STANDARD LOAN
UNIVERSITY OF GLAMORGAN
TREFOREST LEARNING RESOURCES CENTRE
Pontypridd, CF37 1DL
Telephone: (01443) 482626
Books are to be returned on or before the last date below

Biofluid Methods in Vascular and Pulmonary Systems

BIOMECHANICAL SYSTEMS
TECHNIQUES AND APPLICATIONS
VOLUME IV

EDITED BY

CORNELIUS LEONDES

CRC Press

Boca Raton   London   New York   Washington, D.C.

Learning Resources
Centre

Library of Congress Cataloging-in-Publication Data

Catalog record is available from the Library of Congress.

© 2001 by CRC Press LLC

No claim to original U.S. Government works
International Standard Book Number 0-8493-9049-4
Printed in the United States of America 1 2 3 4 5 6 7 8 9 0
Printed on acid-free paper

Preface

Because of rapid developments in computer technology and computational techniques, advances in a wide spectrum of technologies, and other advances coupled with cross-disciplinary pursuits between technology and its applications to human body processes, the field of biomechanics continues to evolve. Many areas of significant progress can be noted. These include dynamics of musculoskeletal systems, mechanics of hard and soft tissues, mechanics of bone remodeling, mechanics of implant-tissue interfaces, cardiovascular and respiratory biomechanics, mechanics of blood and air flow, flow-prosthesis interfaces, mechanics of impact, dynamics of man–machine interaction, and more.

Needless to say, the great breadth and significance of the field on the international scene require several volumes for an adequate treatment. This is the fourth in a set of four volumes, and it treats the area of biofluid methods in vascular and pulmonary systems.

The four volumes constitute an integrated set that can nevertheless be utilized as individual volumes. The titles for each volume are

> Computer Techniques and Computational Methods in Biomechanics
> Cardiovascular Techniques
> Musculoskeletal Models and Techniques
> Biofluid Methods in Vascular and Pulmonary Systems

The contributions to this volume clearly reveal the effectiveness and significance of the techniques available and, with further development, the essential role that they will play in the future. I hope that students, research workers, practitioners, computer scientists, and others on the international scene will find this set of volumes to be a unique and significant reference source for years to come.

The Editor

Cornelius T. Leondes, B.S., M.S., Ph.D., Emeritus Professor, School of Engineering and Applied Science, University of California, Los Angeles has served as a member or consultant on numerous national technical and scientific advisory boards. Dr. Leondes served as a consultant for numerous Fortune 500 companies and international corporations. He has published over 200 technical journal articles and has edited and/or co-authored more than 120 books. Dr. Leondes is a Guggenheim Fellow, Fulbright Research Scholar, and IEEE Fellow as well as a recipient of the IEEE Baker Prize award and the Barry Carlton Award of the IEEE.

Contributors

Sunil Acharya
University of Akron
Akron. Ohio

Gary T. Anderson
University of Arkansas
Little Rock, Arkansas

Joseph P. Archie, Jr.
University of North Carolina
Chapel Hill, North Carolina

Lloyd H. Back
Jet Propulsion Lab
Irvine, California

R.K. Banerjee
Kettering University
Flint, Michigan

Heinrich Brinck
FH Gelsenkirchen
Recklinghausen, Germany

Y.I. Cho
Drexel University
Philadelphia, Pennsylvania

Antonio Delfino
Swiss Federal Institute of Technology
Grolley, Switzerland

Pierre-André Doriot
University Hospital
Geneva, Switzerland

Pierre-André Dorsaz
University Hospital
Geneva, Switzerland

Phillip Drinker
Hood Laboratories, Inc.
Pembroke, Massachusetts

Ding-Yu Fei
Virginia Commonwealth University
Richmond, Virginia

Don Fei
Forest Hills, New York

J.J. Fredberg
Harvard School of Public Health
Boston, Massachuestts

G.M. Glass
Hood Laboratories, Inc.
Pembroke, Massachusetts

Daniel Isabey
INSERM
Créteil, France

M.V. Kaufmann
Silicon Spice, Inc.
Mountain View, California

Clement Kleinstreuer
North Carolina State University
Raleigh, North Carolina

Andre Langlet
L'Universite d'Orleans-Bourges
Bourges, France

Ming Lei
CFD Research Corporation
Huntsville, Alabama

B. Louis
INSERM
Créteil, France

James Moore
Florida International University
Miami, Florida

S. Naili
University of Paris
Créteil, France

Masahide Nakamura
Akita University
Akita, Japan

Christian Ribreau
LGMPB-IUT
Cachan, France

Stanley E. Rittgers
University of Akron
Akron, Ohio

W. Rutishauser
University Hospital
Geneva, Switzerland

B.R. Simon
The University of Arizona
Tucson, Arizona

Marc Thiriet
INRIA
Le Chesnay, France

Jürgen Werner
Ruhr University
Bochum, Germany

Lisa X. Xu
Purdue University
West Lafayette, Indiana

Ryuhei Yamaguchi
Shibaura Institute of Technlogy
Tokyo, Japan

Paul P.T. Yang
Southeast Permanente
    Medical Group
Jonesboro, Georgia

Wen-Jei Yang
University of Michigan
Ann Arbor, Michigan

Contents

1   Hemodynamics Simulations and Optimal Computer-Aided Designs of
    Branching Blood
    Clement Kleinstreuer, Ming Lei, and Joseph P. Archie, Jr. 1-1

2   Techniques in Fluid Dynamical Wall Shear Phenomena and Their
    Application in the Blood Flow
    Masahide Nakamura 2-1

3   A Measurement Method for Wall Shear Stress and Fluid Mechanical
    Application for Vascular Disease
    Ryuhei Yamaguchi 3-1

4   Techniques for Measuring Blood Flow in the Microvascular Circulation
    Lisa X. Xu and Gary T. Anderson 4-1

5   Finite Element Models for Arterial Wall Mechanics and Transport
    B.R. Simon and M.V. Kaufmann 5-1

6   A Three-Dimensional Vascular Model and Its Application to the
    Determination of the Spatial Variations in the Arterial, Venous, and
    Tissue Temperature Distribution
    Jürgen Werner and Heinrich Brinck 6-1

7   Arterial Fluid Dynamics: The Relationship to Atherosclerosis and
    Application in Diagnostics
    James E. Moore, Jr., Antonio Delfino, Pierre-André Doriat, Pierre-André Dorsaz, and
    W. Rutishauser 7-1

8   Computational Fluid Dynamics Modeling Techniques Using Finite
    Element Methods to Predict Arterial Blood Flow
    R. K. Banerjee, L.H. Back, and Y.I. Cho 8-1

9   Numerical Simulation Techniques and Their Application to the Human
    Vascular System
    Ding-Yu Fei, Stanley E. Rittgers, Don Fei, and Sunil Acharya 9-1

10   Flow in Thin-Walled Collapsible Tubes
   M. Thiriet, S. Naili, A. Langlet, and C. Ribreau 10-1

11   Techniques in the Modeling and Simulation of Blood Flows at the Aortic
   Bifurcation with Flexible Walls
   Wen-Jei Yang and Paul P.T. Yang 11-1

12   Airway Dimensions in the Human Determined by Non-Invasive
   Acoustic Imaging
   B. Louis, P. Drinker, G.M. Glass, D. Isabey, and J.J. Fredberg 12-1

Index I-1

1

Hemodynamics Simulations and Optimal Computer-Aided Designs of Branching Blood Vessels

1.1 Introduction 1-1
1.2 Background Information...1-2
    The Biomedical Problem • Current Problem Solutions
1.3 Theory 1-5
    Indicator Equations • Severity Parameters • Flow Waveform
    Parameter • Basic Transport Equations and Auxiliary Condition
1.4 Numerical Method and Model Validation 1-11
    Numerical Method • Grid Generation • Discretized Equations •
    Model Validation
1.5 Results and Discussion 1-24
    Carotid Artery Bifurcation • Bypass Graft-to-Artery
    Anastomosis

Clement Kleinstreuer
North Carolina State University

Ming Lei
CFD Research Corporation

Joseph P. Archie, Jr.
North Carolina State University

1.1 Introduction

Approximately one-quarter million United States citizens each year undergo either carotid endarterectomy for stroke prevention or lower extremity bypass to regain the ability to walk. These arterial reconstructions, as well as others, involve complex flows in branching blood vessels. Both early and late operative failures are due in part to the non-uniform hemodynamics, or disturbed flow related to the reconstruction geometry and its particular flow input waveform. Optimization of the arterial reconstruction of carotid endarterectomies and anastomotic bypass grafts to minimize blood flow disturbances may be crucial in significantly reducing the probability and degree of early postoperative thrombosis and restenosis due to myointimal proliferation and recurrent atherosclerosis. These adverse events can lead to stroke or limb loss and associated disability or mortality. Patient-derived risk factors that influence clinical outcomes of vascular surgical procedures can be favorably modified or controlled, but this lies outside the scope of this review. We have analyzed the effects of geometry and input waveform on disturbed flow in these two arterial systems and have completed work on the theoretical optimization of the branching blood vessel geometries. The methodology has been established, and an extended analysis of computationally derived junction geometries employing available graft material for clinical testing is being initiated.

The primary hemodynamic parameters employed to quantitatively assess disturbed flow patterns include the wall shear stress in general and the temporal and spatial gradients of wall shear stress in particular. Radial blood pressure variations, peak arterial wall stress and strain values, as well as blood particle trajectories and surface depositions can be used as additional indicators of non-uniform hemodynamics and, therefore, susceptible sites to arterial diseases. Optimal reconstructive and anastomotic geometries are computationally designed by iteratively minimizing disturbed flow in three-dimensional branching blood vessels utilizing realistic human hemodynamic input, boundary, and auxiliary conditions. The severity of the potentially harmful hemodynamic factors, in conjunction with the extent of the affected wall surface areas, is expressed in terms of dimensionless groups in order to map and rank various blood vessel configurations and reconstructions as a function of their propensity towards restenosis.

1.2   Background Information

This section summarizes the negative impact of arterial diseases, i.e., thrombosis, atherosclerosis, and hyperplasia, on branching blood vessels, mainly the carotid artery bifurcation as well as the distal end of femoral to popliteal or tibial bypass grafts, and briefly reviews current problem solutions. A detailed review of links between abnormal particle-hemodynamics and the onset of arterial disease processes may be found elsewhere.[1]

The Biomedical Problem

Atherosclerosis is a disease of large and medium-size arteries and is the chief cause of death and disability in the United States and most of the western world. Bypass grafting and endarterectomy have been used for four decades to prevent or delay these events. Carotid endarterectomy for stroke prevention has a relatively low but significant 1% to 4% incidence of early post-operative thrombosis, a 2% to 6% early stroke rate, and a 3% to 10% incidence of significant, i.e., greater than 50%, restenosis due to myointimal hyperplasia in the first few years and later due to recurrent atherosclerosis.[2–8] Similarly, lower extremity synthetic bypass grafts may fail due to restenosis and/or thrombosis. The one-year failure rate is approximately 15%, while 60% to 70% fail within five years.[9,10]

Newer investigative procedures such as balloon angioplasty with or without stenting, laser ablation, or catheter atherectomy are not addressed here, but some of the results presented here may be relevant to those potentially important but unproven reconstruction techniques.

It is generally accepted that critical hemodynamic parameters, particularly related to the wall shear stresses, are directly linked to the localized onset and occurrence of thrombosis, myointimal proliferation, and atherosclerosis. Because each arterial disease process is different, selective hemodynamic factors have to be employed as indicators of locations and extents of susceptible sites in branching arteries. While abnormal hemodynamic factors influence the localization of arterial diseases, their causes are linked to other adverse events including patient-derived risk factors such as smoking, diet, diabetes, hypertension, exercise level, and genetic make-up. The problems being considered are influenced by numerous factors including the arterial reconstruction geometry as well as the type of graft material, placement of the anastomotic junctions, and graft diameter.[11–20] The underlying hypothesis for the initiation of atherosclerosis[21] is that of a response to endothelial injury or dysfunction, leading to high cell turnover and leaky junctions,[22] individual bond rupture of the endothelial cells,[23–25] high blood pressure induced arterial wall stress and strain,[26] and/or prolonged zero-tension state of the endothelial-cells.[27,28,68,86] As a result, at lesion prone sites in branching arteries, endothelial permeability to macromolecules (e.g., low density lipoprotein (LDL)) and to certain cells (e.g., monocytes) is enhanced. In addition to triggering the locally increased lipid flux into the arterial wall as well as foam cell formation, mechanical events may produce changes in gene expression that lead to excessive release of growth factors and subsequent smooth muscle cell proliferation, platelet aggregation, and thrombi formation (Figs. 1.1a and 1.2). Similar events triggered by non-uniform hemodynamic forces can lead to myointimal hyperplasia. In fact, restenosis two years after an arterial reconstruction is considered a continuum transition from hyperplasia to atherosclerosis. This occurs at the toe and heel of the junction of graft-artery bypass configurations along its suture line, and on the bed of the host artery across from the junctions (Figs. 1.1b and 1.2). Hyperplasia followed by atherosclerosis, additional smooth muscle cell proliferation, intra-plaque hemorrhage, and fibrosis all produce restensosis after bypass surgery.[29–33] The anastomotic toe area of the arterial wall most frequently experiences restenosis, which probably leads to graft failure. Even more dangerous is the specter of early post operative complications due to thrombosis. The relations between non-uniform hemodynamics and thrombogenesis have been reviewed,[34] and further discussions may be found elsewhere.[1]

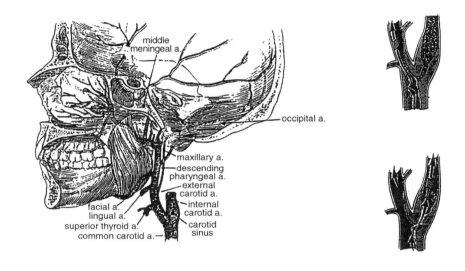

(A)

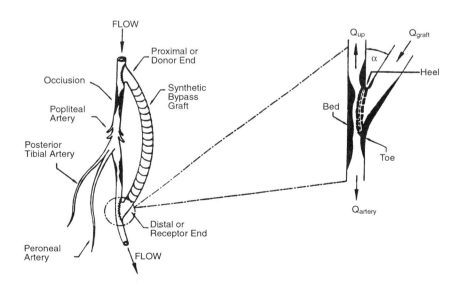

(B)

FIGURE 1.1 Examples of occlusive diseases in branching vessels: (A) atherosclerotic stenosis at the carotid artery bifurcation; (B) femoral to popliteal bypass graft with restenosis in the region of the distal anastomosis.

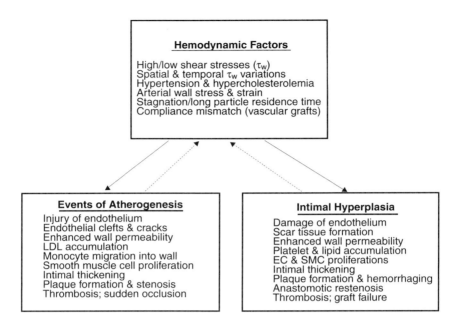

PHYSICO-BIOLOGICAL INTERACTIONS:

Hemodynamic Factors

High/low shear stresses (τ_w)
Spatial & temporal τ_w variations
Hypertension & hypercholesterolemia
Arterial wall stress & strain
Stagnation/long particle residence time
Compliance mismatch (vascular grafts)

Events of Atherogenesis

Injury of endothelium
Endothelial clefts & cracks
Enhanced wall permeability
LDL accumulation
Monocyte migration into wall
Smooth muscle cell proliferation
Intimal thickening
Plaque formation & stenosis
Thrombosis; sudden occlusion

Intimal Hyperplasia

Damage of endothelium
Scar tissue formation
Enhanced wall permeability
Platelet & lipid accumulation
EC & SMC proliferations
Intimal thickening
Plaque formation & hemorrhaging
Anastomotic restenosis
Thrombosis; graft failure

FIGURE 1.2 Non-uniform hemodynamic factors triggering abnormal biological events.

Current Problem Solutions

Significant progress in state-of-the-art computational fluid flow simulations of branching blood vessels has been made in the past decade. For example, Perktold and co-workers[35,36] computed the flow field in a normal carotid artery for the transient three-dimensional flow of a non-Newtonian fluid. Giddens et al.,[16] Nerem,[15] Lou and Yang,[37] and Xu and Collins[38] reviewed relevant experimental and numerical fluid dynamics as well as blood rheology studies related to bifurcating arteries including bypass grafts. Numerous researchers have studied various links between non-uniform hemodynamics and abnormal biological events, but only a few provided quantitative correlations and/or systematically analyzed vessel geometries which may mitigate atherosclerotic plaque formation and/or excessive hyperplasia.[27,39–49]

There are some generally accepted criteria for the selection of the size and composition of prosthetic grafts, such as diameter, the thrombogenicity of flow surfaces, biocompatibility, suturability, porosity, and compliance match.[50,51] However, there are no identified optimal criteria for the geometric design of graft-artery anastomoses. Since graft failure can be caused by myointimal hyperplasia and atheroma developments in the anastomotic region, the geometric design of the junction is very important. For example, based on an energy loss model, Strandness and Sumner[52] suggested that the graft-to-artery angle should be as acute as possible, and the graft diameter should be slightly larger than the distal artery but slightly smaller than the proximal artery to which it is sutured. Except for some of the studies cited above, little effort has been made to improve the geometrical design of graft-artery junctions.

Currently, most lower extremity arterial bypasses are performed on an ad hoc basis depending on the varying experience and decisions of the surgeon, and the patient's arterial anatomy. However, the relationship between disturbed flow and stenosis is gaining recognition, and a few surgeons are investigating the role of the anastomotic geometry and hemodynamics on bypass graft patency rates for coronary bypasses and limb bypasses. Examples include the Taylor patch, the St. Mary's boot, composite prosthetic vein grafts, synthetic cuffs, hoods, etc. One of the more promising graft-to-artery anastomotic geometries is the Taylor-patch.[10] However, the surgical technique is complicated, resulting in varying geometries and mixed results. Similarly, there has been a concentrated effort during the past decade to improve the clinical outcomes of carotid endarterectomies. This effort has been centered around the increased use of vein or synthetic patch reconstruction.[2,3,53] We have achieved encouraging results with customized carotid reconstruction based on operative findings[54,55] and preliminary hemodynamic simulation results,[49,56,57] however there is ample room for significant improvement in outcome.

1.3   Theory

Optimal computer-aided design of branching blood vessels mitigating arterial diseases requires one or more representative hemodynamic predictors based on sound physico-biological hypotheses. A correct hypothesis regarding atherogenesis or intimal hyperplasia must provide a reasonable biological explanation of the mechanisms of these disease processes. In other words, a hypothesis without biological basis is very likely a wrong hypothesis no matter how closely it may match particular distribution patterns of intimal thickening.[1] In the present review focusing on the mitigation of restenosis due to myointimal hyperplasia and/or renewed atheroma, criteria for disturbed flow include very low or very high values of the wall shear stress (WSS), low oscillatory shear, and sustained spatial wall shear stress gradients (WSSG). Specific indicator equations for the localization of atherosclerosis or intimal hyperplasia are described below in conjunction with the basic transport equations, constitutive equations, and auxiliary conditions.

Indicator Equations

In general, the nine-component stress tensor reduces at the wall to a two-component shear stress vector, i.e.,

$$\vec{\vec{\tau}} \Rightarrow \vec{\tau}_w = \left(\tau_{w,m}, \tau_{w,n} \right) \ \text{with} \ \tau_w = \left| \vec{\tau}_w \right| \qquad (1.1a,b)$$

where $\tau_{w,m}$ acts in the mean temporal direction of $\vec{\tau}_w$, and $\tau_{w,n}$ is perpendicular to $\tau_{w,m}$ (Fig. 1.3a), while τ_w is the magnitude of the local instantaneous wall shear stress in N/m². Time averaging over the input cycle of duration T produces

$$\left| \overline{WSS} \right| = \frac{1}{T} \int_0^T \left| \vec{\tau}_w \right| dt \qquad (1.1c)$$

Near-zero WSS regions are associated with longer residence times of blood particles (e.g., platelets, monocytes, etc.) and micro-emboli formation.

The oscillatory shear index (OSI) has been defined[39] as

$$\text{OSI} = \frac{\int_0^T \left| \vec{\tau}^* \right| dt}{\int_0^T \left| \vec{\tau}_w \right| dt} \qquad (1.2a)$$

and more recently[58] as

$$\text{OSI} = \frac{1}{2} \left(1 - \frac{\left| \int_0^T \vec{\tau}_w dt \right|}{\int_0^T \left| \vec{\tau}_w \right| dt} \right) \qquad (1.2b)$$

where Eq. 1.2b is easier to use without giving significantly different contours. Here, T is the period of the input pulse, $\vec{\tau}^{*}$ is the wall shear stress acting opposite to $\overline{\overline{\vec{\tau}}}_{w}$, the mean wall shear stress, and $\vec{\tau}_{w}$ is the instantaneous wall shear stress vector. OSI values vary between 0 and 0.5, with OSI = 0.5 indicating strong sustained oscillatory wall shear effects which previously have been correlated with atherosclerotic lesion developments in carotid bifurcations.[39]

The wall shear stress gradient (WSSG) captures the aggravating impact of temporal and spatial changes in wall shear on the lumen surface. It was suggested that when compared to clinical observation, high sustained, i.e., time averaged WSSG values can be correlated with the onset of atherosclerosis or, following different biochemical mechanisms, with hyperplasia.[23-25,28–35] Specifically,

$$\overline{|WSSG|}_{nd} = \frac{1}{T}\int_{0}^{T}\frac{|WSSG|}{WSSG_{0}}dt \tag{1.3a}$$

where

$$|WSSG| = \left[\left(\frac{\partial\tau_{w,m}}{\partial m}\right)^{2} + \left(\frac{\partial\tau_{w,n}}{\partial n}\right)^{2}\right]^{1/2} \tag{1.3b}$$

and $WSSG_0$ is an assumed reference value. The wall shear stress gradient is obtained by taking the spatial derivatives to the wall shear stress vector, $\vec{\tau}_{w} = (\tau_{m}, \tau_{n})$, which results in an asymmetrical two-dimensional tensor, $\nabla\vec{\tau}_{w}$:

$$\nabla\vec{\tau}_{w} = \begin{bmatrix} \dfrac{\partial\tau_{m}}{\partial m} & \dfrac{\partial\tau_{m}}{\partial n} \\ \dfrac{\partial\tau_{n}}{\partial m} & \dfrac{\partial\tau_{n}}{\partial n} \end{bmatrix} \tag{1.4}$$

The four components of the wall shear stress gradient differ in their impact upon the endothelial cells. The diagonal components,

$$\frac{\partial\tau_{m}}{\partial m} \text{ and } \frac{\partial\tau_{n}}{\partial n},$$

generate intercellular tensions, while the off-diagonal components,

$$\frac{\partial\tau_{n}}{\partial m} \text{ and } \frac{\partial\tau_{m}}{\partial n}$$

generate intercellular shearing forces (Fig. 1.3b). The former may cause the widening and shrinking of the cellular gaps; the latter causes relative movement of adjacent cells. It is assumed that the components influencing the intercellular tension, which leads to dysfunction and elevated permeability of the endothelium, are most important. The use of absolute values takes care of both the positive and negative components of the WSSG. The threshold or reference value for the negative WSSG to trigger the cell response toward enhanced permeability may be different from that of the positive WSSG. However, since there is no experimental data available yet, this combination is probably the simplest and most effective way for calculating the WSSG. Our studies indicated that the positive and negative WSSGs have similar distribution patterns in the distal anastomoses. Furthermore, Kleinstreuer et al.[46] showed that integration of |WSSG| over a given flow input cycle filters out minor temporal changes and yields sustained spatial WSSG distributions.

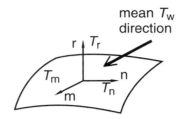

(A)

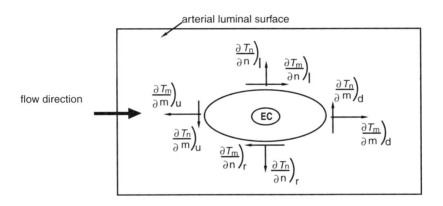

(B)

FIGURE 1.3 Wall stress and wall shear stress gradients: (a) stresses on an arterial surface; (b) forces generated by non-zero WSSGs.

Severity Parameters

Hemodynamic severity parameters indicate the relative impact and extent of arterial diseases. For example, regions of high wall shear stress, important in the thrombi initiation phase, as well as near-zero wall shear stress and long particle residence time, important in the thrombi formation phase, can be combined in a dimensionless form as

$$S_{TH} = \sum \left\{ \frac{A_{\tau,high}\,\tau_{w,high}}{A_0 \tau_0} + \kappa\, \frac{A_{\tau=0}}{A_0} \right\} \qquad (1.5)$$

where $A_{\tau,\,high}$ is the area of high wall shear stress, $\tau_{w,\,high}$ is the high wall shear stress, A_0 is a reference area such as the junction lumen surface, τ_0 is the Poiseuille shear stress at the mean Reynolds number in the inlet tube (based on inlet tube diameter; $Re = UD/\nu$), $A_{\tau=0}$ is the area of the near-zero wall shear stress, and κ is a weighting factor.

The severity parameter to assess restenosis, e.g., neo-intimal hyperplasia and renewed atheroma formation, is based on the wall shear stress gradient and particle residence time (PRT):

$$S_{RE} = \sum \left\{ |\overline{WSSG}|_{nd} \frac{A_{WSSG}}{A_0} + \frac{A_{PRT}}{A_0} \right\} \qquad (1.6)$$

where A_{WSSG} is the impact area based on high sustained WSSG values, and A_{PRT} is equivalent to the area of the near zero wall shear stress.

Equation 1.2 could be used as an alternative severity parameter for restenosis assessment.

Flow Waveform Parameter

The degree of disturbed flow and, therefore, the likelihood of endothelial dysfunction are strongly dependent upon the arterial geometry and the flow input waveform. In order to obtain a quantitative correlation between a flow waveform and its effects, a dimensionless parameter, N, based on the Womersley number, is proposed, which captures the main features of a given input pulse,

$$N = \alpha \, Wo \, \sqrt{\frac{Re_{area}}{Re_{max} - Re_{min}}} \qquad (1.7)$$

Here, α is the number of positive and negative flow portions of the input pulse, e.g., $\alpha = 1$ for the exercise pulse without reverse flow, and $\alpha = 3$ for the resting pulse with a reverse flow portion. $Wo = r_0(2\pi f/\nu)^{1/2}$ represents the Womersley number, a ratio of transient inertia vs. viscous forces, with r_0 representing a reference inlet radius, f representing the pulse frequency, and ν representing the fluid's kinematic viscosity. The extremes of an input pulse are given by Re_{max} and Re_{min}, while Re_{area} is the time averaged value of $|Re_{in}(t)|$:

$$Re_{area} = \frac{1}{T} \int_0^T | Re_{in}(t) | \, dt \qquad (1.8)$$

Clearly, the flow waveform parameter N represents the main characteristics of a given input pulse in terms of disturbed flow impact, including the oscillatory nature and the intensity of the pulse. In general, the larger the value of N, the more complex the flow waveform, possibly resulting in non-uniform hemodynamics in particular junction areas of branching blood vessels.

Basic Transport Equations and Auxiliary Condition

Human blood is generally incompressible and non-Newtonian. Since turbulence rarely appears in the normal artery system,[15,39,59] blood flow is assumed to be laminar and unsteady. Furthermore, it is assumed that the wall is rigid, although it could be permeable, for example, to low density lipoprotein (LDL) transport. The rigid-wall assumption is conservative, i.e., distensible surfaces typically generate lower wall shear stresses.

Assuming the above, the governing equations for blood flow can be written as:[60]

1. Continuity:

$$\nabla \cdot \vec{v} = 0 \qquad (1.9)$$

2. Momentum Transfer:

$$\frac{D\vec{v}}{Dt} = \frac{1}{\rho}\left(-\nabla p + \nabla \cdot \vec{\vec{\tau}}\right) \tag{1.10}$$

where $\vec{v} = (u, v, w)$ or (v_1, v_2, v_3) is velocity vector, p is pressure, t is time, ρ is constant blood density, D/Dt is a substantial derivative

$$\left(\frac{D}{Dt} = \frac{\partial}{\partial t} + \left(\vec{v} \cdot \nabla \vec{v}\right)\right),$$

and ∇ is the spatial gradient operator. The blood rheology is represented by the stress tensor $\vec{\vec{\tau}}$, with the following relations:

1. Power-law fluid

$$\vec{\vec{\tau}} = 2m\left(II_D\right)^{n-1}\vec{\vec{D}} \tag{1.11}$$

2. Casson fluid

$$\vec{\vec{\tau}} = 2\eta\left(II_D\right)\vec{\vec{D}} \tag{1.12a}$$

Here,

$$\vec{\vec{D}} = \frac{1}{2}\left[\nabla\vec{v} + \left(\nabla\vec{v}\right)^T\right]$$

is the rate-of-strain tensor, m = 0.1101 dyn-s^n/m^2, and n = 0.7073 for the power-law fluid. The apparent viscosity, η, is a function of the shear rate given by an extended Casson model[61] as:

$$\eta\left(II_D\right) = \frac{1}{2\sqrt{II_D}}\left[C_1\left(Ht\right) + C_2\left(Ht\right)\sqrt{2\sqrt{II_D}}\right]^2 \tag{1.12b}$$

where Ht is the hematocrit and II_D is the second scalar invariant of $\vec{\vec{D}}$, i.e.,

$$II_D = \frac{1}{2}\left[\left(tr\vec{\vec{D}}\right)^2 + tr\left(\vec{\vec{D}}^2\right)\right] \tag{1.13}$$

Writing out the right hand side in a component form, we have:

$$II_D = D_{11}D_{22} + D_{11}D_{33} + D_{22}D_{33} - D_{12}D_{21} - D_{13}D_{31} - D_{23}D_{32} \tag{1.14}$$

The coefficients C_1 and C_2 were determined for 40% hematocrit as $C_1 = 0.2$ (dyn/cm^2)$^{1/2}$ and $C_2 = 0.18$ (dyn s/cm^2)$^{1/2}$ based on Merrill's experimental data.[62] Fig. 1.4 shows the variation of blood viscosity η (Eq. 1.12) with shear rate

$$\dot{\gamma}\left(= 2\sqrt{II_D}\right).$$

The asymptotic value of the viscosity is $\eta_\infty = 0.0324$ dyn s/cm^2 for 40% hematocrit which, according to Merrill, is appropriate as the shear rate approaches infinity. This asymptotic viscosity is often taken as an approximate Newtonian viscosity, μ. However, the asymptotic value given by Merrill is higher than the value predicted by the Casson model in the range of high shear rates ($1000 < \dot\gamma < 3000$) which also is the upper limit of shear rates in this study. Thus, for computations at moderately high shear rates, values of $\mu = 0.0348$ dyn s/cm^2, i.e., $\nu = 0.033$ cm^2/s, are assumed to guarantee a smooth transition from the Casson model to the limiting Newtonian behavior. At the lower end of the shear rate range, this model is only suitable for $\dot\gamma > 1$ s^{-1}; therefore, $\eta = \eta_{\dot\gamma=1} = 0.1444$ (dyn s/cm^2) when $\dot\gamma < 1$ s^{-1}, which is the zero shear rate condition.

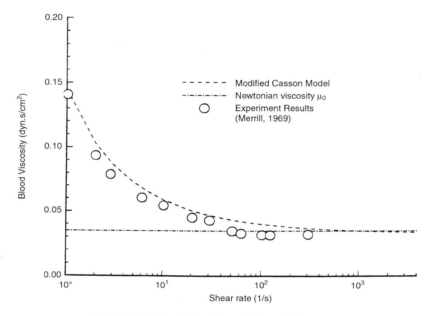

FIGURE 1.4   Blood viscosity variation with shear rate.

The boundary conditions for solving these equations are given below.

Inlet Conditions

A transient parabolic velocity profile is specified at the inlet. The inlet length is made sufficiently long ($l_e > 10\,d_a$, where d_a is the diameter of the artery) so that the inlet velocity profile has no obvious influence on the flow pattern in the junction area. Therefore, the inlet conditions can be expressed as:

$$-\vec{\mathrm{v}} \cdot \hat{n} = u_i(t) \tag{1.15}$$

$$\vec{\mathrm{v}} \cdot \hat{s} = 0 \tag{1.16}$$

where $\hat{n}$ is the outward unit normal vector of the surface at the inlet, $\hat{s}$ is the unit tangential vector, and u_i is a given velocity distribution, varying with time.

Outlet Conditions

A nominal Neumann boundary condition is applied to the velocity field:

$$\frac{\partial \vec{\mathrm{v}}}{\partial n} = 0 \tag{1.17}$$

Usually, there are two outlets, one toward downstream and one toward upstream (Figs. 1.5–1.7). Varying flow partition between these two outlets is specified. The given flow rate ratio must be satisfied through a global mass balance. The length of the downstream outlet is made longer than the upstream outlet ($l_d \geq 11\ d_a$ vs. $l_u \geq 7\ d_a$) since a large portion of the flow goes through the downstream outlet. Zero-pressure condition is applied to one outlet, leaving the other outlet free.

Wall Boundary Conditions

At the impermeable wall, the following velocity conditions are imposed:

$$\vec{v} \cdot \hat{n} = 0 \tag{1.18}$$

$$\vec{v} \cdot \hat{s} = 0 \tag{1.19}$$

1.4 Numerical Method and Model Validation

Investigation of the hemodynamics in branching vessels and the effects of hemodynamic factors influencing atherogenesis and/or neo-intimal hyperplasia can be carried out through *in vitro* fluid flow and cell culture experiments, computer simulations, animal studies, and clinical testing. For engineers, information is obtained mainly through laboratory experiments and computer simulations. During the past two decades, considerable progress has been made in both model bifurcation flow measurements and *in vitro* cell culture experiments. Laser Doppler anemometers have played an important role in obtaining quantitative measurements of velocity profiles *in vitro* model studies.[63–65] The photochromic tracer technique has been recently used to indirectly determine the wall shear stress at selected locations as a function of time.[66] Doppler ultrasound technique is a powerful tool for the measurement of detailed flow patterns *in vivo*.[67] Cell culture experiments provide us with a lot of useful information about endothelial cell response to changes in the hemodynamic environment and, therefore, greatly advance our knowledge of the arterial wall-blood flow interactions. However, experiments have their limitations.[1] Numerical simulation is an alternative approach to obtain quantitative information about flow patterns and wall shear stresses at bifurcations. It becomes more and more important in hemodynamics since it can provide details of the entire flow field and calculate values of disease indicators such as the wall shear stress and wall shear stress gradient quite easily and accurately, especially for complicated geometries under pulsatile flow conditions. Another major advantage of computational study is its flexibility in parametric studies. Recent developments of computer technologies and CFD tools make three-dimensional pulsatile flow simulations possible in branching blood vessels such as the carotid artery bifurcation and femoral end-to-side anastomosis.

In this section, we will describe the numerical method used to solve the basic equations governing blood flow, the key aspects of mesh generation, and numerical code validation.

Numerical Method

All the numerical calculations in this chapter have been carried out with a finite volume-based CFD package, FLOW3D, now CFX from AEA Technologies (Bethel Park, PA). The features, enhancements, and usage of this package are described by Lei.[68] A short overview of the finite volume method (FVM) or control volume method (CVM) is provided below.

The momentum and any scalar transport equations can be rewritten in a general form as:

$$\frac{\partial(\rho\phi)}{\partial t} + \nabla \cdot (\rho\vec{v}\phi) - \nabla \cdot (\Gamma\nabla\phi) = \sum S \tag{1.20}$$

where ϕ is any dependent variable, Γ is the general diffusion coefficient which may be a function of the flow field, and ΣS represents source or sink terms.

This generalized equation can be expressed again in normal form as:

$$L\big(\phi\big) = 0 \qquad (1.21)$$

where $L(\cdot)$ is a nonlinear operator. The purpose of conducting numerical calculations is to find an approximate solution of ϕ. Let ϕ^* be the unknown approximate solution. Substituting ϕ^* into Eq. 1.21 yields:

$$L\big(\phi^*\big) = R \qquad (1.22)$$

where R is called the residual of $L(\phi^*)$. It is evident that $R \to 0$ if ϕ^* approaches ϕ. So, we require:

$$\int_\Omega WRd\Omega = 0 \qquad (1.23)$$

where W is a weighting function, and the integration is performed over the entire computational domain Ω. We solve Eq. 1.23 for a suitable ϕ^* expression as an approximation of ϕ instead of solving Eq. 1.21 directly. This is called the method of weighted residuals (MWR).[69] The type of weighting function W selected determines the type of MWR used. For example, in the Galerkin finite element method, the domain Ω is discretized employing finite elements, and the weighting functions are chosen to be the basis functions of the trial solution. With respect to the finite volume method, we divide the calculation domain Ω into finite, non-overlapping subdomains or control volumes V_i ($i = 1, 2,....., N$), and set the weighting function to be unity over one subdomain at a time and zero everywhere else. This way, we obtain N algebraic weighted-residual equations which can be used to find the value of ϕ^* at each control volume, i.e.,

$$\int_{V_i} R_i dV = 0 \qquad i = 1,2,\dots N \qquad (1.24)$$

Equation 1.24 implies that the integral of the residual over each control volume must become zero, i.e., the integral conservations of mass flow and momentum are satisfied over each control volume. This is the essence of the finite (or control) volume method.

Grid Generation

Proper mesh generation is one of the most challenging tasks in computational fluid dynamics analysis. In this study, a body-fitted coordinate system has been used (Fig. 1.5), and a multi-block grid structure has been implemented (Fig. 1.6). Each block is a topologically rectangular grid and has a local coordinate system (i,j,k) assigned in addition to the global coordinate system (x,y,z). A two-component surface coordinate system (m,n) has been used for the calculation of wall values (Fig. 1.5). The surface coordinate system varies from point to point with m being chosen to follow the local temporal mean wall shear stress direction and n being the direction perpendicular to m. A multi-block grid is generated by merging (or gluing) all the blocks together (Fig. 1.6). A special block arrangement has been employed to avoid finite volume distortion especially at the bifurcation corners. All present geometries are assumed to be symmetrical about a vertical symmetry plane, therefore, only half of each flow domain is simulated. A typical number for the total elements is 16,940 for the streamline connector shown in Fig. 1.7. Further mesh refinement may be necessary in order to ensure mesh-independence.

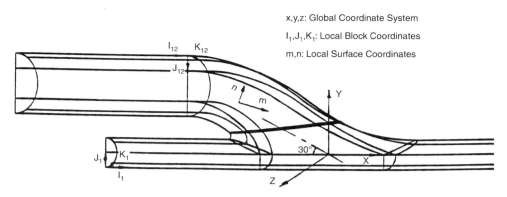

FIGURE 1.5 Coordinate systems.

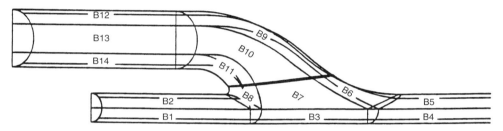

FIGURE 1.6 Multi-block structures for three-dimensional distal anastomosis.

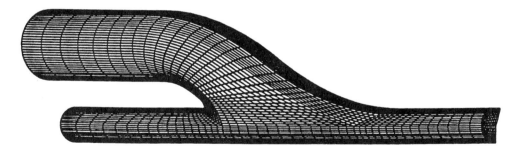

FIGURE 1.7 Computational grid for distal anastomosis.

Discretized Equations

As indicated previously, the finite volume method (FVM) is a kind of conservative finite-difference method, where the physical domain is discretized into multiple control volumes. Then, with all variables defined at the center of the control volumes, each equation is integrated over each control volume to obtain a discrete equation which connects the variable at the center of a control volume with its neighbors. The integration of the generalized governing equation (Eq. 1.20) with the continuity equation being excluded) results in:

$$\int \frac{\partial(\rho\phi)}{\partial t}dV + \int\left(\rho\phi\bar{v}\cdot\hat{n}\right)dA - \int\left(\Gamma\nabla\phi\cdot\hat{n}\right)dA = \int SdV \qquad (1.25)$$

All terms in the equation are discretized in space using a second-order centered differencing scheme, except for the advection terms. For the time derivative terms, a fully implicit backwards difference time stepping procedure has been implemented. For the advection terms, the hybrid differencing scheme[70] is used and the convection coefficients are obtained by using the Rhie-Chow interpolation formula.[71] With the use of the Rhie-Chow algorithm, a staggered grid approach is not necessary, which saves memory when using body-fitted coordinates. The major achievement of the Rhie-Chow algorithm lies in the fact that it provides a prescription for implementing the standard primitive variable algorithms, such as SIMPLE, using a non-staggered grid, while avoiding the well known problem of checker-board oscillations associated with the use of non-staggered grids. The source term for each equation is written as:

$$\int S dV = S_u + S_p \phi_p \qquad (1.26)$$

where S_p is non-positive and S_u is positive. The S_p term enhances the diagonal of the matrix giving more diagonal dominance and better solution behavior. The final discretization formulations have the form

$$A_p \phi_p = \sum_{nb} A_{nb} \phi_{nb} = S_u \qquad (1.27)$$

Here, subscript nb denotes each of the six neighboring nodes, i.e., U (up), D (down), N (north), S (south), E (east), and W (west); A_p and A_{nb} are coefficients. The diagonal coefficient of the matrix, A_p, is given as:

$$A_p = \sum_{nb} -S_p + C_U - C_D + C_N - C_S + C_E - C_W + \frac{\rho V}{\Delta t} \qquad (1.28)$$

where C_U and the other components of Eq. 1.28 are the convection coefficients at corresponding interfaces denoted by the subscripts; the last term stands for the transient effect carried from the previous time step, in which V is the control volume. In the steady-state case, the sum of the convection coefficients in this expression describes the conservation of mass and thus goes to zero in a converged solution.

The continuity equation is replaced by the pressure-correction equation derived by using the SIMPLE or SIMPLEC algorithm.[70,72] The resulting equation has the form

$$b_p P'_p = \sum_{nn} b_{nn} P'_{nn} + S' - m_p \qquad (1.29)$$

where P' is the corrected pressure, b_{nn} and $b_p = \sum_{nn} b_{nn}$ are coefficients, m_p is the residual mass source, and S' is the additional source term due to non-orthogonality.

The discretized transport equations (1.27) are solved iteratively. There are two levels of iteration: an inner iteration to solve for the spatial coupling (non-linearity) for each variable, and an outer iteration to solve for the coupling between variables. Thus, each variable is taken in sequence, while we regard all other variables as fixed. The coefficients of the discretized equations are always reformed, using the most recently calculated values of the variables, before each inner iteration. Stone's method has been used to solve the linearized transport equations for each variable. The SIMPLEC velocity-pressure coupling algorithm, which has been proven to be less sensitive to the selection of under-relaxation factors and therefore is less demanding, has been chosen. The convergence of the iterations is controlled by the mass source residual (the error in continuity) alone. For the transient problems, varying time steps are used depending on the magnitude of the Reynolds number and the slope of the input pulse. More time steps are used for the deceleration phase than for the acceleration phase. Doubling the time steps results in only an approximate 3% difference in the maximum value of the calculated wall shear stress gradient with the same accuracy of final convergence. The mass flux boundary condition is updated each iteration. To eliminate the start-up effect in transient flow, the computations continue over at least four periods.

The computational work has been carried out on a DEC Alpha 3000 machine (Model 300 LX, 96 MB RAM). The convergence criterion was a mass residual of less than 1.0×10^{-6} kg/s or 1.0×10^{-3} g/s with maximum outer iterations limited to 200 to 300 during the start-up periods. Fig. 1.8 shows the convergence processes and relative errors of several sample cases and demonstrates that 200 outer iterations are enough for most of the time steps. Usually, whether the process at a time step converges or not can be determined after approximately 60 iterations. After approximately 100 iterations, the mass residual approaches a steady value which is different for different time steps, but depends on the Reynolds number at a specific time step and its increment (ΔRe) as well as its location on the input curve. In general, the exercise femoral pulse has the fastest convergence rate since the slope of the curve and the difference between the maximum and minimum Reynolds numbers are the smallest. The value of the under-relaxation has no obvious influence on the final convergence, as long as it is not too aggressive.

Model Validation

For both the particular bypass junctions and the deviations from the natural carotid artery bifurcations considered in this study, detailed quantitative experimental data are lacking. Thus, especially for the wall shear stress gradient (WSSG), direct comparisons with experimental measurements cannot be made at this time. Nevertheless, in order to test the accuracy and validity of the finite volume code for the hemodynamics simulations in branching arteries, we have conducted several model validation runs.

Test on Inlet Length

In general, the velocity profile at an arbitrary entrance point may not be parabolic at any time for a branching artery segment under pulsatile flow conditions. However, when we add a sufficiently long conduit to the flow inlet and specify a time-varying parabolic velocity profile at the extended inlet, correct input conditions can be restored for the junction flow. Alternatively, measured transient inlet profiles at a fixed location could be employed as inlet conditions. To determine the minimum length required for an inlet conduit, three-dimensional pipe flow with the standard and resting input pulse for the femoral artery has been selected as an example. Fig. 1.9 demonstrates that the fully developed velocity profiles are not parabolic at most of the time steps even if a parabolic profile is prescribed at the inlet, which is quite different from steady flow results. Fig. 1.10 presents the wall shear stress distributions in the tube at different time steps. When the flow is fully developed, the wall shear stress distribution should be a horizontal line with zero spatial gradient. The inlet length, represented here by $8d_a$, is determined by this condition. In summary, one can conclude the following:

- It is easier and better to use an extended inlet length and specify parabolic inlet velocity profiles as a function of time.
- When a time-varying parabolic velocity profile is specified at the inlet, there exists a minimum inlet length necessary to establish the correct flow field.
- The minimum inlet length is not determined by the maximum Reynolds number, but by the shape of the flow input waveform. Usually, a longer inlet length is required for the decelerating flow since it is less stable than the accelerating flow and needs a longer distance to become fully developed.
- The inlet length requirements are quite similar for two- and three-dimensional flows.

It is worth noting that the inlet lengths analyzed are quite different from the entrance lengths displayed in fluid dynamics textbooks, where l_e is the distance for steady uniform inlet flow to become fully developed in a channel or pipe.

Test on the Mesh Size and Error Estimation

There are two types of errors which determine the accuracy of numerical simulations. One is called round-off error, which is introduced by the floating-point representation of numbers in computers. The other is called truncation error, which is introduced by the discretization of governing equations and the flow domain. The former is determined by the arithmetic precision of computers, so it is referred to as

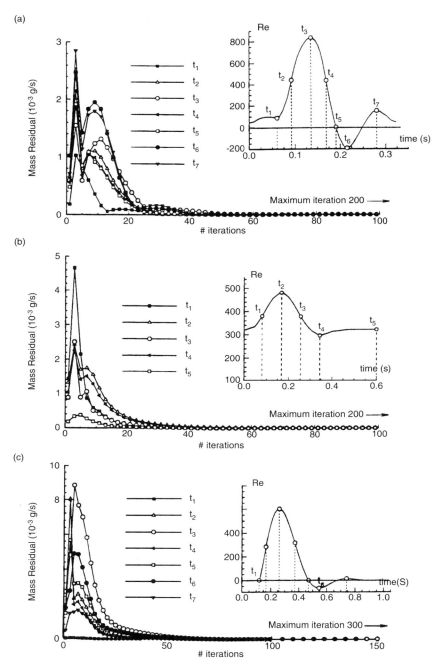

FIGURE 1.8 Convergence processes for sample cases: (a) three-dimensional aorto-celiac junction; (b) 1:1 base case distal anastomosis; (c) 2:1 S-connector.

machine error. Proper programming can reduce the accumulation and propagation of this error, but cannot eliminate it. Fortunately, this round-off error is always much smaller than the truncation so that its presence has no obvious influence on the accuracy of numerical simulations. Truncation error is the major source of inaccuracy. It is determined by the mesh size, degree of element distortion or mesh transformation, the discretization method (numerical scheme), as well as the convergence criterion if an

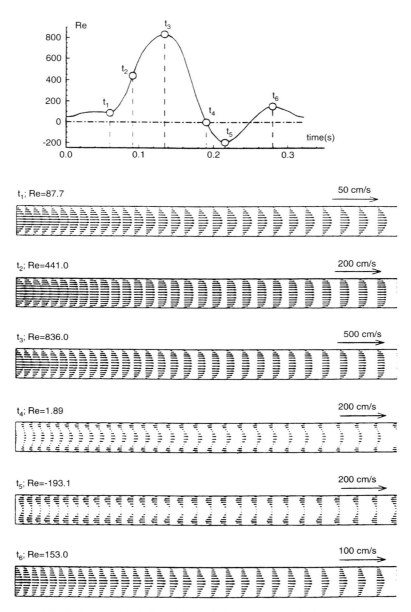

FIGURE 1.9 Flow field variations during an aorto-celiac input pulse.

iterative method is chosen to solve the matrix equations. We can always choose the best numerical schemes and most suitable convergence criterion based on our knowledge, but there is no best mesh size since the truncation error is proportional to it. Usually, the smaller the mesh size, the smaller the truncation error. However, at a certain mesh density, as the mesh becomes smaller and smaller, the round-off error may play an important role in determining the accuracy of numerical calculations. The mesh density is often limited due to the memory and speed limitations of the computer. Thus, not only the accuracy, but also the cost and the availability of computing resources is essential to the determination of an ideal mesh size, as pointed out by Lei,[68] who conducted test runs on the effect of mesh size in terms of cost and accuracy, in a three-dimensional T-junction model.

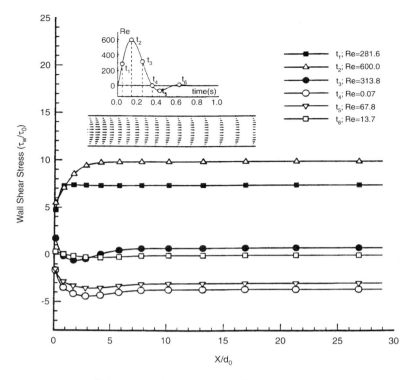

FIGURE 1.10   Wall shear stress distributions at different time steps in a pulsatile flow.

Comparison with LDA Measurements

In order to test the validity of the numerical code for pulsatile flow simulations, we compared numerical results with the LDA measurements of Khodadadi et al.[64] in a plane 90° T- junction with sinusoidal input flow waveforms. Fig. 1.11 shows the grid and the input pulse with a mean Reynolds number of 106. The branch-to-main channel flow rate ratio is set equal to 0.7. The comparisons of velocity profiles at several locations between numerical results and experimental data are shown in Figs. 1.12 to 1.14. Most of these locations are in the disturbed flow regions. It is postulated that if the velocity profiles match the experimental results in the disturbed flow regions, the whole flow field will match. From these graphs, we can see that the numerical results are in agreement with the experimental results.

Comparison with Particle Tracer Study

By using the photochromic tracer technique, Ohja[73] quantitatively measured the temporal variations of the velocity field, from which he computed the wall shear stresses at several locations in a 45° distal anastomosis model with the proximal end of the host vessel completely occluded. Unfortunately, there are no quantitative velocity profiles presented. We carried out three-dimensional numerical simulation in the same geometry with the same input pulse (Fig. 1.15a). The input pulse can be expressed in terms of the Reynolds number as

$$Re = Re_0 \sin(\omega t) + Re_m \tag{1.30}$$

where $\omega = 2\pi f$, f is the pulse frequency, and Re_m is the mean Reynolds number. The values of the parameters are provided in Table 1.1. The results are shown in Fig. 1.15. The mid-plane velocity vector plot at one time step is included to show the flow condition. The wall shear stress variations at the three locations are compared with Ohja's measurements in Fig. 1.15b, which shows there is agreement between the numerical results and experimental measurements.

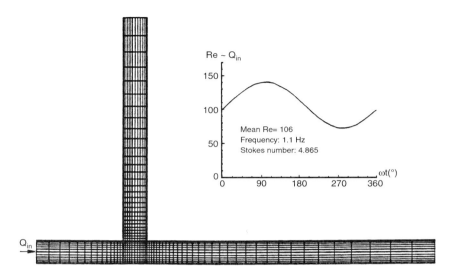

FIGURE 1.11   Sinusoidal input pulse and computational grid for a plane T-junction.

Validation Run for Carotid Artery Bifurcation (CAB)

A velocity profile comparison is made between the experimental measurements of Rindt et al.,[74] who constructed a rigid two-dimensional plexiglass model of the normal CAB, and those of the current numerical model. The dimensions of the plexiglass model are similar to those given by Bharadvaj et al.,[75] with the major difference being the bifurcation angle and the entrance region of the external carotid artery (ECA). The angle created by the axis of the ECA and the horizontal in the plexiglass model of Rindt et al.[74] is 30°, as opposed to an angle of 25.1° in the model of Bharadvaj et al.[75] Also, unlike the model of Bharadvaj et al.,[75] the diameter of the ECA remains constant for a short distance after the ECA entrance, then quickly converges to the distal diameter of the ECA as given by Bharadvaj et al.,[75] whereas the present model and the model of Bharadvaj et al.[75] have a gradual taper of the diameter beyond the entrance to the ECA. Velocity measurements in the model were taken using LDA. In the experimental model, the dimensions of Bharadvaj et al.[75] are scaled up 2.5 times to increase the relative size of the model to the sample volume of the laser. The fluid used by Rindt et al.[74] was a water/sterilized milk mixture (7000:1); since the water/milk volume ratio is so high, the density and viscosity of water at normal room temperature were used without significantly affecting the accuracy of the experiment. Finally, the flow rate ratio between the internal carotid artery (ICA) and ECA was 65/35, as determined by integrating the velocity profiles in each branch.

Unlike the experimental model, the current numerical model is not scaled up from the dimensions of Bharadvaj et al.,[75] and blood, instead of the water/milk mixture, is a non-Newtonian fluid. The scaling of the model and the use of a different fluid does not create a problem because the dynamic similitude is assured by using the same input Reynolds number vs. the time pulse used by Rindt et al.[74] The purpose of the input pulse used by Rindt et al.,[74] shown as an inset to Fig. 1.16, was used for validation of the numerical model. The ICA to ECA flow rate division was set constant at 65/35 throughout the pulse to match the boundary conditions of the experimental model.

Fig. 1.16 depicts some results of the validation run. The solid lines show the velocity profiles as predicted by the present numerical model at the snapshot in time, S8, as indicated in the inset, while the circles indicate the velocity profiles as measured by Rindt et al.[74] There are slight differences in the profiles at the entrances to the ECA, ICA, and the two profiles downstream of the entrance to the ICA. At the entrance to the ICA, the predicted numerical velocities near the divider wall are greater than the measured experimental velocities, while at the entrance to the ECA, the opposite trend is present. The other profiles in the ICA also show numerical velocities greater than the experimental velocities at some locations in the profile — near the divider wall for the profile at the mid-sinus, and near the non-divider wall for

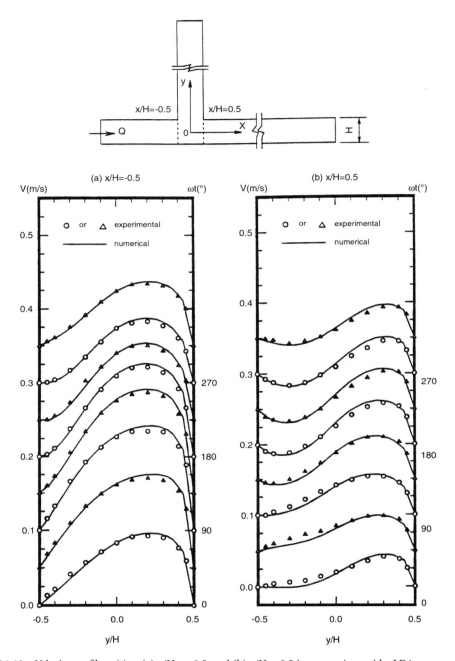

FIGURE 1.12 Velocity profiles u(t) at (a) x/H = −0.5; and (b) x/H = 0.5 in comparison with LDA measurements.[64] (*Source:* Khodadadi, J.M., Vlachos, N.S., Liepsch, D., and Moravec, S., *J. Biomech. Eng.*, ASME, 110, 129–136, 1988. With permission.)

the profile at the end of the sinus. This implies that the flow rate ratio between the ICA and the ECA does not exactly equal 65/35 at this time level, which is believable due to the extreme difficulty in forcing a constant flow rate ratio in an experimental pulsatile flow. Other sources of discrepancies between the numerical and the experimental velocities could be the difficulty in measuring the near wall velocities with LDA techniques, round-off error in the numerical calculations, and the fact that the experimental velocities are obtained by measuring them from graphs.[74] Considering these possible sources of error,

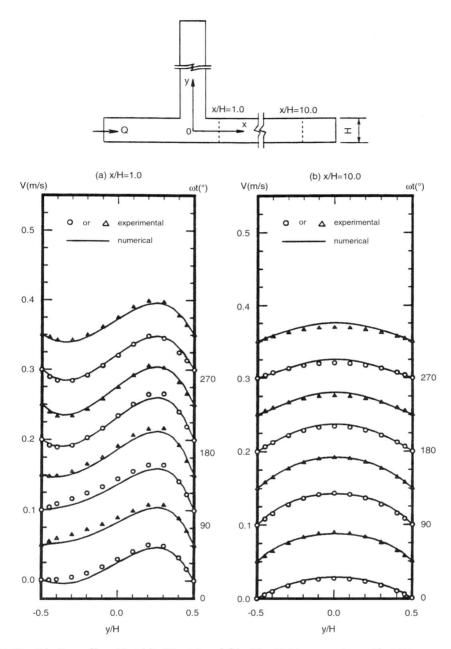

FIGURE 1.13   Velocity profiles u(t) at (a) x/H = 1.0; and (b) x/H = 10.0 in comparison with LDA measurements.[64] (*Source:* Khodadadi, J.M., Vlachos, N.S., Liepsch, D., and Moravec, S., *J. Biomech. Eng.*, ASME, 110, 129–136, 1988. With permission.)

the small discrepancies between the velocity profiles shown in Fig. 1.16 indicate that the current numerical model can accurately predict the transient velocity field in the carotid artery bifurcation.

In summary, based on the model validation studies, we are confident that the numerical code used is sufficiently accurate for flow simulations in branching arteries and graft-bypass junctions. The data comparisons also instill confidence that the computed wall shear stress gradients are sufficiently accurate for: (1) the simulation study for the rabbit aorto-celiac junction; (2) computer-aided design of near optimal graft-artery connectors; and (3) hemodynamics simulations of carotid artery bifurcations.

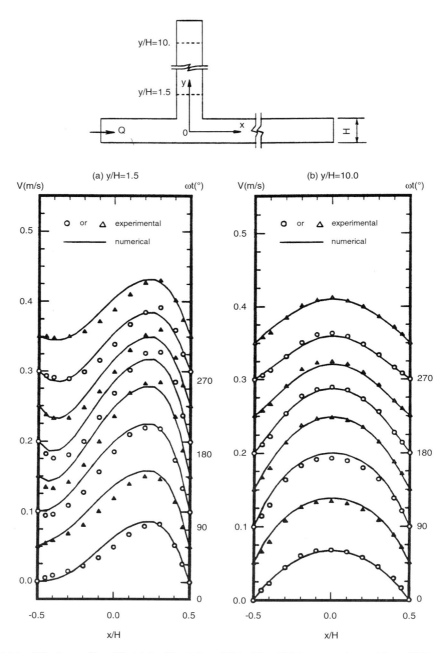

FIGURE 1.14 Velocity profiles v(t) at (a) y/H = 1.5; and (b) y/H = 10.0 in comparison with   LDA   measurements.[64] (*Source:* Khodadadi, J.M., Vlachos, N.S., Liepsch, D., and Moravec, S., *J. Biomech. Eng.*, ASME, 110, 129–136, 1988. With permission.)

TABLE 1.1   Flow Parameters for Ohja's Experiment[73]

D [cm]	ρ [g/cm³]	μ [cP]	Re₀	Reₘ	f (Hz)	Wo
0.51	0.75	1.43	360	575	2.9	7.9

(*Source:* Ohja, M., *J. Biomech.*, 26, 1377, 1993. With permission from Elsevier Science.)

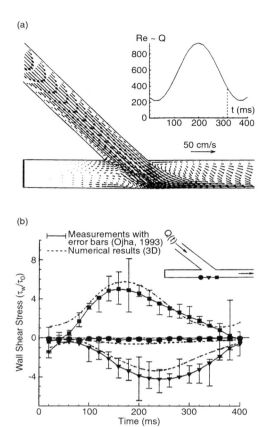

FIGURE 1.15 (a) Mid-plane velocity vector field at t = 318 ms in an occluded distal anastomosis model; (b) comparison between measured wall shear stress variations and predicted results.[73] (*Source:* Ohja, M., *J. Biomech.*, 26, 1377, 1993. With permission from Elsevier Science.)

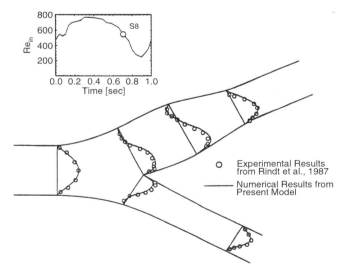

FIGURE 1.16 Comparison of several velocity profiles as computed in the current numerical model and as measured by Rindt et al.[74] at time level S8 as shown in the inset. (*Source:* Rindt, C.C.M., von Steenhoven, A.A., and Reneman, R.S., *J. Biomech.*, 21, 985, 1988. With permission from Elsevier Science.)

1.5   Results and Discussion

Two types of branching blood vessels have been selected for optimal design considerations, i.e., the carotid artery bifurcation and the distal end femoral graft-artery bypass. Both theoretical designs are based on the minimization by trial-and-error of the time-averaged wall shear stress gradient, assuming that a complicated terminal graft segment merges perfectly with an idealized, i.e., circular, artery. Additional disturbed flow predictors to be minimized and more formal nonlinear multivariable optimization methodology to determine the best junction geometries are discussed as future work.

Carotid Artery Bifurcation

As mentioned earlier, significant progress in state-of-the-art computational fluid flow simulations of branching blood vessels has been made in the past decade. For example, Perktold,[35] Perktold et al.,[36] and Lee and Chiu[76] computed the flow fields in a normal carotid artery for the transient three-dimensional flow of a non-Newtonian fluid. Giddens et al.,[16] Nerem,[15] Lou and Yang,[37] and Xu and Collins[38] reviewed other relevant experimental and numerical fluid dynamics as well as blood rheology studies related to branching blood vessels. Numerous researchers have studied various links between non-uniform hemodynamics and abnormal biological events, but only a few provided quantitative correlations and systematically analyzed vessel geometries which may mitigate renewed atherosclerotic plaque formation.[36,39,41,47]

Currently, most carotid endarterectomy reconstructions are performed on an *ad hoc* basis depending on the varying experience and decisions of the surgeon, and the patient's arterial anatomy. There has been a concentrated effort in the past decade to improve the clinical outcomes of carotid endarterectomies. This effort has been centered around the increased use of vein or synthetic patch reconstruction.[2–3,53] Encouraging results have been achieved with customized carotid reconstruction based on operative findings[54,55] and preliminary hemodynamic simulation results.[47]

A typical flow input waveform and three geometric configurations of the carotid bifurcation studied are demonstrated in Figs. 1.17 a–d, of which the base case geometry (Fig. 1.17b) has been taken from Bharadvaj et al.[75] For the computational analysis, the arteries were divided into control-volume meshes within which the numerical solutions for the pressures, velocities, and the associated stresses are obtained. In zones that were shown to have disturbed flow, the finite volumes were made smaller in order to obtain more accurate results. Final mesh-independence of all results was successfully tested. All three carotid bifurcations had the same inlet and outlet diameters. The bifurcation angles in the three geometries were: 50° for the natural geometry, 30° for the patch reconstructed geometry, and 25° for the smoothly tapered geometry. The same input velocity waveform was used for all three configurations. The computational analysis was transient two-dimensional, and the walls were assumed rigid. While the wall compliance effect is generally very small and negligible in older patients, the two-dimensional assumption requires some elaboration. For flow conduits with a symmetry plane as assumed for the carotid artery bifurcation, mid-plane velocity fields are basically the same between two-dimensional and three-dimensional flows[74,75] because the circumferential velocity component is zero in the bifurcation plane. Thus, the effects of geometry changes discussed in this chapter refer to the blood vessels' mid-plane shapes.

As discussed in the Theory Section, transient laminar incompressible flow was assumed. The basic fluid mechanics equations and suitable boundary conditions were used to describe the flow.[60] In this analysis, blood was treated as both a Newtonian fluid and power-law fluid in order to demonstrate the impact, if any, of fluid rheology. At the common carotid inlet, a time-dependent parametric velocity profile[75] was assumed with a maximum Reynolds number of about 800 and a minimum of about 200. The reference length for the instantaneous Reynolds number is the common carotid inlet diameter of 8 mm and an average blood viscosity of 0.0324 dyn s/cm². The usual assumption of no-slip or zero velocity at the wall was utilized. In order to simplify the analysis of the results, the wall shear stresses and shear stress gradients were non-dimensionalized and time-averaged over the input cardiac cycle.[57] Time-averaging does not change the wall shear stress or wall shear stress gradient results along the walls

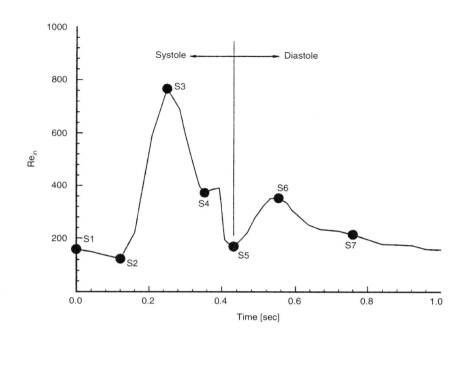

(A)

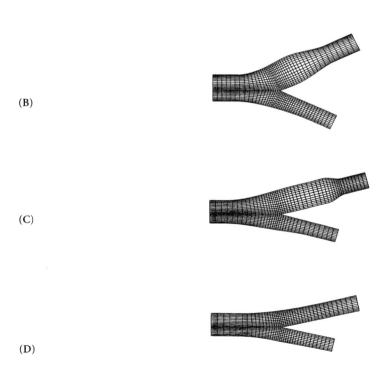

(B)

(C)

(D)

FIGURE 1.17 (A) The flow pulse in the Common Carotid Artery and control volume meshes for the (B) base case; (C) patch reconstruction; and (D) smoothly tapered geometry.

because the integration process gave average sustained magnitudes not very different from their major transient values.[57] The time for one cardiac cycle was arbitrarily chosen to be one second.

Mid-Plane Velocity Vector Profiles

The velocity field in a complex flow system is a good quantitative indicator of disturbed flow structures. Except at the systolic peak (time level S_3 in Fig. 1.17a), reverse flow can be observed in the normal carotid artery bifurcation (NCAB) at all other critical time levels, especially in the carotid bulb (Figs. 1.18a-d). Flow separation and reversal occur because of inlet flow deceleration coupled with adverse pressure gradients caused by conduit widening. Both overcome the relatively small inertia forces present near the wall.

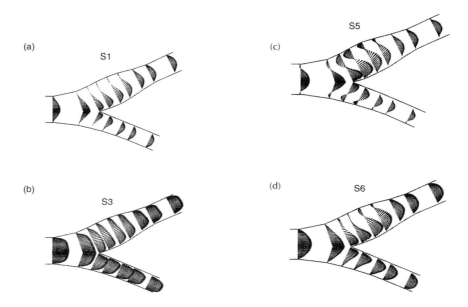

FIGURE 1.18 Velocity vector plots in the NCAB assuming blood to be a power law fluid for (a) time level S_1; (b) time level S_3; (c) time level S_5; and (d) time level S_6.

Wall Shear Stress Distributions

The magnitude and distribution of the time-averaged and non-dimensionalized wall shear stresses for both Newtonian and non-Newtonian fluids are shown in Figs. 1.19a-c for the three carotid bifurcation geometries, respectively. The differences generated by the blood rheology are most notable for the normal carotid bifurcation. The non-Newtonian fluid predicts recirculating zones with high and low wall stresses in the outer wall of the internal carotid artery of both the normal carotid bulb and the patch reconstructed carotid artery. Wall shear stress peaks and troughs appear when blood vessel geometry changes occur. As expected, the junction of the internal and external carotid arteries (walls II and III) exhibits similar behavior with a rapid decrease in the shear stresses at the stagnation point.

Wall Shear Stress Gradients (WSSG)

The magnitude and distribution of the time-averaged and non-dimensionalized WSSG for both Newtonian and non-Newtonian fluids are shown in Figs. 1.20a-c for the three carotid bifurcation geometries, respectively. As expected, all three carotid artery geometries have a high WSSG at the internal-external carotid junction (walls II and III) that rapidly decreases. Peak wall shear stress gradients are bimodal on the lateral walls (walls I and IV) or the normal carotid bifurcation. The extent of susceptible sites for atherosclerotic developments compares favorably with data from angiograms. The peak WSSG in the patch reconstructed carotid artery is somewhat higher and located at the distal end of the patch on walls I and II (Fig. 1.20b). However, in the smoothly tapered carotid bifurcation, peak WSSGs are quite low and primarily occur on the lateral or divider walls near the bifurcation (Fig. 1.20c).

Summary

These results predict a marked dependency of both wall shear stresses and wall shear stress gradients on the geometry of the carotid artery bifurcation. Specifically, large sustained stress gradients occur locally, which may trigger a cascade of biological events leading to the development of atherosclerotic lesions. The normal or primary reconstructed carotid artery with a bulb segment is predicted to have relatively high wall shear stress gradients in the bulb, a site of atherosclerotic developments.[11,77,78] Reconstruction with a patch decreases the flow disturbances and wall shear stress gradient in the bifurcation area, but transfers the peak values downstream to the distal end of the patch. A smoothly tapered reconstruction with a lower bifurcation angle and eliminated bulb segment appears to be the most favorable hemodynamic configuration in terms of minimal wall shear stress gradients. The bifurcation or stagnation point has high shear stress gradients in all three geometries. This is to be expected and probably is not clinically important because of the minimal amount of wall surface involved and special apex cell structure encountered.

To extrapolate these results to carotid endarterectomy reconstruction is interesting. The optimal reconstruction technique and geometry remain to be defined, but these results may give direction. Some surgeons believe patch reconstruction is unnecessary except perhaps in selected cases, while others prefer routine patch reconstruction. Both methods have been shown to give excellent short- and long-term results. Perhaps the optimal reconstruction is a smoothly tapered transition from the common carotid artery to the internal carotid, such as that produced by inter-positioning grafting with autologous saphenous veins. However, the problem of technical errors produced in small arteries, residual proximal and distal atherosclerotic disease, and a thrombogenic endarterectomized surface remain.

In spite of these problems, carotid endarterectomy in the hands of a technically skilled and experienced surgeon has a very low incidence of early post-operative neurological complications from thrombosis and/or embolisms. Optimal reconstruction may further improve outcome. Specifically, consideration should be given to reconstructive techniques that are smoothly tapered with a minimal endarterectomized surface and a complete endarterectomy. Plication shortening of the internal carotid artery endarterectomized segment with a gradually tapered patch or primary closure may give an optimal hemodynamic geometry. In summary, optimal reconstruction may include elimination of the carotid bulb, reduction of the bifurcation angle, and avoidance of acute diameter changes such as that produced at the distal end of a patch.

Bypass Graft-to-Artery Anastomosis

In general, the graft-artery bypass has two types of configurations, i.e., end-to-side (Fig. 1.21a) and end-to-end (Fig. 1.21b). The first configuration, like branching arteries, may have different branching angles (α and β), graft-to-artery diameter ratios (d_g/d_a), tapering of the host artery ($d_{a1} > d_{a2}$), junction curvatures, and placement of the graft relative to an arterial branch. There is still much work necessary to sort out the impact of various geometric and material variables on the hemodynamics around the junctions under different flow conditions. The second configuration, which has varying angle and diameter ratio, is outside the scope of our concern.

Figs. 1.22a-c show the geometric configurations of typical end-to-side anastomoses. In this study, we will focus on the distal (recipient) end anastomosis because most of the graft failure is caused by the distal end intimal hyperplasia development, especially for synthetic grafts.[13,30] Conventional 30° distal anastomoses commonly seen in vascular bypass surgeries with graft-to-artery diameter ratios of 1:1 and 2:1 and a standard flow rate ratio $Q_g:Q_a:Q_u = 100:80:20$ have been considered as base cases.[68] The 2:1 or larger diameter ratio anastomoses are often used in arteries with small diameters such as below-knee popliteal and tibial arteries.[10,12] It is worth noting that the suture line of the anastomosis usually does

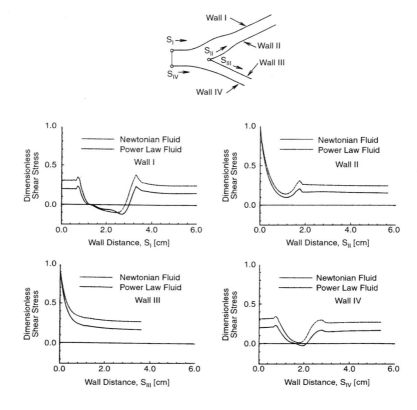

(A)

FIGURE 1.19 The time-averaged, non-dimensionalized wall shear stress along all four walls: (a) NCAB; (b) patch reconstructed geometry; (c) tapered geometry.

not meet with the horizontal diametric plane of the artery due to the nature of common surgical procedures, which causes a slight cross-sectional area reduction and shape change (from circular to elliptic) of the graft near the suture line.[12,45,79] The 2:1 graft-to-artery anastomosis has an even larger graft area reduction and shape change as it approaches the suture line. Based on these two configurations, a variety of geometric changes including the Taylor patch,[10] the effects of branching angle, graft-to-artery diameter ratio, the junction surface curvature, and the position of suture line, etc., have been investigated, and near optimal designs of the distal end anastomosis are suggested based on our WSSG predictor model. Additional effects such as blood rheology, flow rate ratios, and input flow waveforms have been investigated elsewhere.[68]

Two basic input flow waveforms are used in this study for the distal end anastomoses in the femoral artery region. One is called the resting pulse, which is taken from patients at a quiescent state as illustrated in Fig. 1.23a.[80] The frequency for this pulse (f) is 60 beats/min with the maximum and mean Reynolds numbers (Re_{max} and Re_{mean}) equal to 600 and 113.5, respectively. Another waveform is called the exercise pulse, which is taken from patients under moderate exercise conditions as shown in Fig. 1.23b.[81] The frequency for this pulse f is equal to 100 beats/min with $Re_{max} = 480$ and $Re_{mean} = 356.5$. The mean flow rate under the exercise pulse is about three times larger than that under the resting pulse. The resting pulse can be approximated by the following formula:[82]

$$Re = Re_0 e^{-at} \sin(\omega t) \tag{1.31}$$

where the Reynolds number Re can be replaced by the flow rate Q since $Re = 4\rho Q / \pi \mu d_a$, and the angular frequency ω can be a function of time, e.g., $\omega = \omega_0 t^{n-1}$. In Fig. 1.23a, n = 1.5, $\omega_0 = 15.17$, a = 8.357, and

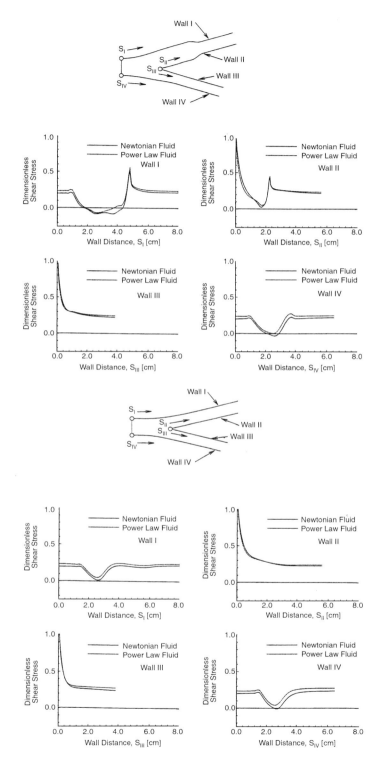

FIGURE 1.19 (CONTINUED)

$Re_0 = 2708.7$. By properly choosing the parameters Re_0, a, and ω (t), the features of the curve, i.e., the maximum value of Re and its location t_{max} (time at Re_{max}), the main and secondary forward flow periods, and the period and magnitude of the reverse flow can be manipulated to fit the physiological flow curves at most of the locations in the artery system.[80,83,84]

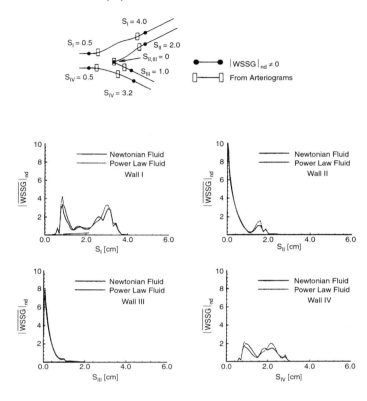

(A)

FIGURE 1.20 The time-averaged, non-dimensionalized wall shear stress gradients along all four walls: (a) NCAB; (b) patch reconstructed geometry; (c) tapered geometry.

Since the diameter of the arteries of interest varies from location to location and from one person to another, we normalize it to $d_a = 1.0$ cm for our numerical analyses, which results in a scaling factor $S_f = d_a/d_a^*$ between the numerical model and the actual physical model, with d_a^* being the original arterial diameter. All other geometric dimensions are multiplied by this factor. By keeping the Reynolds number unchanged, the velocity in the scaled model becomes $\vec{v} = \vec{v}^*/S_f$, and the flow rate increases by the factor S_f. As a result, the wall shear stress and wall shear stress gradient calculated are

$$\tau_w = \tau_w^* / S_f^2 \text{ and WSSG} = WSSG^* / S_f^3,$$

where the superscript * denotes the values in the prototype. Therefore, the actual values for the wall shear stress and wall shear stress gradient should be the calculated values multiplied by S_f^2 and S_f^3 respectively. In the following, the data presented are the calculated values. When comparing with experimental results in different size arteries of different individuals or species, one should take into account the effects of these scaling factors.

Base Case

The $d_{graft}:d_{artery} = 1:1$, $30°$ distal anastomosis (i.e., the base case) is the most common configuration encountered in vascular graft-bypass surgery.[12] The lowest point of the suture line is $1/4 d_a$ above the

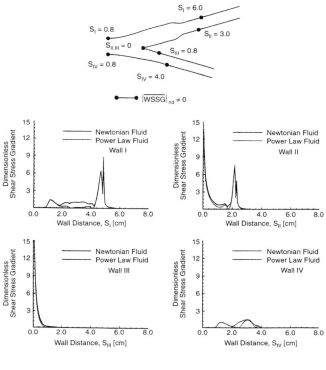

(B)

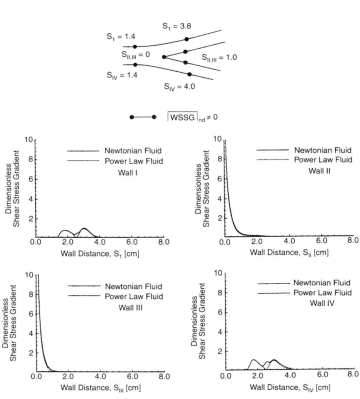

(C)

FIGURE 1.20 (CONTINUED)

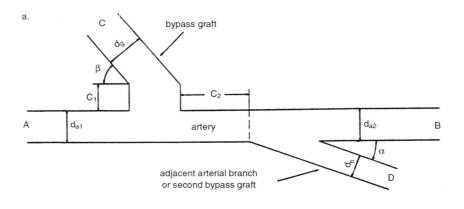

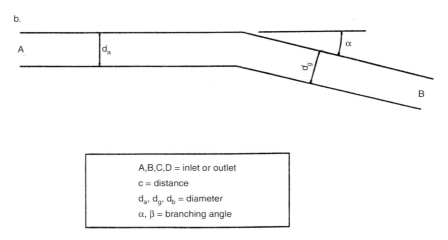

FIGURE 1.21   General artery-graft bypass configurations.

horizontal arterial mid-plane. We take $Q_g:Q_a:Q_u = 100:80:20$ as the standard flow rate ratio, i.e., 20% of the blood coming from the bypassing graft flows back toward the upstream artery section since there are often small arterial branches in that region which need blood supply. Other flow rate ratios are taken as variations to this standard ratio. The two flow waveforms shown in Fig. 1.23a and b, representing both resting and active states of the human body, are applied to this configuration as the input pulses because they envelope a large spectrum of flow waveforms in the femoral artery. The results are presented below, including the transient flow fields, transient and time-averaged wall shear stresses, and wall shear stress gradients. The oscillatory shear index (OSI) as defined in Eq. 1.2a is plotted for comparison purpose.

Flow Field

Figs. 1.24a-c show the velocity vector fields in the vertical mid-plane and horizontal arterial mid-plane (A-A view) for the base case I distal anastomosis at three typical time levels of the resting pulse, representing the flow acceleration phase (t_1), the flow deceleration phase (t_2), and the reverse flow phase (t_3). During the acceleration phase (Fig. 1.24a), the flow is relatively smooth around the junction. No obvious flow recirculation can be seen in both the vertical and horizontal mid-planes. There is a flow stagnation point on the arterial floor, or bed, close to the upstream corner of the junction. During the deceleration phase (Fig. 1.24b), the flow stagnation point on the arterial floor moves downstream to the

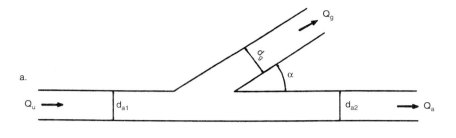

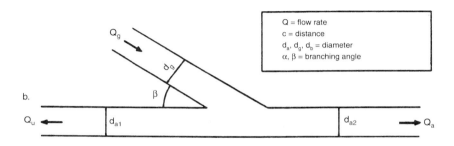

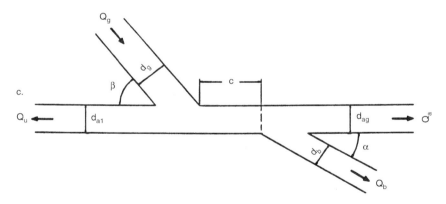

FIGURE 1.22   Typical end-to-side anastomoses.

center of the junction. A large recirculating zone appears just upstream of the stagnation point in the host artery. The velocity is increased in the center region of the flow duct and retarded near the wall. A small quasi-static flow region on the artery side near the toe can be observed in the vertical mid-plane. In the horizontal mid-plane (A-A view), an obvious U-shaped forward-flow/back-flow boundary can be detected with the zero-flow point moving upstream towards the center and downstream near the wall. The secondary flow in the graft (B-B view) is not significant, but in the artery (C-C view) it forms two vortices rotating in the opposite directions. During the reverse flow phase (Fig. 1.24c), the velocity is large around the heel and toe and small in the center of the flow duct. Vortices appear in both the artery and the graft. The stagnation point on the arterial floor moves further downstream and separates the two large vortices in the host artery. Secondary flow is again very small in the branch and relatively large in the host artery, which means the flow in the artery is more disturbed. In summary, the flow is smoother

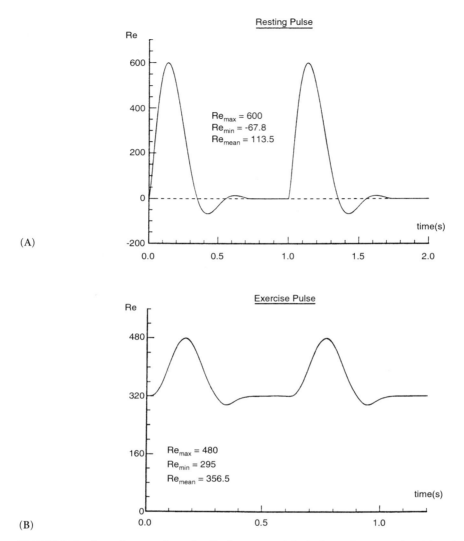

FIGURE 1.23 Input flow waveforms for distal anastomosis in the femoral artery region: (a) resting pulse; (b) exercise pulse.

in the acceleration phase and more disturbed in the deceleration and reverse flow phases. The velocity is always small in the upstream artery section due to the low flow rate ratio.

Figs.1.25a and b show the velocity vector fields of the base case I configuration at two typical time levels of the exercise pulse, i.e., the peak flow phase (t_1) and the minimum flow phase (t_2). The peak flow phase is at the end of the flow acceleration, which shows a recirculating zone upstream of the stagnation point (Fig. 1.25a). The flow pattern in the horizontal mid-plane is similar to the flow pattern in the deceleration phase of the resting pulse (Fig. 1.25b). The minimum flow phase (Fig. 1.25b) is at the end of the flow deceleration, showing a remarkable recirculating region in the artery near the junction heel, which is similar to the deceleration phase of the resting pulse. However, the flow is much more skewed in the downstream artery section in the vertical mid-plane, and it forms M-shaped velocity profiles in the horizontal mid-plane. Therefore, a stronger secondary flow is expected in the downstream artery (C-C view). There is no secondary flow in the graft (B-B view).

The stagnation point on the arterial floor is moving all the time as illustrated in Fig. 1.26. With the resting pulse, it moves from heel to toe during forward flow phases, covering a wide range of arterial

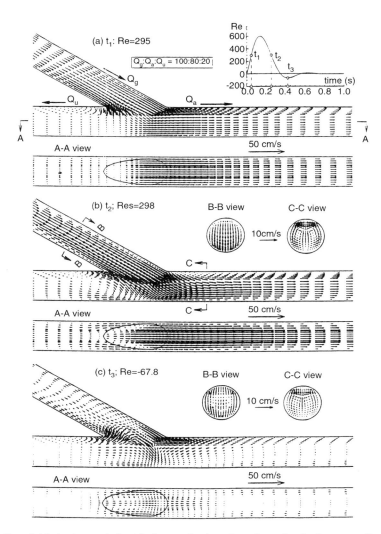

FIGURE 1.24 Flow fields in the vertical mid-plane and other view planes for the base case distal anastomosis at three typical time levels with the resting pulse.

floor at the junction. At the end of the flow deceleration and during reverse flow phase, the stagnation point is either non-existent or not well defined. While under the exercise pulse condition, the stagnation point moves only a short distance. It is almost stationary at the middle of the junction. It appears closer to the heel when the Reynolds number is large and farther when the Reynolds number is small.

Wall Shear Stress Vectors
Figs. 1.27a-c and 1.28a and b show the three-dimensional surface plots of the wall shear stress vectors at different time levels with the resting and exercise pulses, respectively. The τ_w vector plots provide a different view for the blood flow behavior around the anastomosis. These graphs demonstrate that during the flow acceleration phase (Fig. 1.27a), the wall shear stress has a very uniform distribution pattern. The only low shear stress region is around the stagnation point and extends into the whole upstream portion of the artery because of the small upstream flow rate. The relatively high shear stress regions appear around the heel and toe. During the deceleration phases (Figs. 1.27b and 1.28b), nearly the entire graft surface experiences low shear stresses although the Reynolds number is the same as during the acceleration phase. The heel and toe areas of the junction are still relatively high shear stress regions, but the extent

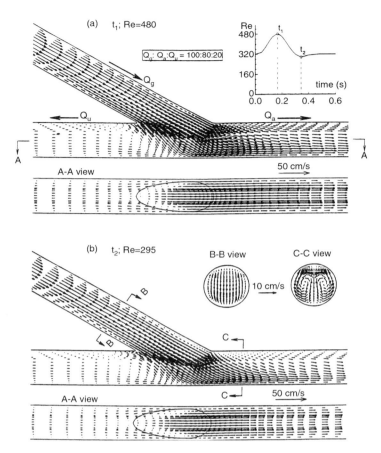

FIGURE 1.25   Flow fields in the vertical mid-plane and other view planes for the base case distal anastomosis at two typical time levels with the exercise pulse.

of the stress is greatly reduced. Another phenomenon evident is that the wall shear stress downstream of the stagnation point increases significantly compared with the acceleration phase. However, beyond the heel, τ_w drops rapidly. In comparison, the reverse flow phase (Fig. 1.27c) has a relatively smoother τ_w distribution pattern than the deceleration phase. There is a small low shear region on the arterial floor opposite the toe. The peak flow interval of the exercise pulse (Fig. 1.28a) has a τ_w distribution pattern intermediate to the acceleration and deceleration phases. The wall shear stress distribution in the graft is quite smooth, similar to the acceleration phase of the resting pulse; however, the distribution in the artery is similar to the deceleration phase. Again, according to the wall shear stress distribution patterns, we can conclude that more flow disturbances are introduced during the flow deceleration phase.

Comparison of the WSSG with Other Physical Indicators

Figs. 1.29a-c and 1.30a-c show the time-averaged wall shear stress and wall shear stress gradient as well as the OSI around the base case distal anastomosis with the resting and exercise pulses, respectively. For both pulses, the wall shear stress contours show high shear regions ($\tau_w > 1.5\ \tau_0$) at the heel and on both sides of the stagnation point zone (Figs. 1.29a and 1.30a). There are low shear regions ($\tau_w < 0.5\ \tau_0$) in the stagnation point zone, the toe area, and at the middle of the suture line. Except for these regions, nearly the entire graft surface and the upstream artery section are in the low shear region. The high shear stress region in the downstream artery also extends far from the junction area. Therefore, the high/low shear stress models and the safe τ_w band width concept cannot pinpoint the susceptible sites for intimal hyperplasia development. The OSI distribution pattern is different for the two pulses. For the resting

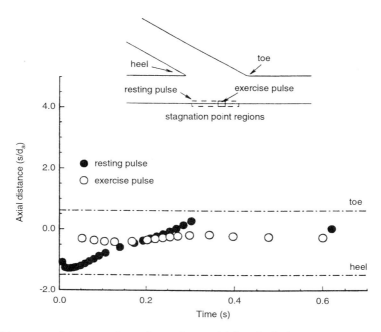

FIGURE 1.26 Movement of the stagnation point on the arterial floor in the base case distal anastomosis for both the resting and exercise flow waveforms.

pulse (Fig. 1.29c), large OSIs appear in the stagnation point region, the upper wall of the upstream artery a short distance from the heel, and along the suture line. However, for the exercise pulse (Fig. 1.30c), the large OSIs appear only at the center of the suture line and the upper wall of the upstream artery. The heel, toe, and arterial floor all experience low OSI values, which does not match the clinical observations about the intimal hyperplasia distribution patterns.[13] In contrast, the WSSG contours consistently show large non-zero WSSGs around the critical areas for both input pulses and match the *in vivo* intimal hyperplasia distribution patterns (Figs. 1.29b and 1.30b). These graphs also demonstrate that the maximum WSSGs are proportional to the mean Reynolds number of the input pulse.

Effects of Suture Line Position
The suture line of the base case distal (recipient) anastomosis does not match the width of the horizontal arterial mid-plane due to the slit-like incision and the contraction of the cutting edge (the oval on the A-A view plane vs. the arterial boundaries in Figs. 1.24 and 1.25). The width of the oval (projection of the suture line onto the A-A plane) is dependent on the graft material. The softer the graft, the narrower the oval. What will happen if the anastomotic orifice is not squeezed and the suture line descends to meet the horizontal arterial mid-plane? The following graphs show the results in a modified 1:1 30° distal anastomosis with the suture line matching the width of the artery and with smoothed wall curvatures at the junction heel and toe.

Figs. 1.31a and b show the velocity vector plots for the modified base case configuration at two typical time levels of the exercise pulse. The flow fields in both vertical and horizontal mid-planes are obviously smoother than for the original geometry, especially at the minimum flow point (t_2) (Figs. 1.31a and b). The secondary flow in the artery is substantially reduced. Other noticeable changes of the flow field are the velocity increase at the toe and the decrease at the heel, both in the artery. Consequently, the low shear stress region at the toe on the artery side is gone, and the high shear region upstream of the stagnation point region is greatly reduced (Fig. 1.32a). Other features of the time-averaged wall shear stress contours remain. However, the time-averaged WSSG contours show a great reduction of the maximum WSSG value (from 171.2 in Fig. 1.30b to 52.7 in Fig. 1.32b). The WSSG value around the toe area is below the first contour value (6.0). For an artery with diameter larger than 5 mm ($S_f = 2$), the

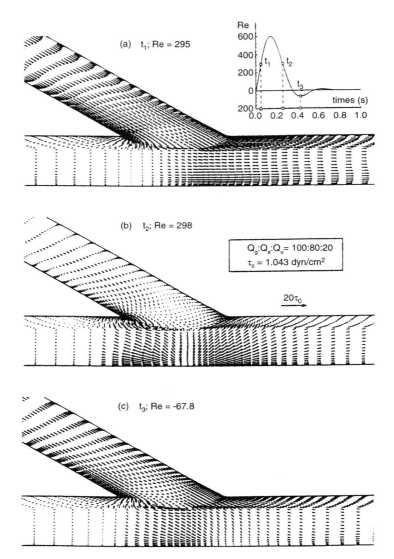

FIGURE 1.27 Three-dimensional surface plots of the wall shear stress vectors for the 1:1 30° distal anastomosis at three typical time levels with the resting pulse.

actual $|\overline{\text{WSSG}}|$ there is below 48 dyn/cm³, which is possibly lower than the critical value of WSSG (e.g., $\text{WSSG}_{\text{ltv}} = 60$ dyn/cm³, a value suggested for New Zealand Rabbits). This means that intimal hyperplasia could be prevented in the toe area for arteries with diameters larger than 5 mm with the modified 1:1 base case geometry. However, for arteries with diameters less than 5 mm, further geometric improvements are needed.

Taylor Patch Connector

The Taylor patch is a surgical technique developed by Dr. Taylor and his medical group for synthetic graft artery bypasses.[10] The basic idea is to use a vein patch connecting the synthetic graft and the artery (Fig. 1.33) so that the compliance mismatch between the graft and the artery can be mitigated and a smaller branch angle can be achieved. The patency rate associated with the use of the Taylor patch connector is relatively higher than the use of the basic 30° connectors. We include this anastomotic configuration in our model studies for comparison purposes. Since the shape and diameter ratio of the

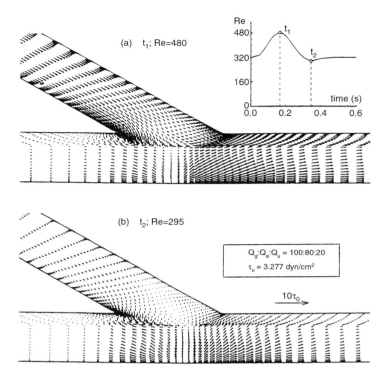

FIGURE 1.28   Three-dimensional surface plots of the wall shear stress vectors for the 1:1 30° distal anastomosis at two typical time levels with the exercise pulse.

Taylor patch vary from time to time and from location to location, depending on the size of artery and the surgeon's personal experience, we selected the 2:1 graft-to-artery diameter ratio and made the shape as shown in Fig. 1.33 according to the description and sketches of Taylor et al.[10] For simplicity, the entire anastomosis is assumed to be rigid. The resting pulse is used as the input flow waveform and the standard flow rate ratio, i.e., $Q_g:Q_a:Q_u = 100:80:20$, is assumed for this geometry. The results are presented below.

Figs. 1.34a-c are the velocity vector plots for the Taylor patch connector at three typical time steps of the resting pulse. During the flow acceleration phase (t_1), the flow is relatively smooth in the whole area with increased velocity at the exit of the junction due to the area reduction. The stagnation point on the arterial floor is close to the heel. There is no noticeable recirculating zone. The secondary flow in the downstream artery is very weak. During the deceleration phase (t_2), an obvious recirculating zone appears in the upstream artery near the junction heel. The secondary flow in the C-C view plane becomes much stronger than during the acceleration phase although the magnitude of the velocity is much smaller than that of the main flow. During the flow reverse phase (t_3), low velocities are everywhere in the center regions of the junction, the graft, and the artery. Relatively high velocities appear at the heel and toe. The secondary flow is much weaker than the deceleration phase.

Figs. 1.35a-c show the various indicators for the Taylor patch under resting conditions. The time-averaged wall shear stress contours (Fig. 1.35a) show two tiny high shear stress regions with one at the heel and another on the arterial floor opposite the toe. Nearly the entire remaining area, except near the toe, is exposed to low shear stresses. According to low shear stress theories, this implies the whole junction area, except the toe area, is susceptible to intimal hyperplasia development. This does not match clinical observations. The time-averaged WSSG contours are quite simple, which demonstrates two distinct areas of high WSSG values, one at the heel and one at the toe. The small area of high WSSG on the arterial floor around the stagnation point region, which appeared in the 2:1 base case anastomosis, does not appear for this configuration. The WSSG values at both the heel and toe are larger than those in the 2:1

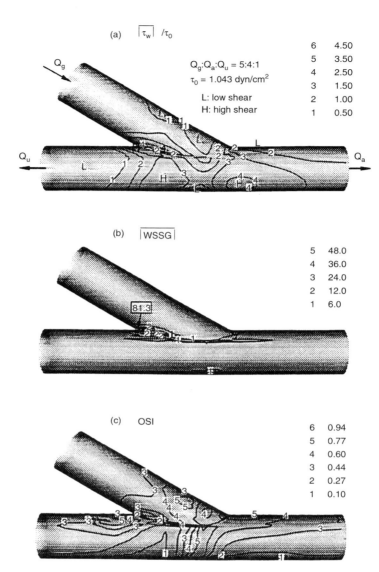

FIGURE 1.29   Three-dimensional contour plots of (a) time-averaged wall shear stress; (b) time averaged WSSG; and (c) OSI for the 1:1 30° distal anastomosis with the resting pulse.

base case due to the relatively sharper geometric changes at those two corners. However, the high WSSG-values are confined to the two small areas. The remaining area around the junction is below the minimum WSSG value (1.5). The severity parameter (S_{RE2}) (Eq. 1.6) for the Taylor patch is about 1.126, which is less than that for the 2:1 base case (1.318) under the same flow conditions. The OSI contours are similar to the 2:1 base case with one additional high OSI region in the curved hood area of the junction. If the critical OSI value is set to 0.3 or lower, the predicted susceptible sites will be too large. If the cut-off value is set to 0.45, the whole toe area will be a low OSI region, which means that the toe area will be free of intimal hyperplasia development. However, the toe area is in fact the most susceptible and critical site for intimal hyperplasia. Only the WSSG model can give the correct prediction here.

Optimal Design of the Distal End Anastomotic Connector

In the previous sections, we saw that geometric changes had substantial effects on the flow and WSSG distribution patterns around the anastomotic area. As shown by Lei,[68] a 30° S-shaped connector constituted

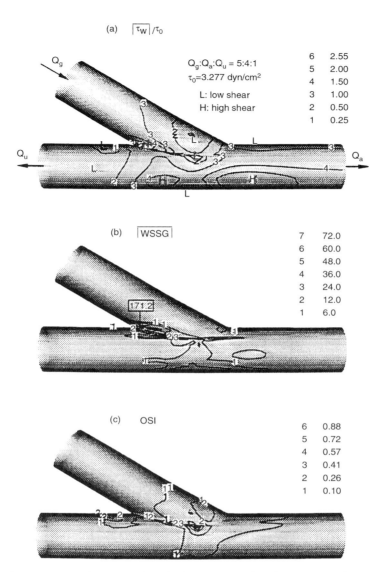

(a) $|\tau_w|/\tau_0$

Q_g

$Q_g{:}Q_a{:}Q_u = 5{:}4{:}1$
$\tau_0 = 3.277$ dyn/cm^2

L: low shear
H: high shear

Q_u

Q_a

6	2.55
5	2.00
4	1.50
3	1.00
2	0.50
1	0.25

(b) |WSSG|

7	72.0
6	60.0
5	48.0
4	36.0
3	24.0
2	12.0
1	6.0

(c) OSI

6	0.88
5	0.72
4	0.57
3	0.41
2	0.26
1	0.10

FIGURE 1.30 Three-dimensional contour plots of (a) time-averaged wall shear stress; (b) time averaged WSSG; and (c) OSI for the 1:1 30° distal anastomosis with the exercise pulse.

a remarkable improvement over all other geometric configurations based on the reduction of WSSGs over the entire junction area. Further improvement of the S-connector can be achieved by using variable anastomotic angles at the graft end so that the curvatures of the junction at the heel and toe can be determined separately based on the reduction of the WSSGs at these locations. The sketch and control parameters for the new S-connector designs are illustrated in Figs. 1.36a and b. The starting angle (α_1) of the S-connector at the heel, which determines the junction curvature there, is set to 45° because this value is proven to be the best angle at the upstream corner of the junction.[52] The second angle (α_2), which is defined as the angle of the tangential line at the inflection point of the downstream hood curve, should be as small as possible depending on the diameter ratio and the cutting length (l_{cut}) required. The smaller the angle, the longer the cutting length.

As an example, we present two S-connector designs here. One has a 2:1 diameter ratio with $\alpha_1 = 45°$ and $\alpha_2 = 35°$, which results in a cutting length of $l_{cut} \sim 7.766\ d_a$ (Fig. 1.37a). If the second angle (α_2) is

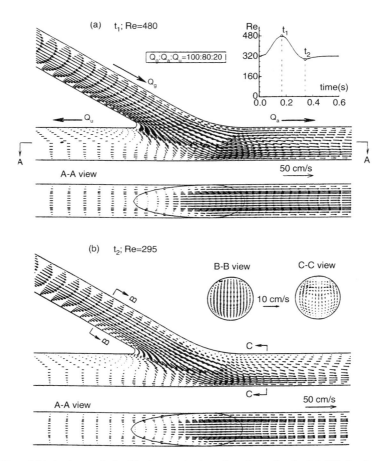

FIGURE 1.31 Flow fields in the vertical mid-plane and other view planes for the 1:1 30° distal anastomosis with a modified suture line position under the resting pulse.

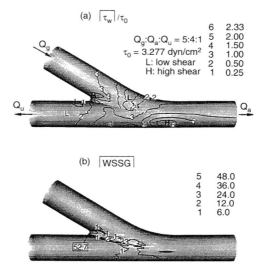

FIGURE 1.32 Three-dimensional contour plots of (a) time-averaged wall shear stress; and (b) time-averaged WSSG for the 1:1 30° distal anastomosis with a modified suture line position under the exercise pulse.

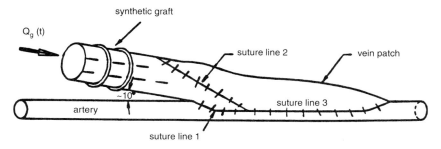

FIGURE 1.33 The Taylor patch (redrawn).[10]

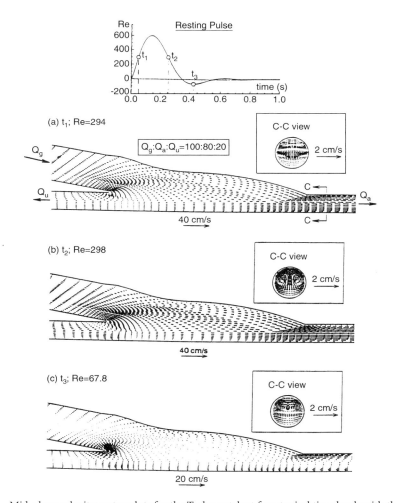

FIGURE 1.34 Mid-plane velocity vector plots for the Taylor patch at four typical time levels with the resting pulse.

set to 25° with the same diameter ratio, the cutting length will be about 11.6 d_a, which is too long. The second design reduces the graft-to-artery diameter ratio (d_g/d_a) to 1.6 and sets $\alpha_2 = 25°$, which requires a cutting length of about 9.79 d_a (Fig. 1.37b). Both designs show remarkable improvement on the reduction of WSSG values around the junction (Figs. 1.37a and b). The maximum WSSG values at the heel are greatly reduced with both designs compared to the 30° S-connector[68] even though a higher frequency (100/min vs. 60/min) exercise pulse is used. In the next section, we will show that a higher frequency generates larger WSSGs. The second design (1.6:1 S-connector) with a smaller downstream angle (α_2) shows a further reduction of the WSSG value in the toe area. Although the 2:1 S-connector with a 45°/35° combination shows a wide area of non-zero WSSGs in the toe region due to a larger angle α_2, the magnitude of the WSSG is small, i.e., smaller than both the Taylor patch connector and the 30° S-connector.[68] Under the resting pulse condition (Figs. 1.38a and b), even smaller WSSG are obtained at the toe area. A small non-zero WSSG region appeared on the arterial floor for the 1.6:1 S-connector with the exercise pulse but vanished with the resting pulse. A slight increase of the WSSG at the heel makes no significant difference since the location of the high WSSG area is within the graft, which contributes less to the intimal hyperplasia development compared with the WSSGs at the toe.

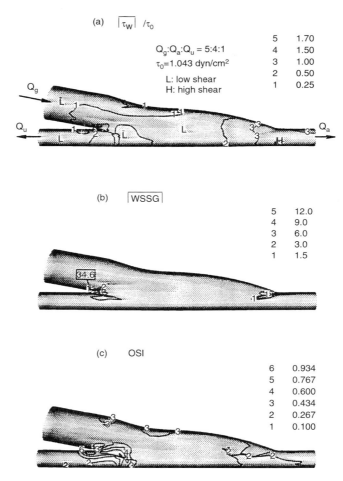

FIGURE 1.35 Three-dimensional contour plots of (a) time-averaged wall shear stress; (b) time-averaged WSSG; and (c) OSI for the Taylor patch with the resting pulse.

Figs. 1.39a and b show the velocity vector fields for these two S-connectors with the exercise pulse. No significant difference is observed. The velocity in the graft is larger in the 1.6:1 S-connector than in the 2:1 S-connector, which is expected. The secondary vortices in the downstream artery in the first design are slightly stronger than the second design. Both of them are much weaker than the Taylor patch (Fig. 1.34b). The location and movement of the stagnation point on the arterial floor are similar for these two designs. Figs. 1.40a and b show the movement of the stagnation point under the resting pulse for

(a)　　　　Sketch of the S-connector

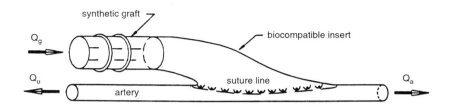

(b)　　　　Design parameters of the S-connector

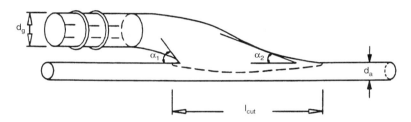

FIGURE 1.36　Sketch and design parameters for the new S-connector.

the two S-connectors. The stagnation point region is closer to the heel for the 1.6:1 S-connector, but the trend of stagnation point movement is similar.

Figs. 1.41a and b show the wall shear stress vector plots for these two designs at a typical time step of the exercise pulse. A relatively uniform wall shear stress distribution over the junction area is achieved.

The temporal variations of the wall shear stress and wall shear stress gradient at ten critical points in the 2:1 S-connector configuration with both the exercise and resting pulses are plotted in Figs. 1.42a and b and 1.43a and b. From Figs. 1.42a and b, one can see that although the wall shear stress differs at different points, its variations with time at all points almost exactly follow the input flow waveforms which are represented by the wall shear stress variations at the control points (artery inlet with the same amount of flow; dotted lines). Note that the absolute values of the wall shear stresses are used here. For the exercise pulse (Fig. 1.42a), relatively large wall shear stress variations occur at the junction heel (points 1, 10) and toe (points 4-6). The magnitude of τ_w at the toe is larger than the control point, however, the temporal gradient is comparable with the control point. Therefore, according to the temporal wall shear stress gradient hypothesis, the S-connector configuration should be basically spared of intimal hyperplasia development under the exercise pulse condition. However, with the resting pulse (Fig. 1.43b), the junction heel and toe as well as all other locations have quite large temporal wall shear stress gradients. Hence, the prediction is quite dependent on the input flow waveform. It can be expected that a flat input flow waveform with a large flow rate will generate zero temporal τ_w gradient in the entire junction area. As discussed for the aorto-celiac junction, the temporal τ_w gradient is completely determined by the magnitude of the local mean wall shear stress and the flow waveform. Therefore, it cannot be used as an independent predictor.

The temporal variations of the wall shear stress gradient also basically follow the input flow waveforms at all points (Fig. 1.43a and b). This time, the heel (point 1) has the largest WSSG variation instead of the toe for both pulses. The stagnation point region (point 7), which has the smallest τ_w variation, has

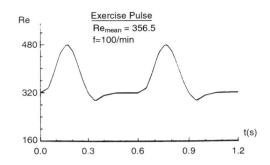

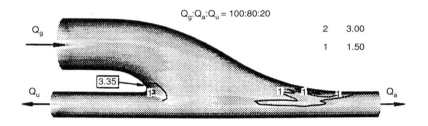

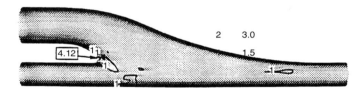

FIGURE 1.37   Three-dimensional contour plots of the time-averaged WSSG for the S-connector designs under exercise pulse conditions: (a) 2:1 S connector with 45°/35° angle combination; (b) 1.6:1 S-connector with 45°/25° angle combination.

a relatively large WSSG variation. Some low shear stress locations (points 9,10) also have relatively large WSSG variations. However, like the wall shear stress variation, the temporal WSSG variation is determined by the time-averaged WSSG and the flow waveform. It is not very useful for the prediction of possible intimal hyperplasia development sites.

Effects of Flow Input Waveforms

The blood flow waveform varies from one location to another in the artery system and from resting to exercise following human activities. It is also different from person to person and may change during sickness. In order to accommodate these facts, we investigated the effects of input flow waveform on the WSSG distribution pattern by varying the frequency, mean Reynolds number (or mean flow rate), and shape of the flow waveform in the 2:1 S-connector (45°/35°) and the 2:1 anastomosis.

Figs. 1.44a and b demonstrate the effect of frequency on the WSSG distribution around the 2:1 S-connector based on the resting pulse. Fig. 1.44a is the result under the standard resting pulse with a

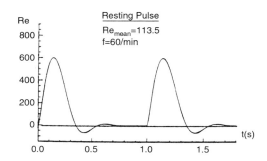

(a)   2:1 S-connector (45°/35°)

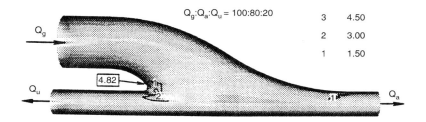

(b)   1.6:1 S-connector (45°/25°)

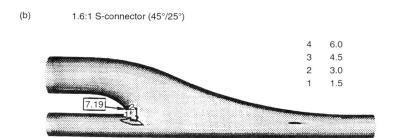

FIGURE 1.38 Three-dimensional contour plots of the time-averaged WSSG for the S-connector designs under resting pulse condition: (a) 2:1 S-connector (45°/35°), (b) 1.6:1 S-connector (45°/25°).

frequency f = 60/min, which is a duplicate of Fig. 1.38a shown here for easy comparison. When the frequency is increased to 100/min with the shape and Re_{mean} unchanged, the WSSG values are increased at both the junction heel and toe (Fig. 1.44b). The extent of non-zero WSSG regions are also enlarged. Figs. 1.45a and b show the effect of the mean Reynolds number based on the exercise pulse. When Re_{mean} is lowered from 356.5 (Fig. 1.45a) to 113.5 (Fig. 1.45b) by keeping the frequency and waveform unchanged, the WSSG value is dramatically reduced. Note that the lowest contour line value is reduced to 1.0 since the WSSG is very small under the low Re_{mean} flow conditions.

Figs. 1.46a-c show the WSSG distribution patterns in the 2:1 base case anastomosis under different input flow waveforms, the exercise pulse (f = 100/min, Re_{mean} = 356.5), the resting pulse (f = 60/min, Re_{mean} = 113.5), and the Inokuchi pulse (f = 60/min, Re_{mean} = 142.0). The results show that the high frequency, large mean flow input pulse yields the largest WSSG values and affected regions. The Inokuchi pulse has a larger Re_{mean} than the resting pulse but generates smaller WSSGs.

The results presented above demonstrate that the WSSG distribution depends not only on the frequency and mean Reynolds number of the input flow waveform, but also on the shape of the waveform.

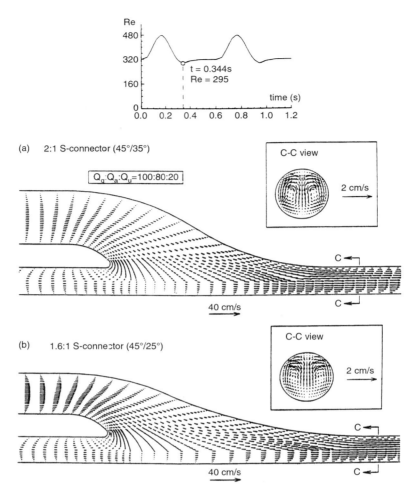

FIGURE 1.39   Velocity vector fields for the S-connector designs: (a) 2:1 S-connector (45°/35°); (b) 1.6:1 S-connector (45°/25°).

To quantitatively assess the impact of the input flow conditions on the WSSG distribution at the anastomotic site, the severity parameter defined in Eq. 1.6 is calculated and correlated to the characteristic flow waveform parameter N defined in Eq. 1.7 for these two geometries. The threshold value, $WSSG_0$, is assumed to be 1.5 dyn/cm^3 for the calculation. The results are shown in Table 1.2 and plotted in Fig. 1.47. The graph establishes a roughly linear correlation, i.e.,

$$S_{RE} = mN + b \tag{1.32}$$

Here the coefficients m and b are functions of the geometry. Different geometries may have different correlation coefficients. For the S-connector, we have approximately $m = 1.05$ dyn/cm^3 and $b = -2.99$ dyn/cm^3. For the 30° 2:1 anastomosis geometry, $m \approx 8.4$ dyn/cm^3 and $b \approx -74.0$ dyn/cm^3. Since S_{RE} is always greater than zero, this linear regression is valid for $N \geq N_{cr}$ only, where N_{cr} is a critical value of N. When N is less than N_{cr}, a new form of correlation should be established. We can approximately set N_{cr} equal to 5 for the S-connector and 9.5 for the 30° 2:1 anastomosis geometry. From Fig. 1.47, one can conclude that the severity of the 30° 2:1 anastomosis geometry is strongly influenced by the flow input waveform, whereas the newly designed S-connector is much less sensitive to the flow waveform. This is a big advantage of the S-connector over the conventional graft-artery bypass configurations. Although

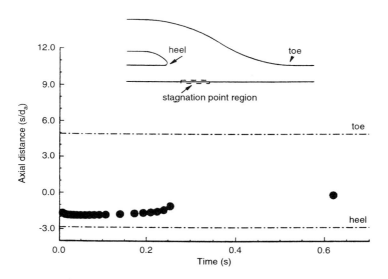

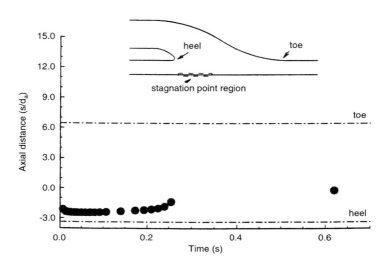

FIGURE 1.40 Movement of the stagnation point for the S-connector designs under resting pulse conditions.

the results presented here may not be exact, the trend predicted is correct. One could use these results as a guide in the design of new geometric configurations or in the evaluation of the severity of natural arterial branches or surgical reconstructions.

Summary

In this chapter, the transient three-dimensional flow fields in different distal end-to-side anastomoses have been numerically analyzed with two basic input flow waveforms, the resting pulse and the exercise pulse. The wall shear stress (WSS) and wall shear stress gradient (WSSG) as well as the oscillatory shear

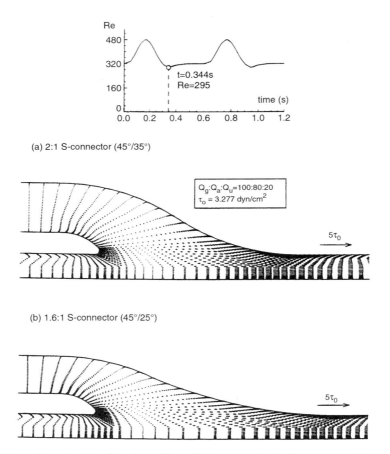

(a) 2:1 S-connector (45°/35°)

(b) 1.6:1 S-connector (45°/25°)

FIGURE 1.41   Three-dimensional surface plots of the wall shear stress for the S-connector designs at a typical time level with the exercise pulse.

index (OSI) have been calculated and their distribution patterns have been compared with published clinical observations of restenosis, e.g., intimal hyperplasia and renewed atheroma developments in the graft-artery junction area. The WSSG appears to be a logical indicator of non-uniform hemodynamics in the anastomotic junction and, therefore, a suitable predictor for intimal hyperplasia and/or atherosclerotic lesion developments. The WSSG has many advantages over other physical disease indicators such as OSI and high/low WSS models. First, the high WSSG values are always located around the junction area in the common regions of intimal hyperplasia developments, i.e., the junction heel and toe, the suture line, and the arterial floor (bed). Second, the WSSG distribution pattern is unique and does not vary significantly with time, i.e., the time-averaged WSSG has the same predictive power as the transient one. Third, the WSSG appears to be the best simple parameter for the representation of disturbed flows in the branching blood vessels. For example, a geometric configuration which generates large secondary flow or recirculating flows is always associated with large WSSGs, while those configurations with smooth flow fields always have small WSSGs. Finally, the WSSG concept has a solid biological background. It can be explained by the zero-tension hypothesis which encapsulates existing theories on atherosclerotic lesion formation and neo intimal hyperplasia development.[68] The high/low shear stress models, the OSI and the temporal variations of the wall shear stress and wall shear stress gradient were all somewhat inadequate for the prediction of sites susceptible to restenosis. However, variations in radial pressure gradients, arterial wall stress and strain, as well as blood particle depositions will be analyzed in future works.

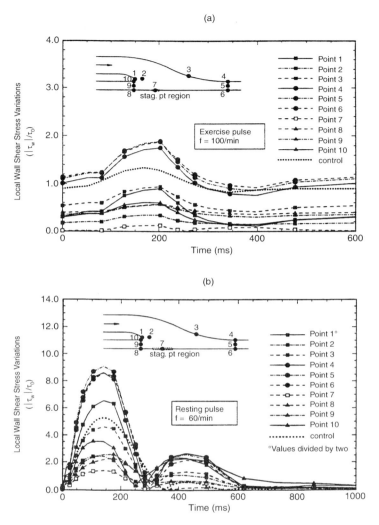

FIGURE 1.42   Temporal variations of the wall shear stress at ten critical points in the 2:1 S-connector: (a) exercise pulse; (b) resting pulse.

In turn, the WSSG predictor model has also been used as a guide for the theoretically optimal design of the distal end graft-artery bypass configurations aiming at the elimination or at least mitigation of intimal hyperplasia and/or atheroma. The basic design parameters and the direction for possible further improvements of the junction geometry have been established. From the simulation results, the suggested near-optimal designs show a significant reduction of WSSG values at the junction. Except for the heel, the whole junction area, including the toe and arterial floor, exhibits low WSSGs. Complete elimination of WSSGs at the junction heel seems impossible for a flow division with backflow. Hence, a pre-manufactured biocompatible insert which maintains all optimal geometric parameters would be a good solution for many reasons. The detailed three-dimensional geometric data sets may be found in Lei.[68] The results presented here indicate that hemodynamically optimal designs are feasible and should be viewed qualitatively.

Extrapolating these results to clinical practice is interesting. If an optimal terminal graft design is manufacturable with several sizes to match arterial diameters and is as easily sutured as a standard end-to-side anastomosis, it may be clinically effective in decreasing the incidence of graft failure. While the thrombogenicity of graft surfaces continues to be a major problem with synthetic materials, an optimally

(a)

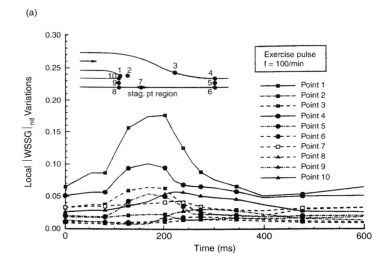

(b)

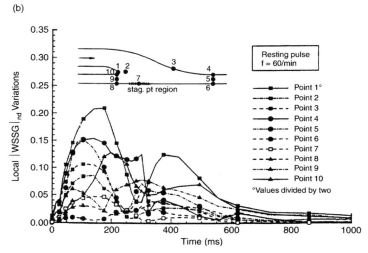

FIGURE 1.43 Temporal variations of the WSSG at ten critical points in the 2:1 S- connector: (a) exercise pulse; (b) resting pulse.

designed terminal geometry may offer some protection against not only myointimal developments, but also the initiation of thrombosis. It is well known that high shear rates from markedly disturbed flows are harbingers of the onset of thrombosis.[85] The proposed hemodynamically optimal terminal graft geometry may reduce the incident of thrombosis.

The finding that a 1.6 to 1 graft-to-artery diameter ratio optimizes hemodynamics is of interest. However, a 2:1 ratio with a simpler input waveform and sub-optimal geometry[56] was comparable. This suggests that standard 6 mm and 8 mm diameter synthetic grafts can be used for femoro-popliteal bypasses, as the ratios are within this window. The results of using these standard grafts for smaller tibial arteries and higher ratios have not been investigated by us.

Since presently no optimal synthetic graft material or geometric configuration is available for clinical use, it is interesting to extrapolate our results to the present technique. We suggest that distal end-to-side anastomosis be performed in a way that allows a smooth transitional curvature at the heel if possible. While the graft-to-artery bevel angle described herein is not of major importance because of

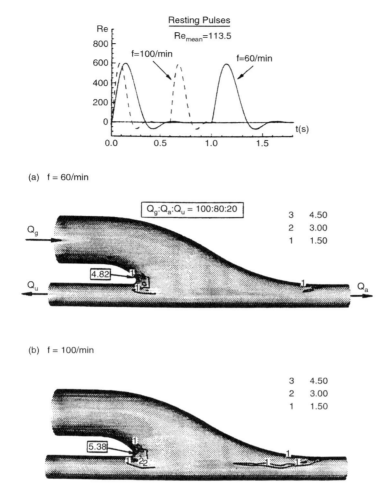

FIGURE 1.44 Three-dimensional contour plots of the time-averaged WSSG for the 2:1 S-connector with the resting pulse of different frequencies.

the optimization of geometry with a smooth transition, it may be advisable to construct the heel with a 10° to 15° heel angle between the graft and artery. The former may be achievable with a cuff-type anastomosis, and the latter by cutting the graft at an appropriate small bevel angle. More importantly, the toe region should be a gradual transition in curvature and cross-sectional area. This recommendation is very similar to the one designed to minimize wall shear stress gradients at the distal end of a carotid endarterectomy patch.[47] This is perhaps best achieved by modifying the cut of the distal end of the graft so that the resulting shape, while still holding the graft flat, looks like a lazy S with a length three or four times the graft diameter, a sharp angle of incidence at the heel relative to the longitudinal axis, and a tapered narrowing toe segment that looks very much like a tapered carotid endarterectomy patch.

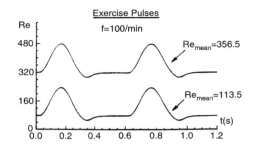

(a) Re_{mean}=356.5

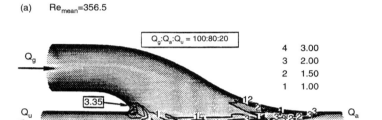

(b) Re_{mean}=113.5

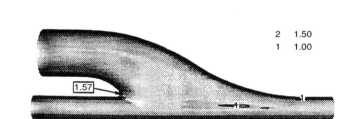

FIGURE 1.45   Three-dimensional contour plots of the time-averaged WSSG for the 2:1 S-connector with the exercise pulse of different mean Reynolds numbers.

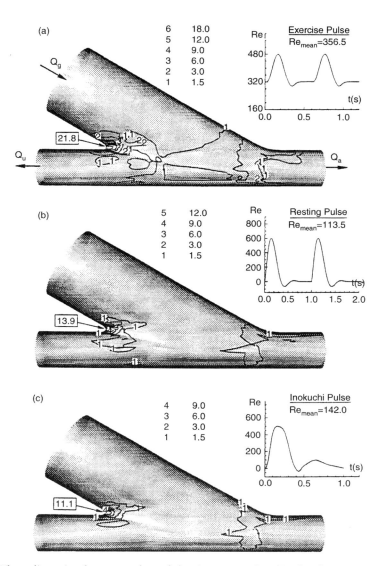

FIGURE 1.46 Three-dimensional contour plots of the time-averaged WSSG for the 2:1 30° anastomosis with different input flow waveforms.

TABLE 1.2   Correlation Between Characteristic Flow Waveform Parameter N and Severity Parameter S_{RE} for 2:1 S-Connector (45°/35°) and the 2:1 30° Anastomosis Geometry

Input Pulse	Exercise		Resting		Inokuchi[84]
f (min^{-1})	100		60	100	60
Re_{mean}	356.5	113.5	113.5		142.0
Re_{max}	480		600		501
Re_{min}	295		−67.8		−38.4
Re_{area}	356.5		217.3		145.0
α	1		3		3
Wo	8.91		6.90	8.91	6.90
N	12.36	6.98	11.81	15.24	10.72
$SP_S{}^a$ [dyn/cm^3]	10.38	4.34	9.48	12.39	—
$SP_a{}^a$ [dyn/cm^3]	33.90	—	22.20	—	18.80

^{a} S = S-connector (45°/35°); a = 2:1 30, anastomosis.

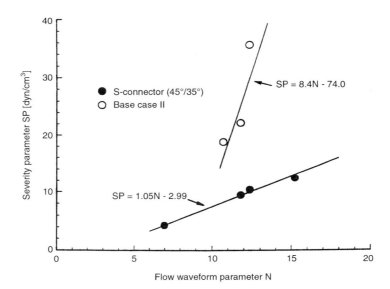

FIGURE 1.47   Correlation between characteristic flow waveform parameter N and severity parameter S_{RE2} for the 2:1 S-connector (45°/35°) and the 2:1 30° anastomosis.

Acknowledgment

The authors wish to thank Cray Research for the financial support of ML in 1993-1995. Use of the software CFX from AEA Technology, Bethel Park, PA is gratefully acknowledged.

References

1. Kleinstreuer, C., Buchanan, J.R., Jr., Lei, M., and Truskey, G.A., in *Biomechanical Systems Techniques and Applications*, Leondes, C., Ed., CRC Press, New York, 2001.
2. Archie, J.P., *Stroke*, 17, 901, 1986.
3. Hertzer, N.R., Beven, E.G., O'Hare, P.J., and Krajewski, L.P., *Ann. Surg.*, 206, 628, 1987.
4. North American Symptomatic Carotid Endarterectomy Trial Collaborators, *New Engl. J. Med.*, 325, 445, 1991.
5. European Carotid Surgery Trialists' Collaborative Group, *Lancet*, 337, 1235, 1991.
6. National Institute of Neurological Disorders and Stroke, *Stroke*, 25, 2523, 1994.
7. Mills, J.L., Fujitani, R.M., and Taylor, S.M., *J. Vasc. Surg.*, 17, 195, 1993.
8. Rosenthal, D., Archie, J., Jr., Garcia-Rinaldi, R. et al., *J. Vasc. Surg.*, 12, 326, 1990.
9. Archie, J.P., *Ann. Vasc. Surg.*, 8, 475, 1994.
10. Taylor, R.S., Loh, A., McFarland, R.J., Cox, M., and Chester, J.F., *Br. J. Surg.*, 79, 348, 1992.
11. Texon, M. *Hemodynamic Basis of Atherosclerosis*, Hemisphere Publishing Corp., New York, 1980.
12. Strandness, D.E., Didiseim, P., Clowes, A.A., and Watson, J.T., Eds., *Vascular Disease: Current Research and Clinical Applications*, Grune and Stratton, Orlando, FL, 1987.
13. Sottiurai, V.S., Vao, J.S.T., Baston, R.C., Sue, S.L., Jones, R., and Nakamura, Y.A., *Ann. Vasc. Surg.*, 3, 26, 1989.
14. McIntire, L.V., *Bioeng. Sci. News*, 15, 51, 1991.
15. Nerem, R.M., *J. Biomech. Eng.*, 114, 274, 1992.
16. Giddens, D.P., Zarins, C.K., and Glagov, S., *J. Biomech. Eng.*, 115, 588, 1993.
17. Friedmann, M.H. and Fry, D.L., *Atherosclerosis*, 104, 189, 1993.
18. Davies, P.F. and Tripathi, S.C., *Circ. Res.*, 72, 293, 1993.
19. Uchiyama, S., Yamazaki, M., Maruyama, S., Handa, M., Ikeda, Y., Fukuyama, M., and Itagaki, L., *Stroke*, 24, 1547, 1994.
20. Barbee, K.A., Davies, P.F., and Lal, R., *Circ. Res.* 74, 163, 1994.
21. Ross, R., *New Engl. J. Med.*, 314, 488, 1986.
22. Weinbaum, S. and Chien, S., *J. Biomech. Eng.*, 115, 602, 1993.
23. Satcher, R.L., Jr. and Dewey, C.F., Jr., *ASME Adv. Bioeng.*, 20, 595, 1991.
24. Satcher, R.L., Jr., Bussolari, S.R., Gimbrone, M.A., Jr., and Dewey, C.F., Jr., *J. Biomech. Eng.*, 114, 309, 1992.
25. DePaola, N., Gimbrone, M.A., Jr., Davies, P.F., and Dewey, C.F., *Arteriosclerosis and Thrombosis*, 12, 293, 1992.
26. Thubrikar, M.J. and Robicsek, F., *Ann. Thorac. Surg.*, 59, 1594, 1995.
27. Lei, M., Kleinstreuer, C., and Truskey, G.A., *J. Biomech. Eng.*, 117, 350, 1995.
28. Buchanan, J.R., Jr., Kleinstreuer, C., Truskey, G.A., and Lei, M., *Atherosclerosis*, 143, 27, 1999.
29. Clowes, A.W., Gown, A.M., Hanson, S.R., and Reidy, M.A., *Am. J. Pathol.*, 118, 43, 1985.
30. Chervu, A. and Moore, W.S., *Surg. Gyn. Obst.*, 171, 433, 1990.
31. Painter, M.A., *Artif. Org.*, 15, 42, 1991.
32. Painter, M.A., *Artif. Org.*, 15, 103, 1991.
33. Bassiouny, H.S., White, S., Glagov, S., Choii, E., Giddens, D.P., and Zarins, C.K., *J. Vasc. Surg.*, 15, 708, 1992.
34. Slack, S.M., Cui, W., and Turitto, V.T., in *Advances in Cardiovascular Engineering*, Hwang, N.H.C. et al., Eds., Plenum Press, New York, 1992, 91.
35. Perktold, K., in *Biofluid Mechanics*, Liepsch, D., Ed., Springer-Verlag, Heidelberg, 1990, 471.

36. Perktold, K., Thurner, E., and Kenner, T., *Med. Biol. Eng. Comput.*, 32, 19, 1994.

37. Lou, Z. and Yang, W.J., *Crit. Rev. Biomed. Eng.*, 19, 455, 1992.

38. Xu, X.Y. and Collins, M.L.W., *J. Eng. Med., Part H*, 204, 205, 1990.

39. Ku, D.N., Giddens, D.P., Zarins, C.K., and Glagov, S., *Atherosclerosis*, 5, 293, 1985.

40. Figueras, C., Jones, S.A., Giddens, D.P., Zarins, C., Bassiouny, H.S., and Glagov, S., in *1991 Advances in Bioengineering*, Vanderby, R., Ed., ASME Press, New York, 1991.

41. Kleinstreuer, C., Nazemi, M., and Archie, J.P., *J. Biomech. Eng.*, 113, 330, 1991.

42. Lei, M., Kleinstreuer, C., and Archie, J.P., Proceedings of the 11th Southern Biomedical Engineering Conference, Memphis, 1992, 124.

43. White, S.S., Zarins, C.K., Giddens, D.P., Bassiouny, H., Loth, F., Jones, S.A., and Glagov, S., *J. Biomech. Eng.*, 115, 104, 1993.

44. Fei, D.Y., Thomas, J.D., and Rittgers, S.E., *J. Biomech. Eng.*, 116, 331, 1994.

45. Staalsen, N.H., Vlrich, M., Winther, J., Pederson, E.M., How, T., and Nygaard, H., *J. Vasc. Surg.*, 21, 460, 1995.

46. Kleinstreuer, C., Lei, M., and Archie, J.P., *J. Biomech. Eng.*, 118, 506, 1996.

47. Wells, D.R., Archie, J.P., and Kleinstreuer, C., *J. Vasc. Surg.*, 23, 667, 1996.

48. Smedby, O., Nillson, S., and Bergstrand, L., *J. Biomech.*, 29, 543, 1996.

49. Lei, M., Kleinstreuer, C., and Archie, J.P., *J. Vasc. Surg.*, 25, 637, 1997.

50. Abbott, W.M. and Cambria, R.P., in *Biological and Synthetic Vascular Prostheses*, Stanley, J.C., Burkel, W.E., Lindenauer, S.M., Barlett, R.H., and Turcotte, J.G., Eds., Grune and Stratton, New York, 1982, 189.

51. Phifer, T.J. and Hwang, N.H.C., in *Advances in Cardiological Engineering*, Hwang, N.H.C. et al., Eds., Plenum Press, New York, 1992, 385.

52. Strandness, D.E., Jr. and Sumner, D.S., *Hemodynamics for Surgeons*, Grune and Stratton, New York, 1975.

53. Eikenboom, B.C., Ackerstaff, R.G.A., Hoeneveld, H. et al., *J. Vasc. Surg.*, 7, 240, 1988.

54. Archie, J.P., *J. Vasc. Surg.*, 17, 141, 1993.

55. Archie, J.P., *J. Vasc. Surg.*, 23, 932, 1996.

56. Lei, M., Kleinstreuer, C., and Archie, J.P., *J. Biomech.*, 29, 1605, 1996.

57. Wells, D.R., Master's thesis, North Carolina State University, Raleigh, NC, 1995.

58. He, X. and Ku, D.N., *J. Biomech. Eng.*, 118, 74, 1996.

59. Stein, P.D., Sabbah, H.N., Anbe, D.T., and Walburn, F.J., *Biorheology*, 16, 249, 1979.

60. Kleinstreuer, C., *Engineering Fluid Dynamics – An Interdisciplinary Systems Approach*, Cambridge University Press, New York, 1997.

61. Macosko, C.W., *Rheology: Principles, Measurements and Applications*, VCH Publishers, New York, 1994.

62. Merrill, E.W., *Physiol. Rev.*, 49, 863, 1969.

63. Ku, D.N. and Giddens, D.P., *J. Biomech.*, 20, 407, 1987.

64. Khodadadi, J.M., Vlachos, N.S., Liepsch, D., and Moravec, S., *J. Biomech. Eng.*, 110, 129, 1988.

65. Keynton, R.S., Rittgers, S.E., and Shu, M.C.S., *J. Biomech. Eng.*, 113, 458, 1991.

66. Ohja, M., *J. Biomech.*, 26, 1377, 1993.

67. Jones, C.J.H., Lever, M.J., Ogasawara, Y., Parker, K.H., Hiramatsu, O., Mito, K., Tsujioka, K., and Kajiya, F., *Am. J. Physiol.*, 31, H1592, 1992.

68. Lei, M. Computational Fluid Dynamics and Optimal Design of Bifurcating Blood Vessels, Ph.D. thesis, North Carolina State University, Raleigh, NC, 1995.

69. Finlayson, B.A., *The Method of Weighted Residuals and Variational Principles*, Academic Press, New York, 1972.

70. Patankar, S.V., *Numerical Heat Transfer and Fluid Flow*, Hemisphere, Washington, D.C., 1980.

71. Rhie, C.M. and Chow, W.L., *AIAA J.*, 21, 1527, 1983.

72. Anon., User Manuals for CFDS-FLOW3D (Ver. 3.2, 3.3) and CFX (Ver. 4.1), AEA Technology Engineering Software, Inc., Pittsburgh, PA, 1991, 1993, and 1995.

73. Ohja, M., *J.Biomech.*, 26, 1377, 1993.

74. Rindt, C.C.M., van Steenhoven, A.A., and Reneman, R.S., *J. Biomech.*, 21, 985, 1988.

75. Bharadvaj, B.K., Mabon, R.F., and Giddens, D.P., *J. Biomech.*, 15, 349, 1982.

76. Lee, D. and Chiu, J.J., *J. Biomech.*, 29, 1, 1996.

77. Wylie, E.J. and Ehrenfeld, W.K., *Extracranial Occlusive Cerebrovascular Disease: Diagnosis and Management*, W.B. Sanders, Philadelphia, PA, 1970.

78. DeBakey, M.E., Lawrie, G.M., and Glaser, D.H., *Ann. Surg.*, 201, 115, 1985.

79. Chandran, K.B. and Kim, Y.H., *IEEE Eng. Med. Biol.*, Aug./Sept. 1994, 517.

80. McDonald, D.A., *Blood Flow in Arteries*, 2nd ed., Williams and Wilkins, Baltimore, MD, 1974, 356.

81. Archie, J.P., *J. Surg. Res.*, 48, 211, 1990.

82. Lei, M. and Kang, Z.H., *Appl. Math. Mech.*, 7, 955, 1986.

83. Kusaba, A., Kina, M., Watanabe, T., Furuyama, M., Okadome, K., Muto, Y., Kamori, M., and Inokuchi, K., *Jap. J. Surg.*, 11, 232, 1981.

84. Inokuchi, K., Okadome, K., Ohtsuka, K., Muto, Y., Kamori, M., Miyazaki, M., and Takahara, H., *J. Vasc. Surg.*, 1, 787, 1984.

85. Badimon, J. and Badimon, L.L., in *Advances in Cardiovascular Engineering*, Hwang, N.H.C. et al., Eds., Plenum Press, New York, 1992, 175.

86. Lei, M., Kleinstreuer C., Truskey, G.A., and Archie, J.P., in *1996 Advances in Bioengineering*, BED Vol. 33, ASME, New York, 1996, 211.

2

Techniques in Fluid Dynamical Wall Shear Phenomena and Their Application in the Blood Flow

2.1 Introduction 2-1
2.2 Wall Shear Stress of a Fully Developed Pulsatile Pipe Flow under the Blood Flow Conditions through the Aorta..2-2
    Introduction • Methods • Results and Discussion
2.3 Wall Shear Stress of the Pulsatile Flow through a Two-Dimensional Channel with an Oscillating Wall.......2-9
    Introduction • Methods • Results and Discussion • Application I • Application II
2.4 Wall Shear Stress of a Non-Newtonian Fluid through an Axisymmetric Diffuser 2-24
    Introduction • Methods • Results and Discussion
2.5 Concluding Remarks 2-30

Masahide Nakamura
Akita University

2.1 General Introduction

Wall shear stress is one of the most significant parameters of fluid mechanics. This parameter plays an important role in evaluating the pressure loss of fluids through the pipeline as well as the force received from the fluids. Therefore, many experimental and theoretical studies were performed to evaluate the behavior of wall shear stress under engineered conditions. Wall shear stress is very significant in biomechanics as well. That is, many experimental and theoretical results suggest that wall shear stress may have a strong relationship to atherogenesis.[1] The detailed mechanics of atherosclerosis have not yet been established, but it has been suggested that atherosclerosis will more readily occur in regions where wall shear stress is relatively low.[1] The study of wall shear stress may therefore be useful in the prevention of some arterial lesions such as atherosclerosis. However, the evaluation of wall shear stress of blood flow through the artery involves great difficulty, and the behavior of wall shear stress *in vivo* is complicated because:

1. The blood flow through the artery is a pulsatile flow.
2. The geometry of the arterial wall changes with time.
3. The blood has a remarkable non-Newtonian effect in the low shear rate region.

Of course, there are many other factors which contribute to the difficulty in evaluating wall shear stress, but these factors are beyond the scope of present consideration. In this chapter, we will introduce the experimental method of evaluating the wall shear stress of a fully developed pulsatile flow through a straight pipe.[2] Next, we will introduce the theoretical method of evaluating the wall shear stress of an unsteady flow through a two-dimensional channel with an oscillating wall.[3] The effective range of these methods is limited, but the methods have some very interesting features and will provide useful information. Finally, we will introduce the numerical results concerning the effect of a non-Newtonian fluid on the distribution of wall shear stress. The experimental and theoretical results presented here will be important as the basic data necessary to understand the behavior of wall shear stress *in vivo*.

2.2   Wall Shear Stress of a Fully Developed Pulsatile Pipe Flow under the Blood Flow Conditions through the Aorta

Introduction

The most simplified model of blood flow in an artery is a fully developed pulsatile flow through a straight pipe. Many studies were performed to evaluate the flow structure of a fully developed pulsatile flow through a straight pipe. For example, the analytical solution in a laminar flow region was derived by Womersley[4] and Uchida.[5] The fully developed pulsatile flow in a turbulent region was investigated systematically by Ohmi et al.[6] However, the condition of blood flow through the aorta lies in the laminar-turbulent transitional region. Therefore, it is expected that the blood flow through the aorta will have a very complicated structure. There is still much more to be studied about the transitional pulsatile flow. For this reason, an experimental investigation of the fully developed sinusoidal pulsatile flow in a straight pipe at a transitional Reynolds number is performed to evaluate the relationship between section averaged axial velocity, pressure gradient, and wall shear stress.[2]

Of course, the value of wall shear stress can be measured by the velocity distribution in the vicinity of the wall, however, the measurement of unsteady velocity distributions in the vicinity of the wall is difficult. So, wall shear stress is often measured by the electrochemical method[7] or the flush-mounting probe.[8] The basis of the electrochemical method is that the wall shear stress is proportional to the cube of electric current flowing between each electrode and counterelectrode.[7] Using the flush-mounting probe, the wall shear stress is determined from the relationship between the wall shear stress and the heat flux from the probe to the fluid.[8] Although the calibration of these methods is not easy, the methods are widely applied, and the reliability of these methods has been demonstrated by many researchers.

The following section introduces an easier method[2] which can be applied to the measurement of wall shear stress in a fully developed unsteady flow through a straight pipe. The application range of the present method is narrow but still very effective for solving our problem.

Methods

In order to understand the principle of our method, we considered the fully developed unsteady pipe flow through a straight pipe, as shown in Fig. 2.1. The equation governing the flow is written as follows:

$$\rho \frac{\partial u}{\partial t} = -\frac{\partial p}{\partial z} + \frac{1}{r}\frac{\partial}{\partial r}\left(r\tau\right) \tag{2.1}$$

where ρ denotes the density of the fluid, u denotes the axial component of the velocity, t denotes the time, p denotes the pressure, τ denotes the shear stress in the fluid, and (r,z) denotes the coordinate components of radial direction and axial direction. In the case of a laminar flow region, the value of τ is written as follows:

$$\tau = \mu \partial u / \partial r \tag{2.2}$$

where μ denotes the viscosity coefficient. In the case of a turbulent flow region or transitional flow region, the reliable equation corresponding to Eq. 2.2 is unknown, but one should note that Eq. 2.2 is not necessary for this discussion. The following equation can be obtained by integrating Eq. 2.1 along the radial direction from 0 to $D/2$.

$$\rho \frac{d\bar{u}}{dt} = -\frac{\partial p}{\partial z} - \frac{4\tau_w}{D} \tag{2.3}$$

where $\bar{u}$ denotes the section averaged axial velocity, τ_w denotes the wall shear stress, and D denotes the diameter of the pipe. Eq. 2.3 is the desired result, and this equation represents the balance of force acting on the fluid. This equation is valid for the fully developed pipe flow in either a laminar flow region or a turbulent flow region. Using this equation we can easily obtain the wall shear stress τ_w from the section averaged axial velocity and the pressure gradient.

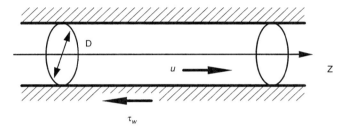

FIGURE 2.1   Model of a fully developed unsteady flow in a straight pipe.

The basic behavior of wall shear stress in a transitional region was investigated on the basis of this method. The experimental apparatus is shown in Fig. 2.2. As the non-Newtonian effect of blood does not play as important a role in a high shear rate region as it does concerning bllod flow in the aorta, water was used as a substitute for blood in this study. Water was pumped from the reservoir to the constant head tank. The flow rate from the constant head tank to the test section is controlled by the valve (2), and the pulsation is generated by the piston (3), which is driven by the scotch-yoke mechanism. The motion of the piston can be closely approximated according to the sinusoidal curve. The back pressure tank (4) is set up to decrease the load of the piston rod. The inner diameter of the test section (plexiglas tube) is 14 mm, and the length of the straight pipe upstream of the test section is about $200D$, which is much longer than the length of the entrance region. So, the fully developed flow is considered established in the test section.

The pressure gradient and the section averaged axial velocity were measured by a variable reluctance-type pressure sensor (Celesco Co., model P7D) and an electromagnetic flowmeter (Yamatake-Honeywell Co., MagneW-3000), respectively. When measuring section averaged axial velocity and pressure gradient, we had to pay attention to the frequency response of these apparatus, and the calibration of these apparatus had to be thoroughly performed. The frequency response of an electromagnetic flowmeter for industrial use is generally not that good. In addition, the natural frequency of the pressure measuring system is low compared to the natural frequency of the pressure sensor. One should note that the natural frequency of the pressure measuring system is strongly affected by the mount-method of the pressure sensor. To improve the frequency response of the pressure measuring system, we increased the diameter of the pressure hole and decreased the volume of the cavity (Fig. 2.3).

The experimental conditions were as follows:

$$<\text{Re}> = 2 \times 10^3 - 1.2 \times 10^4$$
$$\omega^* = 50. - 100. \tag{2.4}$$
$$\Delta u^* = 0.1 - 0.8$$

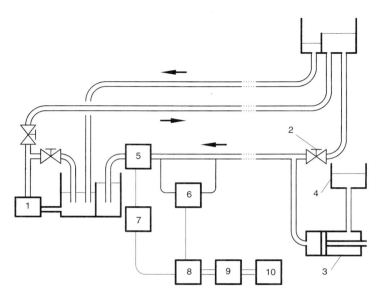

FIGURE 2.2 Experimental apparatus: (1) pump; (2) diaphragm valve; (3) piston; (4) back pressure tank; (5) electromagnetic flowmeter; (6) variable inductance-type pressure transducer; (7) DC canceller; (8) low-pass filter; (9) DC amp.; (10) data recorder.

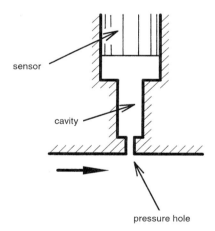

FIGURE 2.3 Installation of a pressure transducer (cavity type).

where <Re> denotes the time-averaged Reynolds number in a cycle, ω^* denotes the non-dimensional angular frequency, and Δu^* denotes the non-dimensional amplitude of the section averaged axial velocity. These parameters are defined as follows:

$$< \mathrm{Re} > = \rho < \bar{u} > D/\mu, \quad \omega^* = D^2 \rho \omega / 4\mu, \quad \Delta u^* = \Delta u / < \bar{u} >$$

where $<\bar{u}>$ denotes the time-averaged section averaged axial velocity in a cycle, ω denotes the angular frequency, and Δu denotes the amplitude of the section averaged axial velocity.

An approximated analysis was performed to compare the calculated results with the experimental results. This analysis was performed on the basis of Eq. 2.3 and the following sinusoidal waveform of the pressure gradient.

$$\partial p/\partial z = <\partial p/\partial z>\left(1+\sin\omega t\right) \qquad (2.5)$$

where $<\partial p/\partial z>$ denotes the time-averaged pressure gradient in a cycle. The wall shear stress was not calculated from Eq. 2.3 directly, but the wall shear stress in a steady pipe flow can be obtained using the following equations:

$$\tau_{ws} = \frac{1}{8}\lambda_s\rho\bar{u}_s^2 \qquad (2.6)$$

where the subscript s denotes the value of the steady flow. In Eq. 2.6, the value of λ_s can be calculated using the following equations (Poiseuille's law and Blasius's law):

$$\lambda_s = \begin{cases} 64/\mathrm{Re}s & \left[\mathrm{Re}s < 2100\right] \\ 0.0305 + 1.37\times10^{-5}\left(\mathrm{Re}s - 2100\right) & \left[2100 < \mathrm{Re}s < 3000\right] \\ 0.3164\,\mathrm{Re}s^{-0.25} & \left[3000 < \mathrm{Re}s\right] \end{cases} \qquad (2.7)$$

where Re_s denotes the Reynolds number of the steady flow ($= \rho\bar{u}_sD/\mu$). In this section, the approximated analysis was performed based on the assumption that these equations are effective under unsteady flow conditions (quasi-steady approximation). On the basis of this assumption, the wall shear stress in Eq. 2.3 is written as follows:

$$\tau_w = \frac{1}{8}\lambda(t)\rho\bar{u}(t)^2 \qquad (2.8)$$

The value of $\lambda(t)$ can be obtained by replacing Re_s in Eq. 2.7 with $\rho\bar{u}(t)D/\mu$. By substituting Eq. 2.8 into Eq. 2.3, the ordinary differential equation of $\bar{u}(t)$ is derived, and the change of the section averaged axial velocity and the wall shear stress can be calculated numerically.

We expected that the characteristics of the flow are strongly affected by the ratio of three terms in Eq. 2.3. So, we introduced the non-dimensional parameters which denote the ratio of these three terms. In this section, the following non-dimensional parameters are adopted.[6]

$$\Phi_{t,1} = \rho\omega\Delta u/\left|\varepsilon<\partial p/\partial z>\right|, \quad \Phi_{z,1} = 4\Delta\tau_w/\left|\varepsilon<\partial p/\partial z>\right| \qquad (2.9)$$

where $\Delta\tau_w$ denotes the amplitude of the wall shear stress in a cycle. The first parameter in this equation denotes the ratio of the amplitude of the inertial term to the amplitude of the pressure gradient term, while the second parameter denotes the ratio of the amplitude of the viscous term to the pressure gradient term. The characteristics of the wall shear stress can be discussed based on $\Phi_{z,1}$ in Eq. 2.9.

Results and Discussion

The experimental results of $\Phi_{z,1}$ are shown in Figs. 2.4 and 2.5. In these figures, the time-averaged Reynolds number is approximately 1.1×10^4, and the pulsatile frequency is 0.30Hz and 0.19Hz, respectively. The abscissa of these figures is the non-dimensional amplitude of section averaged axial velocity. The calculated $\Phi_{z,1}$ based on the quasi-steady approximation is shown by the dot-dash line and the calculated $\Phi_{z,1}$ based on Womersley's solution is shown by the dotted line.. The following findings were obtained from these figures:

1. Non-dimensional amplitude $\Delta u^\star$ does not have a strong effect on $\Phi_{z,1}$.
2. The experimental results lie between the results calculated using the turbulent quasi-steady approximation and the results calculated using Womersley's solution.

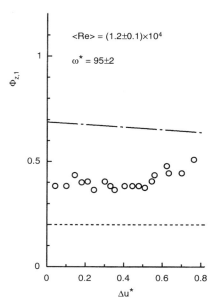

FIGURE 2.4 Experimental results at $\langle Re \rangle = (1.2 \pm 0.1) \times 10^4$, $\omega^* = 95 \pm 2$. Circles represent experiments, the intermittent dashed line represents results calculated based on the turbulent quasi-steady approximation, and the dotted line represents results calculated based on Womersley's solution. (*Source:* Nakamura, M., Sugiyama, W., and Haruna, M., *Trans. ASME, J. Biomech. Eng.* 115, 412–417, 1993. With permission.)

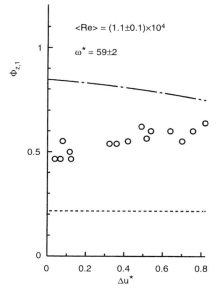

FIGURE 2.5 Experimental results at $\langle Re \rangle = (1.2 \pm 0.1) \times 10^4$, $\omega^* = 59 \pm 2$. Circles represent experiments, the intermittent dashed line represents results calculated based on the turbulent quasi-steady approximation, and the dotted line represents results calculated based on Womersley's solution. (*Source:* Nakamura, M., Sugiyama, W., and Haruna, M., *Trans. ASME, J. Biomech. Eng.* 115, 412–417, 1993. With permission.)

Knowledge of the flow laminarization is essential to understand the pulsatile flow at a transitional Reynolds number. The laminarized pulsatile flow was studied by many researchers whose studies showed that the pulsation changed the flow at $Re = 1.0 \times 10^4$ from the fully turbulent state to the conditionally turbulent state.[9] If the pulsation weakens the turbulence of the flow, the characteristics of the pulsatile flow will lie at an intermediate state between the laminar flow and turbulent flow. The experimental results in Figs. 2.4 and 2.5 are explained by this discussion. Of course, the laminarization effect is not included in the quasi-steady approximation.

Figs. 2.6 and 2.7 show the experimental results when <Re> is approximately 8.5×10^3. Fig. 2.6 shows that the value of $\Phi_{z,1}$ increases with the increase in Δu^*, and the value of $\Phi_{z,1}$ in a high Δu^* region is greater than the results calculated using the turbulent quasi-steady approximation. However, this characteristic is not evident in Fig. 2.7. To understand these results, the bursting frequency (f_B) was introduced. The characteristic of the pulsatile flow at a transitional Reynolds numbers depends on the ratio of the pulsatile frequency to the bursting frequency.[10] Narasimha and Narayanan represent the bursting frequency with the following equation:

$$f_B = <\overline{u}>/5D \tag{2.10}$$

The pulsatile velocity components exhibited an overshoot in the Stokes layer (near the wall) when the ratio of the pulsatile frequency to the bursting frequency was of the order of unity.[10] Since the pulsatile frequency of Fig. 2.6 is closer to the bursting frequency than the pulsatile frequency of Fig. 2.7, the overshoot effect in Fig. 2.6 will be clearer than the one in Fig. 2.7. Since the overshoot of the pulsatile velocity component derives the increase in the amplitude of the wall shear stress, the increase in $\Phi_{z,1}$ can be explained by the overshoot effect. In addition, Fig. 2.6 suggests that the increase in the pulsatile amplitude strengthens the overshoot effect of the pulsatile velocity component.

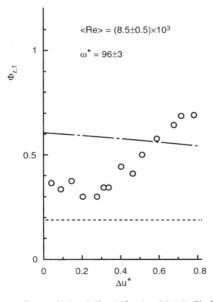

FIGURE 2.6 Experimental results at <Re> = $(8.5 \pm 0.5) \times 10^3$, $\omega^* = 96 \pm 3$. Circles represent experiments, the intermittent dashed line represents results calculated based on the turbulent quasi-steady approximation, and the dotted line represents results calculated based on Womersley's solution. (*Source:* Nakamura, M., Sugiyama, W., and Haruna, M., *Trans. ASME, J. Biomech. Eng.* 115, 412–417, 1993. With permission.)

Figs. 2.8 and 2.9 show the experimental results of $\Phi_{z,1}$ when <Re> is approximately 4000. Fig. 2.8 shows that the value of $\Phi_{z,1}$ is greater than the results calculated using the turbulent quasi-steady approximation through the whole range of the non-dimensional amplitude. This result means that the overshoot effect of the pulsatile velocity components near the wall plays an important role in the whole range of Fig. 2.8. This result is explained by the approach of the pulsatile frequency to the bursting frequency.

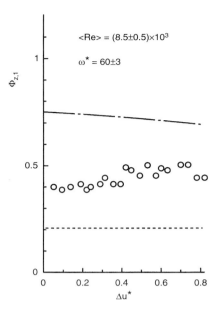

FIGURE 2.7   Experimental results at $<Re> = (8.5 \pm 0.5) \times 10^3$, $\omega^* = 60 \pm 3$. Circles represent experiments, the intermittent dashed line represents results calculated based on the turbulent quasi-steady approximation, and the dotted line represents results calculated based on Womersley's solution. (*Source:* Nakamura, M., Sugiyama, W., and Haruna, M., *Trans. ASME, J. Biomech. Eng.* 115, 412–417, 1993. With permission.)

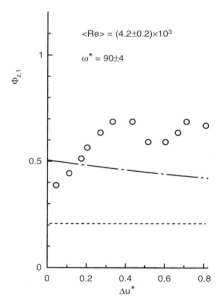

FIGURE 2.8   Experimental results at $<Re> = (4.2 \pm 0.2) \times 10^3$, $\omega^* = 90 \pm 4$. Circles represent experiments, the intermittent dashed line represents results calculated based on the turbulent quasi-steady approximation, and the dotted line represents results calculated based on Womersley's solution. (*Source:* Nakamura, M., Sugiyama, W., and Haruna, M., *Trans. ASME, J. Biomech. Eng.* 115, 412–417, 1993. With permission.)

The effect of the Reynolds number on the value of $\Phi_{z,1}$ is shown in Fig. 2.10. In this figure, the shadowed portion shows the range of the time-averaged Reynolds number in the human aorta.[11] Moreover, one should note that the value of ω^* in the human aorta is approximately 100. This figure shows that the parameters of the aorta lie in the region corresponding to a maximum of $\Phi_{z,1}$. Since $\Phi_{z,1}$ is the ratio of

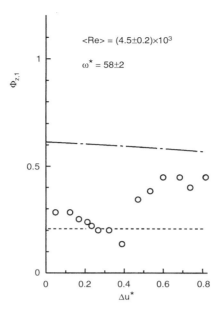

FIGURE 2.9 Experimental results at $<Re> = (4.2 \pm 0.2) \times 10^3$, $\omega^* = 58 \pm 2$. Circles represent experiments, the intermittent dashed line represents results calculated based on the turbulent quasi-steady approximation, and the dotted line represents results calculated based on Womersley's solution. (*Source:* Nakamura, M., Sugiyama, W., and Haruna, M., *Trans. ASME, J. Biomech. Eng.* 115, 412–417, 1993. With permission.)

the amplitude of the wall shear stress to the amplitude of the pressure gradient, this result suggests that the parameters of the aorta are suitable to generate the pulsatile flow with a large amplitude of the wall shear stress. Many studies concerning the relationship between the wall shear stress and some arterial lesions (for example, atherosclerosis) have been performed and suggest that an increase in wall shear stress prevents the development of arterial lesions.[1] Our experimental results demonstrate the important role of blood flow parameters in the aorta. Of course, this is only a hypothesis that explains the role of blood flow parameters in the aorta; more detailed studies should be performed in the future.

The wall shear stress can be obtained accurately if we obtain the accurate flow rate and pressure difference. The measurements of flow rate and pressure difference are much easier to obtain than the measurement of wall shear stress in general. The advantage of this method is clear. Additionally, we should note that Eq. 2.3 is effective for a non-Newtonian fluid. So, our method can be applied to the measurement of the wall shear stress of a non-Newtonian fluid in the future.

2.3 Wall Shear Stress of the Pulsatile Flow through a Two-Dimensional Channel with an Oscillating Wall

Introduction

The flow pattern in the vicinity of the oscillating wall is essential in understanding the blood flow through the artery. Therefore, this flow attracted the interest of many researchers,[3,12] and the flow structure in the vicinity of the oscillating wall was gradually revealed by their research. On the other hand, we expect that the oscillation of the wall has a strong influence on the wall shear stress, but investigations into the relationship between the wall oscillation and wall shear stress are limited compared to investigations into the relationship between the flow pattern and the vicinity of the oscillating wall.[3] As the experimental measurements of wall shear stress in the vicinity of the oscillating wall will involve great difficulty, the experimental techniques are beyond the scope of this section.

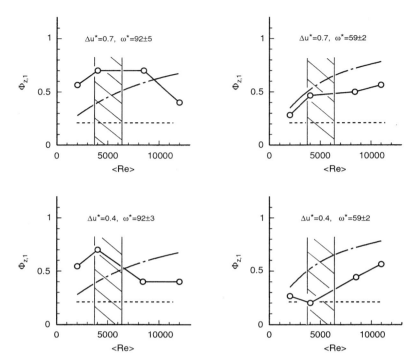

FIGURE 2.10 Effect of the Reynolds number on the value of $\Phi_{z,1}$. Circles represent experiments, the intermittent dashed line represents results calculated based on the turbulent quasi-steady approximation, and the dotted line represents results calculated based on Womersley's solution. (*Source:* Nakamura, M., Sugiyama, W., and Haruna, M., *Trans. ASME, J. Biomech. Eng.* 115, 412–417, 1993. With permission.)

Methods

The model used in this calculation is shown in Fig. 2.11. The laminar pulsatile flows through a two-dimensional channel constructed of a fixed wall (upper wall) and an oscillating wall (lower wall) were analyzed numerically. The height of the lower wall $F(x,t)$ is given by:

$$F\left(x,t\right)=-h_m\exp\left[-5\left(x/D_0-2\right)^2\right]\cos\left(2\pi t/T_0\right) \tag{2.11}$$

where t denotes the time, x denotes the coordinate component (see Fig. 2.11), D_0 denotes the channel width, h_m denotes the amplitude of the wall oscillation, and T_0 denotes the period of the wall oscillation. This equation shows that the center of oscillation lies at $x/D_0 = 2.0$, and the oscillation range of the wall is limited to the range of $x/D_0 = 1.0$ to $x/D_0 = 3.0$. Additionally, the flow rate at the inlet is demonstrated by the following equation:

$$Q=Q_0-\varepsilon Q_0\cos\left[2\pi\left(t-\delta\right)t/T_0\right] \tag{2.12}$$

where Q_{10} denotes the time-averaged flow rate, ε denotes the non-dimensional amplitude of the flow rate, and δ denotes the phase lag between the flow rate and the wall oscillation. This equation was used to determine the boundary condition at the inlet.

The basic equations are written as follows:

$$\frac{\partial\omega}{\partial t}+\frac{\partial\phi}{\partial y}\frac{\partial\omega}{\partial x}-\frac{\partial\phi}{\partial x}\frac{\partial\omega}{\partial y}=\nu\nabla^2\omega \tag{2.13}$$

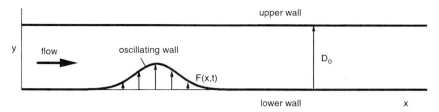

FIGURE 2.11 Two-dimensional channel with an oscillating wall.

$$-\omega = \nabla^2 \phi \qquad (2.14)$$

where ω denotes the vorticity, ϕ denotes the stream function, ∇^2 denotes the two-dimensional Laplacian, and ν denotes the kinetic viscosity coefficient. These equations are very popular in the numerical calculation of a two-dimensional flow. We then introduced the following coordinate transformation from x, y, t to ξ, η, τ.[13]

$$\xi = x, \quad \eta = D_0 \left(y - F\right) / \left(D_0 - F\right), \quad \tau = t \qquad (2.15)$$

Using this transformation, the positions of the upper and lower walls were transformed from $y = d_0$ and $y = F$ to $\eta = D_0$ and $\eta = 0$. That is, the moving boundary problem was transformed into the fixed boundary problem. The importance of this coordinate transformation is clear. In the transformation from (x, y, t) to (ξ, η, τ), the following relationships are effective:

$$\partial / \partial x = \partial / \partial \xi + q_2 \partial / \partial \eta, \quad \partial / \partial y = q_1 \partial / \partial \eta, \quad \partial / \partial t = \partial / \partial \tau + q_4 \partial / \partial \eta \qquad (2.16)$$

where

$$q_1 = \frac{D_0}{D_0 - F}, \quad q_2 = -\frac{D_0 - \eta}{D_0 - F} \frac{\partial F}{\partial x} \qquad (2.17)$$

$$q_4 = -\frac{D_0 - \eta}{D_0 - F} \frac{\partial F}{\partial t} \qquad (2.18)$$

The boundary conditions of our calculations were as follows: the fully developed pulsatile flow pattern through the two-dimensional channel was used as the inlet boundary condition ($x/D_0 = 0$). A value of zero-pressure was used as the boundary condition at the outlet ($x/D_0 = 10.0$). In addition, natural boundary conditions were used for all other variables at the outlet. At the upper wall of the channel ($y = D_0$), it is obvious that the boundary condition of ϕ is $\phi = \phi_0$ (constant). Additionally, the boundary value of ω at the upper wall was determined from Wood's condition.[14] At the lower wall ($y = F$), the boundary condition for ϕ was determined from the following equations:

$$\partial \phi / \partial y \big|_{y=F} = 0, \quad \partial \phi / \partial x \big|_{y=F} = -\partial F / \partial y \qquad (2.19)$$

The boundary value of ω at the lower wall was determined from Wood's condition, too.

The non-dimensional parameters which characterize the flow pattern of our analysis are the Reynolds number, the Strouhal number, the non-dimensional amplitude of section averaged axial velocity, the non-dimensional height of the wall oscillation, and the non-dimensional phase lag between the wall oscillation and the flow pulsation. The Reynolds number and the Strouhal number are defined by the following equations:

$$\text{Reynolds number (Re)} = u_0 D_0 / v \qquad (2.20)$$

$$\text{Strouhal number (St)} = D_0 / \left(u_0 T_0 \right)$$
$$\left(u_0 = Q_0 / D_0 \right) \qquad (2.21)$$

Of course, the non-dimensional amplitude of the section averaged axial velocity is shown by the value of ε in Eq. 2.12 and the non-dimensional amplitude of the wall oscillation is defined by h_m/D_0 in Eq. 2.11. In addition, the non-dimensional phase lag between the wall oscillation and the flow pulsation is defined by δ/T_0 in Eq. 2.12.

In our analysis, the calculated region was transformed into the rectangular region as mentioned above. Therefore, we adopted the finite difference method for the numerical calculation. As the method of discretization, the fully implicit scheme was adopted to stabilize the numerical calculation. The final sparse non-symmetric linear system resulting from this procedure is denoted by $Ax = b$, where A denotes the $n \times n$ nonsymmetric matrix (known), b denotes the vector (known), and x denotes the solution vector of this linear system. Since most CPU time is dissipated by the calculation of this solution, we selected the effective numerical algorithm for this linear system. The numerical solution methods for linear equations are divided into the two groups: direct method and iterative method. A direct method has an algorithm with a finite number of steps, while an iterative method generates a sequence of solution vectors which converge to the solution. It is well known that compared to a direct method, an iterative method is advantageous for the calculation of large linear equations. We adopted the CGS (conjugate gradient squared) method.[15] It has been suggested that the CGS method is one of the most effective iterative methods for solving non-symmetric linear equations.[16] The algorithm of the CGS method is written as follows:[15]

Step 1: Let $m = 0$. Choose x_0 and compute $p_0 = e_0 = r_0 = b - Ax_0$.
Step 2: Compute:

$$\alpha_m = \left(r_0, r_m \right) / \left(r_0, Ap_m \right)$$
$$h_{m+1} = e_m - \alpha_m Ap_m$$
$$r_{m+1} = r_m - \alpha_m A \left(e_m + h_{m+1} \right)$$
$$x_{m+1} = x_m + \alpha_m \left(e_m + h_{m+1} \right)$$
$$\beta_m = \left(r_0, r_{m+1} \right) / \left(r_0, r_m \right)$$
$$e_{m+1} = r_{m+1} + \beta_m h_{m+1}$$
$$p_{m+1} = e_{m+1} + \beta_m \left(h_{m+1} + \beta_m p_m \right)$$

Step 3: Let $m = m + 1$.
Step 4: Repeat steps two and three until convergence is reached.

The distribution of stream function and vorticity can be calculated from these procedures, and the velocity vector (u,v) can be easily obtained from the stream function.

$$u = q_1 \partial \phi / \partial \eta, \quad v = -\left(\partial \phi / \partial \xi + q_2 \partial \phi / \partial \eta \right) \qquad (2.22)$$

The wall shear stress can be calculated from the velocity distributions in the vicinity of the wall. Additionally, the pressure distributions can be calculated from the velocity vectors. The Navier-Stokes equations in the (ξ, η, τ) system are written as follows:

$$\rho\left(\frac{\partial u}{\partial \tau} + q_4\frac{\partial u}{\partial \eta} + \frac{\partial}{\partial \xi}\left(u^2\right) + q^2\frac{\partial}{\partial \eta}\left(u^2\right) + q_1\frac{\partial}{\partial \eta}\left(uv\right)\right) = -\frac{\partial p}{\partial \xi} - q_2\frac{\partial p}{\partial \eta} - \mu q_1\frac{\partial \omega}{\partial \eta}$$

$$\rho\left(\frac{\partial v}{\partial \tau} + q_4\frac{\partial v}{\partial \eta} + \frac{\partial}{\partial \xi}\left(uv\right) + q^2\frac{\partial}{\partial \eta}\left(uv\right) + q_1\frac{\partial}{\partial \eta}\left(v^2\right)\right) = -q_1\frac{\partial p}{\partial \eta} + \mu\left(\frac{\partial \omega}{\partial \xi} + q_2\frac{\partial \omega}{d\eta}\right)$$

where μ denotes the viscosity coefficient. The pressure distribution is easily calculated from these equations and the velocity vector (u,v).

Results and Discussion

Fig. 2.12 shows the calculated stream lines at Re = 500, St = 0.03, $h_m/D_0 = 1/3$, $\varepsilon = 0.22$, and $\delta/T_0 = 0.725$. Because the flow pattern on the downstream side of the oscillating wall is very interesting, the stream lines between $x/D_0 = 2.0$ and $x/D_0 = 7.0$ are also shown. In addition, the channel width of these figures is enlarged in order to more easily understand the characteristics of the flow pattern. Fig. 2.12a shows the stream lines at $t/T_0 = 0.0$, i.e., the height of the oscillating wall has a minimum value. Therefore, the dip forms in the lower wall, and the separation bubble is located in this dip. Fig. 2.12b shows the stream lines at $t/T_0 = 0.25$, i.e., the height of the oscillating wall becomes zero at this time. However, the stream lines are not parallel to the wall because of the unsteady effect. Fig. 2.12c shows the stream lines at $t/T_0 = 0.50$, i.e., the height of the oscillating wall has a maximum value at this time. So, stenosis develops at this time, and a large separation bubble is evident. In addition, we should note the existence of a separation bubble in the vicinity of the upper wall. Fig. 2.12d shows the stream lines at $t/T_0 = 0.75$, i.e., the height of the oscillating wall becomes zero at this time.

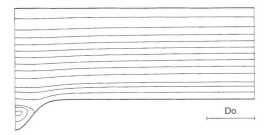

FIGURE 2.12a   Stream lines for an oscillating wall (Eq. 2.11). $t/T_0 = 0.00$; Re = 500; St = 0.03, $hm/D_0 = 1/3$: $\varepsilon = 0.22$; $\delta/T_0 = 0.725$.

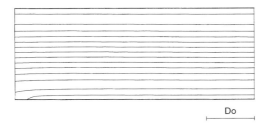

FIGURE 2.12b   Stream lines for an oscillating wall (Eq. 2.11). $t/T_0 = 0.25$; Re = 500; St = 0.03, $hm/D_0 = 1/3$; $\varepsilon = 0.22$; $\delta/T_0 = 0.725$.

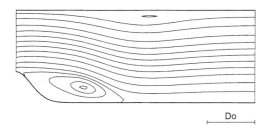

FIGURE 2.12c   Stream lines for an oscillating wall (Eq. 2.11). $t/T_0 = 0.50$; Re = 500; St = 0.03, $hm/D_0 = 1/3$; $\varepsilon = 0.22$; $\delta/T_0 = 0.725$.

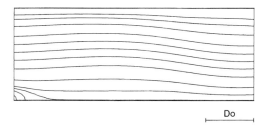

FIGURE 2.12d   Stream lines for an oscillating wall (Eq. 2.11). $t/T_0 = 0.75$; Re = 500; St = 0.03, $hm/D_0 = 1/3$; $\varepsilon = 0.22$; $\delta/T_0 = 0.725$.

It is apparent that the length of the separation bubble is affected by flow conditions. Therefore, we examine the relationship between the length of separation bubble L_s and phase lag δ/T_0. The length of separation bubble L_s at $t/T_0 = 0.5$ is shown in Table 2.1. In this table, D_s denotes the height of the oscillating wall at $t/T_0 = 0.5$, and "steady" denotes the results under steady flow conditions with the same Reynolds number (Re = 500). Table 2.1 demonstrates that the length of a separation bubble can be controlled by the phase lag; the important role of phase lag to the flow patterns is obvious.

TABLE 2.1   Length of Separation Bubble at $t/T_0 = 0.5$

Phase Lag (δ/T_0)	Ls/Ds
−0.025	6.8
0.225	5.9
0.475	6.0
0.725	6.8
Steady	6.3

Next, we considered the distributions of wall shear stress. The distributions of wall shear stress at the upper wall and lower wall are shown in Figs. 2.13a-d. In those figures, the wall shear stress at the upper wall is represented by a dotted line, and the wall shear stress at the lower wall is by a solid line. Fig. 2.13a shows the distributions of wall shear stress at $t/T_0 = 0.0$, i.e., the height of the oscillating wall has a minimum value. Due to the existence of a separation bubble, the negative wall shear stress region appears in the lower wall. Fig. 2.13b shows the distributions of wall shear stress at $t/T_0 = 0.25$. Since the height of the oscillating wall is zero at this time, the value of the wall shear stress is almost constant. Fig. 2.13c shows the distributions of wall shear stress at $t/T_0 = 0.50$. A sharp peak of wall shear stress is evident at the lower wall. One should note that the peak position lies in the upstream side of stenosis ($x/D_0 < 2.0$), which is not in conflict with the results of the fixed wall.[17] Fig. 2.13d shows the distribution of wall shear stress at $t/T_0 = 0.75$. This figure demonstrates the fluctuations in the distribution of wall shear stress far away from the stenosis. This result will be explained by the flow of the separation bubble.

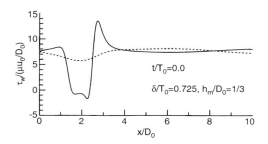

FIGURE 2.13a   Wall shear stress for an oscillating wall (Eq. 2.11). The solid line represents the lower wall, and the dotted line represents the upper wall. $t/T_0 = 0.00$; Re = 500; St = 0.03, $hm/D_0 = 1/3$; $\varepsilon = 0.22$; $\delta/T_0 = 0.725$.

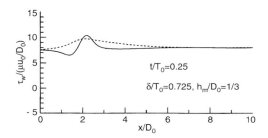

FIGURE 2.13b   Wall shear stress for an oscillating wall (Eq. 2.11). The solid line represents the lower wall, and the dotted line represents the upper wall. $t/T_0 = 0.25$; Re = 500; St = 0.03, $hm/D_0 = 1/3$; $\varepsilon = 0.22$; $\delta/T_0 = 0.725$.

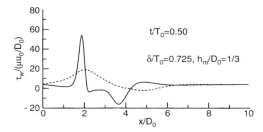

FIGURE 2.13c   Wall shear stress for an oscillating wall (Eq. 2.11). The solid line represents the lower wall, and the dotted line represents the upper wall. $t/T_0 = 0.50$; Re = 500; St = 0.03, $hm/D_0 = 1/3$; $\varepsilon = 0.22$; $\delta/T_0 = 0.725$.

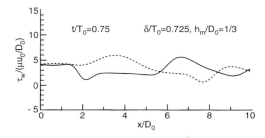

FIGURE 2.13d   Wall shear stress for an oscillating wall (Eq. 2.11). The solid line represents the lower wall, and the dotted line represents the upper wall. $t/T_0 = 0.75$; Re = 500; St = 0.03, $hm/D_0 = 1/3$; $\varepsilon = 0.22$; $\delta/T_0 = 0.725$.

Fig. 2.14 shows the minimum and maximum wall shear stress of the upper and lower wall at each time. In this figure, the labels max and min denote the maximum and minimum value, and the labels UP and LO denote the upper and lower wall, respectively. Fig. 2.14 shows that the wall shear stress of the lower wall reaches its minimum and maximum values at about $t/T_0 = 0.60$. On the other hand, the height of the oscillating wall reaches its maximum value at $t/T_0 = 0.50$. The mechanics of this difference are explained by the growth time of the separation bubble.

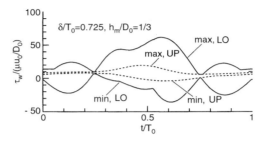

FIGURE 2.14   Maximum and minimum wall shear stress vs. time for an oscillating wall (Eq. 2.11). Max = maximum; min = minimum; LO = lower wall; UP = upper wall. Re = 500; St = 0.03, $hm/D_0 = 1/3$; $\varepsilon = 0.22$; $\delta/T_0 = 0.725$.

It is apparent that the phase lag between the wall oscillation and the pulsation of the bulk flow has a strong influence on the distribution of wall shear stress. The maximum and minimum wall shear stress in a cycle are shown in Table 2.2 as a function of phase lag. In this table, the values of wall shear stress are normalized by $\mu u_0/D_0$.

TABLE 2.2   Influence of Phase Lag on the Maximum and Minimum Wall Shear Stress in a Cycle

Phase Lag (δ/T_0)	Maximum Wall Shear Stress		Minimum Wall Shear Stress	
	UP	LO	UP	LO
−0.025	28.3	86.1	−3.240	−36.4
0.225	24.1	84.4	0.929	−36.3
0.475	14.8	63.3	0.467	−35.2
0.725	19.4	61.7	−3.910	−35.7
Steady	21.2	72.6	−0.927	−35.9

This table shows that phase lag has a strong effect on wall shear stress, and the reversal of wall shear stress at the upper wall can be prevented by selecting the appropriate phase lag. In other words, we can prevent the flow separation in the vicinity of the upper wall by selecting the appropriate phase lag. It is well known that the flow separation has a strong influence on the pressure loss or generation of some arterial lesions.[18] Therefore, this result has important consequences for engineering as well as biomechanics. Additional detailed studies should be performed in the future.

Application I: Pressure Loss of Pulsatile Flow through a Two-Dimensional Channel with an Oscillating Wall

From a practical viewpoint, pressure loss is one of the most important factors, so we considered the pressure loss of the pulsatile flow through a two-dimensional channel with an oscillating wall. The relationship between the wall oscillation and the pressure loss has attracted the interest of many researchers in relation to the drag-reduction mechanism of dolphins and the pressure loss of blood flow in the artery. However, the detailed numerical analysis has not yet been performed, and there is still much more to be studied.

Since the calculation models and methods are almost identical to those discussed earlier in this section, we will present the calculation results directly. Of course, pressure is a function of x, y, t (see Fig. 2.11), but it is apparent that the value of the y-component does not have a strong effect on the pressure. Therefore, we introduced the average pressure along the y-direction within a channel:

$$\bar{p} = \frac{1}{D_0 - F} \int_F^{D_0} p \, dy \tag{2.23}$$

Fig. 2.15 shows the calculated distribution of $\bar{p}$ at Re = 500, St = 0.01, $h_m/D_0 = 0.3$, $\varepsilon = 0.61$, and $\delta/T_0 = 0.695$. The ordinate there is non-dimensional pressure, and the value of the pressure at the inlet ($x/D_0 = 0$) is set at zero. The sharp change of the pressure distribution was found in the vicinity of $x/D_0 = 2.0$. Of course, this sharp change is explained by the pattern of the wall oscillation (see Eq. 2.11).

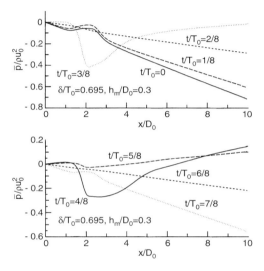

FIGURE 2.15   Pressure distribution for an oscillating wall (Eq. 2.11). Re = 500; St = 0.01, $hm/D_0 = 0.3$; $\varepsilon = 0.61$; $\delta/T_0 = 0.695$.

The most interesting result of this analysis is the relationship between the pressure drop and the wall oscillation from a practical viewpoint. Therefore, we defined the following parameter:

$$\Delta \bar{p} = \bar{p}\left(x/D_o = 10.0\right) - \bar{p}\left(x/D_o = 0.0\right) \tag{2.24}$$

The calculated $\Delta \bar{p}$ is shown in Figs. 2.16a-d. In those figures, the value of the non-dimensional phase lag δ/TI_0 is −0.055, 0.195, 0.445, and 0.695, respectively, and the non-dimensional amplitude of the wall oscillation lies in 0.1–0.4. Figs. 2.16a, b, and d show that the pattern of the pressure drop is strongly distorted by the increase in h_m/D_0. However, Fig. 2.16c shows that the pattern of the pressure drop is hardly affected by the increase in h_m/D_0, thereby demonstrating the importance of the phase lag between the pulsation of the bulk flow and the wall oscillation. To understand the meaning of the wall oscillation, the pressure drop of the pulsatile flow through the channel with a fixed stenosis was calculated for reference. The calculation conditions were the same as the conditions of Figs. 2.16a-d except for the form of the lower wall, which is represented by the following equation:

$$F(x) = h_m \exp\left[-5\left(x/D_0 - 2\right)^2\right] \tag{2.25}$$

The calculated pressure drop based on Eq. 2.25 is shown in Fig. 2.17, which also demonstrates that the absolute value of the pressure drop at the fixed wall is greater than the absolute values of the pressure drop at the oscillating wall (Fig. 2.16). In addition, the pattern of the pressure drop in Fig. 2.17 (fixed wall) resembles the sine curve as compared to the results of Fig. 2.16 (oscillating wall) with one exception (Fig. 2.16c). Next, we considered the blood flow through the artery as an application of these results. The flow through the oscillating stenosis is easily found in the cardiovascular system.[19] Many studies were performed to evaluate the behavior of blood flow through the oscillating stenosis. Among them, a very interesting model study was performed by Rabinovitz et al.[20] They studied the interaction between the oscillating stenosis and the pulsatile blood flow, and they found that the oscillating stenosis had the strong effect of distorting the waveform of the pressure curve compared with the fixed stenosis. In addition, they found that this result was related to some arterial disease (tetralogy of Fallot). Clearly, our numerical results do not conflict with their experimental results, and our investigation theoretically supports their results.

We next introduce the non-dimensional coefficient of an additional pressure drop due to the wall oscillation. First, we divided the cycle-averaged pressure drop into two terms:

$$< \Delta \bar{p} > = \Delta p_0 + \Delta p' \tag{2.26}$$

where $< >$ denotes the cycle-averaged value, and Δp_0 denotes the cycle-averaged pressure drop of the pulsatile flow through a two-dimensional channel with parallel walls ($F(x,t) = 0.0$). We easily calculated the value of Δp_0 using Womersley's two-dimensional solution. Therefore, the additional pressure drop due to the wall oscillation is denoted by $\Delta p'$. On the basis of this result, the non-dimensional coefficient of the additional pressure drop due to the wall oscillation was defined as follows:

$$\Delta p' = -\lambda \frac{1}{2} \rho u_0^2 \tag{2.27}$$

The purpose of our study was to evaluate the basic behavior of the value λ. The calculated results are shown in Fig. 2.18, which demonstrates that an increase in h_m/D_0 increases the value of λ, and we can suppress the increase in the additional pressure drop by selecting the phase lag between the pulsation of the bulk flow and the wall oscillation. In addition, Fig. 2.18 shows that the value of λ satisfies the following inequality:

$$\lambda \geq 0 \tag{2.28}$$

These results are worthy of attention and will be important when evaluating the role of wall oscillation.

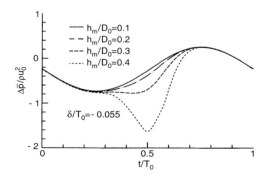

FIGURE 2.16a Pressure difference between x/D_0 0.0 and $x/D = 10.0$ for an oscillating wall (Eq. 2.11). Re = 500; St = 0.01; $\varepsilon = 0.61$; $\delta/T_0 = -0.055$.

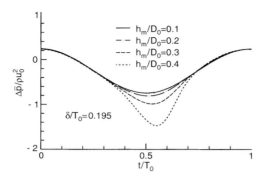

FIGURE 2.16b Pressure difference between x/D_0 0.0 and $x/D = 10.0$ for an oscillating wall (Eq. 2.11). Re = 500; St = 0.01; $\varepsilon = 0.61$; $\delta/T_0 = 0.195$.

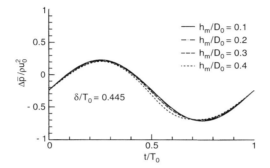

FIGURE 2.16c Pressure difference between $x/D_0 = 0.0$ and $x/D = 10.0$ for an oscillating wall (Eq. 2.11). Re = 500; St = 0.01; $\varepsilon = 0.61$; $\delta/T_0 = 0.445$.

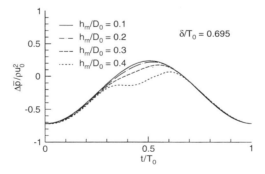

FIGURE 2.16d Pressure difference between $x/D_0 = 0.0$ and $x/D = 10.0$ for an oscillating wall (Eq. 2.11). Re = 500; St = 0.01; $\varepsilon = 0.61$; $\delta/T_0 = 0.695$.

Next, we investigated the effect of the period of wall oscillation by calculating the pressure drop based on the following:

$$F(x,t) = -h_m \exp\left[-5(x/D_0 - 2)^2\right]\cos(4\pi t/T_0) \tag{2.29}$$

The calculated results of λ based on Eq. 2.29 are shown in Fig. 2.19. The calculation conditions there were the same as the calculation conditions shown in Fig. 2.18 except for the period of the oscillating wall. Fig. 2.19 shows that the phase lag between the pulsation of the bulk flow and the wall oscillation does not have a strong effect on the value of λ. In addition, Fig. 2.19 shows that the calculated results of λ satisfy Inequality 2.28. Figs. 2.18 and 2.19 show that the wall oscillation described by Eqs. 2.11 and 2.29 does not decrease the pressure drop through the channel. The relationship between the wall oscillation and the pressure drop through the channel was studied by many researchers, some of whom suggested that the wall oscillation could decrease the pressure drop through the channel.[21] Clearly, this is in conflict with the calculation results described above. The reason for this contradiction can be explained by the form of the wall oscillation. It is apparent that the actual pattern of the wall oscillation has a more complicated form as compared to the form shown in Eqs. 2.11 and 2.29. A numerical study based on the other wall oscillation patterns should be performed in the future.

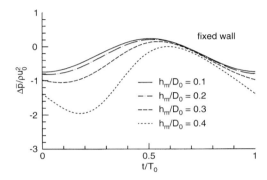

FIGURE 2.17   Pressure difference between $x/D_0 = 0.0$ and $x/D = 10.0$ for a fixed wall (Eq. 2.25). Re = 500; St = 0.01; $\varepsilon = 0.61$.

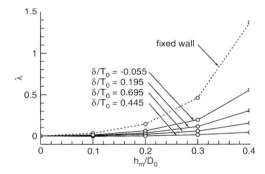

FIGURE 2.18   The value of λ for an oscillating wall (Eq. 2.11) and a fixed wall (Eq. 2.25). Re = 500; St = 0.01; $\varepsilon = 0.61$.

Application II: Influence of the Wall Oscillation on the Structure of the Separation Bubble

In the numerical analysis mentioned above, the geometry of the flow domain was a two-dimensional channel with parallel walls when the oscillation of the wall did not occur ($F(x,y) = 0.0$). So, the separation bubble did not exist when the oscillation of the wall did not occur. We considered the flow through a two-dimensional channel which experienced stenosis and found the existence of the separation bubble possible even if the wall oscillation does not occur. It is clear that the flow pattern in the area of stenosis is strongly affected by the oscillation of the wall. The purpose of this section is to evaluate the change of the separated flow region due to the wall oscillation.

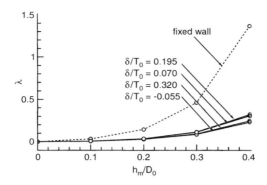

FIGURE 2.19   The value of λ for an oscillating wall (Eq. 2.29) and a fixed wall (Eq. 2.25). Re = 500; St = 0.01; $\varepsilon = 0.61$.

The calculation model is shown in Fig. 2.20. In this model, the geometry of the lower wall $F(x,t)$ is demonstrated by the following equations:

$$
\left.
\begin{aligned}
F(x,t) &= G_0(x) + G_1(x,t) \\
G_o(x) &= 0.2D_0 \exp\left[-5.0\left(x/D_0 - 2.0\right)^2\right] \\
G_1(x,t) &= 0.5h_w\left\{1.0 - \cos\left(2\pi t/T_w\right)\right\}\exp\left[-5.0\left(x/D_0 - \beta\right)^2\right]
\end{aligned}
\right\}
\tag{2.30}
$$

where $G_0(x)$ denotes the geometry of fixed stenosis, and $G_1(x,t)$ denotes the oscillating component of the wall. The constant β denotes the position of the oscillating region. In order to simplify the problem, the pulsation effect of the bulk flow is not taken into consideration. So, the inlet flow rate Q is represented by $Q = Q_f$ (constant). The basic equations and the numerical calculation methods are almost identical to the basic equations and numerical calculation methods mentioned above. Therefore, the detailed explanations of those methods are omitted here and the calculated results are presented directly.

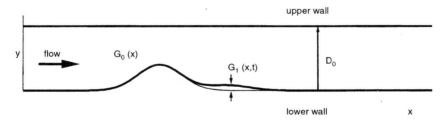

FIGURE 2.20   Two-dimensional channel with fixed stenosis and an oscillating wall.

First we investigated the distributions of wall shear stress. The distributions of wall shear stress in the vicinity of the oscillating region are shown in Fig. 2.21, where τ_w denotes the wall shear stress, and the values of β and h_w/D_0 are set at 3.0 and 0.02, respectively. The Reynolds number and Strouhal number are used to denote the effect of the bulk flow rate and the period of wall oscillation. These parameters are defined as follows:

$$
Re = u_f D_0 / \nu, \quad St = D_0 / (u_f T_w), \quad (u_f = Q_f / D_0)
$$

where the value of Re is 500 and the value of St is 0.1. This figure shows that the oscillation of the wall decreases the separated flow region ($\tau_w < 0$ region) even if the amplitude of the wall oscillation is much smaller than the height of the stenosis or the channel width. This result is worthy of attention. Fig. 2.22 shows the calculated distribution of wall shear stress when the amplitude of the oscillating wall h_w/D_0 is 0.04. The calculation conditions of this figure are the same as the conditions of Fig. 2.21 except for the amplitude of the oscillating wall. The decrease in the separated flow region in Fig. 2.22 is much more noticeable than the decrease in the separated flow region in Fig. 2.21. This result is easy to understand. The length of the separated flow region is shown in Fig. 2.23 as a function of time. In this figure, L_s denotes the length of the separation bubble. The relationship between time and the length of the separation bubble can be determined from this figure.

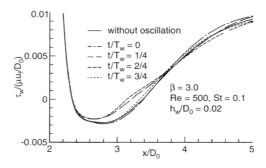

FIGURE 2.21   Wall shear stress for an oscillating wall with fixed stenosis (Eq. 2.30). Re = 500; $\beta = 3.0$; St = 0.1; $h_w/D_0 = 0.02$.

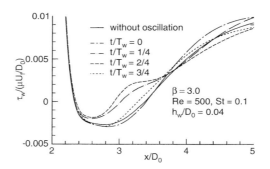

FIGURE 2.22   Wall shear stress for an oscillating wall with fixed stenosis (Eq. 2.30). Re = 500; $\beta = 3.0$; St = 0.1; $h_w/D_0 = 0.04$.

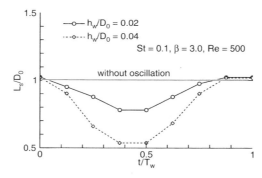

FIGURE 2.23   Length of separated flow region for an oscillating wall with fixed stenosis (Eq. 2.30). Re = 500; $\beta = 3.0$; St = 0.1.

Next we examined the effect of oscillating position on the length of the separated flow region. The position of the oscillating region is characterized by the value of β (Eq. 2.30). The calculated length of the separated flow region is shown in Fig. 2.24 as a function of time. The values of β = 3.0, 3.5, and 4.0 are used here. Figure 2.24 shows that the decrease in the separated flow region is noticeable when the value of β is set at 3.0. However, the length of the separated flow region does not decrease when the value of β is set at 4.0. Fig. 2.21 demonstrates that the separated flow region lies in the range of $2.4 \le x/D_0 \le 3.6$ if wall oscillation does not occur. Clearly, wall oscillation in the vicinity of the separated flow region does decrease the length of the separated flow region.

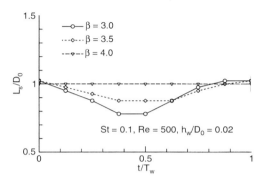

FIGURE 2.24   Length of separated flow region for an oscillating wall with fixed stenosis (Eq. 2.30). Re = 500; St = 0.1; $h_w/D_0 = 0.02$.

Next we examined the effect of wall oscillation period T_w. This parameter is characterized by the Strouhal number (St). The calculated results are shown in Fig. 2.25, demonstrating that the Strouhal number has a strong effect on the length of the separated flow region.

These figures and calculations allowed us to determine the conditions necessary to decrease the length of the separated flow region. These conditions have never been discussed in the literature. As mentioned above, flow separation is significant to the fields of engineering and biomechanics. The implication of these results will become clear in the future.

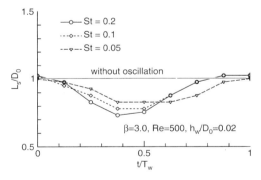

FIGURE 2.25   Length of separated flow region for an oscillating wall with fixed stenosis (Eq. 2.30). Re = 500; β = 3.0; $h_w/D_0 = 0.02$.

2.4   Wall Shear Stress of a Non-Newtonian Fluid through an Axisymmetric Diffuser

Introduction

The fact that blood is a non-Newtonian fluid is very significant in a low shear rate region.[22] The mean shear rate of blood flow was evaluated by some researchers who found that the mean shear rate of the venous flow was lower than that of the arterial or capillary flows. Therefore, it was necessary to consider the non-Newtonian effect of blood when we analyzed the blood flow through the vein. One of the most famous features of venous blood flow is the large amplitude self-excited oscillation between the blood flow and the venous tube.[23] Many laboratory experiments were performed, and the existence of this self-excited oscillation was confirmed *in vitro*. When this large amplitude self-excited oscillation occurs, the divergent and convergent parts of the tube are developed instantaneously. It is clear that the flow separation occurs in the divergent part of the collapsed tube. Many researchers suggested that this flow separation plays an important role in the self-excited oscillation.[24] Therefore, investigation of the separation phenomena in the divergent part of the collapsed tube becomes a very important subject for the clarification of self-excited oscillation.

We numerically studied the laminar steady flow of a non-Newtonian fluid through an axisymmetric diffuser. The steady flow through an axisymmetric diffuser was studied by many researchers, none of whom accounted for the non-Newtonian property of the fluid. The purpose of our calculation was to describe the numerical techniques for a non-Newtonian fluid and to obtain detailed information of the steady separation flow of a non-Newtonian fluid.

Methods

Generally, the Casson model[22] is used as a constitutive equation for blood. The schematic flow curve of a Casson fluid is shown in Fig. 2.26, which demonstrates that the apparent viscosity of a Casson fluid diverges when the shear rate becomes zero. This is not compatible with the measured data.[25] Therefore, the modified Casson model which possesses the upper limit of the apparent viscosity was used. This equation is written as follows:

$$\mu_a = \left[\left(\mu^2 J_2\right)^{1/4} + \left(p_y / 2\right)^{1/2}\right]^2 J_2^{-1/2} \tag{2.31}$$

$$J_2 = \left(e_{rr}^{\ 2} + e_{\varphi\varphi}^{\ 2} + e_{zz}^{\ 2}\right)/2 + \left(e_{r\varphi}^{\ 2} + e_{zr}^{\ 2} + e_{r\varphi}^{\ 2}\right) \tag{2.32}$$

$$\text{if } J_2 < \left(\beta p_y / \mu\right)^2 \quad \text{then } J_2 = \left(\beta p_y / \mu\right)^2 \tag{2.33}$$

where μ_a denotes the apparent viscosity of Casson fluid, μ denotes the constant corresponding to the viscosity coefficient, p_y denotes the yielding stress, e_{ij} denotes the shear rate of the i,j component $(i, j = r, \varphi, z)$, and β denotes the non-dimensional constant related to the upper limit of the apparent viscosity. In these equations, the cylindrical coordinate system (r, φ, z) is used. Eq. 2.32 agrees with the Casson model within the limits of $\beta \to 0$. The value of β is set at 0.01 by considering the upper limit of the apparent viscosity of blood.[25] If the value of β becomes twice as large as this value, the calculated results change by about 1%. Clearly the calculated results are hardly affected by the value of β. Hereafter, this modified Casson model is referred to simply as the Casson model.

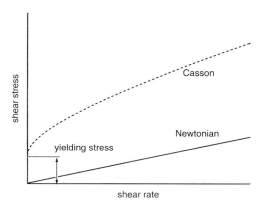

FIGURE 2.26 Schematic flow curve of a Casson fluid.

Basic equations for the axisymmetric incompressible steady flows are written as follows:

$$\rho\left(u\frac{\partial u}{\partial r}+v\frac{\partial u}{\partial z}\right)=-\frac{\partial p}{\partial r}+\frac{\partial}{\partial r}\tau_{rr}+\frac{\partial}{\partial z}\tau_{rz}+\frac{\tau_{rr}-\tau_{\varphi\varphi}}{r} \tag{2.34}$$

$$\rho\left(u\frac{\partial v}{\partial r}+v\frac{\partial v}{\partial z}\right)=-\frac{\partial p}{\partial z}+\frac{\partial}{\partial r}\tau_{rz}+\frac{\partial}{\partial z}\tau_{zz}+\frac{\tau_{rz}}{r} \tag{2.35}$$

$$\frac{1}{r}\frac{\partial(ru)}{\partial r}+\frac{\partial v}{\partial z}=0 \tag{2.36}$$

$$\tau_{rr}=2\mu_a\frac{\partial u}{\partial r},\quad \tau_{rz}=\mu_a\left(\frac{\partial u}{\partial z}+\frac{\partial v}{\partial r}\right) \tag{2.37}$$

$$\tau_{rz}=2\mu_a\frac{\partial v}{\partial z},\quad \tau_{\varphi\varphi}=2\mu_a u/r \tag{2.38}$$

where u denotes the r-component of the velocity vector, v denotes the z-component of the velocity vector, and ρ denotes the density of the fluid.

The following constants are used in this calculation; these constants are determined to simulate the blood flow:

$$\mu=3.5\times10^{-3}\text{ Pa s},\quad \rho=1050\text{ kg/m}^3$$
$$p_y=0.01\text{ Pa (Casson fluid)},\quad 0\text{ Pa (Newtonian fluid)}$$

The geometry of the axisymmetric divergent tube is shown in Fig. 2.27. The geometry of this model is determined by the following three values: n, θ, and D_0. The straight tube is arranged on the upstream and downstream sides of the divergent tube. The straight tube on the downstream side shows the uncollapsed venous tube. The boundary conditions are shown by the following equations:

(1) inlet:

$$u = 0 \qquad (2.39)$$

$$v = \begin{cases} \dfrac{1}{4\mu}\dfrac{\partial p}{\partial z}\left(1+\beta^{-1/2}\right)^{-2}\left(r^2 - r_{cc}^2\right) + v_{0c}, & \left(r \le r_{cc}\right) \\[2ex] \dfrac{1}{4\mu}\dfrac{\partial p}{\partial z}\left[\left(r^2 - R_u^2\right) - \dfrac{8}{3}r_c^{1/2}\left(r^{3/2} - R_u^{3/2}\right) + 2r_c\left(r - R_u\right)\right], & \left(r \ge r_{cc}\right) \end{cases} \qquad (2.40)$$

where

$$r_c = 2p_y / |\partial p / \partial z|, \quad r_{cc} = r_c\left(1 + \sqrt{\beta}\right)^2$$

$$v_{0c} = \frac{1}{4\mu}\frac{\partial p}{\partial z}\left[\left(r_{cc}^2 - R_u^2\right) - \frac{8}{c}r_c^{1/2}\left(r_{cc}^{3/2} - R_u^{3/2}\right) + 2r_c\left(r_{cc} - R_u\right)\right]$$

In these equations, R_u denotes the pipe radius of the upstream side. These equations represent the fully developed velocity profile of a Casson fluid in a straight pipe. A depiction of the shape of the collapsed tube[24] showed that the convergent angle was much smaller than the divergent angle. Therefore, the velocity profile at the entrance to the divergent tube is similar to the fully developed velocity profile in a straight pipe. The reason for using the fully developed velocity profile in a straight pipe as the inlet boundary condition is easily understood by the following:

(2) outlet:

$$u = 0, \quad f_z = 0, \quad p = 0 \qquad (2.41)$$

where f_z denotes the axial component of boundary force. These boundary conditions are similar to the ones used in the calculation of entry flow[26] and the condition of pressure is used for the convenience of numerical calculation.

(3) wall:

$$u = v = 0 \qquad (2.42)$$

(4) center:

$$u = 0, \quad \partial v / \partial r = 0 \qquad (2.43)$$

Many numerical methods have been applied to the calculation of viscous flow. We adopted the finite element method, which has an advantage compared to the finite difference method when solving for the viscous flow in complex geometry. The finite element used in this calculation is triangular. In addition, second and first order interpolation functions are used for the calculation of velocity and pressure distributions, respectively. By integrating the product of the Navier-Stokes equation (Equations 2.34 and 2.35) and the second order interpolation function over one element, we obtained the basic algebraic equations. The remaining equation was obtained by integrating the product of the mass conservation equation (Eq. 2.36) and the first order interpolation function over one element. Detailed explanations of these calculations are provided elsewhere.[17]

The calculation conditions correspond to the venous flow conditions of man and dog. The Reynolds number lies in the range of 70–900, and the diameter D_0 lies in the range of 0.6–2.0 cm. The mean shear

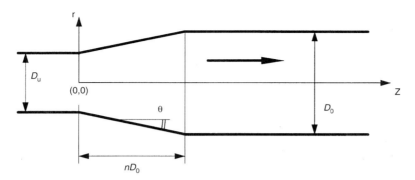

FIGURE 2.27   Model of a divergent tube.

rate of the venous flow lies in the range of 30–100 s^{-1}, which is a low value compared to the artery or capillary flows.

Results and Discussion

The separated flow region is one of the most important parameters for biomechanics, and many important results (e.g., behavior of wall shear stress) can be estimated using this parameter. The calculated results are shown in Figs. 2.28–2.30. In these figures, the longitudinal axis represents the Reynolds number, and the latitudinal axis represents the non-dimensional axial direction. In addition, the separated flow region is denoted by the inner part of these curves. The Reynolds number is defined as follows:

$$\mathrm{Re} = \rho \bar{v} D_0 / \mu \tag{2.44}$$

where $\bar{v}$ denotes the section averaged axial velocity in the downstream side of the diffuser. Fig. 2.28 shows the separated flow region at $n = 1, D_0 = 2.0$ cm, and $\theta = 6$ deg. One should note that flow separation does not occur in the case of a Casson fluid. Fig. 2.29 shows the separated flow region at $n = 1, D_0 = 2.0$ cm, and $\theta = 8$ deg. Figs. 2.28 and 2.29 illustrate that the effect of a non-Newtonian fluid is a decrease in the length of the separated flow region. Fig. 2.30 shows the separated flow region at $n = 1, D_0 = 2.0$ cm, and $\theta = 10$ deg. This figure demonstrates that the increase in the divergent angle decreases the difference in the separated flow region between a Newtonian fluid and a Casson fluid. As mentioned above, the apparent viscosity of a Casson fluid increases with a decrease in the shear rate. So, the apparent viscosity of a Casson fluid in the vicinity of the separation and reattachment points becomes very large. Therefore, it is clear that this increased apparent viscosity prevents the reverse flow. On the other hand, it is obvious that an increase in the divergent angle causes the flow to undergo a sudden change and decreases the low shear rate region. So, the increase in the divergent angle will weaken the effect of a non-Newtonian fluid. The above calculations are explained by these considerations. As mentioned above, the flow separation has a strong relationship to the self-excited oscillation of the venous tube. Many researchers have suggested that flow separation is essential to self-excited oscillation.[24] Therefore, our calculation results suggest that the non-Newtonian property of blood will have the effect of restraining the self-excited oscillation of the venous tube. This demonstrates the important role of the non-Newtonian property of blood. Of course, we cannot draw conclusions from this theory without further study.

In the industrial world, a diffuser is used to convert kinetic energy to pressure. In addition, some researchers have suggested that this energy conversion rate is strongly related to the self-excited oscillation of the venous tube.[24] Therefore, we examined the energy conversion rate of a diffuser. We first calculated the pressure recovery based on Bernoulli's formula. We easily obtained the following equation from Bernoulli's formula:

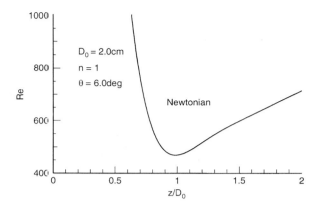

FIGURE 2.28   Separation and reattachment points. $n = 1$; $\theta = 6$ deg; $D_0 = 2.0$ cm.

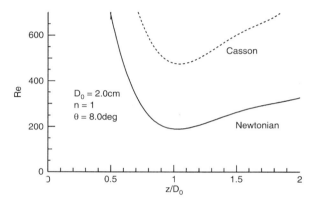

FIGURE 2.29   Separation and reattachment points. $n = 1$; $\theta = 8$ deg; $D_0 = 2.0$ cm.

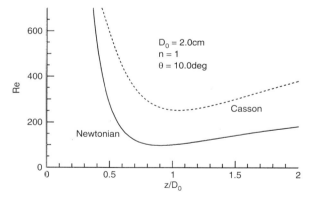

FIGURE 2.30   Separation and reattachment points. $n = 1$; $\theta = 10$ deg; $D_0 = 2.0$ cm.

$$\Delta\overline{p_b} = \overline{p_b}(z = nD_0) - \overline{p_b}(z = 0) = \frac{1}{2}\rho\overline{v}^2\left[\left(D_u/D_0\right)^4 - 1\right] \qquad (2.45)$$

where $\Delta\overline{p_b}$ denotes the section averaged pressure difference between $z = nD_0$ and $z = 0$ (obtained from Bernoulli's formula), $\overline{p_b}$ denotes the section averaged pressure obtained from Bernoulli's formula, and D_u denotes the diameter of the upper tube (Fig. 2.27). Of course, the value of the practical pressure recovery is small compared with the value of Eq. 2.45. So, we define the energy conversion rate η with the following equation:

$$\eta = \Delta \overline{p} / \Delta \overline{p}_b \qquad (2.46)$$

where $\Delta \overline{p}_b$ denotes the practical section averaged pressure recovery between $z = nD_0$ and $z = 0$. The calculated energy conversion rate is shown in Figs. 2.31 and 2.32. In these figures, the longitudinal axis represents the energy conversion rate, and the latitudinal axis represents the angle of divergence (Fig. 2.27). These figures show that the non-Newtonian property of a fluid increases the energy conversion rate. This result can be explained by the decrease in the separated flow region due to the non-Newtonian property of fluid. In addition, these figures show that an increase in the Reynolds number causes a decrease in the energy conversion rate. This result can be explained by the increase in the separated flow region due to the increase in the Reynolds number.

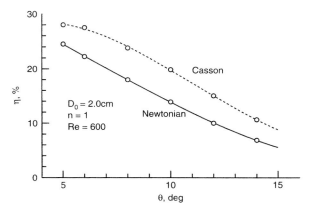

FIGURE 2.31   Efficiency of the pressure conversion rate. $n = 1$; $Re = 600$; $D_0 = 2.0$ cm.

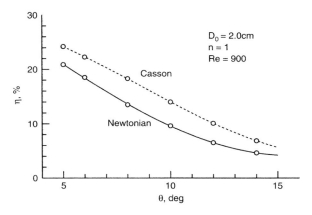

FIGURE 2.32   Efficiency of the pressure conversion rate. $n = 1$; $Re = 900$; $D_0 = 2.0$ cm.

These calculation results prove that a non-Newtonian fluid has the effect of restraining the flow separation and increasing the energy conversion rate. These results clearly illustrate the excellent properties of a non-Newtonian fluid and suggest that the non-Newtonian property of blood plays an important role in the maintenance of life.

2.5   Concluding Remarks

This chapter describes some experimental and numerical techniques for evaluating the distribution of wall shear stress. The effective range of these techniques is limited, but the techniques have some very interesting features. In addition, the results obtained based on these techniques were discussed in detail, we defined the blood flow parameters in the aorta, characterized wall oscillation, and identified the non-Newtonian property of blood. Our results suggest the dexterous nature of blood flow.

As mentioned above, the role of wall shear stress on the arterial lesions has not yet been established, and many problems remain unsolved in the field of biomechanics. Therefore, more detailed studies should be performed in the future. Specifically, the study of blood flow from a biological viewpoint is essential.

References

1. Giddens, D.P., Zarins, C.K., and Glagov, S., *Trans. ASME J. Biomech. Eng.*, 115, 588, 1993.
2. Nakamura, M., Sugiyama, W., and Haruna, M., *Trans. ASME J. Biomech. Eng.*, 115, 412, 1993.
3. Nakamura, M., Kiba, M., and Sato, A. *Trans. JSME (Ser. B)*, 59-567, 3308, 1993 (in Japanese).
4. Womeresley, J.R., *Phil. Mag.*, 46, 199, 1955.
5. Uchida, S., *ZAMP*, 7, 403, 1956.
6. Ohmi, M. and Iguchi, M., *Bull. JSME*, 23-186, 2013, 1980.
7. Yamaguchi, R. and Kohtoh, K., *Trans. ASME J. Biomech. Eng.*, 116, 119, 1994.
8. Tu, S.W. and Ramaprian, B.R., *J. Fluid Mech.*, 137, 31, 1983.
9. Iguchi, M. and Ohmi, M., *Bull. JSME*, 27-231, 1873, 1984.
10. Rao, K.N., Narsimha, R., and Badai Naryanan, M.A., *J. Fluid Mech.*, 48, 339, 1971.
11. Oka, S., *Biorheology*, 2nd Ed., 31, 1974 (in Japanese).
12. Nakao, S., *J. Jap. Hydraulics and Pneumatics*, 15-2, 116, 1984 (in Japanese).
13. Ralph, M.E. and Pedley, T.J., *J. Fluid Mech.*, 190, 87, 1988.
14. Japanese Society of Mechanical Engineers, *Numerical Simulation of Fluid Flows*, 1st Ed., Corona, Tokyo, 1988, 56 (in Japanese).
15. Van der Vorst, H.A., *J. Computational Physics*, 44, 1, 1981.
16. Notera, T., *BIT Special Suppl.*, 19-13, 44, 1987 (in Japanese).
17. Nakamura, M. and Sawada, T., *Trans. ASME J. Biomech. Eng.*, 110, 137, 1988.
18. Ku, D.N., Giddens, D.P., Zarins, C.K., and Glagov, S., *Arteriosclerosis*, 5-3, 292, 1985.
19. Anayiotos, A.S., Jones, S.A., Giddens, D.P., Glagov, S., and Zarins, C.K., *Trans. ASME J. Biomech. Eng.*, 116, 98, 1994.
20. Rabinovitz, R., Degani, D., Guttinger, C., and Milo, S., *Trans. ASME J. Biomech. Eng.*, 106, 309, 1984.
21. Takemitsu, N. and Matsunobu, Y., *Nagare*, 5, 254, 1986 (in Japanese).
22. Oka, S., *Biorheology*, 2nd Ed., 49, 1974 (in Japanese).
23. Conrad, W.A., *IEEE Trans. Biomed. Eng.*, BME 16, 284, 1969.
24. Cancelli, C. and Pedley, T.J., *J. Fluid Mech.*, 157, 375, 1985.
25. Skalak, R., Keller, S.R., and Secomb, T.W., *Trans. ASME J. Biomech. Eng.*, 103, 102, 1981.
26. Crochet, M.J., Davies, A.R., and Walters, K., *Numerical Simulation of Non-Newtonian Flow*, 1st Ed., Elsevier, Amsterdam, 1984, 234.

3

A Measurement Method for Wall Shear Stress and a Fluid Mechanical Application for Vascular Disease

Ryuhei Yamaguchi
Shibaura Institute of Technology

3.1 Introduction..3-1
3.2 Electrochemical Method for Measuring
Wall Shear Stress...3-2
Concept • Principle • Measurement Method •
Creation Process of a Test Electrode •
Effective Surface Area • Frequency Characteristics and
Limitations
3.3 Application...3-17
Physiological View • Stenosis Model • 45° Bifurcation Model

Velocity, pressure, and wall shear stress are important factors in fluid mechanics. Particularly, wall shear stress, i.e., skin friction, which acts as the interface between fluids and the boundary wall, is the most fundamental fluid-dynamic factor. In spite of its importance, there are few effective measurement methods for wall shear stress. This chapter describes an effective method for measuring wall shear stress and its application to an arterial model.

3.1 Introduction

There are several measurement probes of wall shear stress, such as the preston tube, the hot film probe, the pitot tube, and the hot wire probe. Using the preston tube, the wall shear stress at a smooth surface is calculated from the difference between the total pressure and the static pressure which is measured by the surface pitot tube attached to the boundary wall. The hot film probe calculates the wall shear stress with the analogy of heat transfer at the boundary wall. Using the pitot tube and the hot wire probe, the wall shear stress is calculated from the velocity gradient at the boundary wall after measuring the velocity profile. The methods described above only allow indirect measurement of wall shear stress by calculating pressure, heat transfer rate, and the velocity profile at the boundary wall. Furthermore, the flow field around the measurement probe is disturbed by these probes. Generally, the wall shear stress is a very sensitive flow factor. Therefore, direct measurement of wall shear stress is desirable.

The electrochemical method, which is based on mass transfer, is a direct measurement method for wall shear stress. Chemical engineers, who are primarily interested in operations involving liquids, most frequently use this method of electrochemical reaction. Under diffusion-controlling conditions, this method enables the measurement of the mass transfer rate and the wall shear stress at a liquid-solid interface. Although the electrochemical method is very effective for measuring the wall shear stress in a liquid flow, it should be noted that there are several limitations of this method. First, it is limited to liquid flow. Accordingly, the data of mass transfer are limited for high Schmidt numbers. Second, only a few special chemical agent-water mixtures in which a diffusion-controlling electrolytic reaction occurs can be employed as the working solution. Third, after an experiment, we cannot drain the working solution into the sewage channel, because the working fluid contains strong acidic or alkaline components with ionized heavy metal ions. Furthermore, we cannot use this method for a velocity higher than the critical velocity at which the reaction resistance at the cathode becomes relatively significant compared to the decreasing resistance of diffusion. Although there are a few disadvantages to this method, the electrochemical method enables the direct measurement of the distribution of shear stress along the wall in series, and makes it easy to clarify the fluid dynamical and physiological implications of the distribution of wall shear stress.

Furthermore, the laser Doppler velocimeter is the strongest weapon for measuring velocity. Although this equipment is characterized by quick time response and high space resolution, it is very expensive. Although the laser Doppler velocimeter has several advantages, it is not a direct measurement probe, i.e., the velocity gradient, the wall shear stress, is calculated only after obtaining the velocity profile in the vicinity of boundary wall.

On the other hand, the wall shear stress, which is a hemodynamic factor, is considered important for the initiation and development of arterial disease. Atherosclerotic originates in strongly disturbed regions such as areas of stenosis, the aortic arch, and the arterial branch. The development of atherosclerosis may be closely related to the physiological response of the arterial wall to shear stress on a localized basis. Therefore, wall shear stress is the most important factor affecting the interface between blood flow and the arterial wall.

3.2   Electrochemical Method for Measuring Wall Shear Stress

Concept

It is difficult to directly measure the velocity gradient near a boundary wall because the boundary layer is so thin that any measurement probe disturbs the flow. However, the electrochemical method, which uses polarization at a small electrode smoothly embedded in the boundary wall, can be applied to obtain the velocity gradient at the wall, i.e., the wall shear stress.[1-3]

The electrochemical method. by which wall shear stress is measured. is based on mass transfer. In electrode reactions, concentration polarization and chemical polarization exist in series, i.e., the ions transfer from the bulk of the solution to the surface of the electrode where the chemical and physical changes occur. In measuring the mass transfer rate by using electrochemical reactions, it is better and more common to make the chemical polarization negligible because the mass transfer coefficients are most easily obtained from the limiting current when the concentration at the liquid-solid interface can be measured as zero.

The ions are transferred from the bulk of the solution to the surface of the test electrode principally by migration due to the electrical potential field, diffusion due to the concentration difference, and convection due to the flow movement. Assuming the transfer is steady and the fluid flows parallel to the electrode surface, the effect of convection vanishes because there is no bulk flow perpendicular to the electrode surface. Migration will not occur if a large excess of a supporting electrolyte is added to the solution. If such electrolytes, which do not react at the test electrode, exist in the solution in relatively high concentration and have high conductivity compared with the reacting ions, the migration current becomes negligible. The electrolysis current develops from the reaction of ions which reach the electrode surface. This reaction is called the diffusion-controlling reaction. The ion, i.e., mass, transfer flux N is given by the following equation which is the analogy between heat and mass transfer.

$$N = \alpha\left(\theta_\infty - \theta_w\right)$$
(3.1)

where θ_∞ and θ_w are the concentration in the bulk solution and at the surface of the cathode, respectively, and α denotes the mass transfer coefficient.

The current i is related to the mass (ion) transfer flux N according Farady's law:

$$i = AcFN$$
(3.2)

where Ac and F are the cathode area and Faraday's constant 65,900 c/M, respectively. Substituting Eq. 3.1 into Eq. 3.2, we obtain the following:

$$i = Ac\, F\, \alpha\left(\theta_\infty - \theta_w\right)$$
(3.3)

When the migration in the electrical potential field is negligible as mentioned above, the mass transfer is analogous to the heat transfer at a wall with a constant temperature and may be correlated to the result of the heat transfer.

The limiting current can be obtained from the potential-current curve as shown in Fig. 3.1. As the applied negative potential of the cathode is made higher, the current increases exponentially due to the concentration decrease θ_w at the electrode surface. Under conditions of limiting current, the ions are transferred from the bulk of the solution to the surface of the electrode principally by diffusion due to the concentration gradient. Thereafter, it approaches a constant current, i.e., a limiting current, asymptotically, because the concentration at the surface of the electrode becomes zero from 100 mV to −400 mV (Fig. 3.1). Under conditions of limiting current, the increasing potential does not result in an increase in the rate of the reaction. As the ion concentration θ_w of the electrode surface becomes zero at a limiting current, Eq. 3.3 is simplified as follows:

$$i = Ac\, F\, \alpha\theta_\infty$$
(3.4)

Under the same conditions, the limiting current is exactly proportional to the bulk concentration. As distinguished in Fig. 3.1, a further increase of potential over the limiting current region causes a steep increase in current due to the discharge by a secondary reaction such as hydrogen evolution on the cathode. Finally, the cathode surface will be dissolved and damaged.

Principle

The basic electric circuit used in the electrochemical method is shown in Fig. 3.2. The electrolytic solutions employed to measure the mass transfer rate are the redox systems. There are two redox systems which are suitable to the electrochemical method: one is the mixture solution composed of potassium ferricyanide $K_3Fe(CN)_6$ and potassium ferrocyanide $K_4Fe(CN)_6$, and the other is the copper sulfate solution $CuSO_4$.

The former electrolytic solution is composed of potassium ferrocyanide $K_4Fe(CN)_6$, potassium ferricyanide $K_3Fe(CN)_6$, and potassium hydroxide KOH (or sodium hydroxide NaOH) as a supporting electrolyte. These chemical agents are ionized in water as follows:

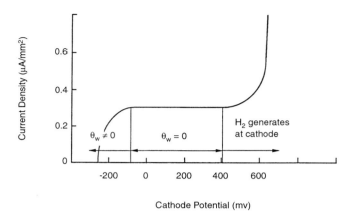

FIGURE 3.1   Limiting current.

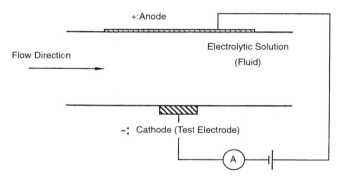

FIGURE 3.2   Electric circuit for electrochemical method.

$$K_3Fe(CN)_6 \rightarrow 3K^+ + Fe(CN)_6^{3-}$$

$$K_4Fe(CN)_6 \rightarrow 4K^+ + Fe(CN)_6^{4-}$$

$$KOH \qquad\quad \rightarrow K^+OH^-, \ or$$

$$NaOH \qquad\quad \rightarrow Na^+ + OH^-$$

In the polarization of this electrolytic solution, only two ferricyanide $Fe(CN)_6^{3-}$ and ferrocyanide ions $Fe(CN)_6^{4-}$ react at the cathode and anode as follows:

$$Fe(CN)_6^{3-} + e^- \rightarrow Fe(CN)_6^{4-} \qquad \text{at cathode} \qquad\qquad (3.5)$$

$$Fe(CN)_6^{4-} + e^- \rightarrow Fe(CN)_6^{3-} \qquad \text{at anode} \qquad\qquad (3.6)$$

In this reaction, the presence of dissolved oxygen influences the mass transfer measurement. It is better to remove oxygen from the test solution and seal the surface of this solution with an inert gas such as the nitrogen gas N_2.

The latter solution is composed of copper sulfate $CuSO_4$ and sulfuric acid H_2SO_4 as a supporting electrolyte. In polarization, the following reaction occurs at the cathode and anode:

$$Cu^{2+} + 2e^- \rightarrow Cu \quad \text{at the cathode}$$

$$Cu \rightarrow Cu^{2+} + 2e^- \quad \text{at the anode}$$

When this solution is employed, we should pay attention to the condition of the cathode surface, i.e., the cathode surface will change with time due to copper deposition during polarization.

When measuring the wall shear stress and the mass transfer rate, the former electrolytic solution is preferred as a typical working fluid, because the shape of the cathode surface has not changed with time during measurement, and there is no effect on the flow pattern around the cathode surface. It is recommended that metals such as platinum or nickel, which have as high a hydrogen overvoltage as possible and are inert in acidic solutions, are selected as the electrode material. Furthermore, an anode with a large surface area and a cathode with a much smaller surface area should be coupled in order to maintain the reductive reaction controlled at the cathode. The addition of an excess of a supporting electrolyte to the solution has an advantage for eliminating the migration in practical operation—it makes the electric resistance of the solution negligible.

In the electrolysis cell, the oxidative and the reductive reactions occur at the anode and cathode as expressed in Eqs. 3.5 and 3.6, respectively. Through this oxidative-reductive reaction, the ferricyanide ion $Fe(CN)_6^{3-}$ is transferred from the bulk of the solution to the cathode surface and converted to the ferrocyanide ion $Fe(CN)_6^{4-}$ at the cathode as expressed in Eq. 3.5, i.e., the ion transfer rate is equivalent to the current i flowing between the anode and cathode.

For the sake of simplicity, we consider a two-dimensional steady flow over the flat plate as shown in Fig. 3.3. The mass transfer flux N is also expressed in Fick's diffusion law as follows:

$$N = D\frac{\partial \theta}{\partial y}\bigg|_{wall} \tag{3.7}$$

where D is the diffusion coefficient. As the Schmidt number $Sc \approx 10^3$ is very high in the present solution, the thickness δ_c of the mass transfer boundary layer is much thinner than that δ of the viscous layer in the present electrolytic solution, i.e., $\delta_c = \delta(S_c)^{-1/3} \cong \delta/10$, the velocity profile within the mass transfer boundary layer is linearly proportional to the distance y from the wall. The diffusion in the transverse direction z is neglected, the same as the velocity v in the direction y. Then, the diffusion equation is governed by the following expression:

$$u\frac{\partial \theta}{\partial x} = D\frac{\partial^2 \theta}{\partial y^2} \tag{3.8}$$

where the flow velocity u near the wall is replaced by κy, κ being the velocity gradient

$$\partial u/\partial y\big|_{wall}$$

at the wall. The boundary conditions associated with Eq. 3.8 are as follows:

$$\theta = \theta_w, \quad y = 0 \left(\text{at wall}\right) \tag{3.9a}$$

$$\theta = \theta_\infty, \quad y = \infty \left(\text{in bulk solution}\right) \tag{3.9b}$$

Using the coordinate transformation variable, $\beta = y(\kappa/9Dx)^{1/3}$, the following ordinary differential equation is obtained under the boundary conditions:

$$\frac{d^2\theta}{d\beta^2} + 3\omega^2 \frac{d\theta}{d\beta} = 0 \tag{3.10}$$

At the boundary,

$$\theta = \theta_w, \quad \beta = 0 \tag{3.11a}$$

$$\theta = \theta_\infty, \quad \beta = \infty \tag{3.11b}$$

Eq. 3.10 can be easily solved for the concentration profile:

$$\frac{\theta - \theta_w}{\theta_\infty - \theta_w} = \frac{1}{\Gamma(4/3)} \int_0^\beta e^{-\beta^3} d\beta \tag{3.12}$$

where Γ denotes the gamma function, $\Gamma(4/3) = 0.8930$. The mean mass transfer flux N across the length ℓ of the cathode is obtained from Eqs. 3.7 and 3.12 as follows:

$$\begin{aligned} N &= \frac{1}{\ell} \int_0^\ell D \frac{\partial \theta}{\partial y}\bigg|_{wall} dx \\ &= \frac{\sqrt[3]{3}}{2\Gamma\left(\dfrac{4}{3}\right)} (\theta_\infty - \theta_w) \left(D^2 \frac{\kappa}{\ell} \right)^{1/3} \end{aligned} \tag{3.13}$$

Since the wall shear stress is expressed as follows,

$$\tau_w = \mu \left(\frac{du}{dy} \right)_{wall} = \mu\kappa \tag{3.14}$$

the wall shear stress is exactly associated with the current i using Eqs. 3.2, 3.13, and 3.14.

$$\tau_w = B\,i^3 \tag{3.15}$$

where

$$B = \frac{8}{3} \frac{\mu\ell}{D^2} \left\{ \frac{\Gamma(4/3)}{(\theta_\infty - \theta_w) Ac\,F} \right\}^3 = 1.9 \frac{\mu\ell}{D^2} \left\{ \frac{1}{(\theta_\infty - \theta_w) Ac\,F} \right\}^3 \tag{3.16}$$

The coefficient B depends on the shape of the cathode, the property of electrolytic solution, and the temperature. Therefore, the wall shear stress τ_w is proportional to the cube of current i under the limiting current.

Measurement Method

The general concept of the electric circuit and the flow circuit employed in the measurement is illustrated in Fig. 3.4. The electrolytic solution composed of 0.01 M potassium ferrocyanide $K_4Fe(CN)_6$, 0.01 M

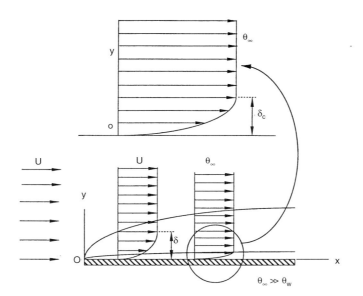

FIGURE 3.3 Velocity and concentration profile in boundary layer.

potassium ferricyanide $K_3Fe(CN)_6$, and 1.0 M potassium hydroxide KOH (or sodium hydroxide NaOH) as a supporting electrolyte is employed as the working fluid. In order to remove the oxygen dissolved in the working fluid, the nitrogen gas is continuously injected into the electrolytic solution in the reservoir. In order to measure the natural potential of the flow circuit, which is an absolute potential against the ground, the electrolytic solution within the flow circuit is connected to the reference electrode through the Luggin capillary. When the potential voltage is applied between the cathode and the anode by means of the potentiostat, the oxidative and reductive reactions occur at the anode and the cathode, respectively, and the current i corresponds to the ion transfer rate between the anode and the cathode.

The typical polarization curve, i.e., the current density vs. the cathode potential, is shown in Fig. 3.5. The current density is constant from 100 mV to –600 mV, i.e., the increasing potential does not result in an increase of current because the concentration θ_w at the cathode surface vanishes as expressed in Eq. 3.4. This constant current is called the limiting current. By increasing the flow rate under the same conditions of electrolysis, the value of the limiting current rises and the flat portion of the polarization curve disappears above a certain upper limit of the flow rate as shown in Fig. 3.6. In such situations, the limiting current is no longer indicated since the reaction is too slow to remove all ions reaching the electrode surface. In other words, the thickness δ_c of the mass transfer boundary layer will be comparable to that δ of the viscous layer as described earlier. This upper limit of the flow rate is called the critical flow rate. Therefore, the electrochemical method cannot apply beyond the critical flow rate.

One measurement example of the measurement of the limiting current vs. the absolute potential is shown in Fig. 3.7a. This limiting current is measured in the fully developed laminar steady flow through the parallel rectangular channel with a width of 4 mm and at the cathode area of 10×30 mm² embedded in the channel wall. As the mean velocity U increases, the flat portion of the limiting current becomes narrower. Another example is the measurement of limiting current vs. reference potential in the fully developed laminar steady flow through a straight circular tube with a diameter of 24 mm as shown in Fig. 3.7b. The cathode, which has a diameter of 0.5 mm, is embedded in the tube wall perpendicular to the tube axis. The natural potential of this flow circuit is about 280 mV, i.e., this value corresponds to zero potential of the reference potential. The flat portion of the limiting current is measured from 150 mV to –150 mV.

The relationship between the current i and the wall shear stress τ_w is shown in Fig. 3.8. The current is measured in the fully developed laminar steady flow through the straight circular tube with a diameter of 24 mm as described in Fig. 3.7b. When measuring, the reference potential is set up at –30 mV. The

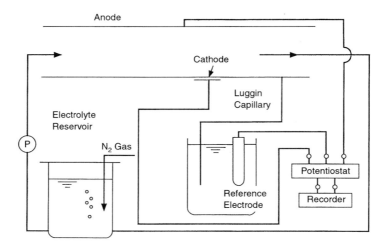

FIGURE 3.4   Electric and flow circuit.

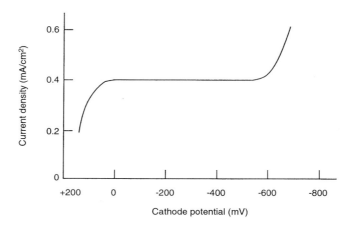

FIGURE 3.5   Typical limiting current.

wall shear stress τ_w is theoretically estimated from the velocity profile of the Poiseuille flow using the flow rate Q. In the laminar flow region, the wall shear stress τ_w is exactly proportional to the cube of the current i up to Re = 1200. The results mentioned in the application section below are calculated from this calibration line, and the high value beyond the region shown in Fig. 3.8 is extrapolated. The application limit of the electrochemical method is shown in Fig. 3.9. Open and closed symbols denote the wall shear stress measured using the electrochemical method and the wall shear stress estimated from the velocity profile obtained by the laser Doppler velocimeter, respectively. In Fig. 3.9, we have adopted the data measured using the electrochemical method as a true value. The value measured by the laser Doppler velocimeter is inaccurate because the velocity near the tube wall is sensitive and depends on the accuracy of the measurement position. The wall shear stress measured using the electrochemical method coincides with the wall shear stress estimated from the velocity profile up to a Reynolds number of 4000. Although the wall shear stress might be measured above Re = 4000 by using the electrochemical method, the result would be overestimated compared to the true value. Actually, above RE = 6000, the wall shear stress measured by the electrochemical method is twice as large as the wall shear stress estimated from the velocity profile.

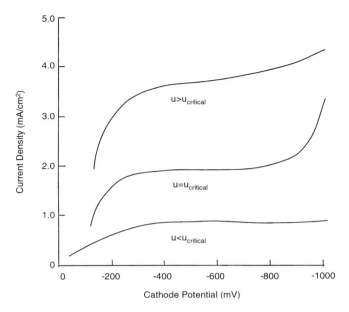

FIGURE 3.6 Critical flow rate.

Creation Process of a Test Electrode

The material of the electrode, the flow channel, the adhesive agent, and the creation process are described in this section.

Electrode

There are two metals suitable for creating the test electrode, i.e., nickel and platinum. The properties of nickel and platinum are tough and mild, respectively. Although platinum is expensive, it is easy to process and makes it possible to polish the surface of the electrode by hand. Therefore, platinum is more popular as the test electrode material because the polishing process is easier for platinum than for nickel.

Flow Channel

There are two types of flow channel, i.e., a circular channel drilled into the acrylic rectangular block and the acrylic circular tube as shown in Figs. 3.10 and 3.11, respectively.

Adhesive Agent

Epoxy adhesive and acrylic cement are the two possible adhesive agents. Although it is easy to embed the platinum wire electrode using the epoxy adhesive, the epoxy resin is apt to absorb the water within the electrolytic solution causes the surface shape around the wire electrode to deform. The deformation of the surface around the wire electrode disturbs the flow pattern on the electrode. The wall shear stress cannot be exactly measured since the wall shear stress is very sensitive to the flow disturbance. Therefore, acrylic cement is frequently employed as the adhesive agent.

How to Embed a Wire Electrode

One half-channel plate and the cross-sectional view of the wire electrode with a diameter of 0.5 mm embedded in the channel wall is also shown in Fig. 3.10. When experimenting, another half-channel plate is symmetrically assembled with it.

In order to embed the wire electrode, the slot is cut by a milling cutter with the radius of the circle 1 mm thick. After milling, there will definitely be some residual stress around the slot. In order to remove the residual stress, an annealing process needs to be performed, i.e., the channel plate is placed in a furnace and kept at a constant temperature of 70°C for at least three hours. After annealing, the wire

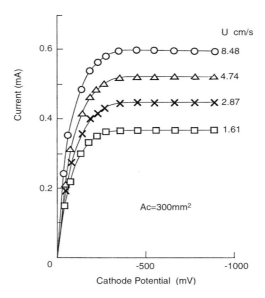

FIGURE 3.7a　Limiting current measured at an electrode with area of 10×30 mm^2 in a rectangular channel with a width of 4 mm.

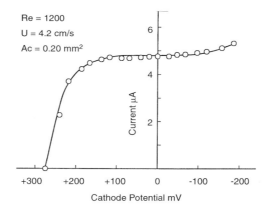

FIGURE 3.7b　Limiting current measured at an electrode with a diameter of 0.5 mm in a circular tube with a diameter of 24 mm.

electrode is glued into the slot using acrylic cement. After glueing, residual stress occurs again because the acrylic cement contracts during the hardening process. So, the acrylic model with the wire electrode embedded must be again placed in the constant temperature furnace for three hours to undergo the annealing process. If the residual stress caused by machining and gluing is not removed, many cracks will appear around the test electrode. The appearance of cracks results in the increase of the surface area of the test electrode. After several annealing processes, the wall surface with the embedded wire electrode is polished by hand using sand paper and emery paper until the wall surface approaches a smooth, mirror-like surface.

In order to embed the wire electrode in the circular tube as shown in Fig. 3.11, the hole is drilled perpendicular to the tube axis. In the same process as described above, the wall surface with the embedded wire electrode is polished by hand.

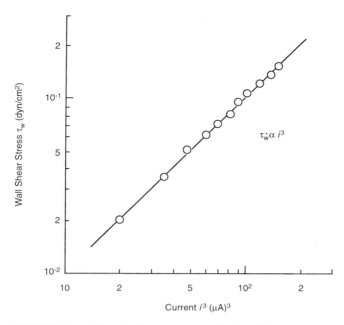

FIGURE 3.8 Calibration between current i and wall shear stress τ_w.

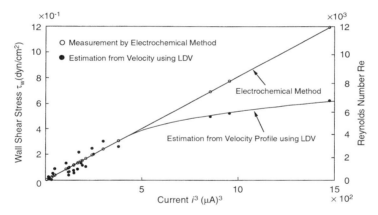

FIGURE 3.9 Comparison of wall shear stress measured with electrochemical method with that estimated from the velocity profile measured by laser Doppler velocimeter.

Effective Surface Area

This section describes a voltage-sweep polarographic method[4] for the effective measurement of the surface area of an electrode. The real surface area of a metal electrode is generally different from its projected or geometrical area because of surface curvature or an irregular geometry. This is particularly true for the round platinum wire electrode embedded at the curvature wall. This situation may arise where the surface of the electrode has an irregular geometry which makes mechanical or optical measurements difficult. Actually, it is difficult to measure the surface area of an electrode embedded along the stenosis inside the circular tube employed in our recent work. The polarographic method is a simple technique which overcomes such difficulty.

We considered a stationary test electrode whose circular cross-sectional surface was exposed in a redox solution at rest. If the initial potential was zero, ferricyanide was not reduced at the electrode. As the

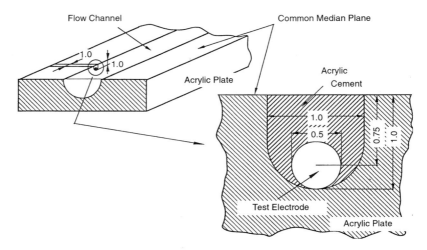

FIGURE 3.10   Side view of a test electrode embedded in an acrylic plate.

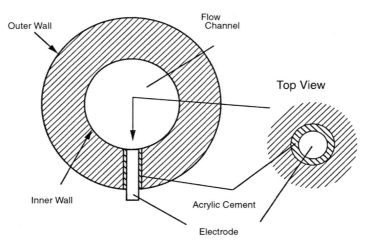

FIGURE 3.11   Electrode embedded in a circular tube.

negative potential was continuously increased, a charging current developed until it reached plateaus as shown in Fig. 3.12. Then, the reductive current in the system increased as the ferricyanide ion diffused from the bulk solution to the cathode and reacted at the cathode surface. At a higher negative potential, the system became concentration diffusion limited since reduction at the cathode surface was relatively fast compared to the transfer rate of the ferricyanide ion to the cathode. The diffusion rate of the ferricyanide ions toward the cathode did not depend on the cathode potential because the electric field was saturated with a high concentration of non-reducible ions. A region of depletion of ferricyanide ion extended further and further from the cathode surface and the current decreased. Thus, the current reached a maximum value called the peak current i_p. Mathematical analysis of the diffusion equations for this system has been presented by Meites[5] as follows:

$$i_p = \left(Kn^{3/2} D^{1/2} \theta_\infty \right) A_c v^{1/2} \tag{3.17}$$

where A_c represents the cross-sectional area of the electrode, D represents the diffusion coefficient of the electrolytic solution, θ_∞ represents the bulk concentration, v represents the voltage ramp rate, n represents the number of electron exchanges per molecule, and K represents the numerical constant.

The original solution was derived for one-dimensional diffusion to a recessed electrode surface through a stagnant fluid, which does not accurately describe our experimental system. Therefore, for our system, the equation constant was decided empirically, and the voltage-sweep polarographic method was performed by checking the dependence of peak current i_p on two parameters, electrode area A and voltage ramp rate v.

When experimenting with the voltage-sweep polarographic method, four electrodes with wire diameters of d = 0.5 mm embedded along the surface of the stenosis model were used as shown in Fig. 3.13. The axes of the four electrodes were perpendicular to the axis of the stenosis. Although the diameter of the four electrodes is same, the surface area exposed to the electrolytic solution is different, i.e., the shape of the standard electrode E_0 is circular, and the shapes of electrodes E_1, E_2, and E_3 are elliptical. Therefore, the surface area of electrodes E_1, E_2, and E_3 embedded at the stenosis are larger than that of the standard electrode E_0. The minor diameter of the ellipse embedded in the stenosis is approximately the same as the diameter of the wire electrode, and the major diameter d_s is longer than the diameter d = 0.5 mm.

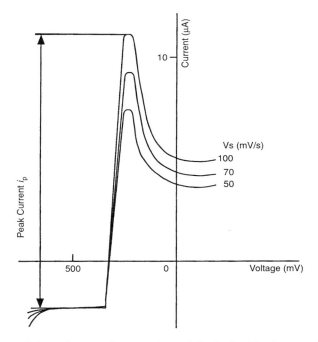

FIGURE 3.12 Typical chart of current from test electrode E_2 obtained by the potential sweep method.

Several voltage ramp rates between 50 and 100 mV/s are applied to each of four electrodes E_3, E_1, E_2, and E_3, and the peak current is measured. The data from a typical test for the electrode E_0 showing electrode current and applied potential at three different voltage ramp rates, are shown in Fig. 3.12. The peak current i_p, measured from the plateau to its maximum, is about 13.5 μA at a voltage ramp rate v = 100 mV/s. As the applied voltage proceeds from an initial value of +300 mV toward more negative voltages, the current rises and falls after reaching a peak value i_p. The peak current is plotted vs. the voltage ramp rate in Fig. 3.14 for four electrodes. The functional dependence of peak current on voltage ramp rate is clarified, i.e., the peak current is proportional to the square root of the voltage ramp rate.

Using this functional dependence, the unknown areas of electrodes E_1, E_2, and E_3 are geometrically shown in Fig. 3.15 at the voltage ramp rate of 70 mV/s. Since the area of electrode E_0 is known, the

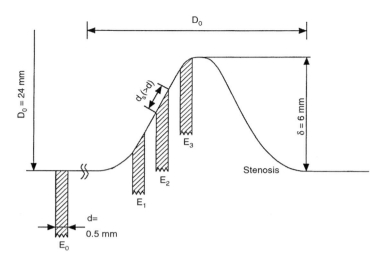

FIGURE 3.13   Electrode embedded around stenosis.

unknown areas of electrode E_1, E_2, and E_3 are easily estimated from Fig. 3.15. The real areas of electrode E_1, E_2, and E_3 are 1.09, 1.32, and 1.06 times as large, respectively, as the area of the standard electrode E_0.

Frequency Characteristics and Limitations

The flow in the large human artery is pulsating with a wave form which is expressed in at least four harmonic components. In order to employ the electrochemical wall shear probes for the measurement of instantaneous wall shear stress in pulsating flow models, the frequency response of the probes must be exactly known. Several theoretical investigations have been made to determine the response of wall shear probes subjected to nonreversing unsteady flows, and an attempt to extend the analysis to reversing flows has also been made.[6,7] Several experimental studies of the frequency response of local electrochemical wall shear probes have been performed.[8–10]

The objective of this section was to experimentally assess the frequency response of electrochemical mass transfer wall shear probes. Attention was restricted to laminar flow, because under normal conditions, the flow in the human circulatory system is laminar. Only nonreversing flow was studied, because no satisfactory theory exists for the reversing flow despite the effort which has been made to consider the reversing flow problem. In order to establish the frequency characteristics of wall shear probes, we considered the case of a fully developed laminar pulsating flow through a straight circular tube with a sinusoidal component superimposed on a steady mean flow, since the instantaneous wall shear stress for this flow can be exactly evaluated.[11,12]

The experiments were carried out using the same electrolytic solution and the same flow circuit as in Fig. 3.4 except for stenosis, i.e., the inner diameter of the straight circular tube is $D_0 = 24.5$ mm, and the diameter of the platinum test electrode is d = 0.5 mm. The wall shear stress was measured in the fully developed laminar pulsating flow with an amplitude A of flow rate. The pulsating flow rate is expressed as the following equation:

$$Q/Q_0 = 1 + A \sin \omega t \qquad (3.18)$$

where Q, Q_0, and ω are the instantaneous, the mean flow rates, and the angular velocity of pulsating frequency, respectively. The measurement was conducted using a mean Reynolds number of Re = 800, a flow amplitude of A = 0.3 and several non-dimensional frequencies of $\alpha = R_0 \, (\omega/\nu)^{1/2} \leq 10$.

The comparison of the wave form of measured wall shear stress in the straight circular tube during one period to the exact solution is shown in Figs. 3.16a, b, and c at $\alpha = 3.3$, 6.6, and 10.4, respectively.

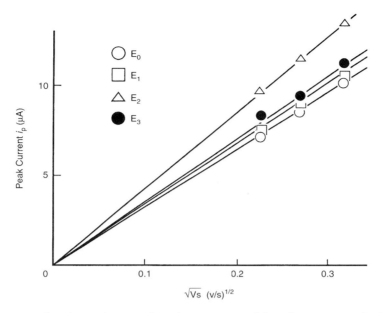

FIGURE 3.14 Data replotted as peak current i_p vs. the square root of the voltage ramp rate for four electrodes.

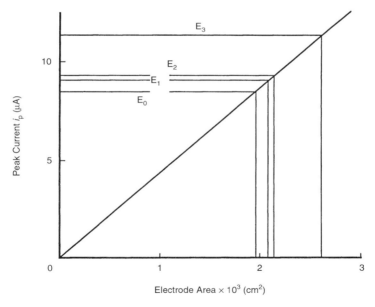

FIGURE 3.15 Data replotted as peak current i_p vs. electrode area for four electrodes.

The wave form of the flow rate is also indicated in these figures. The measured wall shear stress is estimated from the result calibrated in a steady Poiseuille flow as shown in Fig. 3.9, i.e., $\tau_w = Bi^3$. In these figures, the wall shear stress is normalized by the Poiseuille flow, i.e., $\tau_w = 2.0$ in a mean steady flow.

At $\alpha = 3.3$, the measured wall shear stress closely corresponds to that of the exact solution. At $\alpha = 6.6$, the amplitude of measured wall shear stress is still similar to that of the exact solution. However, the phase lead appears at $\alpha = 6.6$. At a higher frequency of $\alpha = 10.4$, the amplitude of the measured wall shear stress becomes smaller than that of the exact solution, and the phase lead becomes much larger. The

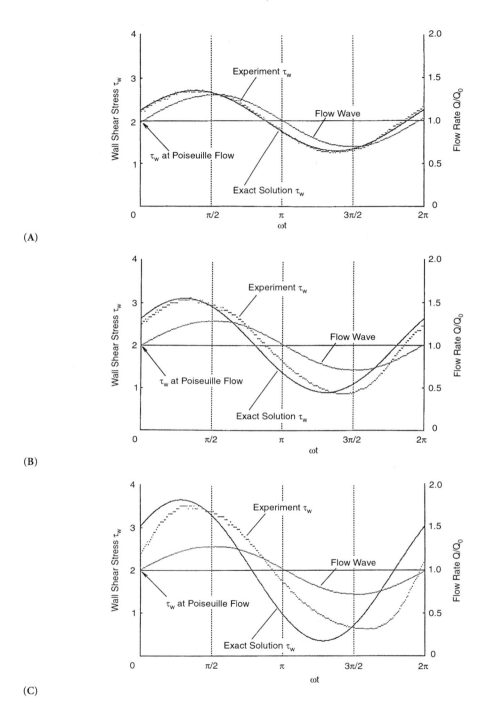

FIGURE 3.16 Comparison of experimental and exact solution of wall shear stress in a pulsating flow: (a) $\alpha = 3.3$, $A = 0.30$; (b) $\alpha = 6.6$, $A = 0.30$; (c) $\alpha = 10.4$, $A = 0.30$.

frequency characteristics are shown in Fig. 3.17. The amplitude ratio σ is defined as the ratio of the amplitude of the measured wall shear stress to the amplitude corresponding to the quasi-steady flow. The phase lead δ is also defined as the phase difference between the time $\omega t = \pi/2$ at a maximum flow rate and the time at a maximum wall shear stress. The phase lead δ reaches a maximum at $\alpha = 7.0$, and

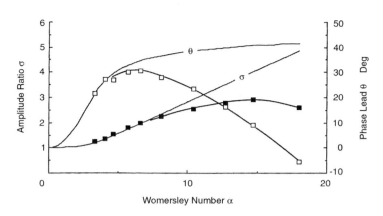

FIGURE 3.17 Frequency characteristics of wall shear stress in a pulsating flow.

the amplitude ratio σ is the same as the exact solution until $\delta = 7.0$. There are several reports concerning the frequency characteristics of the electrochemical wall shear probe.[7–9] According to the definition by Mao and Hanratty,[9] the criterion parameter $\omega^+ N^{1/3} L^{+2/3}$ is about 0.38 in the present study of Re = 800, $\alpha = 6.6$, and A = 0.30. And the criterion parameter λ defined by Pedley[7] is about 0.30. This parameter lies on the boundary between the quasi-steady flow and the fully developed pulsating flow. Therefore, this method is considered valid in the range $\alpha < 7$ for the pulsating flow of Re = 800 and A = 0.30.

3.3 Application

Physiological View

Flow behavior through the artery is closely associated with vascular disease such as atherosclerosis. From the viewpoint of hemodynamics, it is well known that vascular disease occurs in highly disturbed regions of stenosis, branching arteries, and large curvature wall.[13–15] Concerning the generation and development of atherosclerotic disease, there are two major hypotheses,[16,17] i.e., low shear stress and high shear stress. Furthermore, it has been reported that the shape of an endothelial cell is distorted in the direction of its action by the wall shear stress.[18,19] Clearly, fluid mechanical factors such as wall shear stress and vortex may determine the sites where atherosclerotic disease will generate and develop.

The arterial branch is well known as one of the predilection regions where atherosclerotic disease frequently generates and develops. There are many experimental reports describing the flow situation through the branch involving flow pattern, velocity profile, wall shear stress, and vortex. Generally, it is assumed that in the two-dimensional branch flow, the wall shear stress is high around the apex and low along the proximal wall of the daughter tube. Lutz et al.,[20] and El Masry and Feuerstein[21] measured the wall shear stress in the three-dimensional mesenteric branch with a daughter tube bifurcated from a parent tube at 45°. They suggested that the wall shear stress along the proximal wall of the daughter tube is not necessarily low and varies with magnitude, and the variation of wall shear stress might be associated with the site of vascular disease. Karino et al.[22,23] visualized the flow in the T-branch and observed that, at the proximal wall of the daughter tube, there are small reversing flows, not stagnating flows, and there are a pair of helical flows within the daughter tube. In the internal carotid artery, Ku et al.[24] measured the wall shear stress at several locations and clarified that the wall shear stress is low and the intimal thickness becomes thicker at the outer wall of the internal carotid artery, where the flow separates. Concerning the wall shear stress in casts of a human aortic branch with flexibility, Duncan et al.[25] measured the wall shear stress at 19 locations. They clarified that, due to pulsation, the cast compliance reduced shear rates at the outer wall, while the shear rate was increased at the inner wall along the flow divider.

The flow field in several branch models was numerically estimated using the two-dimensional numerical analysis.[26,27] These results show that the wall shear stress near the apex becomes extremely high but is low downstream of the upstream corner of the daughter tube. Recently, the flow pattern through the three-dimensional symmetrical branch was numerically calculated using a finite control volume method. Perktold et al. indicated that, in the flow field through the T-branch in a pulsating flow, the regions of high and low wall shear stress are close to each other.[28]

Concerning atherosclerotic disease at the branch of the artery, experimenting on animals and pathological observation has been very revealing. The observed results of pathological research show that atherosclerosis exists around the apex of the branch, i.e., the region of high shear stress,[29,30] in small animals. In contrast, atherosclerotic disease occurs at the outer wall of the daughter tube, i.e., the region of low shear stress,[31,32] in large animals. The site of atherosclerotic disease varies in each experiment, which may be related to the different feeding method of cholesterol lipids, the flow pattern, and the flow distribution ratio in each tube. Yamaguchi et al.[33] indicated in their recent investigation that atherosclerosis in the inferior mesenteric artery proliferates at the proximal wall where the wall shear stress is lower than that around the apex. This inferior mesenteric artery site resembles the mesenteric branch model studied by Lutz et al., and El Masry and Feuerstein.[20,21] However, since the measuring region covered in the experiment was not wide enough, the behavior of wall shear stress in the daughter tube is not adequately clarified.

It is also interesting to examine the flow situation through the stenosis. In the flow field through the stenosis, high wall shear stress occurs in the vicinity of low wall shear stress, i.e., high wall shear stress appears just upstream of the stenosis, and low wall shear stress appears downstream of the stenosis where the flow separates.[34]

Stenosis Model

In this section, the behavior of wall shear stress in the developing process of the occlusion in an arterial wall, such as stenosis, has been experimentally described in a laminar pulsating flow.[35] The wall shear stress has been locally measured using the electrochemical method, which is based on the reduction of the ferricyanide ion on an electrode embedded in the tube wall. Furthermore, this flow has been numerically analyzed using a finite element method.

Experiment and Numerical Approach

Apparatus

The configuration of axisymmetric stenosis expressed by a cosine curve is shown in Fig. 3.18. In the experiment, three stenosis ratios were employed, i.e., $\delta/R = 0.1, 0.2$, and 0.3, where δ and $R = 12$ mm are the height of the stenosis and the radius of the straight tube, respectively. The length of the stenosis is the same as the diameter of the straight tube. In order to measure the wall shear stress, 34 platinum test electrodes with diameters of 0.5 mm were embedded in the bottom wall upstream and downstream of the stenosis.

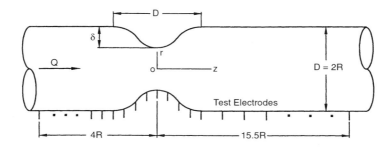

FIGURE 3.18   Sketch of flow field and arrangement of electrodes.

As shown in Fig. 3.19, the working fluid flows from the constant head tank into the model stenosis through the inlet straight circular tube with a length of 2 m. Through an electromagnetic flowmeter, the working fluid flows into the flow control orifice and returns to the reservoir₁, within which the working fluid was maintained at a constant temperature. The control orifice had a slit aperture port with a width of 10 mm and a height of 1 mm. The aperture area sinusoidally changes around a mean area in the direction of width by an eccentric cam. Therefore, the flow rate is expressed as a summation of the steady and pulsating flow rates. The experiment was carried out in the sinusoidal laminar pulsating flow of a non-dimensional pulsating frequency of $\alpha = R(\omega/v)^{(1/2)} = 7.0$ with the mean flow rate of $Re = 2RU/v = 560$ and the non-dimensional amplitude of flow rate of $A = 0.3$, where Re, U, v, and ω are the Reynolds number, the mean velocity, the kinetic viscosity, and the pulsating angular velocity, respectively. The wall shear stress was also measured in a steady flow corresponding to the maximum, mean, and minimum flow rates in a pulsating flow.

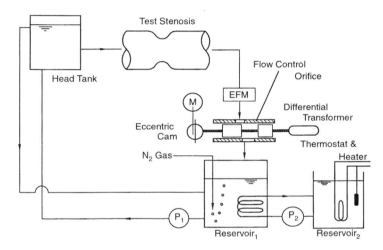

FIGURE 3.19 Experimental apparatus and flow circuit.

Measurement
In measuring the wall shear stress, the electrolytic solution composed of 0.01 M potassium ferricyanide, 0.01 M potassium ferrocyanide, and 1.0 M potassium hydroxide as a supporting electrolyte was employed as the working fluid. The property of the working fluid was approximately the same as that of distilled water, i.e., $\rho = 1.044$ g/cm³, and $v = 0.839 \times 10^{-6}$ m²/s at a temperature of 303 K. The wall shear stress was measured using the electrochemical method as previously described.

Numerical Analysis
The flow pattern and the wall shear stress through the axisymmetric stenosis were also calculated using a finite element method.[36] The fluid was assumed to be Newtonian. The governing equations were the axisymmetric momentum equation and the continuity equation. The number of rectangular finite elements was 637, and the number of grids was 2745. The boundary condition for velocity in the upstream inlet section was provided by the fully developed laminar pulsating flow,[11,12] and the fluid was no-slip on the tube wall.

Results

Steady Flow
The distribution of the wall shear stress around the stenosis in a steady flow of $\delta/R = 0.3$ is shown in Fig. 3.20. The Reynolds numbers indicated in this figure correspond to the maximum, mean, and minimum flow rates in a pulsating flow as described below. The ordinate denotes the non-dimensional wall shear stress normalized by the wall shear stress of the Poiseuille flow at Re = 560 in the upstream

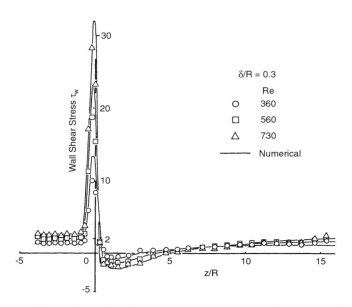

FIGURE 3.20   Distribution of wall shear stress around stenosis in a steady flow.

straight tube, i.e., the wall shear stress τ_w in the upstream straight tube at Re = 560 is 2.0. The abscissa is normalized by the tube radius R. The solid line denotes the numerical result.

The wall shear stress reached its maximum just upstream of the narrowest section of the stenosis and increased with the Reynolds number. Except for $\delta/R = 0.1$, the flow separated, and the wall shear stress was negative downstream of the stenosis and its absolute value was small. Therefore, although the wall shear stress was small downstream of the stenosis, the gradient of wall shear stress with respect to the tube axis became markedly large at the narrowest section.

Pulsating Flow

The instantaneous wall shear stress was calculated from the relationhip $\tau_w = Bi^3$ based on the steady flow as shown in Fig. 3.9. The instantaneous wall shear stress at the maximum and minimum flow rates in a pulsating flow of Re = 550, $\alpha = 6.6$, and A = 0.34 is shown in Figs. 3.21a, b, and c. This pulsating flow was laminar and turbulence never occurred. The wall shear stress τ_w in this figure is normalized by the wall shear stress of the Poiseuille flow corresponding to the mean flow rate, i.e., the wall shear stress in the steady flow corresponding to the mean flow rate is $\tau_w = 2.0$. The gradient of wall shear stress with respect to the tube axis becomes markedly large downstream of the narrowest section, i.e., around the separation point. The magnitude of this gradient increases at a period of increasing flow rate. Although not indicated at $\alpha = 3.3$, the gradient of wall shear stress near the separation point becomes much larger with the pulsating frequency.

The oscillation shear index (OSI)[24] is shown in Fig. 3.22 for the above results. Although not indicated, the OSI in $\delta/R = 0.1$ is zero at any location because there is no separation flow behind the stenosis as shown in Fig. 3.21a. In $\delta/R = 0.2$ and 0.3, the OSI is not zero around the separation point just downstream of the narrowest section of the stenosis or the reattachment point behind the stenosis. Therefore, the flow direction periodically changes at both the separation point and the reattachment point.

Remarks

The gradient of wall shear stress with respect to the tube axis becomes very large just downstream of the narrowest section of the stenosis although the wall shear stress is low near this point. Furthermore, the OSI has a finite value near the separation point. Therefore, around the separation point just downstream of the narrowest section, the flow direction periodically changes with the large gradient of wall shear stress.

45° Bifurcation Model

In this section, the wall shear stress and the flow pattern through the 45° branch model in laminar steady and pulsating flows have been experimentally clarified by the electrochemical method and by flow visualization. The daughter tube bifurcates from the parent tube at 45°. This model simulates the artery site from the abdominal aorta to the mesenteric artery.[10,20,21,33] The experiments were performed for two flow distribution ratios downstream of each tube.

Experiment

Branch model

The schematic view of the experimental apparatus and the flow circuit is shown in Fig. 3.23. Through the inlet tube with the length of 2.2 m, the working fluid flowed from the constant head tank into the test branch. It then returned to the reservoir₁, within which the working fluid was maintained at a constant temperature (± 0.2 K). The flow rate downstream of each tube was measured by an electromagnetic flowmeter. In our experiment, the sinusoidal pulsating flow was produced by the change of peripheral resistances, i.e., the constriction of the flow control orifice. In the case of an even flow distribution ratio, the flow control orifice as shown in Fig. 3.23 has two slit aperture ports with a width of 10 mm and a height of 1 mm. These aperture ports, through which the working fluid issues in the atmosphere, sinusoidally change around a mean area in the direction of width by an eccentric cam. So, the flow rate in both the downstream parent tube and the daughter tube is a sinusoidal pulsating flow superposed on the steady flow. Therefore, the flow rate through the upstream parent tube is expressed as a summation of the flow rate in both downstream tubes.

The configuration of the branch model is shown in Fig. 3.24. The daughter tube (diameter $D_1 = 14$ mm) bifurcates at 45° from the parent tube (diameter $D_0 = 24.0$ mm) with a radius of curvature of $R_D = 3.5 D_0$ at the proximal wall. The cross-sectional area ratio $D_1^2/(D_0^2+D_1^2)$ is 0.25. The branch model is comprised of two separate acrylic plates that are 30 mm thick. These two plates, fixed by tightening eight screws, are milled by 24 mm and 14 mm drills in the parent tube and in the daughter tube, respectively. After milling, two separate half-branch plates are also cut along the proximal wall with a radius of curvature $R_D = 3.5 D_0$ using the ball end mill with a diameter of 12.5 mm. The junction with a discontinuous ridged surface between parent and the daughter tubes is smoothly polished by hand. The one half-bifurcation plate employed in the measurement of wall shear stress and the cast made of plaster are shown in Figs. 3.25a and b, respectively. As shown in Fig. 3.25, the junction has a smooth surface connection between the parent and daughter tubes.

The coordinate system and the measuring position of wall shear stress are also shown in Fig. 3.24. The origin at which the y-axis of the daughter tube intersects with the x-axis of the parent tube is defined as the branch point. The plane which includes both axes x and y is defined as the common median plane and is installed horizontally. The electrode A_i is along the outer wall of the parent tube, B_i is along the proximal wall, E_i is along the distal wall of the daughter tube, and F_i is along the inner wall of the parent tube. Approximately 130 platinum test electrodes (diameter 0.5 mm) are embedded in the common median plane. A cross-sectional view of the electrodes A_i, B_i, E_i, and F_i embedded in the common median plane is shown in Fig. 3.10. The center of these electrodes deviates from the common median plane by 3 and 6 degrees in the parent tube and daughter tube, respectively.

Wall Shear Stress and Flow Visualization

When measuring the wall shear stress, the electrolytic solution composed of 0.01 M potassium ferricyanide $K_3Fe(CN)_6$, 0.01 M potassium ferrocyanide $K_4Fe(CN)_6$, and 1.0 M potassium hydroxide KCH as a supporting electrolyte were employed as the working fluid. The property of the working fluid was approximately the same as that of distilled water, i.e., $\rho = 1.044$ g/cm³, $\nu = 0.839 \times 10^{-6}$ m²/s at a temperature of 303 K. The wall shear stress is measured using the electrochemical method. The flow pattern is visualized by a dye injected from the parent tube upstream of the branch point.

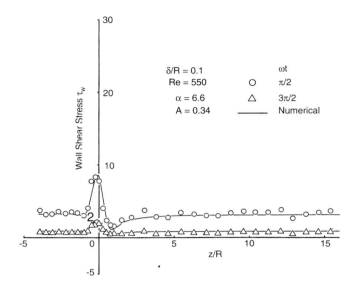

(A)

FIGURE 3.21 Instantaneous wall shear stress at maximum flow rate ($\omega t = \pi/2$) and minimum flow rate ($\omega t = 3\pi/2$) around stenosis in a pulsating flow: $\delta/R = 0.1$; $\delta/R = 0.2$; $\delta/R = 0.3$.

Results

In this experiment, the author aimed to simulate the flow situation through the 45° branch model such as the flow from the artery site to the mesenteric artery from the abdominal aorta. The effects of both the flow distribution ratio into each downstream tube and the pulsation on the wall shear stress have been studied. The measurements were carried out in the following range: the Reynolds number is approximately $Re = UD_0/\nu = 800$, where U is the mean velocity in the upstream parent tube. The flow distribution ratios are the geometrical flow ratio ($Q_D/Q_T = 0.30$) approximately equal to the cross-sectional area ratio $D_1^2/(D_0^2 + D_1^2)$ such as rest and the even flow ratio ($Q_D/Q_T = 0.51$) as another typical flow condition such as exercise. The non-dimensional pulsating frequency is $\alpha = 6.6$ and the non-dimensional flow amplitude in the sinusoidal pulsating flow is A = 0.30.

Generating System of Pulsating Flow

It is also necessary to examine the generating system of the pulsating flow. In order to examine the flow wave at the entrance of the junction, the wave forms of wall shear stress in A_1 ($x/R_0 = -9.6$), at the outer wall of the parent tube, and B_1 ($x/R_0 = -9.6$), at the proximal wall, are shown in Fig. 3.26. The pulsating flow is produced by the displacement of the flow control orifice. This displacement corresponds to the change of aperture area in the flow control orifice. Therefore, the flow wave is expressed by the displacement of the flow control orifice. It is assumed that these wave forms of wall shear stress at A_1 and B_1 sinusoidally change in the sine wave. The magnitude of the wall shear stress at B_1 is a little larger than that at A_1, because the core flow in this section is beginning to deflect to the daughter tube. Therefore, it is considered that the flow rates in both the parent and the daughter tubes sinusoidally change in the sine wave, and the flow distribution ratio Q_D/Q_T does not change with time in a pulsating flow. This is an important characteristic in the current generating system of the pulsating flow.

Flow Visualization

The flow pattern visualized by a dye stream in a steady flow is shown in Fig. 3.27 for $Q_D/Q_T = 0.51$. The shutter speed in the following photos is 1/60 second. Dye injected from the outer wall of the parent tube separates near the branch point as shown in Fig. 3.27a. As it is displayed in this figure, one of the dye streams flows downstream and the other turns along both the top and bottom walls of the parent tube.

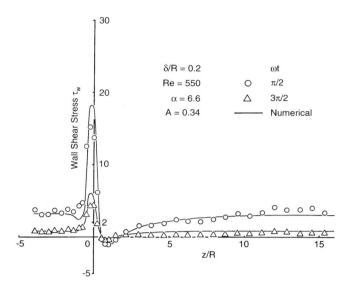

(B)

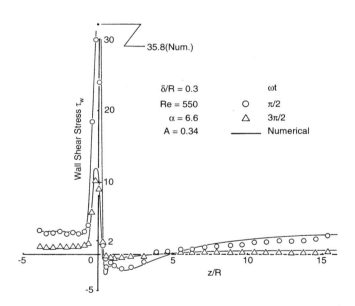

(C)

FIGURE 3.21 (CONTINUED)

Then, the dye flows into the daughter tube as a pair of helical flows after impinging the proximal wall near B19 (y/R1 = 3.5). The dye injected from the core flow above the common median plane flows into the daughter tube with skewing, and then it approaches the proximal wall near B26 (y/R1 = 8.3) as shown in Fig. 3.27b. These two points B19 and B26 correspond to the position of two relative maxima of wall shear stress at the proximal wall as discussed in the next section.

In the case of $Q_D/Q_T = 0.30$, the separation flow at the outer wall of the parent tube appears more clearly as shown in Fig. 3.28a. Although a pair of helical flows go downstream with skewing in the daughter tube, there is no separation flow at the proximal wall as shown in Fig. 3.28b. It is assumed that the separation flow does not occur in this model at the daughter tube with a large radius of curvature,

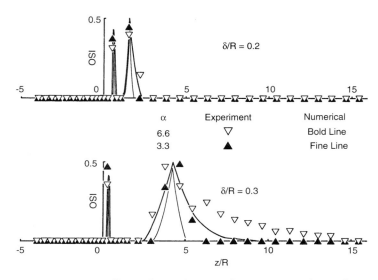

FIGURE 3.22   Oscillation shear index around stenosis in a pulsating flow.

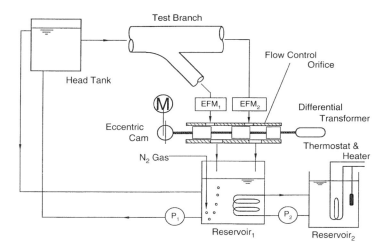

FIGURE 3.23   Experimental apparatus and flow circuit.

because the flow region along the proximal wall is compensated by a pair of helical flows transferred into the daughter tube from the separation region of the parent tube.

Wall Shear Stress in Steady Flow

The distributions of wall shear stress for even and geometrical flow distribution ratios in a steady flow are shown in Figs. 3.29 and 3.30, respectively. Left and right side figures indicate the distribution at the parent tube and daughter tube, respectively. The abscissas are taken so that x is in the direction of parent tube, y is in the direction of daughter tube, and they are normalized by the tube radii R_0 and R_1, respectively. The wall shear stress τ_w is normalized by the Poiseuille flow upstream of the parent tube, i.e., $\tau_w (= \tau_T) = 2.0$ denotes the wall shear stress of the Poiseuille flow upstream of the parent tube. The asymptotic values τ_T, τ_P, and τ_D denote the wall shear stress of the Poiseuille flow at each tube. The vertical lines through the symbol in Fig. 3.29c indicate the fluctuation about the time-averaged wall shear stress.

For an even flow distribution ratio $Q_D/Q_T = 0.51$ at Re = 790 as shown in Fig. 3.29b, the wall shear stress along the outer wall (A_i, $\triangle$) of the parent tube deviates from $\tau_T = 2.0$ and decreases downstream

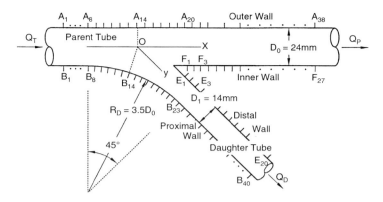

FIGURE 3.24 Flow field and electrode arrangement in 45° bifurcation model.

(a)

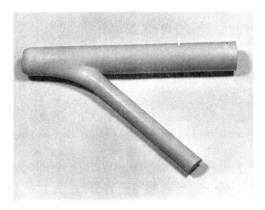

(b)

FIGURE 3.25 (a) Half-bifurcation plate in measurement; (b) Cast of 45° bifurcation model.

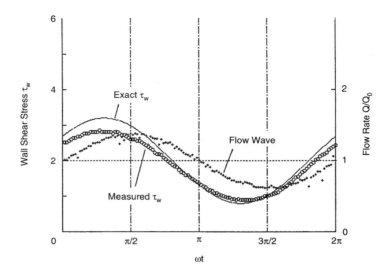

(a)

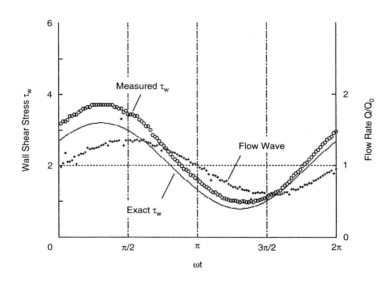

(b)

FIGURE 3.26 (a) Comparison of measured wall shear stress at A_1 ($x/R_0 = -9.60$) and exact wall shear stress in a pulsating flow (Re = 800, $\alpha = 6.6$, A = 0.31); (b) Comparison of measured wall shear stress at B_1 ($x/R_0 = -9.60$) and exact wall shear stress in a pulsating flow (Re = 800, $\alpha = 6.6$, A = 0.31).

of the parent tube. It adopts a negative value for the range of $0 < x/R_0 < 10$ downstream of the branch point, since the separation occurs in this region as evidenced by the flow visualization shown in Fig. 3.27a. Along the inner wall (F_i, $\Diamond$) of the parent tube, the wall shear stress becomes extremely high, and it sharply decreases as going downstream. Along the proximal wall (B_i, $\bigcirc$) within the parent tube, the wall shear stress around B_8 just upstream of the daughter tube is a little larger than $\tau_T = 2.0$. It is presumed that this elevation results from the deflection of the core flow toward the daughter tube. Flowing into the daughter tube, the wall shear stress steeply increases and attains its first relative maximum of about $6 \times \tau_T$ at B_{19} ($y/R_1 = 3.5$). As shown in Fig. 3.27a, position B_{19} corresponds to the first point of the proximal wall where dye injected from the outer wall of the parent tube impinges. Downstream, the wall shear

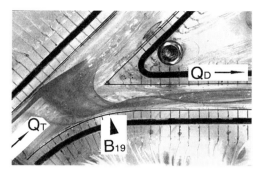

(a)

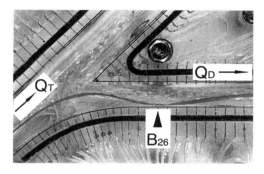

(b)

FIGURE 3.27 Flow visualization at an even flow ratio [Re = 800, Q_D/Q_T = 0.51, O.W. = outer wall, P.W. = proximal wall, $y/R_1(B_{19})$ = 3.5, $y/R_1(B_{26})$ = 8.3]. (a) Dye streamline from outer wall of parent tube; (b) dye streamline from core flow above common median plane.

stress suddenly decreases and then again increases, and reaches its second relative maximum at B_{26} (y/R_1 = 8.3). As shown in Fig. 3.27b, position B_{26} is the second point of the proximal wall where dye injected from the core flow approaches. Afterwards, the wall shear stress along the proximal wall varies significantly with position in the form of a damped sine wave and approaches the asymptotic value τ_D. Along the distal wall (E_i, $\square$), the wall shear stress becomes extremely high around the apex and steeply approaches τ_D. The relative maximum at E_3 corresponds to the relative minimum at B_{23}, and E_6 corresponds to the second relative maximum at B_{26}. Thus, the locations of the extreme values at the distal wall correspond to those at the proximal wall. Therefore, the sinusoidal variation of wall shear stress along the proximal wall results from the secondary helical motion of fluid transferred into the daughter tube from the separation region of the parent tube with skewing. The variation of wall shear stress becomes more remarkable with the increase of the Reynolds number. Due to turbulence, the disturbance occurs downstream of both tubes in the case of Re = 1040 as shown in Fig. 3.29c. The critical Reynolds number for the present model at an even flow distribution ratio is less than 1000.

For a geometrical flow distribution ratio Q_D/Q_T = 0.30 as shown in Fig. 3.30, the general tendency of wall shear stress is similar to that for an even flow distribution ratio although the magnitude is smaller. The wall shear stress along the proximal wall in the daughter tube reaches a single relative maximum around B_{20} (y/R_1 = 4.2). Position B_{20} corresponds to the point on the proximal wall where dye injected from the outer wall of the parent tube impinges, as shown in Fig. 3.28a. The secondary helical velocity becomes smaller with the decrease of the flow rate into the daughter tube. At the same Reynolds number of Re ≈ 800, the wall shear stress at F_1 around the apex in Q_D/Q_T = 0.30 is smaller than that in Q_D/Q_T = 0.51 despite the increase of the flow rate into the parent tube. The fluid velocity approaching the apex in Q_D/Q_T = 0.51 has approximately the maximum velocity of the Poiseuille flow upstream of the parent

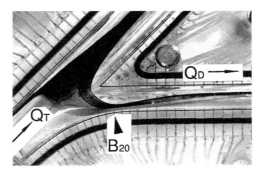

(a)

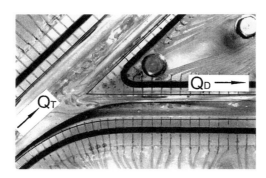

(b)

FIGURE 3.28 Flow visualization at a geometrical flow ratio [Re = 810, Q_D/Q_T = 0.30, O.W. = outer wall, P.W. = proximal wall, $y/R_1(B_{20})$ = 4.2]. (a) Dye streamline from the outer wall of parent tube; (b) dye streamline from the proximal wall of parent tube.

tube, while in Q_D/Q_T = 0.30 it is smaller than its maximum velocity. Therefore, the wall shear stress at F1 around the apex in Q_D/Q_T = 0.30 is smaller than that in Q_D/Q_T = 0.51.

As decided in the flow visualization, there is no separation flow along the proximal wall, so that the relative maxima at B_{19}, B_{26} in Fig. 3.29b, and B_{20} in Fig. 3.30 would be positive.

Wall Shear Stress in Pulsating Flow

The distributions of instantaneous wall shear stress at a maximum flow rate ($\omega t^* = \pi/2$) and a minimum flow rate ($\omega t^* = 3\pi/2$) for an even flow distribution ratio in the sinusoidal pulsating flow of Re = 800, α = 6.6, and A = 0.31 are shown in Figs. 3.31a and b, respectively. The asymptotic values τ_P, τ_P, and τ_D denote the instantaneous wall shear stress at each time in the pulsating flow through each straight tube. The general tendency of wall shear stress is similar to that of a steady flow. The distribution is larger at a maximum flow rate and smaller at a minimum flow rate. Along the proximal wall (B_i, ○) in the daughter tube, the wall shear stress varies significantly with position in the form of a damped sine wave, comparable to what occurs around the apex. Along the distal wall (E_i, □), the wall shear stress also varies with the reverse phase against the proximal wall.

The distribution of the wall shear stress amplitude $\Delta\tau_w$ at an even flow distribution ratio in the pulsating flow is shown in Fig. 3.32a. The asymptotic values $\Delta\tau_P$, $\Delta\tau_P$, and $\Delta\tau_D$ denote the amplitude of wall shear stress at Re = 800, α = 6.6, Q_D/Q_T = 0.51, and A = 0.31 through each straight tube. The general tendency of this amplitude distribution is similar to the distribution of wall shear stress in a steady flow. The amplitude at each downstream tube has a slight scatter which is caused by turbulence occurring downstream of both tubes. The similarity between the distribution of wall shear stress in a steady flow and the amplitude distribution of wall shear stress in a pulsating flow suggests that there is a severe change of wall shear stress along the proximal wall, and its time variation becomes more obvious at the site where the magnitude of the wall shear stress is large. Also, the flow condition of Re = 800, $\alpha = 6.6$, and A = 0.31 might be regarded as the quasi-steady flow. The amplitude of the wall shear stress at a geometrical flow distribution ratio is shown in Fig. 3.32b. Although the tendency is similar to that of an even flow distribution ratio, the variation of the amplitude is smaller.

General Feature of Wall Shear Stress and Physiological Implications

The flow situation through the 45° asymmetric branch model employed in this study is characterized around the branch point, particularly in the daughter tube at even flow distribution ratio. When approaching the daughter tube, the wall shear stress is a little larger than that of the Poiseuille flow upstream of the parent tube. Flowing into the daughter tube, the wall shear stress steeply increases and, at B_{19} ($y/R_1 = 3.5$), reaches the first relative maximum which is several times larger than that of the Poiseuille flow upstream of the parent tube. Downstream, the wall shear stress sharply decreases and then again increases, and reaches its second relative maximum at B_{26} ($y/R_1 = 8.3$). Afterwards, the wall shear stress asymptotically approaches the wall shear stress τ_D. It is worth mentioning that the magnitude of these relative maxima along the proximal wall is comparable to the magnitude around the apex, and the wall shear stress reaches an extremely high value at E_1 just downstream of the apex. Although downstream it sharply decreases and approaches the asymptotic wall shear stress, the wall shear stress sinusoidally varies along the distal wall and reaches a relative maximum at E_3 and a relative minimum at E_6. These sinusoidal variations of wall shear stress along both the proximal and the distal walls relate to the secondary helical motion of fluid transferred into the daughter tube from the separation region of the parent tube with skewing, as shown in Fig. 3.27. At the outer wall of the parent tube, the wall shear stress decreases from τ_T and becomes negative around the branch point where the separation occurs, as depicted by the visualization. Further downstream, it increases and asymptotically approaches the wall shear stress τ_p of the Poiseuille flow downstream of the parent tube. The aforementioned tendencies appear to be severe with the increase of the flow rate into the daughter tube.

The diameter ratio of $D_1/D_0 = 0.58$ in the experimental branch model is different than that of $D_1/D_0 = 0.41$ in the physiological condition.[33] However, the flow situation of this model helps to clarify the flow characteristics through the 45° branch. Furthermore, this study has mainly been conducted in an even flow distribution ratio, although the flow distribution ratio is not proportional to the one in the physiological condition. In the case of Re = 820 and $Q_D/Q_T = 0.30$ in Fig. 3.30, the wall shear stress reaches one maximum value at B_{20} ($y/R_1 = 4.2$) and seems to reach a second maximum value around B_{25} ($y/R_1 = 7.6$). This tendency markedly appears in an even flow distribution ratio. Therefore, the even flow distribution ratio emphasizes the flow behavior in the daughter tube.

The distribution of wall shear stress measured in this section has several physiological implications. In small animals atherosclerosis has been observed around the apex,[29,30] and the wall shear stress high around the apex. In contrast, atherosclerosis in comparatively large animals such as humans has been pathologically observed at the entrance of the daughter tube along the proximal wall.[33] The magnitude of wall shear stress is generally considered small along the proximal wall, and the mass transfer rate dependent on the shear rate would be also smaller in this region. According to the results of this study, the wall shear stress immediately upstream of the entrance to the daughter tube is not necessarily small, and the wall shear stress along the proximal wall of the daughter tube has a magnitude comparable to the wall shear stress around the apex. This stress magnitude becomes several times larger than that of the Poiseuille flow upstream of the parent tube. Certainly, since the electrode E_1 is embedded about 10 mm downstream of the apex, the wall shear stress at the site between the apex and location E_1 would be larger than that at E_1. The wall shear stress along the proximal wall varies significantly with position in the form of a damped sine wave. The behavior of wall shear stress in this region might correspond to the change of wall shear stress indicated by Perktold and Peter,[28] i.e., zones of small and large wall shear stress are close to each other. Therefore, the results of this study suggest that the gradient of wall shear stress with respect to the tube axis and the time variation of wall shear stress are closely associated with the generation and development of atherosclerosis.

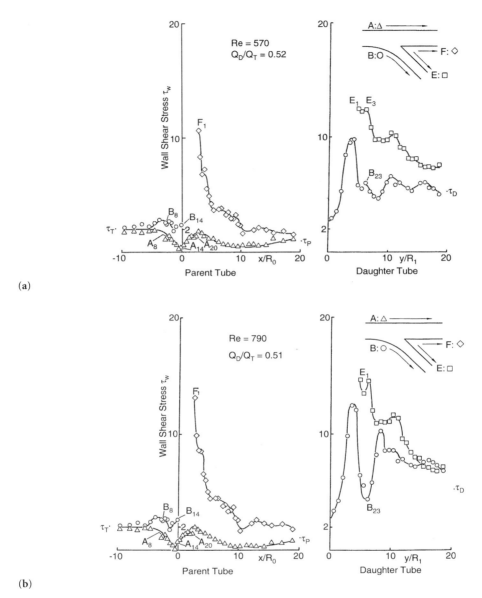

FIGURE 3.29 Distribution of wall shear stress at an even flow ratio in steady flow: (a) Re = 570; (b) Re = 790; (c) Re = 1040.

Remarks

In this section, the wall shear stress in the 45° bifurcation model has been experimentally studied in both steady and pulsating flows. The distribution of wall shear stress is characterized along the wall of the daughter tube. After flowing into the daughter tube, the wall shear stress sinusoidally varies along the proximal wall, and its magnitude is comparable to its magnitude along the distal wall around the apex. For the pulsating flow of Re = 800, $\alpha = 6.6$, $Q_D/Q_T = 0.51$, and A = 0.30, the wall shear stress along the proximal wall markedly changes and its time variation appears more obvious at the site where the magnitude of the wall shear stress is large. These results suggest that the gradient of wall shear stress with respect to the tube axis and the time variation of wall shear stress are closely associated with the generation and development of atherosclerosis.

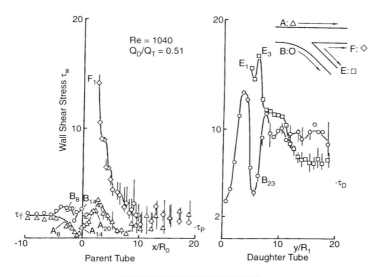

(c)

FIGURE 3.29 (CONTINUED)

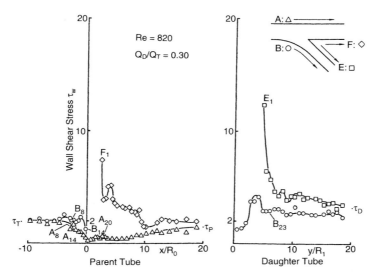

FIGURE 3.30 Distribution of wall shear stress at a geometrical flow ratio in steady flow (Re = 820).

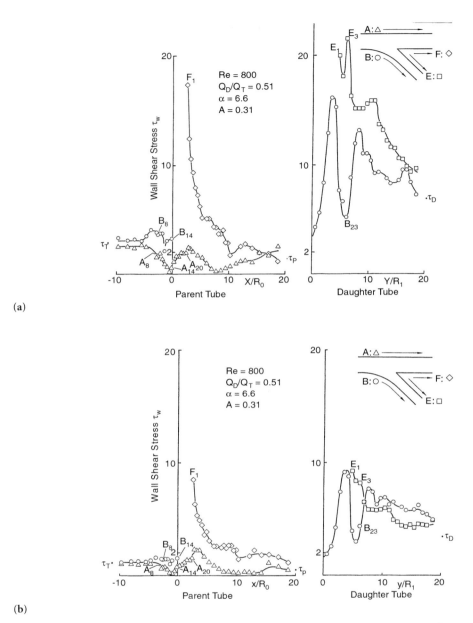

FIGURE 3.31 Distribution of instantaneous wall shear stress at an even flow ratio in a pulsating flow (Re = 800, $\alpha = 6.6$, A = 0.31). (a) Maximum flow rate: $\omega t = \pi/2$; (b) minimum flow rate: $\omega t = 3\pi/2$.

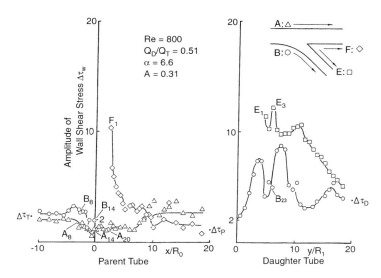

FIGURE 3.32a Amplitude of wall shear stress at an even flow ratio in a pulsating flow (Re ≈ 800, α = 6.6, A ≈ 0.31). $Q_D/Q_T = 0.51$.

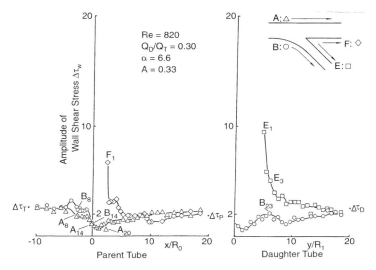

FIGURE 3.32b Amplitude of wall shear stress at an even flow ratio in a pulsating flow (Re ≈ 800, α = 6.6, A ≈ 0.31). $Q_D/Q_T = 0.30$.

References

1. Mizushina, T., *Advances in Heat Transfer*, Vol. 7, Academic Press, New York, 1971, 87.
2. Ranz, W.E., *AIChE J.*, 4, 338, 1958.
3. Lin, C.S. et al., *Ind. Eng. Chem.*, 43, 2136, 1951.
4. Lutz, R.J. et al., *AIChE J.*, 28, 1027, 1982.
5. Meites, L., *Polarographic Techniques*, 2nd Ed., John Wiley & Sons, New York, 1965.
6. Pedley, T.J., *J. Fluid Mech.*, 55, 329, 1972.
7. Pedley, T.J., *The Fluid Mechanics of Large Blood Vessels*, Cambridge University Press, Cambridge, U.K., 1980, 126.
8. Talbot, L. and Steinert, J.J., *ASME J. Biomech. Eng.*, 109, 60, 1987.
9. Mao, Z.X. and Hanratty, T.J., *Exp. Fluids*, 3, 129, 1985.
10. Yamaguchi, R. and Kohtoh, K., *ASME J. Biomech. Eng.*, 116, 119, 1994.
11. Uchida, S., *ZAMP*, 7, 403, 1956.
12. Womersley, J.R., *J. Physiol.*, 127, 553, 1955.
13. Lighthill, M.J., *J. Fluid Mech.*, 52, 475, 1972.
14. Stehbens, W.E., *Quart. J. Exp. Physiol.*, 44, 110, 1959.
15. McDonald, D.A., *Blood Flow in Arteries*, Edward Arnold, London, 1974.
16. Fry, D.L., *Circ. Res.*, 22, 167, 1968.
17. Caro, C.G. et al., *Proc. Roy. Soc. (ser. B)*, 177, 109, 1971.
18. Nerem, R. et al., *ASME J. Biomech. Eng.*, 103, 172, 1981.
19. Dewey, C.F., Jr. et al., *ASME J. Biomech. Eng.*, 103, 177, 1981.
20. Lutz, R.J. et al., *Circ. Res.*, 41, 391, 1977.
21. El Masrey, O.A. and Feuerstein, I.A., *ASME J. Biomech. Eng.*, 104, 290, 1982.
22. Karino, T. et al., *Biorheology*, 16, 231, 1979.
23. Karino, T. and Goldsmith, H.L., *Biorheology*, 22, 87, 1985.
24. Ku, D.N. et al., *Arteriosclerosis*, 5, 293, 1985.
25. Duncan, D.D. et al., *ASME J. Biomech. Eng.*, 112, 183, 1990.
26. Kawaguti, M. and Hamano, A. *J. Physics Soc. Japan*, 49, 817, 1980.
27. O'Brien, V. et al., *J. Fluid Mech.*, 75, 315, 1976.
28. Perktold, K. and Peter, R., *J. Biomed. Eng.*, 12, 2, 1990.
29. Cornhill, J.F. and Roach, M.R., *Atherosclerosis*, 23, 489, 1976.
30. Sinzinger, H. et al., *Atherosclerosis*, 33, 149, 1979.
31. Cornhill, J.F. et al., *Atherosclerosis*, 35, 103, 1980.
32. Friedman, M.H. et al., *Atherosclerosis*, 39, 425, 1981.
33. Yamaguchi, T. et al., *Recent Adv. Cardiovasc. Dis.*, 7, 97, 1986 (in Japanese).
34. Lee, J.S. and Fung, Y.S., *ASME J. Applied Mech.*, 37, 9, 1970.
35. Yamaguchi, R. *7th Inter. Conf. Biomed. Eng.*, 517, 1992.
36. Fung, T.J., *Finite Element Analysis on Fluid Dynamics*, McGraw-Hill, New York, 1978.

4

Techniques for Measuring Blood Flow in the Microvascular Circulation

4.1   Introduction 4-1
4.2   Thermal Clearance Techniques 4-2
        Bioheat Transfer Equation • Thermal Pulse Decay Technique •
        Self-Heated Thermistor Method • Other Heat Dilution
        Methods
4.3   Indicator Techniques 4-21
4.4   Imaging Techniques 4-22
        Single Photon Emission Computed Tomography (SPECT) •
        Positron Emission Tomography (PET) • Ultrafast Computed
        Tomography (CINE) • Magnetic Resonance Imaging
4.5   Doppler Techniques 4-24
        Ultrasound Doppler • Laser and Optical Doppler
4.6   Plethysmography Techniques 4-25

Lisa X. Xu
Purdue University

Gary T. Anderson
University of Arkansas

4.1   Introduction

Blood flow in microcirculation, including the capillary network plus small arterioles and venules of less than 100 μm in diameter, is usually referred to as perfusion, which is defined as the blood flow rate per unit tissue volume. Perfusion plays an important role in the local transport of oxygen, nutrients, pharmaceuticals, and heat throughout the body. Measurements of tissue blood flow are important for a wide variety of clinical areas. Perfusion is an important parameter in determining the viability of tissue and organ transplants and for assessing the prognosis of patients with ischemic heart disease. It can play a significant role in determining brain damage in stroke victims, as well as determining the state of diseased kidneys, livers, pancreas, and lung tissue. Perfusion is also a factor in many drug studies, in thermal dosimetry during hyperthermia, in treatment of cancer, and in laser photocoagulation of tumors. Despite its clinical significance, there is at present no ideal way to measure perfusion. The purpose of this chapter is to briefly describe some of the main techniques for measuring tissue perfusion, with a discussion of the advantages and limitations of each technique. Special attention is given to the thermal pulse decay and self-heated thermistor methods. Other techniques discussed include thermal clearance, indicator, imaging, Doppler methods, and plethysmography.

4.2   Thermal Clearance Techniques

Thermal clearance techniques for measuring blood flow are based on the basic principles of heat transfer in blood perfused tissue. Although it is difficult to evaluate tissue blood flow by heat analysis, the problem becomes less complex if the tissue of interest is far away from the body surface. In that case, heat exchange between the tissue and the environment by condensation, evaporation, and radiation can be neglected, and heat transfer is limited to conduction and convection. Depending on the direction of the associated temperature gradient, conduction causes heat to be either gained or lost due to the direct transfer of intramolecular energies to adjacent tissue at a different temperature. The rate of conduction is dependent on the thermal conductivity of tissue, an intrinsic thermal property of a homogeneous, isotropic solid mass in a thermal steady state. Heat is also transferred in a perfused tissue by convection, as blood, lymph, and other body fluids flow through the tissue. The convective component of heat transfer is proportional to the difference in temperature between the tissue and the fluid flowing through it. Blood flow rate, often called blood perfusion rate, plays an important role in convection in a living tissue. Sometimes, the term effective thermal conductivity is used as a lumped parameter describing the combined thermal effect of conduction and convection in perfused tissue. Thermal clearance techniques all require an initial temperature perturbation and thus involve a transient measurement. However, the induced heat field needs to be weak enough that it does not affect the regional blood flow. [1–3]

Thermal clearance techniques to measure perfusion are based on the premise that heat flux due to conduction can be separated from heat flux due to convection (e.g., blood flow). Either theoretical or empirical models of heat transfer due to blood flow are then used to quantitatively evaluate the perfusion rate in a given tissue. Searching for an accurate model of perfusion has attracted a great deal of attention over the last several decades. [4–7] The well known Pennes bioheat equation[8] was developed in 1948 based on the assumption that blood–tissue heat transfer takes place in capillary beds. This model assumes that the temperature of venous blood is in equilibrium with the local tissue temperature while that of arterial blood is not. It has since been shown that blood–tissue heat transfer actually starts to occur in pre-capillary beds and therefore the basis for the Pennes equation is questionable. [7] However, the Pennes equation-based models have yielded realistic predictions in strong agreement with experimental data under a variety of conditions. [9–15]

In the following sections, an introduction to the theoretical basis and assumptions of the Pennes bioheat transfer equation will be provided. Then, various thermal clearance techniques including the pulse decay method,[12] transient thermistor,[16] and sinusoidal techniques,[17,18] developed based on this and similar equations will be discussed, respectively. Finally, other thermal clearance techniques based on theories of either the Fick relation or effective thermal conductivity will be briefly discussed.

Bioheat Transfer Equation

The mathematical formulation of the Pennes heat transfer equation in blood perfused tissue is given as:[8]

$$\rho_t c_t \frac{\partial T_t}{\partial t} = \nabla k_t \cdot \nabla T_t + \rho_{bl} c_{bl} \omega_b \left(T_a - T_t \right) + q_{mb} \tag{4.1}$$

Originally applied to predict temperature fields in the human forearm, this equation has become well known as the bioheat equation. It is a relatively simple modification of the ordinary heat conduction equation (the first two terms), in which the contribution of moving blood is treated as a volumetric isotropic heat source (the third term) and metabolic heat production (the fourth term). This equation was developed based on the assumption that the major blood–tissue thermal energy exchange occurs in capillaries, where the vessel surface area is large and blood flow velocity is very low. The thermal contribution of blood can then be modeled as if it enters an imaginary pool at the same temperature as that in major supply arteries, T_a. Due to its immediate thermal equilibration with the surrounding tissue, blood exits the pool and enters the venous circulation at the tissue temperature T_t. The total energy transferred by the flowing blood can therefore be described by a nondirectional heat source, the magnitude of which is proportional to the blood perfusion rate (ω_b; ml/s/ml) and the temperature difference between the arterial blood and local tissue. This term is often referred to as the Pennes perfusion term, which neglects all pre- and post-capillary heat exchange between the blood and tissue. Although it has been questioned since its inception over 50 years ago,[4–7,19] the Pennes perfusion term has often been applied successfully in estimations of blood perfusion in tissue far away from large vessels, where the collective treatment of the vasculature in a continuum formulation is valid for blood vessels of terminal arterial branches and small vessels.

Thermal Pulse Decay Technique

Approach

The thermal pulse decay (TPD) technique for the determination of local blood perfusion and tissue thermal conductivity is based on a comparison of measured and simulated temperature decays following a very short and small heating pulse. The technique was initiated by Chen and Holmes[12] and was subsequently enhanced during the following ten years.[20–25] A solution of the Pennes bioheat transfer equation (Eq. 1.1) was used to build the theoretical model for this technique. The model considers a tissue sphere which is small enough to assume both a uniform blood perfusion rate (ω_b; ml/s/ml) and uniform thermal properties, but sufficiently large to simulate the thermistor bead as a point source located at the sphere center. Densities and heat capacities of the blood and tissue are assumed to be equal, and the initial and boundary conditions are:

$$T_t\left(r,t=0\right)=T_{ss}\left(r\right); \quad T_t\left(r=\infty,t\right)=0 \tag{4.2}$$

where T_{ss} is the initial steady state tissue temperature. A heat pulse is deposited locally in tissue through a very small thermistor bead as:

$$q_p\left(r\right)=P\cdot\delta\left(0\right) \quad \text{for } t\le t_p; \quad q_p\left(r\right)=0 \text{ for } t>t_p \tag{4.3}$$

where P is the deposited power, and $\delta(0)$ is the Dirac delta function. The temperature of the arterial blood (T_a) and the metabolic heat source term (q_{mb}) in Eq. 4.1 can be eliminated from the calculation by considering only the decay of transient temperature elevation $\theta = T_t - T_{ss}$ after the pulse heating. Thus, for the tissue surrounding the thermistor bead, Eq. 4.1 becomes:

$$\frac{\partial\theta}{\partial t}=\alpha\cdot\nabla^2\theta-\omega_b\theta+\frac{1}{\rho c}q_p \tag{4.4}$$

$$\theta\left(r,t=0\right)=0; \quad \theta\left(r=\infty,t\right)=0 \tag{4.5}$$

For the limiting case of an infinitesimally small probe with an infinitesimally short heating pulse, the solution of Eq. 4.4 for the interval of temperature decay takes the form:

$$\theta=\gamma\cdot\int_0^{t_p}\left(t-s\right)^{-1.5}\cdot e^{-\omega_b(t-s)}\cdot e^{-r^2/[4\alpha(t-s)]} \, ds \tag{4.6}$$

$$\alpha=P\left(\rho c\right)^{0.5}/8\pi^{1.5}; \quad \gamma=a/k_t^{1.5}/t_p^{0.5} \tag{4.7}$$

Using the relations:

$$\tau = t/t_p; \quad \eta = s/t_p; \quad w = \omega_b t_p; \quad F_o = \alpha \, t_p/r^2 \tag{4.8}$$

Eq. 4.6 becomes nondimensionalized as:

$$\theta = \gamma \cdot \int_0^1 e^{-w(\tau-\eta)} \cdot (\tau - \eta)^{-1.5} \cdot e^{-1/[4 F_o(\tau-\eta)]} \, d\eta \tag{4.9}$$

Due to the fact that the bead diameter is nearly zero compared to the large measuring tissue volume, the temperature of the tissue adjacent to the thermistor bead surface can be approximated using r = 0.0 in Eq. 4.9. The tissue conductivity and blood perfusion rate are calculated simultaneously by comparing predicted and measured temperatures so that the least squares of deviations between them are minimized. A computing algorithm has been developed based on the above mentioned scheme:[23]

Probe Design

In the TPD measuring system, a thermistor bead probe is used first to deposit the heating pulse and afterwards to measure the tissue temperature. Fig. 4.1 shows that each probe consists of one or two small (0.15 mm dia.) thermistor beads situated at the end or near the middle of a glass fiber reinforced epoxy shaft. The diameter of the finished probe is typically 0.3 mm, and the length can vary as desired. Probes are structurally strong and, because the end can be sharpened to a point, they are capable of piercing most tissues with very minimal trauma.

The thermistor microprobe functions as a temperature sensing element when it forms one arm of a resistance bridge (Fig. 4.2). Typically, as described by Arkin et al.,[23] the bead used has a resistance $R_p = 2000$ $\Omega \pm 5\%$ at 25°C. The thermistor resistance–temperature relationship can be expressed as:

$$R_p = R_{pk} \exp\left[\beta\left(\frac{1}{T_p} - \frac{1}{T_k}\right)\right] \tag{4.10}$$

R_{pk} is the bead resistance evaluated at $T_k = 311.15$ K, a value which represents the center of the working range in living tissue. At this temperature, the thermal response time of the thermistor bead is about 10 msec in water. Temperature calibration against a thermometer traceable to NBS standard establishes values for R_{pk} and β for each probe assembly.

Instrumentation Hardware and Software

The TPD bridge unit includes a total of six bridge circuits. Each bridge circuit consists of a computer controlled and relay-activated thermistor heating circuit, an electrically isolated resistance bridge circuit, and an Analog Devices instrumentation amplifier (model 286J) which amplifies (×100) the output of the resistance bridge. A two-conductor and shielded probe interface cable is provided for each bridge channel to serve as a connection between the thermistor bead probe and the TPD bridge unit. One end of the interface cable plugs into the socket on the front panel of the bridge unit, while the other end is designed to receive the probe assembly. When it is connected to the unit, the thermistor bead resistance located in the measurement probe serves as an arm of a resistance bridge. The output voltage channels of the unit are connected to a Burr-Brown PCI–20002M Analog Input Module mounted on a PCI–20041C High Performance Carrier Board located inside the computer. The thermistor pulse heating circuitry within the bridge unit is computer controlled from an I/O port (P5) located on the PCI–20041C board. Changes in the value of the bead resistance result in changes in the measured bridge output voltage, from which the bead temperature can be calculated from Eq. 4.10. As shown in Fig. 4.3, the menu driven supporting software written in FORTRAN [23] permits the user to control the duration of the heating pulse,

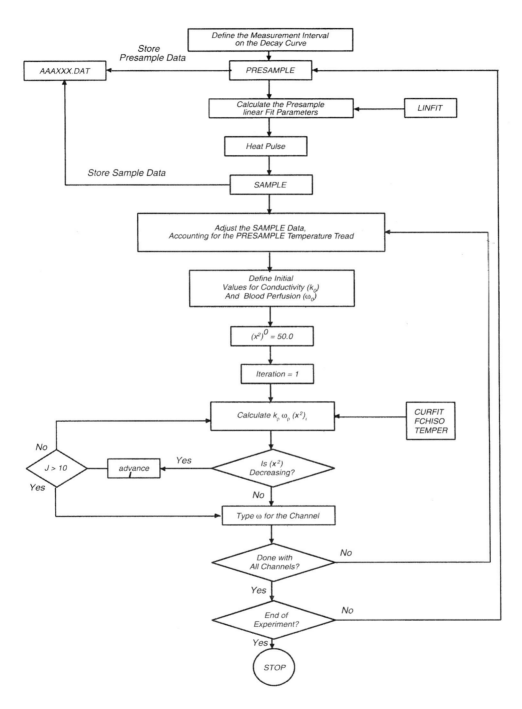

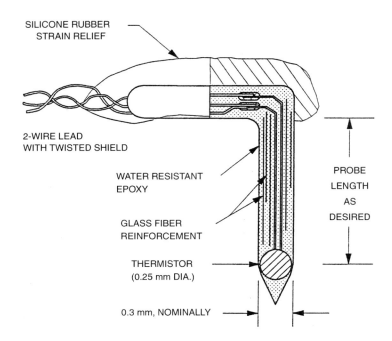

FIGURE 4.1   Thermistor bead microprobe configuration for the TPD system.

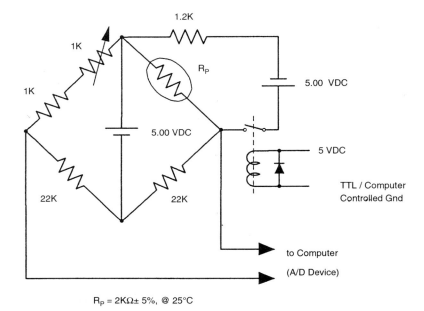

FIGURE 4.2   The resistance bridge for measuring variations of probe resistance.

```
TIME 13:53:43          TPD DATA ACQUISITION SYSTEM      Ver 3.1A        11/19/1996

BP7: 0           BP8: 0        A/D Gain: 10    System Gain: 100
                              1          2          3          4          5          6
1) Samp/Calc/Coup Status     S/C        S          S/C        S          S/C        S
2) Sample Data Filename      KID01001   KID02001   KID03001   KiD04001   KID05001   KID06001
3) Probe Number              40         41         42         43         44         45
4) Heat Pulse (secs)         3.0        0.0        3.0        0.0        3.0        0.0
5) Bridge Resistance         520        740        876        432        589        678
   Probe Reading (mV)        3.531      2.456      1.541      4.321      2.901      1.430
   Probe Temperature ©       22.766     22.578     22.290     22.320     21.997     22.150
   Last Presample Temp. ©    22.189     22.683     21.590     21.091     22.050     22.100
   Last Perfusion            0.012      0.521      1.269      0.001      0.571      1.950
```

Durations (secs)	Readings		
6) Presample 10.00	100	S-Sample	A-Automatic
7) Sample 10.00	100	T-Test Switches	F-Set A/D Gain
8) Alarm OFF Limits: -5.00	5.00	D-Disp Sample Data	P-Print Sample Data
9) Auto Run Settings		B-Calibrate BP	E-Set System Gain
Start Time: 14:05:00		W-Set Calc. Param.	G-Graph Mode
Repetitions: 180		Q-Quit	I-Save Parameters
Inter Run Time (secs): 22			
0) Window OFF Filename:	WIN000		

FIGURE 4.3   The user's interface for the TPD system.

the data acquisition, and storage. The software also provides an on-line calculation of the tissue thermal conductivity (k_t) and the local blood perfusion rate (ω_b) using the iterative algorithm described above.

Calibration of the TPD system involves the thermistor bead probe calibration as described above, trimming the instrumentation amplifier, and balancing the resistance bridge. On the front panel of the TPD bridge unit, the bridge-trim switch is provided to assure that a zero voltage input to the amplifier produces a zero voltage output. After the instrumentation amplifier has been properly trimmed, a phantom thermistor probe with a measured resistance R = 1310 ± 5 Ω is plugged into the end of the probe interface cable. With the indicating dial on the front panel set to 310, the output voltage of the bridge (as viewed on the TPD screen shown in Fig. 4.3) should be 0.000 ± 0.015 VDC. If this condition is not observed, it may be necessary to remove the dial from the shaft of the variable resistance and adjust the position of the shaft to obtain the desired output voltage. This sequence of switch placements and resistor adjustments needs to be repeated for each bridge channel contained within the bridge unit. Bridge elements can be expected to change their properties slightly with time and environmental (room) temperature. Therefore, it is not unreasonable to expect that periodic re-balancing of the bridge circuit may be necessary. With the calibrated system, the accuracy of the temperature measurement is $\epsilon_\theta = \pm 0.003°C$.

Error and Sensitivity Analyses

There are two main types of possible errors in the TPD technique.[20] The first is related to the measuring errors in tissue temperature (± 0.003°C), the supplied power (5 mW ± 15%), and the length of the heating pulse (3 sec ± 0.003). The second type of error may result from the inexact nature of the model, such as neglecting the probe diameter and fluctuations of the baseline temperature measurement. Fig. 4.4 plots the overall maximum error for a typical set of experimental conditions, as well as the specific contribution of each measured parameter to the overall error. Immediately after the cessation of the heating pulse, the primary error is caused by the fact that the thermistor bead has a finite size which is not equal to zero. However, this error diminishes very quickly with time. This is mainly because the sampling tissue volume becomes larger as time increases, so the assumption of zero bead diameter is more realistic. After a long period of time, on the other hand, the main error comes from the imprecise temperature measurement. Although the measurement error is very small ($\pm$ 0.003°C), it is rather constant. The relative temperature rise induced by the heating pulse decays with time, so that the error becomes an increasing fraction of the measured temperature. It is important to note that for the applied set of parameters, the overall maximum error is basically a combination of the error caused by the bead size and the error in measured temperature. Other measured quantities have only secondary contributions. Thus, the optimal measurement interval of time mentioned earlier has to be implemented by compromising these two errors. In practical measurement, the following parameters have been used:[20] t_p = 3 sec; k_t = 0.5 W/m/k; ϵ_θ = $\pm$ 0.003°C; P = 5mW. The analysis assumes the existence of a linear relationship between the dependent variables and the measured parameters in the neighborhood of the point of calculation. Thus, the change of the dependent variable with a small change of each parameter may be estimated by its first partial derivative. For the perfusion rate, $\epsilon_\omega = \epsilon_\theta\,(\partial\omega_b/\partial\theta)$; whereas $\epsilon_k = \epsilon_\theta$ $(\partial k/\partial\theta)$ for the tissue conductivity, while the first derivatives of θ to ω_b or to k can be obtained from Eq. 4.6. Fig. 4.5 compares the overall errors in conductivity and perfusion rate at two different perfusion rates (3 ml/min/gm and 6 ml/min/gm). For a short time after the power is shut off, the primary error is associated with the measurement of the perfusion rate. However, for very high perfusion rates and long sampling times, the predominant error will be associated with tissue conductivity. In other words, the measuring sensitivities for conductivity and perfusion rate vary with respect to time and the local perfusion rate, respectively. From Fig. 4.5, one can anticipate that the sensitivity of the perfusion measurement will decrease at low perfusion rates but increase for the conductivity measurement because of the dominance of conduction. More importantly, Fig. 4.5 reveals an optimal measuring interval of time (MIT) dependent on the local perfusion rate. For example, if the local perfusion rate is 3 ml/min/gm or higher, the MIT was selected to be 3s < t < 6s after the heating pulse. In this case, the sensitivity can be as high as about 99% for the conductivity and 97% for the perfusion rate. On the other hand, with lower perfusion rates and small conduction effects, the MIT needs to be shifted to the right in Fig. 4.5 so that the convection effect can be distinguished. Nevertheless, due to the fact that the relative temperature rise induced by the heating pulse decays with time, the temperature measurement error becomes more dominant in this region. Therefore, for very low perfusion measurements, the TPD technique is not strongly recommended.

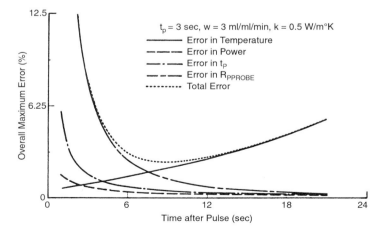

FIGURE 4.4 Contribution of measured parameters to the maximum overall error for a typical set of experimental conditions.[20] (Courtesy of ASME. With permission.)

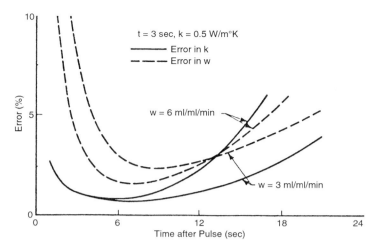

FIGURE 4.5 Typical errors in estimation of conductivity and blood perfusion rate as a function of time for various perfusion rates.[20] (Courtesy of ASME. With permission.)

Self-Heated Thermistor Method

Background

There are several variations of the original self-heated thermistor method to measure perfusion, but most of them are based on the same principle. Chato[26] was the first to suggest that perfusion can be calculated from the amount of power necessary to cause a set increase in thermistor temperature. The basic technique involves placing a miniature thermistor bead in perfused tissue and heating it to a predetermined temperature. As the bead temperature begins to rise, both perfusion and intrinsic tissue conductivity act to carry heat away from the thermistor. The contributions of perfusion and intrinsic tissue conductivity to the total heat transfer can be lumped together into an effective thermal conductivity term, which is the apparent conductivity the tissue would have if one did not know that perfusion was present. From the power required to maintain a set temperature rise in a thermistor, the effective thermal conductivity of the perfused tissue can be determined. Perfusion is then calculated based on either a theoretical or empirical model of the heat transfer due to perfusion. As mentioned earlier, the correct model of perfusion is a hotly debated issue, and appears to be tissue dependent. For example, a model developed by Pennes[8] has been shown to work well for the pig renal cortex[15] and the rat liver,[17] while a small artery model (SAM)[18] works better for the canine kidney. Weinbaum and Jiji[6] and Valvano and Anderson[17,18] developed a general method for determining perfusion models for given tissues based on the vasculature of the tissue.

Several people have contributed to the development of the self-heated thermistor technique. Chato started with a simplified view of the thermistor as a lumped mass and then solved the bioheat equation for the perfused tissue surrounding the thermistor bead. Balasubramanium and Bowman[16] relaxed Chato's simple assumption by modeling the thermistor as a sphere with a distributed mass. Jain[27] developed a model that described the thermistor as having a spherical heat generating center surrounded by a passive glass shell. Afterwards, Valvano et al.[17] obtained a closed form transient solution for measurements of thermal conductivity and diffusivity. Hayes and Valvano[28] confirmed the thermal conductivity measurement equation of Valvano et al. for more realistic thermistor shapes and heating patterns. Valvano[29] and Anderson[30] showed that a thermistor heated with a combination of steady-state and sinusoidal power could simultaneously measure the intrinsic and effective thermal conductivities of perfused tissue, thus simplifying the quantitative evaluation of absolute perfusion rates.

Approach

The theory behind the steady-state self-heated thermistor technique is based on solutions to a bioheat equation. The general form of the bioheat equation is

$$\rho_t c_t \frac{\partial T_t}{\partial t} = k_t \nabla^2 T_t + q_{perf} + q_{ext} + q_{mb} + q_{ves} \tag{4.11}$$

where $\rho_t c_t$ is the heat capacity of tissue, k_t is the conductivity of tissue, and T_t is the temperature rise in tissue. The metabolic heat generation q_{mb} can be ignored if it is spatially and temporally uniform or is much smaller than q_{ext}, the volumetric power applied to the thermistor. q_{ext} can be a steady-state, transient, or combination of steady-state and periodic sources. q_{ves} is the heat transfer due to major vessels in an organ. The following analysis assumes that there are no major vessels in the field of influence of the thermistor, and this term will be ignored.

If the shape of a thermistor is approximated as a sphere, then the equations for a bead embedded in nonperfused tissue become:

$$\rho_b c_b \frac{\partial T_b}{\partial t} = k_b \nabla^2 T_b + q_{ext} \quad r < a \tag{4.12}$$

$$\rho_t c_t \frac{\partial T_b}{\partial t} = k_t \nabla^2 T_t \quad r > a \tag{4.13}$$

where r is radial distance from the thermistor center and a is the radius of the thermistor. Five boundary and initial conditions are in effect. Before heating begins, the tissue and thermistor are at a uniform temperature:

$$T_t = T_b = 0 \tag{4.14}$$

There is no heat flux at the center of the thermistor:

$$\partial T_b / \partial r = 0 \qquad r = 0 \tag{4.15}$$

The temperature at the tissue–thermistor boundary matches:

$$T_b = T_t \qquad r = a \tag{4.16}$$

The heat flux at the tissue–thermistor boundary also matches:

$$k_b \frac{\partial T_b}{\partial r} = k_t \frac{\partial T_t}{\partial r} \qquad r = a \tag{4.17}$$

Since the thermistor is a finite heat source in an infinite medium:

$$\lim_{r \to \infty} T_t = 0 \tag{4.18}$$

A steady-state solution to Eqs. 4.12–4.18 with q_{ext} held constant allows the extraction of thermal conductivity from the ratio of the average thermistor temperature (ΔT) to applied power (P):

$$k_t = \frac{1}{\dfrac{4\pi a \Delta T}{P} - \dfrac{1}{5k_b}} \tag{4.19}$$

where

$$\Delta T \equiv \frac{\int\limits_{0}^{a} 4\pi r^2 T(r)\, dr}{\frac{4}{3}\pi a^3} \tag{4.20}$$

$$P \equiv \int\limits_{0}^{a} 4\pi r^2 q_{ext}\, dr = \frac{4}{3}\pi a^3 q_{ext} \tag{4.21}$$

In perfused tissue, the effects of tissue conductivity and perfusion are combined into an effective conductivity term, k_{eff}:

$$k_{eff} = \frac{1}{\dfrac{4\pi a \Delta T}{P} - \dfrac{1}{5k_b}} \tag{4.22}$$

where

$$k_{eff} = f\left(k_t, \omega_b\right) \tag{4.23}$$

To measure the intrinsic tissue conductivity, either blood flow must be temporarily stopped and Eq. 4.19 used, or a transient heating technique must be employed. Fig. 4.6 shows a typical graph of k_{eff} vs. perfusion rate in an alcohol-fixed canine kidney.[3] The effective conductivity with no perfusion is taken to be the tissue intrinsic conductivity.

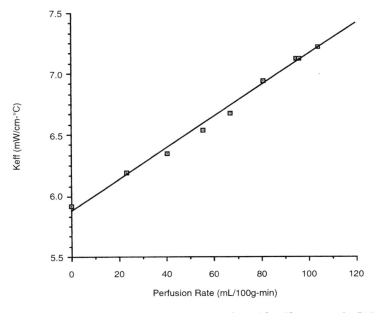

FIGURE 4.6 k_{eff} vs. perfusion in a canine kidney. Data was taken with a Thermometrics P100 thermistor.

Perfusion can be determined from k_{eff} and k_t either empirically or from theoretical considerations. For some perfusion models, most notably the widely used Pennes formulation, perfusion rate can be calculated directly from theory. Pennes[8] proposed that blood–tissue heat transfer takes place in the capillary bed, and that the temperature of venous blood is in equilibrium with the local tissue temperature, while that of arterial blood is not. For Pennes' model:

$$q_{perf} = \omega_b \rho_{bl} c_{bl} \left(T_a - T_t \right) \tag{4.24}$$

where ω_b is perfusion, T_a is the temperature of arterial blood, and T_t is the tissue temperature. Perfusion is then related to k_{eff} by:[32]

$$\omega_b = \frac{\left(k_{eff} - k_t \right)^2}{k_t \rho_{bl} c_{bl} a^2} \tag{4.25}$$

Other models have no simple analytical solution and an experimental relation between the effective thermal conductivity and perfusion must be determined. These empirical determinations require a calibration plot of perfusion vs. $(k_{eff} - k_t)$ for the tissue type of interest. A typical setup for performing this calibration is shown in Fig. 4.7. Tissue is placed in a temperature controlled water bath to minimize temperature drifts in the tissue during the measurements. The tissue is then perfused with a fluid, typically degassed saline or a mannitol-saline solution.[31] Care must be taken not to allow bubbles to flow into the tissue, as they can get trapped in capillaries and block further flow. Flow through the tissue can be varied by changing the pressure, and is measured by weighing the amount of effluent that exits the tissue in one minute. The tissue is then injected with a dye such as Evans Blue and dissected to determine where perfusion occurred. The perfused volumes of tissue are weighed, and perfusion is calculated by dividing the flow rate by the weight of the perfused tissue. A plot of $(k_{eff} - k_t)$ vs. perfusion rate is used as a calibration curve for measurements in live tissue.

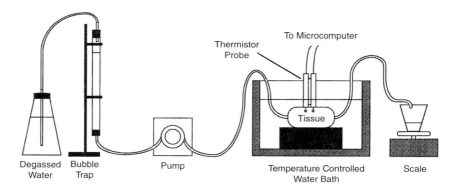

FIGURE 4.7   Setup to determine k_{eff} vs. perfusion curves for different tissues.

Sensitivity Analysis

Once the relationship between perfusion and thermistor temperature is known, it is possible to analyze the sensitivity of the measurement. Several factors affect the measurement of tissue conductivity, including the thermal properties of the thermistor and its size. Intrinsic tissue conductivity, k_t, and effective conductivity, k_{eff}, are both determined by the thermistor temperature rise, ΔT, for a given applied power, P. Because increasing the applied power causes the thermistor temperature to increase, $\Delta T/P$ is a constant for a given intrinsic tissue conductivity k_t (as is indicated by Eq. 4.19). The sensitivity of the conductivity measurement to thermistor temperature can be derived by differentiating Eq. 4.19 with respect to ΔT:

$$\frac{\partial k_t}{\partial \Delta T} = \frac{-100\pi a P k_b^2}{\left[20\pi a \Delta T k_b - P\right]^2} \tag{4.26}$$

Eq. 4.26 shows $\partial k_t / \partial \Delta T$ is approximately proportional to a/a^2, and therefore the sensitivity increases as the thermistor size decreases. As will be discussed later, practical considerations put a limit on the minimum size of the thermistor.

Measurements of perfusion depend on developing a theoretical or empirical relation between perfusion and $k_{eff} - k_t$. Both analytical and empirical models require several input parameters, each of which is known only to a limited accuracy. Thermistor bead radius, thermistor conductivity, and intrinsic tissue conductivity are known only to within a certain percentage error, and vary from one probe or tissue to another. The average temperature rise in the thermistor bead is subject to noise, computer roundoff errors, and other perturbations in the measurement system. If empirical relations are used, then a technique such as factorial analysis[33] must be employed to determine the sensitivity of the measurement to uncertainties in the values of these parameters. However, if relationships such as those given by Eq. 4.19, 4.22, and 4.25 are known, limits on the accuracy of the measurements can be determined analytically by one or both of the following equations:[34]

$$\Delta k_{eff} = \frac{dk_{eff}}{d(parameter)} \times \Delta(parameter) \tag{4.27}$$

$$\Delta \omega_b = \frac{d\omega_b}{d(parameter)} \times \Delta(parameter) \tag{4.28}$$

where (parameter) is thermistor radius, intrinsic tissue conductivity, thermistor conductivity or thermistor temperature rise, and Δ (parameter) is the uncertainty in that parameter. Table 4.1 shows how each uncertainty affects knowledge of the perfusion rate. Uncertainties in tissue and thermistor conductivities are the largest contributors to ambiguity in the measurements, followed by noise in the thermistor temperature. Uncertainties in the perfusion measurement are much smaller for high blood flow rates (100 mL/100 g/min) than for lower rates (10 mL/100 g/min). Although this analysis assumes the Pennes formulation of heat transfer, other models have shown the same relative contribution of each factor to uncertainties in the perfusion measurement.

TABLE 4.1 Sensitivities of Perfusion Measurement to Uncertainties in Probe Parameters for the Pennes Model

Parameter (Units)	Nominal Value	Perturbation (Percent)	$(\Delta\omega/\omega)$ ω = 10ml/100 g/min	$(\Delta\omega/\omega)$ ω = 100 ml/100 g/min
a (cm)	0.050	0.0005 1%	0.02	0.02
k_m (W/cm-°C)	0.005	0.0001 2%	0.69	0.23
k_b (W/cm-°C)	0.001	0.00002 2%	0.67	0.21
ΔT (°C)	4.000	0.02 0.5%	0.34	0.11

Practical Considerations

The above formulation contains several assumptions that are not strictly correct. For example, thermistors are not spherical, but have a somewhat irregular shape. Also, power deposition in a self-heated thermistor is not uniform, but concentrated in the center of the device. For practical reasons, the thermistors used are usually not constructed of a single material, but have a protective glass coating around them. In addition, the thermal properties of the transducer are usually unknown.

To overcome these obstacles, calibrations must be performed on the thermistor probes before perfusion measurements can be taken. During measurements, Eq. 4.19 is replaced with:

$$\frac{1}{k_{eff}} = k_1 + k_2 \frac{P}{\Delta T} \tag{4.29}$$

Calibration coefficients k_1 and k_2 are typically determined by taking thermistor measurements in glycerol and agar-gelled water (at 300°C, $k_{glycerol}$ = 2.86 × 10^{-3} W/cm/°K; k_{water} = 6.13 × 10^{-3} W/cm/°K). Eq. 4.19 demonstrates that k_1 is a function of thermistor conductivity, while k_2 depends on thermistor radius:

$$k_1 = 1/5k_b \tag{4.30}$$

$$k_2 = 4\pi a \tag{4.31}$$

Despite the many deviations of actual thermistors from the theory, Eq. 4.29 has been shown to remain valid[28,34] for non-ideal devices.

Another factor that needs to be considered when taking measurements is temperature drifts over time. The effective conductivity measurement (Eq. 4.22) is quite sensitive to changes in temperature, with as little as a 0.02°C change in temperature causing a 34% error at low perfusion rates. It generally takes a self-heated thermistor about five minutes to reach steady-state, making small drifts in ambient temperature likely. Typically, a probe will consist of two devices: a self-heated thermistor to measure perfusion and a sensing thermistor to measure drifts in tissue temperature over time (Fig. 4.8). The sensing thermistor must be far enough from the heated device to be outside its heating volume of influence, but close enough to minimize spatial differences in temperature between the two thermistors. A separation of 1–2 cm is typically used.

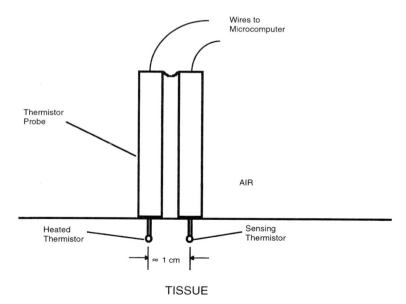

FIGURE 4.8   Diagram of a typical thermistor probe.

The sensitivity of the perfusion measurement to small changes in measured temperature necessitates careful calibration of thermistor resistance vs. temperature curves. The temperature is typically fit to:

$$1/T = a_0 + a_1\left(\ln R_t\right) + a_3\left(\ln R_t\right)^3 \tag{4.32}$$

where R_t is the resistance of the thermistor and a_n are calibration coefficients. A quartz thermometer, accurate to 0.005°C, is often used as a temperature reference. Less accurate references can be used if many reference temperatures are taken and care is used to eliminate systematic errors in data collection.

Many circuits can be used to either measure thermistor resistance or heat the thermistor by applying a voltage across it. The sensitivity of the measurement necessitates the use of low noise and precision elements, however. Care must be taken not to self-heat the sensing thermistor, especially when small diameter devices are used ($a \leq 0.15$ cm). The amount of self-heating incurred can be estimated from:

$$\Delta T = \frac{3P}{4\pi a}\left(\frac{1}{15k_b} + \frac{1}{3k_t}\right) \tag{4.33}$$

where $P = I^2 R$ is the power applied to the thermistor. Assuming a thermistor conductivity $k_b \approx 0.001$ W/cm/°C and a tissue conductivity $k_t \approx 0.005$ W/cm/°C, then $\Delta T \approx 32\ P/a$. For a self-heating effect of less than 0.01°C, $P \leq 0.0003\ a$, where P is measured in watts and a in cm.

The size and type of thermistor play an important role in perfusion measurements. The sensitivity of the tissue conductivity measurement increases as thermistor size decreases. However, this does not necessarily mean the sensitivity of the perfusion measurement also increases with smaller thermistors, as is indicated in Table 4.1. Another factor that must be considered is contact resistance between the probe and tissue, which for small thermistors can be variable. On the other hand, large thermistors can compress the tissue surrounding them, thus affecting the perfusion rate in the immediate vicinity.

Thermistor construction can also affect the perfusion measurements. Bare thermistors are fragile, and must be coated with a waterproof sealant to prevent the electrodes from shorting together. Often, a small amount of moisture will seep into the probe during measurements, thus slightly altering the resistance vs. temperature calibration. For this reason, many thermistors come with a protective glass coating that makes them durable and easy to use. Hayes[16] showed that Eq. 4.29 is still valid for the glass coated thermistors. In the author's experience, glass coated thermistors with a total diameter of 0.09–0.16 cm (0.035–0.06 inches) give the most reliable measurements.

Measurement Volume

The temperature rise of a thermistor will be most affected by the thermal properties of tissue closest to the bead. The volume of tissue that has an effect on ΔT is called the measurement volume. It is possible to determine how large this volume is for a steady-state heating. Balasubramanium[16,32] derived equations to describe the average temperature rise in a thermistor, the temperature field in a thermistor, and the temperature field in homogeneous tissue due to heating a spherical thermistor with a time dependent source. The steady-state form of Balasubramanium's equations can be derived more easily:

$$k_b \nabla^2 T_b + q_b = 0 \tag{4.34}$$

$$k_t \nabla^2 T_t + q_{perf} = 0 \tag{4.35}$$

where $q_b = \Gamma$ is the constant volumetric power applied to the thermistor. Note that Γ is the applied power divided by the volume of the thermistor:

$$\Gamma = \frac{P}{\frac{4}{3}\pi a^3} = \frac{3P}{4\pi a^3} \tag{4.36}$$

where the boundary conditions defined in Eqs. 4.14–4.18 still apply. By integrating Eqs. 4.34 and 4.35 and applying the boundary conditions, a solution to this system can be found. It is assumed that the perfusion term, q_{perf}, is either equal to zero or can be accounted for by modifying the conductivity term, k_m, into an effective conductivity, k_{eff}. The resulting solution allows description of the average temperature rise in a thermistor (ΔT), the temperature field in a thermistor ($T_b(r)$), and the temperature field in homogeneous tissue due to heating a spherical thermistor ($T_t(r)$):

$$T_b(r) = \frac{\Gamma a^2}{6k_b}\left(1 - r^2/a^2\right) + T_s \tag{4.37}$$

$$T_s = \frac{\Gamma a^2}{3k_t} \tag{4.38}$$

where T_s is the temperature at the surface of the bead.

$$T_t(r) = \frac{\Gamma a^3}{3k_t r} \tag{4.39}$$

$$\Delta T \equiv \frac{\int_0^a 4\pi r^2 T_b(r)\,dr}{4/3\pi a^3} = \frac{\Gamma a^2}{15k_b} + T_s \tag{4.40}$$

If a shell of material is placed around the thermistor so that the radius of the thermistor plus the shell is r_{shell}, then Eqs. 4.37 and 4.40 remain valid, but the surface temperature of the thermistor becomes:

$$T_s \equiv \frac{\Gamma a^3}{3}\left[\frac{1}{k_{shell}}\left(\frac{1}{a} - \frac{1}{r_{shell}}\right) + \frac{1}{k_t r_{shell}}\right] \tag{4.41}$$

One can extend the above analysis to include a system with an infinite number of shells of material around a spherical thermistor. It can be shown by induction that Eqs. 4.37 and 4.40 still apply for this system, but the thermistor surface temperature now becomes:

$$T_s = \frac{\Gamma a^3}{3}\left[\sum_{i=0}^{\infty} \frac{1}{k_{shell,i}}\left(\frac{1}{r_{shell,i}} - \frac{1}{r_{shell,i+1}}\right)\right] \tag{4.42}$$

where $k_{shell,i}$ is the thermal conductivity of the material in the shell i and $r_{shell,i}$ is the radius from the thermistor center to the outer edge of shell i. Now set the thickness of each shell equal to Δ so that for all i the following relationship is true:

$$\frac{1}{r_{shell,i}} - \frac{1}{r_{shell,i+1}} = \frac{1}{r_{shell,i}} = \frac{1}{r_{shell,i} + \Delta} \tag{4.43}$$

Substituting Δ into Eq. 4.42 and rearranging terms yields:

$$T_s = \frac{\Gamma a^3}{3}\left[\sum_{i=0}^{\infty}\frac{1}{k_{shell,i}}\left(\frac{1}{r_{shell,i}^2}-\frac{1}{\Delta r_{shell,i}}\right)\right] \tag{4.44}$$

Taking the limit as Δ goes to 0, the equation for T_s becomes:

$$T_s = \frac{\Gamma a^3}{3}\int_0^{\infty}\frac{dr}{k(r)r^2} \tag{4.45}$$

where the tissue conductivity $k(r)$ is heterogeneous in the r direction, but homogeneous in the θ and φ directions. Substituting Eqs. 4.45 and 4.40 into Eq. 4.22 yields:

$$k_{eff} = \frac{1}{a\int_0^{\infty}\frac{dr}{k(r)r^2}} \tag{4.46}$$

where k_{eff} is the apparent conductivity of the heterogeneous tissue as seen by the thermistor. Let b be the distance at which:

$$p = \frac{\int_a^b \frac{dr}{k(r)r^2}}{\int_0^{\infty}\frac{dr}{k(r)r^2}} \quad 0 < p < 100\% \tag{4.47}$$

Now define the measurement volume:

$$\text{Measurement Volume} = \left(4/3\right)\pi\left(b^3 - a^3\right) \tag{4.48}$$

The term p in Eq. 4.47 indicates the amount of the k_{eff} signal that comes from a sphere of tissue with radius b. If the conductivity $k(r)$ in Eq. 4.48 is homogeneous, the steady-state measurement volume of a thermistor depends only on its radius, a. Table 4.2 gives values of the radial measurement b and measurement volume for given values of p. For example, 75% of the signal measured by the thermistor is determined from the tissue properties within 4 thermistor radii, 90% of the thermistor signal comes from within 10 thermistor radii, and 99% of the signal comes from within 100 radii.

TABLE 4.2 Thermistor Measurement Lengths and Measurement Volumes for Various Values of Measurement Volume Parameter p

Percent of Signal p	Measurement Length b	Measurement Volume
25	1.3a	$5.7a^3$
50	2a	$29a^3$
75	4a	$260a^3$
90	10a	$4200a^3$
95	20a	$33{,}000a^3$
99	100a	$4.2 \times 10^6 a^3$

The measurement volume of the probe vs. thermistor diameter is plotted in Fig. 4.9 for $p = 90\%$. Note that the measurement volume of a thermistor is proportional to the cube of its radius. A large thermistor is therefore less sensitive to nonuniformities in the thermal properties of small areas of tissue immediately surrounding it. Such nonuniformities might be caused by tissue disturbance during probe insertion and thermistor–tissue thermal contact resistance.

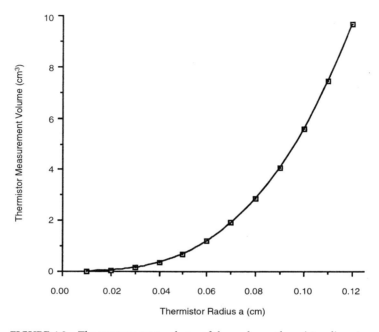

FIGURE 4.9   The measurement volume of the probe vs. thermistor diameter.

Transient Measurements

Eq. 4.22 works well for measuring perfusion, but $\Delta T/P$ takes at least five minutes to reach a steady state value in most situations. However, Chato[26] suggested that it is also possible to estimate the final $\Delta T/P$ from the initial responses of a thermistor to an applied power. In the absence of perfusion, Balasubramanium[32] has demonstrated experimentally and Valvano[17] has demonstrated analytically that the transient power follows $t^{-1/2}$. The situation may be more complex in perfused tissue. Using the Pennes model, Valvano has shown analytically that the transient power does not follow $t^{-1/2}$ exactly, but rather:

$$P = \Gamma + \beta\left[t^{-1/2}e^{-zt} - \sqrt{(z\pi)}\,erfc\sqrt{(zt)}\right] \qquad (4.49)$$

where $z = \omega_b\rho_{bl}c_{bl}/c_t r_t$, while Γ and β are the coefficients of steady-state and transient power, respectively. Balasubramanium,[32] Valvano,[17] and Patel[36] have all used this information to estimate the steady-state response of a thermistor from its transient response to the applied power:

$$\frac{P(t)}{\Delta T(t)} = I + St^{-1/2} \qquad (4.50)$$

where $P(t)$ is the applied power, $\Delta T(t)$ is the temperature rise in the thermistor, t is time, and I and S are coefficients that depend on tissue properties.

Three methods have been used to implement this technique. Bowman[37] and Valvano[17,18] utilized an electronic feedback circuit to maintain the thermistor at a fixed heated temperature, T_h. If the tissue baseline temperature T_s is constant, the probe temperature rise $\Delta T = T_h - T_s$ will also be fixed. The second method involves using a computer control loop to maintain constant power across the thermistor and measuring the temperature response. Finally, a steady voltage can be applied across the thermistor and both power and thermistor temperature rise while being simultaneously measured. In all three cases, $P/\Delta T$ can be calculated from measurements of applied power and/or temperature as a function of time and fit to Eq. 4.50. The steady-state $P/\Delta T$ response is then $1/I$. In practice, these methods work best if the first several seconds of power and temperature information are thrown away before attempting to fit the data to Eq. 4.50.

Simultaneous Measurements of Tissue Conductivity and Perfusion

The above techniques all require knowledge of intrinsic tissue conductivity to calculate an absolute perfusion rate. Usually, conductivity must be measured in the absence of perfusion, making such determinations difficult. Valvano[29,30] proposed a clever method to simultaneously measure both tissue intrinsic conductivity and the effective conductivity of perfused tissue. In this technique the thermistor is heated with a combination of steady state and sinusoidal power:

$$P = A + B \sin(2\pi f t) \tag{4.51}$$

The resulting thermistor temperature rise is fit to the same form equation:

$$\Delta T = C + D \sin(2\pi f t + \phi) \tag{4.52}$$

where ϕ is the phase difference between the applied sinusoidal power and the resulting thermistor temperature. Intrinsic and effective conductivities are calculated from the following empirical equations:

$$1/k_{eff} = k_1 + k_2\, C/A \tag{4.53}$$

$$1/k_{sin} = k_3 + k_4\, D/B \tag{4.54}$$

Coefficients k_1–k_4 are experimentally determined by operating the thermistor probe in media of known thermal conductivities. If a thermistor is assumed to be spherical and all thermistor parameters are known, then k_1 and k_2 can be calculated from Eq. 4.19. Since the size, shape, and thermal properties of thermistors vary, calibration coefficients are used. Eq. 4.53 is simplified from Eq. 4.22. Eq. 4.54 has no analytic basis, but has been verified numerically and experimentally.[30] Experimental results have shown that tissue intrinsic conductivity can be measured in the presence of perfusion by selecting an appropriate sinusoidal heating frequency.

Fig. 4.10 shows how k_{sin}, the conductivity calculated by Eq. 4.54, varies with perfusion for heating periods of 20 and 200 seconds. The data was collected in an alcohol-fixed canine kidney that was perfused using the setup in Fig. 4.7. In this case, k_{sin} is insensitive to perfusion at a heating period of 20 seconds, but a pronounced linear relationship is apparent at a heating period of 200 seconds.

The reason the technique works is that heat is applied and then removed before blood flow can have an effect on the thermistor. Although much of the literature on heat transfer due to perfusion assumes that blood flow in the capillaries plays a major role in determining the temperature fields in tissue, this assumption is questionable. First, the temperature of blood equilibrates with that of its surrounding tissue in the arteries and arterioles long before it reaches the capillaries. There is, therefore, little heat transfer due to differences in temperature between the capillary blood and tissue. Second, the pervasiveness of capillaries in tissue allows blood to flow in several directions at once. Often, the flows in different

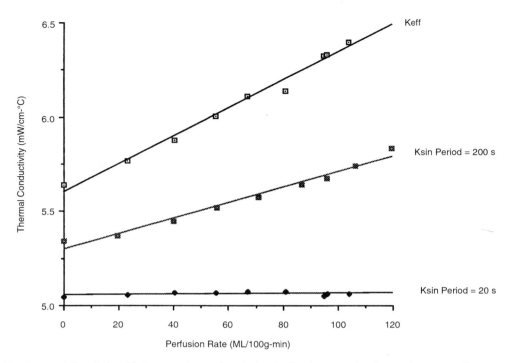

FIGURE 4.10   The relationship between k_{sin} and perfusion for heating periods of 20 and 200 seconds. Data was taken with a Thermometric P60 thermistor.

directions cancel out, leaving little net flow of blood in a given direction. As a result, heat transfer in a perfused volume of tissue is dominated by the largest blood vessels in that volume. Often, these are small arteries and arterioles that are 30–200 microns in diameter. Measurements of tissue intrinsic conductivity require a heating period fast enough to be unaffected by blood flowing in these small arteries.

The lower limit of the sinusoidal heating period is determined by the thermal diffusivity of the tissue. As the heating period decreases, so does the volume of tissue heated by the thermistor. The measured conductivity in the resulting small measurement volume can be strongly influenced by the traumatized tissue immediately surrounding the thermistor. This is illustrated in Fig. 4.10, which shows considerable difference in unperfused tissue ($\omega_b = 0$) between k_{sin} with a heating period of 20 seconds, k_{sin} with a heating period of 40 seconds and k_{eff}. When the thermistor was inserted into the kidney cortex, it squeezed the tissue immediately surrounding it. This had the effect of lowering the thermal conductivity of the compressed tissue. The measurement with the 20 second heating period has the smallest measurement volume and thus is most influenced by the compressed tissue. The steady-state k_{eff} measurement has the largest measurement volume and is least influenced by traumatized tissue. All three measurements were taken with a 0.06 inch diameter thermistor. Smaller thermistors will cause less tissue trauma, but may also produce less consistent measurements.

Other Heat Dilution Methods

Two methods that employ coarser approximations than those employed in the bioheat transfer equation have been used for many years to estimate tissue perfusion. One method follows the Fick relation, which assumes that heat transfer in blood and tissue by conduction may be neglected. In this case, the rate of change of internal energy in response to a bolus or steady infusion (or removal) of heat is balanced by convective transport of heat by the tissue blood supply and local heat generation rate.

As a matter of interest, the simple washout model was originally developed by Gibbs[39] some 60 years ago. Many approximate methods for estimating perfusion by heat clearance have been developed based on this analysis. A thermoelectric probe, supplied with a constant amount of heat, was inserted in a tissue through which blood was flowing. The temperature sensed by the probe varied inversely with the flow of blood. That is, an increased flow of blood would cool the probe, and a decreased flow heated it up. By correlating the probe temperature with the blood flow, an index of the blood flow rate was obtained. Using this technique, measurements were taken in the cat kidney cortex and muscle, respectively.

Perl and Hirsch[9] improved the empirical technique reported by Stow and Scheive.[40] The probe employed in their study consisted of a hypodermic needle with a thermistor placed near the tip. Heat was injected continuously into a small volume of tissue, and the temperature difference between heated and unheated tissue regions was monitored. The blood flow to the tissue was suddenly stopped for a few seconds. A sudden change in the rate of perfusion elicited a discontinuous change in the temporal derivative of temperature, which served to approximate conductive heat flux. A relationship was then produced between blood perfusion rate and the rate of change of internal energy. This measuring technique was tested on the dog and rabbit kidney, and satisfactory results were found at low blood flow. Difficulties in providing a total occlusion of the blood supply and other errors inherent in approximating heat diffusion by the clamping maneuvers limited the successful extension of this method.

Cameron[41] provided a heat clearance method for measuring regional cerebral cortical blood flow, involving the injection of a small amount of cool saline into the cerebral circulation. Blood flow was estimated from the slope of the subsequent thermal clearance curve, with temperature being measured by a small thermistor probe situated under the dura in contact with cerebral cortical tissue. The accuracy of the simple washout method is of concern because the negligence of heat conduction causes an overestimation of perfusion, especially with low blood flow rates. In other words, reasonable accuracy can be achieved only when the convective effect is much stronger than that of the conduction in the measured tissue region.

Another approximation method is based on the measurement of effective thermal conductivity, which has recently been developed to a higher degree.[11,16,42–46] The principle of this method is to measure the temperature difference which is established between a heated and an unheated probe. A thermistor embedded in tissue is heated for a very short time and the temperature difference between sensors located a known distance apart is measured. The system is modeled as a steady, spherically symmetric heat source conducting into an infinite thermal mass. The resulting equation (the well known heat conduction equation) is solved and the effective thermal conductivity includes the intrinsic thermal conductivity (without perfusion) and a convective heat transport contribution. With a no-flow calibration under otherwise identical conditions, the ratio of effective to intrinsic conductivity can then be correlated with the perfusion rate. Similar techniques have been used to measure blood perfusion in brains, kidneys, spleens, and thyroids.[47–50] However, the drawback of this method is its requirement of a no-flow calibration, since it is difficult to fully stop the blood supply to most organs.

4.3 Indicator Techniques

In addition to heat clearance, other indicator dilution techniques have also been developed and used widely for the estimation of tissue blood perfusion.[51–54] The general method follows from the Fick relation, in which the net transport of any indicator into a region of interest is balanced by the transient buildup or removal of the indicator. These techniques have been developed based on the following principal assumptions: (1) when the indicator is injected into a tissue, the blood flow system is assumed to be in a steady state; (2) indicator diffusion equilibrium between tissue and capillary blood is reached before sample collection; (3) the tissue of interest is isotropic; and (4) recirculation of the indicator can be neglected.

Indicators that have been used for organ blood flow analysis include injected radioactive and other diffusible inert gases,[55] and a locally generated hydrogen ion.[56] Perfusion measurements with diffusible tracers are based on conservation of mass as described by Fick's Law.[51,57] The basic idea is that the amount of tracer taken up by tissue equals the quantity supplied by the arteries minus the amount leaving through the veins. Blood flow is normalized by a unit of organ mass (usually 100 grams) to get perfusion rate. It is typically assumed that the venous concentration of tracer is proportional to the tissue concentration:

$$C_0 / C_v = \lambda \qquad (4.55)$$

where C_o is the tissue concentration of the tracer, and C_v is the venous concentration. The change in tissue tracer concentration is then described by:

$$dC_0(t)/dt = \omega_b \left[C_a(t) - C_0(t)/\lambda \right] \qquad (4.56)$$

where ω_b is the perfusion rate (mL/100 g/min). Perfusion is determined from the slope of the tissue concentration washin or washout curve. In certain cases, an integral form of Eq. 4.56 can be used. Difficulties in applying this technique include accurate determination of tracer concentration in tissue, controlling for nonuniform inflow concentrations, recirculation of the tracer during measurements, and nonuniform partitioning of the tracer.

In trapped indicator methods, radio-labeled indicators, fluorescent microspheres, or aggregated albumin are distributed to a capillary bed in proportion to the local flow rate.[58–60] These nondiffusible tracers remain inside the capillary walls and do not enter the tissue. Blood flow can be derived from classical indicator-dilution theory,[61,62] which states that flow in a tube can be calculated by knowing the mass of an injected indicator, its average concentration, and its time course. Perfusion is determined by measuring the concentration of the indicator in a given volume of tissue. In addition to the general assumptions for the indicator techniques stated earlier, this model assumes: (1) complete mixing of the indicator; (2) the volume of the indicator is negligible; (3) there is no extravascular loss of indicator; and (4) the contrast density can be accurately measured. Considering that outflow of the indicator from the tissue is much slower than inflow, perfusion is proportional to the ratio of peak concentration of the indicator in tissue to the area of the indicator clearance curve in the supply artery:

$$\omega_b \propto \frac{\Delta C_t(t_m)}{\int_0^{t_f} C_a(t)\,dt} \qquad (4.57)$$

where $\Delta C_t(t_m)$ is the maximum change in tissue concentration, t_f is the time when all the indicator has cleared the organ, and C_a is the arterial input concentration of the indicator. Alternately, perfusion can be calculated by determining the volume distribution of the tracer (V_d) and its mean transit time (MTT), the average time it takes a molecule of tracer to traverse the tissue:[57,62]

$$\omega_b = V_d / MTT \qquad (4.58)$$

This second method requires knowledge of the tracer distribution volume and the shape of the bolus of tracer injected into the tissue. Both these quantities are difficult to determine, making the use of this technique problematic.

4.4   Imaging Techniques

Imaging methods show great promise for one day being able to provide absolute quantitative information on organ perfusion in a clinical setting. Advances in equipment speed and resolution make this a rapidly advancing field. At present, several imaging methods allow for semi-quantitative evaluation of perfusion. Typically, a tracer that can be monitored is injected into the blood and its concentration over time in the tissue of interest is recorded. The tissue is usually divided into discrete regions of interest (ROI), with perfusion in each ROI assumed to be uniform. Difficulties in evaluating the exact concentrations of the tracer lead to a myriad of different algorithms for extracting perfusion. Often, realistic perfusion rates can be derived only with the use of empirically obtained constants or correction factors, making the measurements semi-quantitative. Problems common to many imaging methods include motion artifacts associated with increasing blood volume with increasing perfusion rate, and nonlinearities in the relationship of tracer concentration and perfusion.

Techniques currently being used or examined for perfusion measurements include Magnetic Resonance Imaging (MRI), ultrafast computed tomography (CINE), Single Photon Emission Computed Tomography (SPECT), and Positron Emission Tomography (PET). The majority of studies in the literature focus on measurements of perfusion in the myocardium or brain, with fewer reports concentrating on tissue blood flow in the lungs, liver, pancreas, skeletal muscle, or kidney. Despite the many hardware options for obtaining images, there are only two different general approaches to measuring perfusion using imaging methods. Both approaches involve the injection and monitoring of tracers in the blood. Diffusible tracers pass through the capillary walls and are exchanged with tissue, while nondiffusible tracers remain inside the blood vessels at all times.

Single Photon Emission Computed Tomography (SPECT)

SPECT has been used for perfusion studies in the myocardium and brain.[63–68] Numerous diffusible radiotracers are used in SPECT perfusion studies, including ^{201}Tl, ^{18}FDG, and ^{99m}technetium-labeled tracers such as ^{99m}Tc-MIBI, ^{99m}Tc-sestamibi, and ^{99m}Tc-teboroxime for evaluating coronary artery disease. Brain investigations have utilized ^{99m}Tc-HMPAO, ^{99m}Tc-ECD, and ^{123}I-IMP. These radioactive tracers are injected into a vein, and perfusion is calculated from the time course of tissue activity counts of radiation from the tissue of interest compared to counts from a reference area, such as the aorta. The sensitivity of SPECT is limited by image artifacts caused by uneven attenuation of the signal in different tissues. Often, a correction factor and empirical linearization constants are used to determine absolute perfusion rates, making this a semi-quantitative technique. The advantage of SPECT is its relatively low cost and the wide availability of both the equipment and tracers needed to carry out perfusion studies.

Positron Emission Tomography (PET)

Like SPECT, PET uses radiolabeled diffusible tracers to measure perfusion.[65,69–71] Typical tracers include ^{15}O-water, ^{82}Rb, and ^{13}N-ammonia. Tissue activity counts are collected for the tissue of interest and a convenient artery. Perfusion is then numerically calculated, taking into account the effects of metabolism on the tracer. PET is currently the imaging method that provides the most reliable quantitative data on perfusion. PET is several times more sensitive to radiotracers than SPECT, has a higher spatial resolution, and is not as susceptible to attenuation artifacts. A wide range of perfusion rates (2 to > 400 mL/100 g/min) have been reportedly measured with PET.[65,71] Its major disadvantage is its high cost and the limited number of institutions that have access to PET hardware.

Ultrafast Computed Tomography (CINE)

CINE uses a nondiffusible iodinated contrast agent to measure perfusion in the heart.[61,62,72–74] The indicator is rapidly injected intravenously, and the exponential time course of the agent is monitored in the myocardium and aorta or left ventricle. Perfusion can be estimated from the ratio of the peak tissue concentration to the arterial output curve. CINE has been used to assess relative myocardial perfusion in humans, but has several limitations. CINE has been shown to underestimate perfusion, especially at high flow rates.[74] CT scanners are prone to image artifacts that produce nonlinearities in CT count measurements.[62] Also, high iodine concentrations can introduce nonlinearities through beam hardening effects.

Magnetic Resonance Imaging

Magnetic resonance imaging is still an experimental method of measuring perfusion, but rapid improvements in MRI equipment make it a method that shows good promise.[57,75-79] Both diffusible and nondiffusible tracers can be used. With diffusible tracers (e.g., $CH^{19}F_3$, D_2O, and ^{17}O-water), a bolus of indicator is rapidly injected into a vein or right ventricle. Arterial input functions are assessed with MRI or other methods, and washout curves in tissue are measured by measuring either pixel intensities or integrated spectral line areas. A commonly used paramagnetic contrast agent is gadolinium diethylenetriamine pentacetic acid (Gd-DTPA). Gd-DTPA is freely diffusible in most tissue, but can be considered nondiffusible in the brain. Perfusion in the brain and heart has been reported using the mean transit time method with Gd-DTPA.[57,76,77] An interesting technique for measuring perfusion in the brain uses the water in blood as a contrast agent. Protons are imparted with saturated spins in the neck region, and their flow through the brain is mapped. Present difficulties with MRI methods include low signal to noise ratios and/or relatively slow imaging times. The introduction of scanners with stronger magnets and more stable hardware is expected to alleviate both these problems.

4.5 Doppler Techniques

Doppler techniques are used to measure blood flow rates based on the shift in frequency of the reflected ultrasound (or laser) caused by moving red cells. If the applied signal has a frequency of f_0 and the detected wave frequency is f, then the change in frequency when the source and detector approach each other with velocity v is,

$$f - f_0 = f_0 \frac{v}{\left(v_s - v\right)} \qquad (4.59)$$

When $v_s \gg v$ and the detector is stationary, the frequency shift is doubled,

$$f - f_0 = \pm 2f_0 \frac{v\cos\theta}{v_s} \qquad (4.60)$$

where $\pm$ is for the red cell moving toward or away from the detector, and θ is the incident angle of the beam.

Ultrasound Doppler

Ultrasound is a term for sound waves with a frequency above the audible range. For diagnostic use, frequencies between 1 and 10 MHz are applied. All Doppler ultrasound systems may be grouped in general as either pulsed wave (PW) or continuous wave (CW). PW systems use a single transducer for transmission and reception and define the measurement volume by timing the echo signals. CW systems, commonly used for peripheral circulation measurements, employ separate transducers for transmission and reception. In these systems, the measurement volume is defined by the intersection of the two transducers' directivity functions.

Ultrasound Doppler has been widely used for non-invasive measurement of the velocity of blood within the body in fields such as cardiology,[80-83] cerebral circulation,[84] obstetrics,[85-98] skin circulation,[99-101] and tumor circulation.[102] However, systems for applications to capillary blood flow estimation are still under development. This is because the frequency shift from the relatively low flow velocity in capillary beds is quite small, and the echoes from the red cells are difficult to separate from those of the surrounding tissue. To overcome these difficulties, a system has been designed based on the modified conventional CW Doppler ultrasound by adding an extra frequency converter, a coherent demodulation, and a spectrum analyzer with fast Fourier transform (FFT).[103] This system has been developed by employing a heterodyne demodulation technique to avoid the attenuation of the low Doppler shift through any filtering.[104] A high frequency pulsed ultrasonic Doppler system was also proposed for tissue perfusion measurement.[105]

The inherent visual appeal of color Doppler flow imaging has caused this field to develop rapidly as a new generation of sonographic imaging devices capable of displaying regional venous flow[106] has emerged. In peripheral vascular imaging, an angle must be created between the face of the linear transducer and blood flowing vessels that are parallel to the skin surface. Color assignment is dependent on the frequency shift determined by both the velocity of the moving red cells and the angle of insonation between the ultrasound beam and the blood stream. In general, a light color hue is used to encode greater frequency shift. Flow rate sensitivity can be varied by changing the pulse repetition rate for the recorded Doppler signal. When the sampling rate is decreased, the system becomes more sensitive to slower flow. Although the image provided by the color Doppler system appeals to the user, this technique for flow measurement is limited to a qualitative rather than quantitative study due to the difficulties involved with calibration and color resolution of the system.

Laser and Optical Doppler

Laser Doppler velocimetry (LDV) has been commonly used for skin perfusion measurements.[107–112] The LDV instrument operates on the same Doppler principle as that used for ultrasound devices, except that light is substituted for sound. The laser light (i.e., from a 5 mW He-Ne @ $f_s = 4.75 \times 10^{14}$ Hz) is transmitted to the skin through a quartz optical fiber. The light is then backscattered from both stationary skin components and moving objects, primarily erythrocytes in capillaries. The backscattered light consisting of mixed frequency waves is transmitted to a photodiode through another optical fiber. The output signal from the photodiode is composed of a spectrum of Doppler-shifted frequencies from which the blood flow rate can be obtained by taking the normalized root mean square bandwidth of the signals. However, this technique has limited application due to the lack of depth of penetration of the laser beam.

An optical Doppler technique using noncoherent light generated by optical arrangements[113] is an alternative to the laser Doppler method for velocity determinations in small vessels. To implement an optical Doppler system, light and dark patterns must be superimposed on the real image of the vessel to be studied before the light focuses on each sensor. Low-capacitance photodiodes are used as optimal sensors to minimize noise and to allow nonelectronic subtraction to complete the differential aspect of the gratings. After optimizing noise figures by bandpass filtering signals before frequency conversion, average blood cell velocities are calculated from the resulted frequency. This method has its limitation in requiring a transparent preparation to allow transmission and reflection of light from the object, in addition to the limited penetration depth also apparent in the laser Doppler method.

4.6   Plethysmography Techniques

The mercury-in-rubber strain gauge has also been used as a means of measuring limb blood flow.[114–116] The technique is based on measurement of changes in limb volume due to blood flow. Calibration of the strain gauge plethysmography can be accomplished either mechanically stretching the gauge or by electrical calibration. The former requires some necessarily bulky apparatus and suffers from inaccuracy due to limb movement. The latter measures the change in resistance of a mercury-in-rubber strain gauge, which is related to its length, and thus the circumference of the limb. Changes in limb volume are therefore directly proportional to changes in resistance, which can be measured by constructing the strain gauge as one arm of a Whetstone bridge.

Limb blood flow can be measured by applying an occlusive cuff and a distal strain gauge on the limb. The cuff is rapidly inflated to a pressure which is sufficient to stop venous return from the limb while still allowing arterial inflow. The output of the plethysmography is recorded and the initial slope of this volume is used to calculate flow rate into the limb. Percentage change in limb volume per unit of time is usually expressed as blood flow per tissue in $cm^3/100 \ cm^3/min$. With this technique, the accuracy of measurements can be influenced by the sensitivity of the instrument and the electrical calibration. The pulsatility of the measured curves caused by systolic and diastolic cycles can make the slope greater or lower than the proper value.

Infrared photoplethysmography also allows a non-invasive monitoring of the cutaneous microcirculation of the finger according to changes of reflected radiation intensity.[117] However, the flow rate can only be examined qualitatively and is affected by the limb position and the degree of venous filling.

Nomenclature

a	thermistor bead diameter
C	tracer concentration
CINE	ultrafast computed tomography
CT	computed tomography
CW	continuous wave
c	specific heat
FFT	fast Fourier transform
f	frequency
I	current
k	thermal conductivity
LDV	laser Doppler velocimetry
MIT	measuring interval of time
MRI	magnetic resonance imaging
MTT	mean transit time
P	volumetric supply power
PET	positron emission tomography
PW	pulsed wave
q	volumetric heat rate
R	resistance
ROI	region of interest
SPECT	single photon emission computed tomography
T	temperature
TPD	thermal pulse decay
t	time
V	volume distribution of tracer
v	velocity

Greek

α	thermal diffusivity
β	coefficient of the thermistor bead or power
ϵ	the smallest measurable value
θ	temperature elevation from the steady state
λ	ratio of the tracer concentrations of tissue and blood
ρ	density
ω	perfusion rate

Subscript and Superscript

a	arterial
b	bead or blood
bl	blood
d	distribution
m	modified
mb	metabolic

p	pulse
s	surface or sound
ss	steady state
t	tissue

References

1. Hyman, C. and Paldino, R.L., *Circ. Res.*, X, 89, 1962.
2. Sekins, K.M. et al., *Arch. Phys. Med. Rehabil.*, 65, 1, 1984.
3. Xu, L.X. et al., *Ann. Biomed. Eng.*, 24, S20, 1996.
4. Chen, M.M. and Holmes, K.R., *Ann. N.Y. Acad. Sci.*, 335, 137, 1980.
5. Chato, J.C., *J. Biomech. Eng.*, 102, 110, 1980.
6. Weinbaum, S. and Jiji, L.M., *J. Biomech. Eng.*, 107, 131, 1985.
7. Charny, C.K., *Advances in Heat Transfer*, 22, Academic Press, San Diego, CA, 1992, 19.
8. Pennes, H.H., *J. Appl. Physiol.*, 1, 93, 1948.
9. Perl, W. and Hirsch, R.L., *J. Theor. Biol.*, 10, 251, 1966.
10. McCaffery, T.V. and McCook, R.D., *J. Appl. Physiol.*, 39, 170, 1975.
11. Johnson, W.R. et al., *ASME J. Biomech. Eng.*, 101, 58, 1979.
12. Chen, M.M., et al., *J. Biomech. Eng.*, 103, 253, 1981.
13. Bowman, H.F., *N.Y. Acad. Sci.*, 335, 155, 1980.
14. Eberhart, R.C., Shitzer, A., and Hernandez, E.J., *Ann. N.Y. Acad. Sci.*, 335, 107, 1980.
15. Xu, L.X. et al., *Advances in Bioengineering*, HTD 189, BED 18, ASME, New York, 1991, 15.
16. Balasubramanium, T.A. and Bowman, H.F., *J. Heat Transfer*, 296, 1974
17. Valvano, J.W. et al., *J. Biomech. Eng.*, 106, 192, 1984.
18. Anderson, G.T. and Valvano, J.W., *J. Biomech. Eng.*, 116, 71, 1994.
19. Baish, J.W., *J. Biomech. Engl.*, 116, 521, 1994.
20. Arkin, H. et al., *J. Biomech. Eng.*, 108, 54, 1986a.
21. Arkin, H. et al., *J. Biomech. Eng.*, 108, 208, 1986b.
22. Arkin, H. et al., *J. Biomech. Eng.*, 108, 306, 1986c.
23. Arkin, H. et al., *J. Biomech. Eng.*, 9, 38, 1987a.
24. Arkin, H. et al., *J. Biomech. Eng.*, 109, 346, 1987b.
25. Arkin, H. et al., *J. Biomech. Eng.*, 111, 276, 1989.
26. Chato, J.C., *Symposium on Thermal Problems in Biotechnology*, ASME, New York, 1968, 16.
27. Jain, R.K., *J. Biomech. Eng.*, 101, 82, 1979.
28. Hayes, L.J. and Valvano, J.W., *J. Biomech. Eng.*, 107, 77, 1985.
29. Valvano, J.W., Badeau, A.F., and Pearce, J.A., *Heat Transfer in Bioengineering and Medicine*, HTD 2995, BED 7, ASME, New York, 1987, 31.
30. Anderson, G.T., Valvano, J.W., and Santos, R.R., *IEEE Trans. Biomed. Eng.*, 39, 877, 1992.
31. Holmes, K.R. and Chen, M.M., *Advances in Bioengineering*, ASME, New York, 1984, 9.
32. Balasubramanium, T.A. and Bowman, H.F., *J. Biomech. Eng.*, 99, 148, 1977.
33. Mason, R.L., Gunst, R.F., and Hess, J.L., *Statistical Design and Analysis of Experiments*, John Wiley & Sons, New York, 1989, 115.
34. White, C.H., Master's Thesis, University of Texas at Austin, Austin, TX, 1989.
35. Anderson, G.T. et al., *Computational Methods in Bioengineering*, BED 9, ASME, New York, 1988, 301.
36. Patel, P.A. et al., *J. Biomech. Eng.*, 109, 330, 1987.
37. Bowman, H.F., in *Heat Transfer in Medicine and Biology*, Shitzer and Eberhart, Eds., Plenum Press, New York, 1985, 193.
38. Valvano, J.W. et al., *Int. J. Thermophys.*, 6, 301, 1985.
39. Gibbs, F.A., *Proc. Soc. Exp. Biol. Med.*, 31, 141, 1933.
40. Stow, R.W. and Scheive, J.F., *J. Appl. Physiol.*, 14, 215, 1959.

41. Cameron, B.D., *Phys. Med. Biol.*, 15, 715, 1970.
42. Grayson, J., *J. Physiol.*, 118, 54, 1952.
43. Hokanson, D.E. et al., *IEEE Trans. Biomed. Eng.*, BME 22(1), 21, 1975.
44. Hensel, H., *Ber. Physiol.*, 162, 360, 1954.
45. Betz, E. and Benzing, H., *Dtsch. Ges. Kreislaufforsch*, 28, 437, 1963.
46. Muller-Schauenburg, W. et al., *Basic Res. Cardiol.*, 70, 547, 1975.
47. Linzell, J.L., *J. Physiol.*, 121, 390, 1953.
48. Mowbray, J.F., *J. Appl. Physiol.*, 14, 647, 1959.
49. Grangsjo, G. et al., *Acta Physiol. Scand.*, 66, 366, 1966.
50. Leatherman, N.E. and Bean, J.W., *J. Appl. Physiol.*, 23(4), 585, 1967.
51. Kety, S.S., *Keio J. Med.*, 43, 9, 1994.
52. Wood, E.H., *Circ. Res.*, 10, 357, 1961.
53. Bassingthwaighte, J.B., *Science*, 167(3923), 1347, 1970.
54. Paradise, N.F. and Fox, I.J., *Univ. Park Press, Baltimore*, 335, 1974.
55. Lassen, N.A., *Acta Med. Scandi. Suppl.*, 472, 136, 1967.
56. Koyama, T. *Univ. Park Press, Baltimore*, 10, 525, 1975.
57. LeBihan, D., *Invest. Radiol.*, 27, S6, 1992.
58. Rudolph, A.M. and Heymann, M.A., *Circ. Res.*, 21, 163, 1967.
59. Buckberg, G.D. et al., *J. Appl. Physiol.*, 31(4), 598, 1971.
60. Austin, R.E. et al., *Am. Physiol. Soc.*, H280, 1989.
61. Rumberger, J.A. et al., *J. Am. Coll. Cardiol.*, 9, 59, 1987.
62. Gould, R.G., *Invest. Radiol.*, 27, S18, 1992.
63. Matsuda, H. et al., *Eur. J. Nucl. Med.*, 19, 195, 1992.
64. Matsuda, H. et al., *Eur. J. Nucl. Med.*, 22, 633, 1995.
65. Schaiger, M., *J. Nucl. Med.*, 35, 693, 1994.
66. Saha, G.B., MacIntyre, W.J., and Go, R.T., *Semin. Nucl. Med.*, 24, 324, 1994.
67. Svrota, A. and Jehenson, P., *Eur. J. Nucl. Med.*, 18, 897, 1991.
68. Delbeke, D. et al., *J. Nucl. Med.*, 36, 2110, 1995.
69. Schuster, D.P. et al., *J. Nucl. Med.*, 36, 371, 1995.
70. Hoffman, J.M. and Coleman, R.E., *Invest. Radiol.*, 27, S22, 1992.
71. Keenan, G.F. et al., *Spine*, 20, 408, 1995.
72. Rumberger, J.A. and Bell, M.R., *Invest. Radiol.*, 27, S40, 1992.
73. Weiss, R.M. et al., *J. Am. Coll. Cardiol.*, 23, 1186, 1994.
74. Ludman, P.F. et al., *Am. J. Card. Imaging*, 7, 267, 1993.
75. Pickens, D.R., *Invest. Radiol.*, 27, S12, 1992.
76. Manning et al., *J. Am. Coll. Cardiol.*, 18, 959, 1991.
77. Keijer, J.T. et al., *Am. Heart J.*, 130, 893, 1995.
78. Patel, P.P., Koppenhafer, S.L., and Scholz, T.D., *Magn. Resonance Imaging*, 13, 799, 1995.
79. Muhler, A., *MAGMA*, 3, 21, 1995.
80. Angelsen, B.A.J. and Brubakk, A.O., *Cardiovasc. Res.*, 10, 368, 1960.
81. Darsee, J.R. et al., *Cardiology*, 46, 613, 1980.
82. Rifkin, M.D. et al., *Ultrasound Med. Biol.*, 11, 341, 1985.
83. Schrope, B. et al., 6th Ann. Symp. Adv. Echocard., Aug. 21-28, 1991.
84. Brar, H.S. et al., *J. Ultrasound Med.*, 8, 187, 1989.
85. Assali, N.S. et al., *Am. J. Obstet. Gynecol.*, 79, 86, 1960.
86. Fitzgerald, D.E. and Drumm, J.E., *Br. Med. J.*, 2, 1450, 1977.
87. Eil-Nes, S.H., *J. Biomed. Eng.*, 4, 28, 1982.
88. Kurjak, A. and Rajhvajn, B., *J. Perinat. Med.*, 10, 3, 1982.
89. Backe, B. et al., *Ultrasound Med. Biol.*, 9, 587, 1983.
90. Eldridge, M.W. and Berman, W., *Serial Measurement of Human Fetal Aortic Blood Flow*, Kisco, New York, 1983, 211.

91. Griffin, D. et al., *Clinics Obstet. Gynecol.*, 10, 565, 1983.
92. Gill, R.W. et al., *Ultrasound Med. Biol.*, 10, 349, 1984.
93. Marsal, K. et al., *Ultrasound Med. Biol.*, 10, 339, 1984.
94. Van Lierde, M. et al., *Obstet. Gynecol.*, 6, 801, 1984.
95. Wladimiroff, J.W. et al., *Recent Advances in Ultrasound Diagnosis 4*, Exerpta Medica, Amsterdam, 1984, 63.
96. Erskine, R.L.A. and Ritchie, J.W.K., *Br. J. Obstet. Gynaecol.*, 92, 600, 1985.
97. Teague, M.J. et al., *Ultrasound Med. Biol.*, 11, 27, 1985.
98. Pearce, J.M. et al., *Br. J. Obstet. Gynaecol.*, 95, 248, 1988.
99. Satumora, S., *J. Acoust. Soc. Am.*, 29, 1181, 1958.
100. Franklin, D.L. et al., *Science*, 132, 564, 1961.
101. Fronek, A. et al., *Am. J. Surg.*, 126, 205, 1973.
102. Golberg, B.B. et al., *Radiology*, 177, 713, 1990.
103. Burns, S.M. and Reid, M.H., *Biomat. Art. Cells Art. Org.*, 17, 61, 1989.
104. Ting, T.H. et al., *Ultrasonics*, 30(4), 225, 1992.
105. Basler, S. et al., *Adv. Exp. Med. Biol.*, 220, 223, 1987.
106. Foley, W.D. and Erickson, S.J., *AJR*, 156, 3, 1991.
107. Nilsson, G.E. et al., *IEEE Trans. Biomed. Eng.*, 27, 597, 1980.
108. Enkema, L. et al., *Clin. Chem.*, 27(3), 391, 1981.
109. Fischer, J.C. et al., *Microsurgery*, 4, 164, 1983.
110. Tenland, T. et al., *Int. J. Microcirc. Clin. Exp.*, 2, 81, 1983.
111. Tur, E. et al., *J. Invest. Derm.*, 81, 442, 1983.
112. Boggett, D. et al., *J. Biomed. Eng.*, 7, 225, 1985.
113. Borders, J.L. and Granger, H.J., *Microvas. Res.*, 27, 117, 1984.
114. Whitney, R.J., *Proc. Phys. Soc.*, 5, 1949.
115. Mason, D.T. and Braunwald, E., *Am. Heart J.*, 64, 796, 1962.
116. Brakkee, A.J.M. and Vendrik, A.J.H., *J. Appl. Physiol.*, 21(2), 701, 1966.
117. Hokanson, D. E. et al., *IEEE Transc. Biomed. Eng.*, BME–22(1), 21–25, 1975.

5

Finite Element Models for Arterial Wall Mechanics and Transport

5.1   Abstract **5**-1
5.2   Introduction **5**-2
5.3   PHETS Theoretical Basis **5**-3
      Fundamental Fields and Preliminary Kinematics •
      Phenomenological Equations • Eularian PHETS Initial
      Boundary Value Problem • Principles of Virtual Velocities •
      Lagrangian PHETS Initial Boundary Value Problem • Material
      Properties in PHETS Formulations • Equivalence of Poroelastic
      and Mixture-Based Biomechanical Models
5.4   Finite Element Models (FEMs) **5**-10
5.5   Application of FEMs to Arterial Structures....................**5**-12
      FEMs of Normal Arteries • FEMs of Diseased Arteries • FEMs
      of the Arterial Wall and CFD Models • Some FEMs of Other
      Arterial Structures
5.6   Conclusion **5**-24

B. R. Simon
The University of Arizona

M. V. Kaufmann
Silicon Spice, Inc.

5.1   Abstract

Arterial structural mechanics have been studied for almost two centuries. This chapter surveys the development and use of finite element models (FEMs) for the analysis of arterial structures, principally addressing the mechanics of the walls of large arteries. We consider representative applications of FEMs for the study of the mechanics of normal and diseased arteries. Special attention is given to the coupled interactions between structural response and transport phenomena in the arterial wall. Some applications of arterial wall FEMs in conjunction with computational fluid dynamics (CFD) are also introduced. Our presentation is based on a general theory that uses a porohyperelastic-transport-swelling (PHETS) model to account for coupled elastic response and the transport of mobile tissue fluid and dissolved species. A soft tissue structure is viewed as a fluid-saturated porous medium (continuum), i.e., a porous solid (fibrous matrix) in which an incompressible fluid (water) flows. The transport of mobile species (solute) in this fluid is accomplished by both diffusion and convection processes that are influenced by large deformations of the porous solid. Arterial FEMs are classified and described according to material law, e.g., we discuss hyperelastic (HE) and porohyperelastic (PHE) models of arteries as special cases of the

PHETS formulation. An equivalence is noted for PHETS models and multiphasic mixture-based models for soft tissues. FEMs developed using commercial programs, such as ABAQUS, and specialized finite element programs are also presented. Normal arterial response is considered with HE-FEMs, and an initial FEM of the tissue remodeling processes (evidenced as "opening angles") is described. Mobile tissue fluid motion in the arterial wall is simulated using PHE-FEMs subjected to quasi-static and cyclic loading conditions. Mobile species transport in the arterial wall (here considering a single uncharged species) is illustrated using the PHETS material law in FEMs of rabbit aortas. FEM results are compared to experimental data where both diffusion and convection of labeled albumin are significant. FEMs using HE theory are described for the study of diseased states associated with atherosclerosis and aneurysms. Combined HE-FEMs and CFD simulations are illustrated as an introduction to the complex fluid-structure interaction problems associated with vascular mechanics. In the future, PHETS models in conjunction with CFD models should provide fundamental quantitative information relating arterial blood flow, wall shear, endothelial damage, transport, and wall mechanics that may, in turn, lead to a better understanding of atherogenesis at specific sites in the arterial tree. Such FEMs and associated experimental studies will also be useful for the fundamental understanding of structural-transport and active tissue remodeling and ingrowth associated with the design of tissue-engineered vascular grafts. PHETS-FEMs will find application in the design and simulation of catheter-based local drug delivery systems where species transport in arterial tissues after angioplasty is of prime importance for the optimal administration of substances that may prevent recurrence of atherogenesis in the arterial wall.

5.2 Introduction

The mechanics of the arterial wall have been studied for nearly two hundred years, beginning with the early efforts of well known scientists such as Robert Hooke and Thomas Young. The structure of the wall of a large artery is complex and serves the primary function of carrying flowing blood from the heart to the microcirculation where the exchange of oxygen and other substances can be accomplished. A large artery is a thick tube composed of soft tissue organized in three layers — the intima, media, and adventitia. The intima is the thin innermost layer composed of endothelial cells, basal lamina, and a subendothelial layer of collagen, elastin, and smooth muscle. The media contains smooth muscle and elastic laminar networks of elastic fibrils and collagen. The adventitia is the outermost layer containing collagen fibers, ground substance, nerves, and blood vessels (vasa vasorum).

A large artery is subjected to pulsatile pressures as well as significant axial tethering forces, and exhibits visible deformation and a highly nonlinear material response. Life scientists and engineers have studied the arterial wall in order to obtain a basic understanding of its mechanics and physiology. Quantitative biomechanical structural analyses should be able to provide predictions of deformation and stresses in normal and diseased arteries. Such analyses should be useful in the investigation of the relation of wall strains and stresses to the observed active remodeling of the arterial wall. In addition, biomechanicians are attempting to understand the interrelationships among blood flow, arterial wall mechanics, fluid-species transport, and atherogenesis. Such information would also be of fundamental importance in the development of catheter-based local drug delivery systems, which introduce substances that are to be transported into the arterial wall after angioplasty. A quantitative description of the active response of arterial wall tissue to mechanical, biological, and chemical conditions is also necessary in order to design and develop modern tissue-engineered vascular grafts and other arterial prostheses.

The structural response of a large artery is characteristically complex and includes the nonlinear, history-dependent response of a nonhomogeneous anisotropic tube undergoing finite deformations. At the outset, the finite strain elasticity theory has provided analytical solutions for arterial mechanical studies. The artery is usually assumed to be an incompressible, hyperelastic material so that the deformation field can be determined analytically. Then the resulting strains and stresses can be calculated using the constitutive law. However, analytical solutions are only available for homogeneous materials and relatively simple boundary conditions and geometries. During the last thirty years, numerical finite element models (FEMs) have been developed for the detailed study of the arterial wall and other biological structures. Such FEMs can provide solutions for both compressible and incompressible materials in non-homogeneous biological structures that are complex in shape and exhibit highly nonlinear material response with finite strains. History-dependence and transport processes can also be included in the FEMs. Thus, the FEM has become the most popular method for structural analysis of soft arterial tissues. A quantitative knowledge of the material constitutive law is a key ingredient to successful finite element modeling of arterial structures. It is not our intent to describe the numerous studies of the material properties of the arterial wall. The reader is referred to the works of Hayashi[1] and Humphrey[2] for critical reviews of research in this area.

The scope of this chapter is limited to the consideration of FEMs of arterial structures that are based on elastic, poroelastic (or biphasic), and porohyperelastic-transport-swelling (or triphasic/multiphasic) theories. No intrinsically viscoelastic models will be considered. Mixture and poroelastic theories have been applied to model cardiac structures (e.g., Huyghe[3]). Both poroelastic[4,5] and biphasic models[6,7] have been developed for the study of soft tissues. The equivalence of poroelastic and biphasic (mixture) models was illustrated by Simon.[8] These models have been extended to the triphasic-mixture model[9] and the porohyperelastic-transport-swelling (PHETS) model.[10,11] An equivalence between PHETS models and multiphasic-mixture models for soft tissues can also be established. A brief summary is presented of the PHETS field theory that serves as the basis for classification of the FEMs described in this chapter. We consider hyperelastic (HE) and porohyperelastic (PHE) models as special cases of the PHETS theory. A PHE soft tissue is viewed as a fluid-saturated porous medium (continuum) composed of an incompressible HE porous solid (fibrous matrix). An incompressible mobile fluid (water) flows through this highly deformable porous solid. In the PHETS soft tissue model, mobile species are considered to be dissolved in this fluid. The experimental basis for the use of poroelastic models for arterial tissues has been discussed by a number of authors (e.g., Kenyon[12,13] and Jayaraman[14]). Chuong and Fung[15] have also described the effects of mobile tissue fluid motion in the arterial wall. The PHETS formulations and FEMs given in this chapter have been discussed by Simon[8] and Kaufmann et al.[11] and were developed fully by Kaufmann.[16] We begin with a brief development of a PHETS formulation that includes the effects of a single, uncharged species, leaving the description of PHETS models including electrical effects to other publications.

5.3 PHETS Theoretical Basis

Fundamental Fields and Preliminary Kinematics

We consider soft tissue to be a continuum composed of an incompressible, porous solid, an incompressible fluid, and one neutral species dissolved in the fluid. The theory is readily extended to multiple dissolved charged species. Unless noted otherwise, summation is implied only for repeated subscripts (i, j, k, l, etc.). The superscript η denotes all constituents, with $\eta = (s, f, c)$, i.e., the solid, fluid, and neutral species, whereas superscripts α, β, and γ denote (f, c). A particle of constituent η is located at X_i^η at time t_0. Its current position at time t is given as $x_i^\eta = x_i^\eta(X_j^\eta, t)$. At time t, the apparent domain of all constituents coincides with the domain of the solid so that $x_i^\eta = x_i^s = x_i$.

In this section we summarize a mixed PHETS formulation in which the fundamental fields are the displacement of the solid, u_i, the pore fluid stress, π^f, and the concentration of the mobile species, c. The displacement of the solid is

$$u_i = x_i - X_i = u_i^s = x_i^s - X_i^s \tag{5.1}$$

and the velocity of the solid is

$$\dot{u}_i = v_i^s = \dot{u}_i^s = \frac{du_i^s}{dt} \tag{5.2}$$

where a superposed dot denotes the material time derivative defined with respect to the solid, i.e.,

$$\dot{(\,)} \equiv \frac{d}{dt}(\,) \equiv \frac{\partial}{\partial t}(\,)\bigg|_{X_i} = \frac{\partial}{\partial t}(\,)\bigg|_{x_i} + v_k^s \frac{\partial}{\partial x_k}(\,) \tag{5.3}$$

The total displacement, u_i, will be represented without the superscript s, whereas the solid velocity, v_i^s, will be represented with the superscript s to distinguish it from the other constituent velocities. The volume strain, J, deformation gradients, F_{ij} and F_{ij}^{-1}, and Finger's strain tensor, H_{ij}, are

$$J = \frac{dV}{dV_0} = \det F_{ij}, \quad F_{ij} = \frac{\partial x_i}{\partial X_j}, \quad F_{ij}^{-1} = \frac{\partial X_i}{\partial x_j}, \quad H_{ij} = F_{im}^{-1}F_{jm}^{-1} \tag{5.4}$$

The apparent relative velocities of the fluid and the species are

$$v_i^{fr} = n\left(v_i^f - v_i^s\right), \quad v_i^{cr} = n\left(v_i^c - v_i^s\right) \tag{5.5}$$

where, for example, v_i^f is the average velocity of the fluid at x_i and t. Here, v_i^f is defined so that the fluid volume flow rate through a unit area perpendicular to the x_i-axis is nv_i^f. A similar view applies to v_i^c. The fluid fully saturates the pores in the solid skeleton, and the volume of the dissolved species is neglected. Then the porosity of the fluid-saturated material, n, is the ratio of the fluid volume, dV^f, to the total volume, dV, i.e.,

$$n = \frac{dV^f}{dV}, \quad dV = dV^s + dV^f \tag{5.6}$$

Assuming the solid material is incompressible, the porosity is given as

$$n = 1 - J^{-1}\left(1 - n_0\right) \tag{5.7}$$

where $n_0 = dV_0^f / dV_0$ is the porosity in the fluid-saturated reference configuration.

The concentration of the species dissolved in the fluid is defined as the mass of the species per unit of fluid volume,

$$c = \frac{dm^c}{dV^f} \tag{5.8}$$

The apparent density of the η constituent (mass per unit of total tissue volume) is

$$\rho^\eta = dm^\eta / dV = \phi^\eta / \left(dm^\eta / dV^\eta\right) = \phi^\eta \rho_T^\eta,$$

where $\rho_T^\eta = dm^\eta / dV^\eta$ is the true density and $\phi^\eta = dV^\eta / dV$ is the volume fraction of the η constituent per unit of total volume. The apparent density of the species (per unit of total tissue volume), $\rho^c = dm^c / dV$, is related to the concentration as $\rho^c = nc$.

Phenomenological Equations

The phenomenological equations relate electrochemical potentials to drag forces and provide a starting point for the theory. The general form of these equations (including electrical effects) can be obtained from entropy inequalities and expressed in terms of absolute velocities[9] or in terms of the relative fluid and species velocities.[16] Here, we specialize the equations of Kaufmann[16] for a single neutral species to the form

$$-n^2 \rho_T^f \frac{\partial \mu^f}{\partial x_i} + n^2 \frac{\partial \pi^f}{\partial x_i} = a_{ij}^{ff} v_j^{fr} - a_{ij}^{fc} v_j^{cr} \tag{5.9}$$

$$-n^2 c \frac{\partial \mu^c}{\partial x_i} = -a_{ij}^{cf} v_j^{fr} + a_{ij}^{cc} v_j^{cr} \tag{5.10}$$

where μ^f and μ^c are chemical potentials, $a_{ij}^{\alpha\beta}$ are drag coefficients, and the pore fluid stress π^f = (pore fluid pressure). We eliminate the relative species velocity in the formulation by solving Eq. 5.10 for v_i^{cr}, i.e.,

$$v_i^{cr} = -\hat{a}_{ij}^{cc} n^2 c \frac{\partial \mu^c}{\partial x_j} + \left(\hat{a}_{ij}^{cc} a_{jk}^{fc}\right) v_k^{fr} \tag{5.11}$$

where

$$\hat{a}_{ij}^{cc} = \left[a_{ij}^{cc}\right]^{-1} \tag{5.12}$$

The phenomenological equation in the form of Eqtation 5.11 can be used to develop either the Eulerian or Lagrangian PHETS formulations of the field equations.

Eulerian PHETS Initial Boundary Value Problem

The kinematic equations include strain and deformation measures and gradients in the fundamental fields. The strain can be characterized using the Eulerian strain tensor, E_{ij}. The spatial velocity gradient $\partial v_i^s / \partial x_j$ is split into the symmetric velocity strain or rate of deformation tensor, D_{ij}, and the anti-symmetric vorticity or spin tensor, W_{ij}. These tensors are

$$E_{ij}^* = \frac{1}{2}\left(\delta_{ij} - F_{ki}^{-1} F_{kj}^{-1}\right), \quad D_{ij} = \frac{1}{2}\left(\frac{\partial v_i^s}{\partial x_j} + \frac{\partial v_j^s}{\partial x_i}\right), \quad W_{ij} = \frac{1}{2}\left(\frac{\partial v_i^s}{\partial x_j} - \frac{\partial v_j^s}{\partial x_i}\right) \tag{5.13}$$

The pore fluid stress gradient, e_i^π, is

$$e_i^\pi = \frac{\partial \pi^f}{\partial x_i} \tag{5.14}$$

and the relative volumetric fluid strain rate is $D_{kk}^{fr} = \partial v_k^{fr} / \partial x_k$. The species concentration gradient, e_i^c, is

$$e_i^c = \frac{\partial c^c}{\partial x_i} \tag{5.15}$$

The conservation equations include conservation of momentum and mass. The quasi-static equations of motion (no body forces) in terms of total Cauchy stress, σ_{ij}, are given as

$$\frac{\partial \sigma_{ji}}{\partial x_j} = 0 \qquad (5.16)$$

Assuming that soft tissue is composed of an incompressible, porous solid material that is saturated by an incompressible fluid, the conservation of solid and fluid mass equation becomes

$$\frac{\partial v_j^{fr}}{\partial x_j} + D_{kk} = 0 \qquad (5.17)$$

The conservation of species mass equation is given in terms of concentration as

$$\frac{\partial}{\partial t}\left(nc\right) + \frac{\partial j_i^{cr}}{\partial x_i} + \frac{\partial}{\partial x_i}\left(ncv_i^s\right) = 0 \qquad (5.18)$$

where $j_i^{cr} = cv_i^{cr}$ is the relative species flux.

The constitutive equations include definitions of the effective stress, the generalized Darcy law, and the relative species flux. The total Cauchy stress is expressed in terms of effective stress, σ_{ij}^e, and pore fluid stress, π^f, as

$$\sigma_{ij} = \sigma_{ij}^e + \pi^f \delta_{ij}, \quad \sigma_{ij}^e = \sigma_{ij}^e\left(E_{kl}^*, c\right) \qquad (5.19)$$

where, again $\pi^f = -$(pore fluid pressure), which is indeterminate subject to the constraint of Eq. 5.17. Eq. 5.11, when substituted into Eq. 5.9, produces a generalized Darcy Law identifying pressure gradients (mechanical and chemical) that act to produce fluid velocity relative to the deforming solid, i.e.,

$$v_i^{fr} = k_{ij}\left(e_j^\pi + \frac{\partial \pi^c}{\partial x_j}\right) \qquad (5.20)$$

where

$$k_{ij} = n^2\left[a_{ij}^{ff} - a_{im}^{fc}\hat{a}_{mk}^{cc}a_{jk}^{cf}\right]^{-1} \qquad (5.21)$$

Here, k_{ij} is the hydraulic permeability and $\partial \pi^c/\partial x_j$ is an osmotic pressure gradient obtained in the generalized Darcy Law and has the form

$$\frac{\partial \pi^c}{\partial x_i} = g_{ij}^{cc}e_j^c + g_{imkl}^{E^*}\frac{\partial E_{kl}^*}{\partial x_m} \qquad (5.22)$$

When Eq. 5.11 for v_i^{cr} is used in the definition of the relative species flux, $j_i^{cr} = c^c v_i^{cr}$, then

$$j_i^{cr} = -d_{ij}^{cc}e_j^c - h_{ijkl}^c\frac{\partial E_{kl}^*}{\partial x_j} + b_{ij}^{cf}cv_j^{fr} \qquad (5.23)$$

The mechanical property functions appearing in Eqs. 5.22 and 5.23 are given below.

Principles of Virtual Velocities

Both the Eulerian and Lagrangian principles of virtual velocities (PVVs) are used as a basis for FEMs, as well as to identify the appropriate correspondence rules for stresses and relative velocities in the formulation of the PHETS field problem. The Eulerian PVVs in mixed form are

$$\int_V \delta D_{ij}\sigma^e_{ji}dV + \int_V \delta D_{kk}\pi^f dV - \int_A \delta v^s_i \sigma_{ij}\hat{n}_j dA = 0 \tag{5.24}$$

$$\int_V \delta\pi^f D_{kk}dV - \int_V \delta e^\pi_i k_{ij} e^\pi_j dV - \int_V \delta e^\pi_i k_{ij}\frac{\partial \pi^c}{\partial x_j}dV + \int_A \delta\pi^f v^{fr}_k \hat{n}_k dA = 0 \tag{5.25}$$

$$\int_V \delta\dot{c}\frac{\partial}{\partial t}(nc)dV - \int_V \delta\dot{e}^c_i j^{cr}_j dV + \int_V \delta\dot{c}\frac{\partial}{\partial x_i}(ncv^s_i)dV + \int_A \delta\dot{c}j^{cr}_i \hat{n}_i dA = 0 \tag{5.26}$$

where $\hat{n}_i$ is the unit normal to the area dA. Using appropriate correspondence rules, Eqs. 5.24–5.26 can be transformed into mixed Lagrangian PVVs of the form

$$\int_{V_0} \delta\dot{E}_{ij}S^e_{ji}dV_0 + \int_{V_0} \delta\dot{E}_{ij}H_{ij}\pi^f JdV_0 - \int_{A0} \delta\dot{u}_i T_{ji}\hat{n}_{0j}dA_0 = 0 \tag{5.27}$$

$$\int_{V_0} \delta\pi^f \dot{E}_{ij}H_{ij}JdV_0 - \int_{V_0} \delta\tilde{e}^\pi_i \tilde{k}_{ij}\tilde{e}^\pi_j dV_0 - \int_{V_0} \delta\tilde{e}^\pi_i \tilde{k}_{ij}\frac{\partial \pi^c}{\partial X_j}dV_0 + \int_{A_0} \delta\pi^f \tilde{v}^{fr}_k \hat{n}_{0k}dA_0 = 0 \tag{5.28}$$

$$\int_{V_0} \delta\dot{c}\frac{d}{dt}(Jnc)dV_0 + \int_{V_0} \delta\dot{\tilde{e}}^c_i \tilde{d}^{cc}_{ij}\tilde{e}^c_j dV_0 + \int_{V_0} \delta\dot{c}\tilde{b}^{cf}_{ij}\tilde{v}^{fr}_j \tilde{e}^c_i dV_0$$
$$+ \int_{V_0} \delta\dot{c}c\frac{\partial}{\partial X_i}(\tilde{b}^{cf}_{ij}\tilde{v}^{fr}_j)dV_0 + \int_{A_0} \delta\dot{c}(\tilde{j}^{cr}_i)_s \hat{n}_{0i}dA_0 = 0 \tag{5.29}$$

where $\hat{n}_{0i}$ is the unit normal to the area dA_0. Note that a specified surface flux, $(\tilde{j}^{cr}_i)_s$, has been introduced in Eq. 5.29 to avoid a spurious reflection of mass transfer in the FEM due to convection at the boundary (see the ABAQUS manual[17]), i.e.,

$$\left(\tilde{j}^{cr}_i\right)_s = \tilde{j}^{cr}_i - c\tilde{b}^{cf}_{ij}\tilde{v}^{fr}_j = -\tilde{d}^{cc}_{ij}\tilde{e}^c_j \tag{5.30}$$

Lagrangian PHETS Initial Boundary Value Problem

In the total Lagrangian formulation, all field variables are considered to be dependent on X_i and t, i.e., the displacement is $u_i = u_i(X_j, t) = x_i - X_i$, $\pi^f = \pi^f(X_j, t)$, $c = c(X_j, t)$, etc. The relations between field variables described in the Eulerian and Lagrangian PVVs allow the identification of correspondence rules. The classical correspondence rules[18] relate Cauchy stress to the first and second Piola-Kirchhoff stresses, which are defined below. The relative fluid and species velocities are also related by correspondence rules; e.g., consider the relative fluid velocity, v^{fr}_i, which will be referred, in the Lagrangian sense, to the reference configuration. We define a Lagrangian relative fluid velocity, $\tilde{v}^{fr}_i$, that corresponds to v^{fr}_i so that the same relative fluid mass flow rate occurs through the faces dA and dA_0, i.e., $\rho^f_T v^{fr}_i \hat{n}_i dA = \rho^f_{T0}\tilde{v}^{fr}_i \hat{n}_{0i}dA_0$.

A similar correspondence rule applies to relative species velocities, $\tilde{v}_i^{cr}$ and v_j^{cr}; i.e., with definitions of Eulerian and Lagrangian relative species flux given as $j_i^{cr} = cv_i^{cr}$ and $\tilde{j}_i^{cr} = c\tilde{v}_i^{cr}$, the correspondence is $j_i^{cr}\hat{n}_i dA = \tilde{j}_i^{cr}\hat{n}_{0i} dA_0$ (equal relative species flux through corresponding faces dA and dA_0). Then, using the Nanson formula[19] for an incompressible fluid ($\rho_T^f = \rho_{T0}^f$) with $c(x_i, t) = c(X_i, t)$, $\tilde{v}_i^{fr}$ corresponds to v_i^{fr} and $\tilde{v}^{cr}$ to v^{cr} as

$$\tilde{v}_i^{fr} = J\frac{\partial X_i}{\partial x_j} v_j^{fr}, \quad \tilde{v}_i^{cr} = J\frac{\partial X_i}{\partial x_j} v_j^{cr} \qquad 5.(31)$$

Lagrangian strain and deformation measures and gradients include Green's strain, E_{ij}, and Green's strain rate, $\dot{E}_{ij}$,

$$E_{ij} = \frac{1}{2}\left(F_{ki}F_{kj} - \delta_{ij}\right), \quad \dot{E}_{ij} = F_{ki}D_{km}F_{mj} \qquad (5.32)$$

and the pore fluid stress gradient, $\tilde{e}_i^{\pi}$,

$$\tilde{e}_i^{\pi} = \frac{\partial \pi^f}{\partial X_i} \qquad (5.33)$$

The Lagrangian relative volumetric fluid strain rate is $\tilde{D}_{kk}^{fr} = \partial\tilde{v}_k^{fr}/\partial X_k$, and the constraint that both solid and fluid materials are incompressible has the form $JH_{ij}\dot{E}_{ij} + \tilde{D}_{kk}^{fr} = 0$. The species concentration gradient, $\tilde{e}_i^c$, is

$$\tilde{e}_i^c = \frac{\partial c}{\partial X_i} \qquad (5.34)$$

The Lagrangian form for the quasi-static equations of motion (no body forces) is

$$\frac{\partial T_{ji}}{\partial X_j} = 0 \qquad (5.35)$$

where the first and second Piola-Kirchhoff stresses are $T_{ij} = JF_{jk}^{-1}\sigma_{ki}$ and $S_{ij} = JF_{im}^{-1}\sigma_{mn}F_{jn}^{-1}$, respectively. The Lagrangian form for conservation of solid and fluid mass becomes

$$\frac{\partial\tilde{v}_j^{fr}}{\partial X_j} + JH_{ij}\dot{E}_{ij} = 0 \qquad (5.36)$$

for the case where both solid and pore fluid materials are incompressible. The conservation of species mass has the Lagrangian form

$$\frac{d}{dt}\left(nJc\right) + \frac{\partial\tilde{j}_i^{cr}}{\partial X_i} = 0 \qquad (5.37)$$

The Lagrangian form of the constitutive equations includes the effective stress, the generalized Darcy law, and the relative species flux. The second Piola-Kirchhoff stress is expressed in terms of effective stress, S_{ij}^e, and pore fluid stress, π^f, as

$$S_{ij} = S_{ij}^e + J\pi^f H_{ij}, \quad S_{ij}^e = \frac{\partial U^e}{\partial E_{ij}}, \quad U^e = U^e\left(E_{ij}, c\right) \tag{5.38}$$

where π^f is indeterminate. The Lagrangian form of the generalized Darcy Law is

$$\tilde{v}_i^{fr} = \tilde{k}_{ij}\left(\tilde{e}_j^\pi + \frac{\partial \pi^c}{\partial X_j}\right) \tag{5.39}$$

where $\tilde{k}_{ij}$ is the Lagrangian hydraulic permeability and, again, $\tilde{e}_j^\pi = \partial \pi^f / \partial X_j$ is the gradient in pore fluid stress. The generalized osmotic pressure gradient, $\partial \pi^c / \partial X_i$, is

$$\frac{\partial \pi^c}{\partial X_i} = \tilde{g}_{im}^{cc}\tilde{e}_m^c + \tilde{g}_{imkl}^E \frac{\partial E_{kl}}{\partial X_m} \tag{5.40}$$

The Lagrangian form for relative species flux, defined as $\tilde{j}_i^{cr} \equiv c\tilde{v}_i^{cr}$, is

$$\tilde{j}_i^{cr} = -\tilde{d}_{ij}^{cc}\tilde{e}_j^c - \tilde{h}_{ijkl}^c \frac{\partial E_{kl}}{\partial X_j} + \tilde{b}_{ij}^{cf} c\tilde{v}_j^{fr} \tag{5.41}$$

We now have corresponding Eulerian and Lagrangian PHETS theories that provide partial differential equations to be solved for u_i, π^f, and c subject to boundary conditions and initial conditions.

Material Properties in the PHETS Formulations

Two corresponding sets of material properties have been identified — one associated with the Eulerian formulation and the other with the Lagrangian formulation. Direct mathematical relations exist between these material properties. By the principle of equipresence, the Eulerian properties are dependent on E_{ij}^* and c, whereas the Lagrangian properties depend on E_{ij} and c. The porous solid material is characterized by a generalized effective strain energy density function, U^e, which defines the effective stress as

$$\sigma_{ij}^e = J^{-1}F_{ik}S_{km}^e F_{jm}, \quad S_{ij}^e = \frac{\partial U^e}{\partial E_{ij}}, \quad U^e = U^e\left(E_{ij}, c\right) \tag{5.42}$$

Isotropic forms for the material properties are noted elsewhere in this chapter. The necessary Eulerian material properties are the effective stress, σ_{ij}^e; chemical potentials, μ^c and μ^f; coupling coefficients, l_{ij}^{cc}; convection coefficients, b_{ij}^{cf}; chemical swelling parameters, g_{ij}^{cc} and $g_{ijkl}^{E^*}$; and permeability, k_{ij}. The remaining material property functions obtained from these basic properties are

$$d_{ij}^{cc} = l_{ij}^{cc}\frac{\partial \mu^c}{\partial c}, \quad h_{ijkl}^c = l_{ij}^{cc}\frac{\partial \mu^c}{\partial E_{kl}^*} \tag{5.43}$$

where the diffusion coefficients are d_{ij}^{cc} and h_{ijkl}^c. The Eulerian and Lagrangian mechanical properties and chemical potentials correspond directly, and the Lagrangian property list is the same as the Eulerian list, with properties identified with a superposed tilde. The scalar chemical potential functions have the same value in both descriptions, but are functions of different variables, e.g., $\mu^c(E_{ij}^*, c) = \mu^c(E_{ij}, c)$, etc. Other relations between Lagrangian and Eulerian material properties are

$$\tilde{k}_{ij} = JF_{im}^{-1}k_{mn}F_{jn}^{-1} \tag{5.44}$$

$$\tilde{l}_{ij}^{cc} = JF_{im}^{-1}l_{mn}^{cc}F_{jn}^{-1} \tag{5.45}$$

$$\tilde{d}_{ij}^{cc} = \tilde{l}_{ij}^{cc}\frac{\partial\tilde{\mu}^{c}}{\partial c} \tag{5.46}$$

$$\tilde{h}_{ijkl}^{c} = \tilde{l}_{ij}^{cc}\frac{\partial\tilde{\mu}^{c}}{\partial E_{kl}} \tag{5.47}$$

$$\tilde{b}_{ij}^{cf} = F_{im}^{-1}b_{mk}^{cf}F_{kj} \tag{5.48}$$

Thus, we have complete sets of anisotropic material property functions, which can be quantified by experimental methods and introduced in appropriate FEMs. Additionally, the distribution of the porosity in the reference configuration, $n_0 = n_0(X_i)$, must also be known so that the current porosity can be calculated using $n = 1 - J^{-1}(1 - n_0)$.

Equivalence of Poroelastic and Mixture-Based Biomechanical Models

Biphasic-mixture theory and FEMs have been used by a number of research groups to study soft tissues.[6,7,20] At this point, we note that PHE models and biphasic models as applied to soft tissues are equivalent.[8] The biphasic models are formulations using absolute fundamental fields (e.g., v_i^f, etc.) whereas *relative* fields [e.g., $v_i^{fr} = n(v_i^f - v_i^s)$, etc.] are used in PHE models. A change in variables provides the relation between the equivalent biphasic and PHE models, i.e.,

$$v_i^f = n^{-1}v_i^{fr} - v_i^s, v_i^s = \dot{u}_i, \hat{\sigma}_{ij}^s = \sigma_{ij} - \hat{\sigma}_{ij}^f, \hat{\sigma}_{ij}^s = (1-n)\sigma_{ij}^s, \text{ and } \hat{\sigma}_{ij}^f = n\pi^f\delta_{ij}.$$

The PHE property $S_{ij}^\varepsilon = \partial U^e/\partial E_{ij}$ is equal to the biphasic property $S_{ij}^{SE} = \rho_0^s(\partial\psi_s/\partial E_{ij}^s)$, with $E_{ij} = E_{ij}^s$, and the PHE permeability, k_{ij}, is equal to the biphasic permeability. Representative functional forms for S_{ij}^{SE} and permeability can be found in Spilker et al.[20] This equivalence means that the ABAQUS finite element program (HE material and pore pressure elements with nonlinear permeability) and our PHETS finite element program should be useful for the analysis of soft tissue structures where either biphasic or poroelastic constitutive theories and material properties are available. Similarly, it can also be shown that the PHETS model is equivalent to the triphasic model for soft hydrated tissues, the former being based on field variables (fluid flow, species flux, etc.) viewed relative to the deforming solid and the latter based on the corresponding absolute field variables. Thus, PHETS FEMs can be used for analysis of soft tissues that are described using equivalent triphasic (or multiphasic) models.

5.4   Finite Element Models (FEMs)

Various FEMs, including displacement, mixed, and hybrid formulations, have been utilized to study soft tissue structures, including arteries. Both Eulerian- and Lagrangian-based FEMs have been developed; e.g., the ABAQUS program is Eulerian-based and can solve both HE and PHE (or equivalent biphasic) field problems. In this chapter we use a total Lagrangian view to illustrate the modeling approach for a PHETS FEM based on a mixed formulation (since when solid and fluid materials are incompressible, π^f is indeterminate). The primary variables are u_i (interpolated using quadratic functions of X_i), π^f, and c (both interpolated using linear functions of X_i) given as[21]

$$\delta\dot{u}_i = N_N^u \delta\dot{u}_{Ni}, \quad \delta\dot{u} = \mathbf{N}^u \delta\dot{\overline{u}}, \quad u_i = N_N^u u_{Ni}, \quad u = \mathbf{N}^u \overline{u} \tag{5.49}$$

$$\delta\pi^f = N_M^\pi \delta\pi_M^f = \mathbf{N}^\pi \delta\overline{\pi}^f, \quad \pi^f = N_M^\pi \pi_M^f = \mathbf{N}^\pi \overline{\pi}^f \tag{5.50}$$

$$\delta c = N_M^c \delta c_M = \mathbf{N}^c \delta\overline{c}, \quad c = N_M^c c_M = \mathbf{N}^c \overline{c} \tag{5.51}$$

Bold characters denote matrices, and barred quantities represent the unknown time-dependent nodal quantities. Here, summation is implied for repeated subscripts N and M, which denote the number of nodes in the finite element, and N_N^u, N_M^π, and N_M^c are interpolation functions of X_i. The unknown nodal quantities are

$$\overline{u}^T = \langle u_{11}\, u_{12}\, u_{13}\dots u_{N1}\, u_{N2}\, u_{N3} \rangle, \quad \overline{\pi}^{fT} = \langle \pi_1^f\, \pi_2^f\dots \pi_M^f \rangle$$
$$\overline{c}^T = \langle c_1 c_2 \dots c_M \rangle \tag{5.52}$$

Symmetric stress and strain tensors are represented as column matrices using the vector-tensor notation of Zienkiewicz,[22] e.g., $\mathbf{E}^T = (E_{11}\ E_{22}\ E_{33}\ E_{12}\ E_{23}\ E_{31})$, etc. The Green's strain can be written in terms of displacements as

$$E_{ij} = \frac{1}{2}\left(\frac{\partial u_i}{\partial X_j} + \frac{\partial u_j}{\partial X_i} + \frac{\partial u_k}{\partial X_i}\frac{\partial u_k}{\partial X_j} \right) \tag{5.53}$$

After interpolation using Eq. 5.49, E_{ij} can be split into linear (small strain) and nonlinear components as

$$\mathbf{E} = \mathbf{E}_0 + \mathbf{E}_{NL} = \left(\mathbf{B}_o^u + \frac{1}{2}\mathbf{B}_{NL}^u \right)\overline{u} = \mathbf{B}^u \overline{u} \tag{5.54}$$

and the rate of Green's strain is interpolated as

$$\dot{\mathbf{E}} = \left(\mathbf{B}_o^u + \mathbf{B}_{NL}^u \right)\dot{\overline{u}} = \mathbf{B}\dot{\overline{u}} \tag{5.55}$$

The gradients in pore fluid stress and concentration are interpolated as

$$e^\pi = \nabla\pi^f = \nabla\mathbf{N}^\pi \overline{\pi}^f = \mathbf{B}^\pi \overline{\pi}^f \tag{5.56}$$

$$e^c = \nabla c = \nabla\mathbf{N}^c \overline{c} = \mathbf{B}^c \overline{c} \tag{5.57}$$

where ∇ denotes the spatial gradient operator matrix that contains $\partial/\partial X_i$.

Using these interpolations with the Lagrangian PVVs produces the discretized elemental conservation equation

$$\hat{\mathbf{c}}\dot{\mathbf{p}} + \mathbf{P}^{int} = \mathbf{P}^{ext} \tag{5.58}$$

where $\mathbf{p}$ is a time-dependent vector of nodal unknowns, i.e.,

$$\mathbf{p}^T = \left(\overline{u}^T\ \overline{\pi}^{fT}\ \overline{c}^T \right) \tag{5.59}$$

and $\hat{\mathbf{c}}$ is the elemental generalized damping matrix. Internal and external nodal forces are

$$\mathbf{P}_{int}^{T} = \langle \mathbf{P}_{int}^{uT} \; \mathbf{P}_{int}^{\pi T} \; \mathbf{P}_{int}^{cT} \rangle \tag{5.60}$$

$$\mathbf{P}_{ext}^{T} = \langle \mathbf{P}_{ext}^{uT} \; \mathbf{P}_{ext}^{\pi T} \; \mathbf{P}_{ext}^{cT} \rangle \tag{5.61}$$

Specific forms for the entries in Eqs. 5.49–5.61 are given in Kaufmann.[16] The global form of Eq. 5.58 after standard assembly procedures is

$$\mathbf{C}\dot{\mathbf{r}} + \mathbf{R}^{int} = \mathbf{R}^{ext} \tag{5.62}$$

For static analysis, the first term in Eq. 5.62 is zero. Our PHETS FEM program solves quasi-static problems using a predictor-corrector method in which the time-stepping options include forward difference, backward difference, or a trapezoidal Crank-Nicholson formula followed by the Newton-Raphson iteration to achieve convergence.[16] A modified Petrov-Galerkin approach based on Yu and Heinrich[23,24] is used to eliminate artificial diffusion and numerical dispersion due to the presence of convection terms in the relative species flux equation. The incremental effective static problem solutions are carried out using a nonsymmetric, banded equation solver.

5.5   Application of FEMS to Arterial Structures

In this section we present a selected review of the literature (as well as some recent results from our ongoing research) in order to illustrate the application of FEMs to the analysis of various arterial structures. A review of multiphase poroelastic FEMs for soft tissue structures was presented by Simon.[8] Some of the literature associated with finite element analysis of large arteries was also reviewed by Simon et al.[25] In this chapter, selected references from those reviews are included and additional works are cited. Again, only HE, PHE, and PHETS FEMs are considered here. However, one of the earliest FEMs of a blood vessel appeared in Wiederhielm et al.[26] and included both the elastic and the viscoelastic material behavior of an arteriole. Many FEMs of arterial structures were not developed using commercial programs. Recently, commercial programs such as ABAQUS, ANSYS, and MARC have been used for both HE and PHE finite element analyses. Following are some illustrations of the use of FEMs to study normal large arteries, diseased arteries, arterial grafts, arterial angioplasty, and arterial aneurysms. We have also included some examples of combined structural FEMs and computational fluid dynamics (CFD) models that have been used to study coupled arterial wall motion and blood flow.

FEMs of Normal Large Arteries

Hyperelastic (HE) FEMs

HE constitutive law has been used by various researchers to develop FEMs of normal arteries. The pore fluid and dissolved species are then omitted in the PHETS formulation above. Early arterial FEMs were developed based on transversely isotropic HE material models.[27,28] Simon et al.[29] used these properties in a FEM of a sector of the arterial wall including a simulated vasa vasorum. Flow in the vasa vasorum calculated from FEM results was comparable to experimental observations. Ayorinde et al.[30,31] presented FEMs based on isotropic HE properties of canine aortas (with and without the influence of anesthetics). Material laws were quantified and stresses calculated in the aortic wall. Vawter[32] developed HE FEMs to study the lung parenchyma, including hemodynamics, surface tension effects, and stress concentrations in the structure. Pleural pressures were found to affect vessel diameter whereas perivascular pressure is more important in determining flow resistance. More recently, Vorp et al.[33] coupled a nonlinear FEM with experimental data using a least-squares algorithm to carry out material identification for canine carotid and aortic tissue. Orthotropic, compressible HE constitutive laws were quantified, and stress distributions were calculated.

HE FEMs have been used to simulate remodeling in arterial tissues. Chuong and Fung,[34] Liu and Fung,[35] and Fung[36] (among other investigators) have described active remodeling effects in arterial tissues evidenced as opening angles. Opening angles are observed when the cross section of an excised vessel is cut through the thickness and placed in a physiological solution. Formulas for the calculation of stresses in blood vessels including opening angles are given by Fung.[36] In the absence of this opening angle, an artery is usually analyzed as a thick-walled HE cylinder where severe stress gradients develop during stretching and pressurization. The opening angle (associated with remodeling) may serve to reduce these gradients and the levels of stress that develop at the intimal surface where fragile endothelial cells reside. We used the ABAQUS program to study the arterial wall deformation and stress distributions associated with opening angles.[16] The material properties were based on the HE part of the PHE aortic models described below. Opening angles were introduced in a FEM of an idealized quarter-sector of an aorta (Fig. 5.1) that was subjected to an internal pressure of 100 mm Hg and held at constant length. The opening angle shown in this figure corresponds to Fung's $\theta_0 = \pi/2$, and FEM results were compared with FEM results from the same model with no opening angle ($\theta_0 = \pi$). As expected, circumferential stresses were reduced with increased opening angles ($\theta_0 = \pi/2$), whereas in the absence of an opening angle ($\theta_0 = \pi$), significant stress gradients developed with highly elevated levels occurring at the internal surface (Fig. 5.2). We are currently developing more detailed FEMs that include opening angles in both PHE and PHETS material models.

ABAQUS

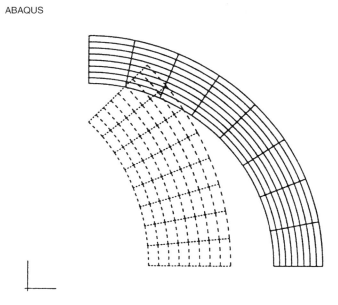

FIGURE 5.1   ABAQUS finite element model to simulate an opening angle ($\theta_0 = \pi/2$) observed in an arterial segment.

Porohyperelastic (PHE) FEMs of Arterial Structures

The PHE theory is a special case of the PHETS theory, above, for which tissue fluid motion is included and the dissolved species are ignored. Early efforts to model arterial tissue characterized as PHE materials were reported by Kenyon[13] and later by Jayaraman[14,37] and Jain and Jayaraman[38] using analytical methods. We considered the PHE response of arteries in Simon et al.[39,40] using representative states, including, (1) the undrained condition, a rapid inflation for which, given the low arterial permeability, the response is incompressible; (2) the drained condition, a steady-state, compressible inflation for which there is no relative fluid motion when the pore pressure becomes uniform; and (3) the intramural flowing condition, a pressurized steady-state (representative of mean *in vivo* pressures) in which the vessel is inflated until motionless, but a steady relative tissue fluid flow remains within the arterial wall. Internal pressures, axial

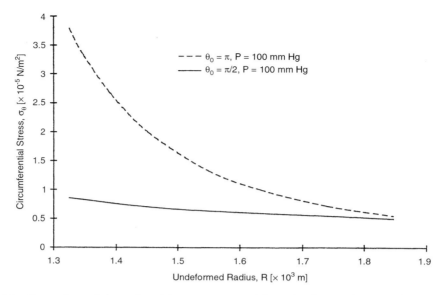

FIGURE 5.2 Comparison of circumferential stress, σ_θ, vs. undeformed radius, R, due to applied pressure, axial force, and an opening angle obtained from the ABAQUS FEM of an arterial segment ($\theta_0 = \pi$, no opening angle; $\theta_0 = \pi/2$, opening angle shown in Fig. 5.1).

stretch, pore fluid pressure, relative tissue fluid flow, and vessel dimensions were calculated using both ABAQUS and our PHE FEMs program. These parameters were also measured in experiments. We characterized mobile tissue fluid flow through the arterial wall by measuring the motion of an air bubble placed in a small-caliber tube in the inflation cannula (Fig. 5.3). The volume flow rate through the arterial wall can be determined from v_B, the velocity of this air bubble. Details of these tests and the experimental apparatus shown in Fig. 5.3 may be found in Simon et al.[40] Some representative results of these experiments and FEMs will be highlighted below.

The first poroelastic FEMs for the arterial cross section used u-π mixed procedures based on a linear isotropic theory and allowed detailed description of both steady-state and transient arterial wall strains and stresses, pore fluid pressure, and relative tissue fluid flow.[41,42] A reversed inward flow of fluid was noted at the external arterial surface for early times after step pressurization. This inward flow was also discussed by Kenyon[13] and appears again in other results given below. These specialized PHE FEMs of Simon et al.[41] were extended by Simon and Gaballa[43] to study rabbit aortas and included the characteristic material nonlinearity and finite-strain response of arterial tissue. A compressible form for U^e (see Chuong and Fung[15]) and simplified forms for permeability were introduced in the FEMs. Pressure-radius curves (for the undrained, drained, and flowing cases) and wall stress and pore fluid gradients were calculated. In related works, we reported more detailed PHE material properties (U^e and strain-dependent permeability) and used FEMs to calculate wall stress gradients and transient pore fluid pressure gradients.[44,45] In Yuan and Simon,[46] we described a PHE FEM based on a penalty method in which large arteries were simulated using an exponential, neo-Hookean PHE form for U^e. Deformations, stresses, pore fluid pressure, and relative fluid motion were calculated for both transient and steady states.

PHE FEMs were used in conjunction with experimental studies on rat carotid arteries with simulated congestive heart failure produced by coronary ligation.[47,48] HE law and hydraulic permeability were introduced into large-strain PHE FEMs. This analysis indicated that heart failure conditions resulted in decreased radial stress gradients and increased circumferential stress gradients. Material response was also determined for these arteries after treatments with the drugs captopril and hydralazine. Captopril treatment resulted in decreased wall stresses, whereas hydralazine treatment showed no effect on wall stresses.

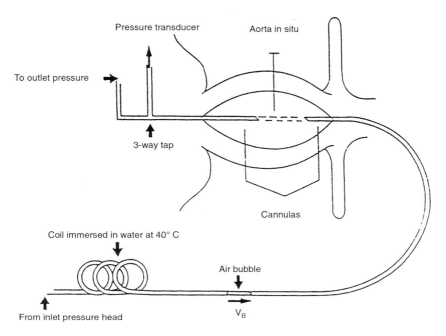

FIGURE 5.3   Experimental apparatus for determination of internal pressure, dimensions, and air bubble velocity, v_B, for *in situ* rabbit aortas. (*Source:* Simon, B.R., *ASME, J. Biomech. Eng.*, 120, 188, 1998. With permission.)

Recently, we used the ABAQUS finite element program to develop FEMs that combine HE material law and pore pressure element capabilities to simulate rabbit aortas.[39] The modeling procedure presented here illustrates the use of such commercial programs for the PHE analysis of soft arterial tissue. The ABAQUS FEM of an axisymmetric segment of an aorta subjected to finite plane strain (axial stretch and internal pressure) is shown in Fig. 5.4. The FEM is composed of 43 eight-noded quadrilateral pore pressure elements (ABAQUS element CAX8P), providing spatially and temporally converged results. ABAQUS utilizes the Eulerian FEM formulation, and time integration is accomplished using a backward-difference scheme with subsequent Newton-Raphson iterations at each time step. Finite displacements were prescribed at the top of the model and suppressed at the bottom to impose finite axial stretch ($\lambda = 1.5$), and the segment was pressurized. The PHE material law was introduced by combining the pore pressure element with the UHYPER option in ABAQUS. Structurally, the material was considered to be homogeneous and isotropic, i.e., the intima and media were assigned the same elastic properties, whereas the intimal and medial permeabilities were different (see below). The relatively loose adventitia was removed in the experiments and was not included in the FEM.

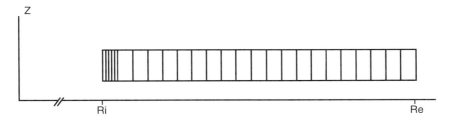

FIGURE 5.4   Axisymmetric plane strain FEM (PHE and PHETS materials) of a rabbit aorta including intima (*I*) and media (*M*). The adventitia was removed in the experiments. (*Source:* Simon, B.R., *Int. J. Sol. Struct.*, 35, 5021, 1998. With permission from Elsevier Science.)

These vessels were also studied experimentally and were subjected to undrained, drained, and steady-state flow conditions in order to determine material parameters in U^e and k_{ij}.[40] The aortas were assumed to be isotropic so that U^e is dependent on $\bar{I}_1, \bar{I}_1$, and J, where the deviatoric invariants are

$$\bar{I}_1 = J^{-2/3}\bar{I}_1 \text{ and } \bar{I}_2 = J^{-4/3}I_2, \text{ with } I_1 = 3 + 2E_{kk} \text{ and } I_2 = 3 + 4E_{kk} + 2\left(E_{ii}E_{jj} - E_{ij}E_{ij}\right).$$

Then the effective stress was derived in a form compatible with ABAQUS as

$$\sigma_{ij}^e = DEV\left[\sum_{N=1.2} J^{-1}F_{ik}\frac{\partial U^e}{\partial \phi}\frac{\partial \phi}{\partial \bar{I}_N}\frac{\partial \bar{I}_N}{\partial E_{km}}F_{jm}\right] + \frac{\partial U^e}{\partial \phi}\frac{\partial \phi}{\partial J}\delta_{ij} \tag{5.63}$$

where DEV is the deviatoric operator. A generalized compressible, isotropic Fung form for U^e was developed (based on experimental data) as

$$U^e = U^e(\phi) = \frac{1}{2}C_0\left(e^\phi - 1\right) \tag{5.64}$$

where ϕ is a function of strain invariants assumed as

$$\phi = C_1\left(\bar{I}_1 - 3\right) + C_2\left(\bar{I}_2 - 3\right) + K'\left(J - 1\right)^2 \tag{5.65}$$

for the isotropic case. The UHYPER option in ABAQUS was used to introduce this form for U^e in the FEMs. The intima was treated as a thin layer with low stiffness and was neglected structurally. The values for the material constants of the media (obtained using our data-reduction procedures and experimental data) are $C_0 = 8133\ N/m^2$, $C_1 = 0.907$, $C_2 = 0.002475$, and $K' = 20$.

Pressure-radius curves for undrained and drained conditions are shown in Fig. 5.5, where FEM results are compared to experimental measurements. The flowing state was characterized by plots of free tissue fluid flow (bubble velocity, v_B) vs. pressure in the aortic segments. Assuming isotropy, the hydraulic permeability is $k_{ij} = \hat{k}\delta_{ij}$. In ABAQUS, $\hat{k}$ is assumed to be a function of e, the voids ratio, where $e = dV^f/dV^s = n/(1-n)$ for a saturated material. The function for $\hat{k} = \hat{k}(e)$ was determined from experimental data from de-endothelialized and intact rabbit aortas. The media was considered to be a PHE material, and our experimental data indicate that medial permeability is nearly constant, i.e.,

$$\hat{k}_{MED} = 5.26 \times 10^{-16}\left[m^4/N - \sec\right].$$

The medial permeability value for $\hat{k}_{MED}$ was comparable to permeability values reported for other soft biological tissues.[6,13] The intima is a relatively thin layer and is *not* a PHE material, but it was represented by PHE elements in order to give an initial approximation to the complex behavior of this layer of the arterial wall. Again based on our experiments, the intimal elements were assigned a nonlinear permeability, $\hat{k}_{INT}$, determined as a function of e from our experimental data.[40]

ABAQUS FEMs were used to simulate two flowing test experiments, one for intact vessels and a second for de-endothelialized vessels. Fig. 5.6 shows the FEM predictions and experimental data for the bubble velocity, v_B, for these conditions. The de-endothelialized test provided the constant value for $\hat{k}_{MED}$, which was then used with the intact test data to determine the function $\hat{k}_{INT} = \hat{k}_{INT}(e)$ for the intima. Fig. 5.7 shows intimal and medial permeabilities vs. voids ratio, e, obtained from our data-reduction procedures.[40] The bubble velocity curve for the intact vessel shown in Fig. 5.6 is indicative of the varying hydraulic resistance associated with the intima, i.e., as pressure increases, the intimal permeability decreases (intimal hydraulic resistance increases). Initially, v_B increases with pressure and then exhibits a characteristic dip

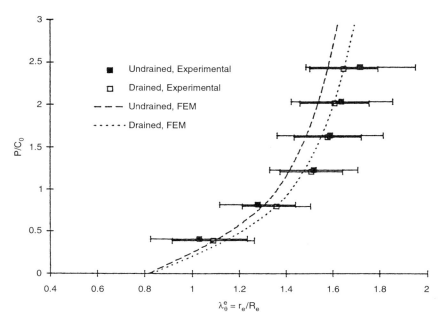

FIGURE 5.5 Experimental and FEM results for internal pressure (P/C_0) vs. external radius ($\lambda_{\theta_e} = r_e/R_e$) data for intact rabbit aortas subjected to undrained (rapid inflation) and drained (steady state, $\pi^f = 0$) conditions. (*Source:* Simon, B.R., *Int. J. Sol. Struct.*, 35, 5021, 1998. With permission from Elsevier Science.)

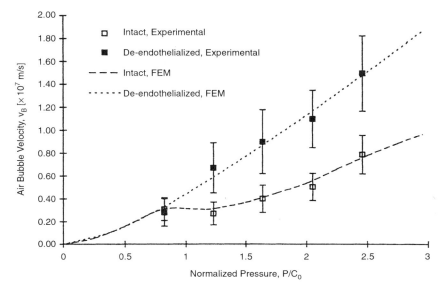

FIGURE 5.6 Experimental and FEM results for air bubble velocity (v_B) vs. pressure (P/C_0) data for intact and de-endothelialized rabbit aortas subjected to steady-state pressurization. (*Source:* Simon, B.R., *Int. J. Sol. Struct.*, 35, 5021, 1998. With permission from Elsevier Science.)

at physiological pressure levels (100 mm Hg) as intimal permeability decreases during further pressurization. Eventually, further increases in pressure result in an increasing flow trend through the wall. Note that further research is needed in order to properly characterize the complex transport properties of the intima.

Fig. 5.8 shows pore fluid pressure distributions in the arterial wall at various times after inflation. At times very soon after the step loading, the fluid pressure takes on significant negative values that produce an inward flow at the intimal and at the external surface. Again, similar results were seen in earlier poroelastic FEMs of arteries[43] and have been noted by Kenyon.[13] There is also a marked drop in the pore fluid pressure across the intimal barrier associated with the relatively low permeability (high resistance) of the endothelium. These pore pressure gradients are related to water and species transport in the arterial wall. A typical wall stress plot (no-opening-angle case) is shown in Fig. 5.2 for an aorta experiencing steady-state flow conditions. Significant gradients develop in the axial and circumferential stresses due to the exponential form of U^e. We note that stress and pore fluid pressure gradients may be significantly altered by the initial bending stresses associated with observed opening angles,[36] as illustrated in Fig. 5.2 above.

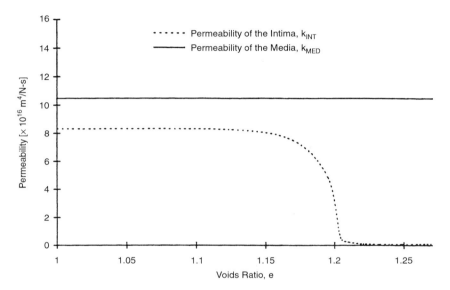

FIGURE 5.7   Hydraulic permeability of the intima, k_{INT}, and the media k_{MED}, vs. voids ratio, e, for the rabbit aorta.

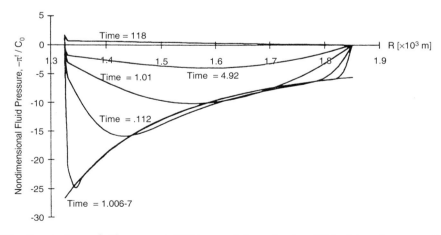

FIGURE 5.8   Transient pore fluid pressure $(-\pi^f/C_0)$ vs. undeformed radius (R) for internal pressure, $P = 100$ mm Hg. Time (sec) after pressurization ranges from 0+ to steady-state conditions.

PHE Analysis of a Simulated Cardiac Cycle Using ABAQUS

The ABAQUS program was used to simulate cyclic pressurization of a large artery composed of PHE material in order to study free tissue fluid motion that may occur during a cardiac cycle. A sawtooth pressure–time function with a frequency of 1 Hz and pressure variation between 80 and 120 mm Hg was applied to the FEM to approximate the cardiac cycles. The FEMs were cycled until a repeating response was produced in each subsequent cycle, and results were then analyzed. Significant spatial variations in pore fluid pressure distribution and the associated fluid flux (relative fluid velocity) were noted within the arterial wall. A representative plot of pore fluid pressure vs. position in the arterial wall is shown in Fig. 5.9 for the systolic part of the simulated cardiac cycle. Fig. 5.10 shows the corresponding relative mobile fluid velocity profile in the arterial wall for the systolic portions of the cardiac cycle. The velocity profiles shown represent the conditions at maximum (120 mm Hg) and minimum (80 mm Hg) pressure levels in each cycle. The relative velocity is nearly zero at the internal surface when $P = 80$ mm Hg, whereas the relative velocity reaches a maximum value at this same location when $P = 120$ mm Hg. Furthermore, at 120 mm Hg, the relative fluid velocity plot indicates that fluid is imbibed at both the internal and external surfaces of the artery. There are also regions within the wall where the tissue fluid flow changes direction, resulting in a churning effect during the cardiac cycle. Similar results were produced in the diastolic phase of the cardiac cycle. Such relative fluid velocity profiles were presented for the first time by Kaufmann[16] and are significant in that they illustrate possible influences of relative mobile fluid motion on species transport that occurs in the arterial wall. These results have implications regarding normal or pathological transport processes in the arterial wall. This fluid flow field in the arterial wall is also important for understanding the transport processes associated with the delivery of drugs or other substances to the interior of the arterial wall during the cardiac cycle. We are currently developing more detailed arterial FEMs based on the PHETS theory in order to study both water and species transport during cyclic loading.

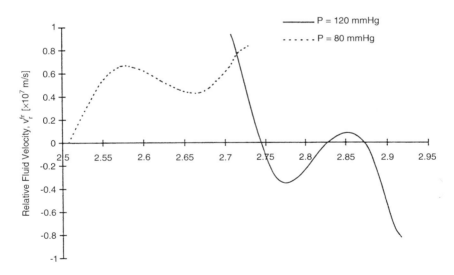

FIGURE 5.9 Fluid pressure $(-\pi^f)$ distribution in the deformed arterial wall (r) at various times $(0 \le t/t_c \le 0.33)$ during systole of a simulated cardiac cycle $(80 \le P \le 120$ mm Hg$)$.

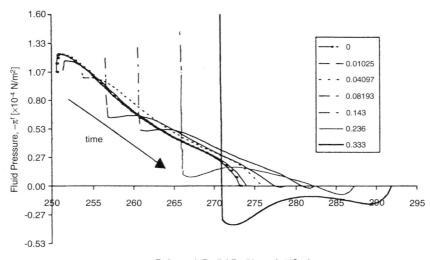

FIGURE 5.10 Eulerian relative fluid velocity (v_r^{fr}) distribution in the deformed arterial wall (r) at two times $(t/t_c = 0$ and $t/t_c = 0.333)$ during a systole, simulated cardiac cycle $(80 \le P \le 120$ mm Hg).

Porohyperelastic-Transport-Swelling (PHETS) FEMs of Arteries

Our first PHETS theory and FEMs were based on the linear forms of the field equations above and did not include convection or osmotic pressure gradient effects.[49,50] The theoretical basis and the extension of the associated FEMs to finite strains, nonlinear materials, and coupled convection effects were introduced by Simon[8] and were fully developed by Kaufmann.[16] A specialized PHETS finite element program for transient analysis of soft tissues was also developed by Kaufmann. This PHETS program allows the study of both HE and PHE models as special cases of the general PHETS approach. The accuracy of this program was verified by comparing the results with HE and PHE analyses obtained analytically and using ABAQUS. In addition, our PHETS FEM results matched results for a rigorous ABAQUS test problem, thereby verifying the PHETS FEM solution using the Petrov-Galerkin method for convection-dominated transport problems. Here we briefly highlight some recent applications of this PHETS FEM program to the study of species and fluid transport in the walls of large arteries.

In Simon et al.,[10] a PHETS theory and FEM were described for soft tissues in general, with specific application to arterial mechanics, and material properties were identified and quantified experimentally. In companion papers, examples were given using PHETS FEMs for mechanical-transport analysis of rabbit aortas.[51,52] These first formulations were limited to small strains, and transient tissue displacements and concentration profiles of a single uncharged species were predicted. We have extended these first transport models to include the finite strain, nonlinear response, and the associated diffusive and convective transport in the arterial wall using the PHETS theory.[11] Our initial nonlinear PHETS FEMs did not include the following effects: the osmotic pressure gradient coupling term, $\partial \pi^c / \partial x_i$, strain dependence in electro-chemical potentials, or applied electrical potentials. Labeled albumin was considered to be a single, uncharged, mobile species. The hyperelastic function U^e was assumed to be independent of albumin concentration and had the exponential form given in Eq. 5.64, with the material constants C_0, C_1, C_2, and K' reported above for $\lambda = 1.5$. PHETS FEM-based generalized least squares data-reduction procedures were developed for determination of the drained bulk modulus parameter (K'), permeability, diffusion, and convection material properties for rabbit aortas needed in the PHETS FEMs.[53] We considered the intima and media of aortas (adventitia removed) and assumed isotropy so that

$$\tilde{k}_{ij} = JkH_{ij}, \ \tilde{d}_{ij} = jdH_{ij}, \text{ and } \tilde{b}_{ij}^{cf} = b\delta_{ij},$$

where k, d, and b are isotropic Eulerian permeability, diffusion, and convection coefficients. Only the radial component of these material properties (e.g., $k = k_{rr}$, etc.) was used in the FEMs. Furthermore, based on experimental observations, we assumed that the medial and intimal properties k_{MED}, d_{MED}, b_{MED}, d_{INT}, and b_{INT} are all constant, whereas k_{INT} is not constant (as noted above). Experimental data were obtained from our laboratory and from the experiments of Tedgui and Lever.[54] The intima is a complex layer that is represented by PHETS finite elements, but k_{INT}, d_{INT}, and b_{INT} are again associated with thin-layer approximations (i.e., resistance in the layer). We determined values for C_0, C_1', and C_2' for the media using the undrained and drained test data given above. The form of k_{INT} and the values for K', k_{MED}, d_{INT}, d_{MED}, b_{INT}, and b_{MED} were determined in the least squares optimization procedure. The procedure was applied separately to (1) determine k_{INT} and k_{MED} from free tissue flow data in the aortic wall; (2) determine the d_{MED} and d_{INT} from the concentration profiles of labeled albumin in a diffusion test at zero internal pressure ($P = 0$); and (3) determine b_{MED} and b_{INT} and based on the steady-state concentration profiles for labeled albumin in Tedgui and Lever's[54] tests ($P = 70$ and 180 mm Hg) with both diffusion and convection. Optimal values determined for all material parameters (C_0, C_1', C_2', d_{MED}, d_{INT}, b_{MED}, b_{INT}, and k_{MED}) and k_{INT} will be presented in other publications. Some typical results are shown in Fig. 5.11, where Tedgui and Lever's[54] steady-state concentration profiles of labeled albumin are compared to our PHETS FEM predictions. In their experiments, rabbit aortas were held at steady internal pressure levels (70 and 180 mm Hg) and the concentration of labeled albumin was elevated in the external bath surrounding each vessel. Thus, the pressurized arterial wall was subjected to diffusion of labeled albumin inward from the external surface in combination with convection associated with water flux outward from the internal surface. This procedure was carried out for intact, as well as damaged (de-endothelialized), vessels ($P = 70$ and 180 mm Hg). The effects of large strains on the shape of the albumin concentration profiles in the arterial wall are clearly seen in Fig. 5.11. The PHETS FEM results agree very well with the experimental measurements. Further details of this recent work are given in Simon et al.[10]

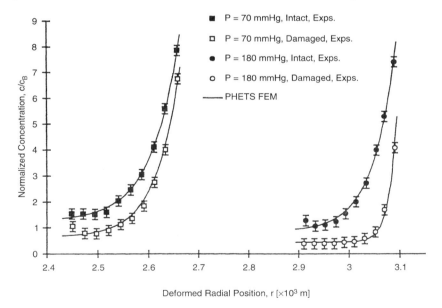

FIGURE 5.11 Steady-state labeled albumin concentration profiles (c/c_B) in the deformed arterial wall: finite element results compared with Tedgui and Lever's[54] experimental data; c_B = bath concentration of albumin. (*Source:* Simon, B.R., *Int. J. Sol. Struct.*, 35, 5021, 1998. With permission from Elsevier Science.)

FEMs of Diseased Arteries

Finite element methods with HE material laws have been used to simulate diseased conditions principally associated with atherosclerosis in large arteries. Chandran et al.[55] studied the effects of local variations of elastic properties in the aortic arch for both normal and arteriosclerotic conditions. Fenton et al.[56,57] analyzed vasoconstrictions at arterial branches and suggested that the flow divider may be a site of disease initiation (due to both flow stresses and structural stresses). It was noted that vasoconstriction and tensile stress actually may tear arterial tissue. Coronary and carotid arteries were described by Keeny and Richardson[58] using FEMs that included possible local mechanical failure in the intima due to the presence of atherosclerotic plaque. MacWilliams et al.[59] related wall deformation and endothelial permeability to lipid transport in rabbit aortas that were experimentally subjected to stenosis. Their FEMs suggest that wall longitudinal shear changed sign rapidly near the stenosis. The occurrence of rounded polygonal endothelial cells in the arterial wall was strongly correlated with regions of elevated longitudinal shear stress. The permeability of such polygonal endothelial cells was thought to be abnormally high, thus increasing the potential for lipoprotein uptake by the arterial wall. Additionally, these structural FEMs were associated with flow models of stenoses.[60,61]

Vito et al.[62,63] developed FEMs of diseased and normal arterial cross sections and found significant changes in stress with variations in plaque elastic modulus, relative size, and medial tears. Thubrikar et al.[64] calculated stresses in bovine arterial branches associated with variations in the branch geometry, wall thickness, orthotropic material properties, and incremental pressure loads. Maximum stresses were independent of elastic properties and occurred at the proximal and distal regions of the ostium. Stresses at the inner surface of the branch were 3 to 4 times greater than stresses in straight segments. Salzar et al.[65] used these same FEMs to study relations between atherosclerosis and wall stress in human carotid arterial bifurcations. Parametric studies varying wall thickness and mechanical properties revealed areas of stress concentration in the arterial walls at the point of bifurcation and the sinus bulb that were correlated with the location of lesion formation. Aoki and Ku[66] developed large-deformation FEMs (using the ANSYS program) to identify stress concentrations and the buckling pressures associated with the collapse of diseased arteries with eccentric cross sections. Their HE FEMs may allow the prediction of locations for plaque-cap rupture due to compressive stresses. Veress et al.[67] modeled pressurized coronary artery walls with FEMs including plaque morphology. Lipid maps were used to define lesion dimensions and wall stress distributions were classified and quantified for four different patient age ranges. The location of the highest stress in the 15–19 age group shifted from the healthy wall region opposite the lesion to the normal intima adjacent to the plaque cap. This trend increased with age, and Veress et al.[67] suggested that age-related development of atherosclerosis may lead to increased localized wall stress and plaque rupture. Hayashi and Imai[68] simulated atherosclerotic walls of large arteries using large-strain FEMs based on nonlinear HE properties of plaque and the media. The isotropic strain energy density functions were based on tensile test data. Elevated stresses in the inner wall of an atheroma were predicted, and the possibility of surface fracture of plaques was suggested.

FEMs of the Arterial Wall and CFD Models

In recent years, structural FEMs have been used in conjunction with computational fluid dynamics (CFD) to study the interaction between flowing blood and the mechanical/transport response of the arterial wall and its possible relations to atherosclerosis. Following are a few examples of this type of combined FEM development. We present these examples to show a starting point for more detailed studies of the interactions between flowing blood and arterial wall mechanics. In particular, we anticipate that a combination of PHETS FEM and CFD models will be developed to link transport processes in flowing blood with associated water and species transport in the wall.

Perktold and Rappitsch[69] surveyed the overall role of coupled structural FEMs and CFD models in understanding the fluid dynamic-induced mechanisms for atherogenesis. The interactions of arterial wall mechanics and local blood flow phenomena in a human carotid arterial bifurcation had been considered earlier by Perktold and Rappitsch.[70] The pulsatile, non-Newtonian flow field (time-dependent Navier-Stokes equations) and the arterial wall (nonlinear thin shell theory, incremental linear elasticity) were both analyzed using coupled FEMs. A significant reduction of flow separation and decrease in wall shear stress were noted when rigid arterial wall models were replaced by distensible wall models. Rappitsch and Perktold[71] studied flow and convection-dominated diffusion oxygen transport processes in local constrictions that simulated stenoses. They demonstrated the effects of wall shear stress and recirculating flow on concentration distribution and mass transfer in the walls of abdominal aortas. Coupled CFD and FEMs were developed to solve the incompressible Navier-Stokes equations (blood is assumed to be a Newtonian fluid) and the convection-diffusion mass transport equation using a streamline upwind procedure for the transport equation and a special sub-element technique. Constant and shear-dependent wall permeabilities were introduced. Permeability was shown to have a strong influence on flux and interfacial concentration profiles along the wall. The flow significantly influenced mass transport; e.g., in the reversed flow region downstream of a stenosis, the oxygen concentration was decreased to 75% of the inlet concentration value. Xu et al.[72] coupled FEMs of blood flow (again using time-dependent Navier-Stokes equations) and the arterial wall (linear elastic, small-displacement model) to predict time-dependent wall displacement and stress fields and the flow field for physiological situations. Heil and Pedley[73] used Lagrangian nonlinear shell FEMs coupled with viscous fluid flow models using lubrication theory to study flow in collapsible arteries. Pre-buckling deformation, structural stability, and nonsymmetric buckling were examined at various axial stretch levels in simulated experimental configurations.

Some FEMs of Other Arterial Structures

FEMs have been developed for a number of other arterial structures that deserve note. Intracranial saccular aneurysms were studied using symmetric and nonsymmetric FEMs by Kyriacou and Humphrey.[74] Representative geometries and two-dimensional HE constitutive laws were developed based on an exponential strain energy function. The maximum stresses increased with an increased neck/height ratio and were always at the fundus, whereas rupture was near the neck for axisymmetric lesions.

Balloon angioplasty was simulated by Oh et al.[75] using the ABAQUS program. Large-strain FEMs of balloon angioplasty procedures on atherosclerotic arteries used HE material properties for both arterial tissue and plaque. Three typical plaque configurations were subjected to balloon dilatation. Intimal splitting, intimal–medial dehiscence, and medial stretching were studied using deformations and stress distributions predicted by FEMs.

Arterial grafts and stents have been considered using FEMs. Chandran et al.[76] studied the effects of pressure loading on the compliance mismatch in an artery–graft anastomosis and found regions of increased compliance and relatively large wall stresses that may be related to a loss of patency in small-diameter vascular grafts. Subsequently, Chandran et al.[77] developed linear isotropic elastic FEMs of end-to-end anastomoses of arteries with vein grafts, Dacron grafts, and polytetrafluoroethylene (PTFE) grafts. They evaluated the distribution of compliance and stresses due to compliance mismatch and found a hypercompliant zone on the arterial side and high wall tensile stresses on the graft side of the anastomosis. Furthermore, hypercompliance was larger in Dacron and PTFE grafts than in vein grafts, but tensile stresses were larger in the vein grafts than in the synthetic grafts. They noted that increasing graft diameter to increase flow may result in a significant increase in hypercompliance on the arterial side of the anastomosis.

Ballyk et al.[78,79] considered the stresses in end-to-side anastomoses using FEMs composed of cubic Lagrangian isoparametric elements (linear elastic material) to model the large-strain shell equations. Stress concentrations and elevated stress gradients were predicted at the arterial-graft junction, supporting the view that intimal hyperplasia is part of a remodeling process that attempts to reduce mechanical stress to physiological levels in the arterial graft structure.

Vascular stents have been studied using FEMs. Borgersen and Sakaguchi[80] used the MARC/MENTAT program (MARC Analysis Research Corp., Palo Alto, CA) to simulate the contact problem associated with the inflation of a stent. Separate three-dimensional structural models of the catheter balloon, barrel stent, and arterial wall were developed, and a three-dimensional contact algorithm was introduced. The balloon and stent were considered to be nonlinear HE, whereas the arterial tissue was considered to be a linear elastic material. A surgical pressurization schedule was simulated and initial results for deformation and stress levels were obtained. During expansion to surgical pressurization levels, both the balloon and the stent contacted the artery. During de-pressurization of the balloon, the stent was allowed to relax and be compressed radially by the arterial wall.

5.6   Conclusion

This chapter has provided an overview of techniques in finite element modeling of the arterial wall that include the coupled effects associated with finite strains and mobile fluid/species transport in this soft biological tissue. These techniques have been classified and described by the material model used in each FEM. Arterial FEMs considered first were HE (hyperelastic) analyses, followed by the PHE (porohyperelastic/biphasic) and PHETS (porohyperelastic-transport-swelling/triphasic/multiphasic) analyses. The emphasis of the chapter is on the use of the latter two models in studies where species and fluid flux (as well as structural response) were determined using FEMs of large arteries based on appropriate experimental data. Applications of the finite element analysis of arterial structures included studies of normal and diseased arteries, as well as aneurysms. Combined FEM and CFD models were also introduced. FEMs of arterial grafts and stents were discussed briefly.

Specific attention was given to the PHETS formulation using a mixed FEM as a general theoretical framework that allowed the consideration of finite strain, fluid transport, and the transport of a single uncharged mobile species. These initial models were based on a simplified isotropic strain energy function. The intima and media were included in the FEMs and experimental models, for which the adventitia had been removed. Water permeabilities and species transport properties were also simplified and considered only for the radial direction for the media and intima. Experimental data were matched by FEM predictions for a number of different conditions including undrained, drained, flowing, and transport of labeled albumin. Of particular note is the close agreement obtained using the PHETS FEM to simulate species transport experiments. The model accurately predicted the concentration profiles of labeled albumin in the wall of both intact and damaged (de-endothelialized) rabbit aortas, where both diffusion and convection were significant.

There are a number of directions for future research and extension of the present PHETS FEM capabilities for arterial mechanics. At the outset, experimental observations are a key ingredient in all future developments. These initial models can be extended to include the characteristic anisotropy of the arterial wall. This can be accomplished by carrying out biaxial and triaxial testing including bending, twisting, or other deformational states that excite the directional coupling in the HE material law of the PHETS models. The anisotropic forms of permeability can be quantified by characterizing the diagonalized form (principal vlaues) of the permeability tensor and permeability functions. The next stage of modeling could be based on the assumption that the radial, circumferential, and axial directions are the principal directions for the permeability tensor. Anisotropic effects in the diffusion and convection properties will require significant theoretical and experimental efforts. The initial results described here provide a beginning for such research.

Various soft tissue research efforts (e.g., Gu et al.[81]) have been developing theories that include the fixed charges in soft tissues and the associated effects on solutes (e.g., NaCl) and water transport. These methods can be included in the extension of the PHETS model to include charged mobile species. The theoretical basis for this extension is given by Kaufmann.[16]

The future goals for any study of arterial wall transport will include the accurate simulation of the transport of relatively large molecules, such as LDL cholesterol. Clearly, the modeling of such a large species or the consideration of multiple interacting charged species will introduce more complexity. The models described in this chapter provide a beginning for combined FEM and experimental procedures, which are essential for the development of the appropriate theory and experiments associated with transport simulations for LDL in arterial tissues.

The role of smooth muscle in the arterial wall should also be investigated. Smooth muscle has been rendered passive in many of the models described here. However, PHETS FEMs can be used for the simulation of both passive and active smooth muscle states, and the resulting structural and transport response can be determined. Further development of the PHETS model and experiments will be necessary in order to characterize the role of arterial smooth muscle in wall mechanics and transport.

We anticipate that tissue remodeling effects (e.g., opening angles) will be included in future PHETS FEMs. We currently measure opening angles in all tests on rabbit aortas. The techniques for simulation of opening angles, initiated here using ABAQUS, can be introduced in the arterial PHETS FEMs in order to establish the effects of such prestress conditions on the overall and local transport of water and mobile species in the deforming arterial wall.

The development of PHETS FEMs subjected to cyclic pressures associated with the cardiac cycle seem especially promising. Our initial results from such cyclic ABAQUS FEMs using only the PHE model (no mobile species) revealed complicated mobile tissue fluid motion in the arterial wall at normal and hypertensive mean pressure levels. The next step will be to add representative mobile species to the FEMs in order to quantify transport of free tissue fluid and various species during the cardiac cycle.

Our first FEMs have considered the intima to be very thin, with average (thin-layer) PHETS material properties. It would be especially interesting to consider the intimal layer (endothelium, elastic membrane, etc.) in greater detail. The PHETS continuum model has provided an initial quantitative view of the overall transport across this thin layer. However, it is now appropriate to extend modeling to the micro-scale of the intima. A recent paper by Huang et al.[82] describes such a model. Micro-scale models could be combined with the PHETS macro-scale simulation of the media. Also, the adventitia deserves more attention in order to complete the modeling of a large artery. The mechanics and transport in this layer await quantitative study that will have significance in the understanding of species transport processes through this outer layer associated with normal function, as well as the design of local drug delivery systems that introduce substances at the external wall surface.

Inclusion of complicated osmotic effects in the electro-chemical-mechanical osmotic driving fluid pressure gradient, $\partial \pi^c/\partial X_i$, can also be initiated as an extension of the PHETS models described in this chapter. This would include the study of electrical potentials and charged species effects using a combination of FEM simulation, experimental testing, and extended PHETS FEM-based data-reduction procedures. Strain-dependent chemical potentials and concentration dependence in U^e can also be studied based on the initial results given in this chapter.

Once the PHETS arterial wall model extensions are underway (including development of fully three-dimensional FEMs), we anticipate that significant new knowledge regarding normal and diseased states in the arterial system will be obtained by combining PHETS FEMs with modern computational fluid dynamics (CFD) models of large arterial structures. The study of fluid-structure interaction, local wall shear and normal stresses, and species transport (LDL, etc.) should lead to important basic information and possible new insights regarding the complex processes associated with atherogenesis. Atherosclerosis has been associated with numerous aspects of the mechanics of flowing blood in the regions where the disease is known to first develop in the arterial tree. But atherosclerosis is a disease of the arterial wall. Thus, detailed PHETS FEMs of the arterial wall coupled with CFD models of blood flow conditions should quantify stress and transport phenomena that could be related to the development of arterial disease in specific sites in the walls of various arterial structures. Micro-scale models for the intima could also be introduced in this study.

One should also be able to use PHETS FEMs to study the effects of known causal factors for atherogenesis, such as hypertension, elevated LDL levels, altered states of smooth muscle tension, alterations in the arterial wall associated with remodeling, damage, etc. These FEMs will allow quantitative prediction of the response of the wall to such factors applied singly or in various combinations to study complex interactions among these causal factors. For example, one might study the effect of elevated blood pressure and elevated LDL in the blood using a PHETS FEM of a bifurcation in a large artery — a site implicated in early stages of atherogenesis.

This chapter has presented a few early FEMs of arterial grafting and stents. The PHETS modeling capability should find further application in the design and development of arterial grafts. The emerging field of tissue engineering of such arterial grafts provides problems in which more generalized PHETS FEMs should be useful. We have begun the development of such models for the study of PTFE materials to quantify the transport of water, blood, and other substances in order to design better arterial grafts. The complex processes associated with tissue remodeling and ingrowth may eventually be quantified and more fully understood using extended PHETS FEMs supported by experimental data.

An important future application of the PHETS model will include the study of local drug delivery systems. A typical delivery system configuration is associated with angioplasty and utilizes the balloon to contact the vascular wall and deliver drugs by diffusive and convective transport. Our models of arterial wall mechanics and species transport can be introduced in PHETS FEMs of the balloon–arterial contact configuration and the resulting delivery of species to wall tissues could be quantified. We have initiated the development of such models, beginning with water transport simulation using ABAQUS, and are currently extending the approach using our PHETS FEMs.

Finally, we point out that the PHETS theory and FEMs described in this chapter should find applications in the quantitative study of many other highly deformable hydrated soft-tissue structures (cardiovascular, articular cartilage, intervertebral disc, etc.). Such studies can be developed with either PHETS material laws or equivalent multiphase material laws (using the direct mathematical relationships between multiphase and PHETS material properties) in PHETS FEMs of biological structures where nonlinear structural response coupled with free-tissue fluid motion and charged species transport are of interest.

Acknowledgments

The authors gratefully acknowledge the support provided by NSF grants BES 9210282, BES 9443135, and BES 9410471.

Nomenclature

$a_{ij}^{\alpha\beta}$	Relative drag coefficients
$\hat{a}_{ij}^{\alpha\beta}$	Function of relative drag coefficients
b_{ij}^{cf}, $\tilde{b}_{ij}^{cf}$	Convection material properties
B	Shape function derivative matrix
C_0, C_1^I, C_2^I	Hyperelastic material constants
C, $\hat{c}$	Damping matrices
c	Concentration of the mobile species (dm^c/dV^f)
D_{ij}	Velocity strain or rate of deformation
D_{kk}^{fr}, $\tilde{D}_{kk}^{fr}$	Relative volumetric fluid strain rate
$d_{ij}^{\alpha\beta}$, $\tilde{d}_{ij}^{\alpha\beta}$	Diffusion coefficients
E_{ij}	Green's strain
E_{ij}^*	Almansi's strain
e	Voids ratio
e_i^c, $\tilde{e}_i^c$	Concentration gradient of the mobile species
F_{ij}, F_{ij}^{-1}	Deformation gradients
FEM	Finite element model
g_{ij}^{cc}, $\tilde{g}_{ij}^{cc}$, $g_{ijkl}^{E^*}$, $\tilde{g}_{ijkl}^{E}$	Osmotic material functions
HE	Hyperelastic
H_{ij}	Finger's strain

$h_{ijkl}^c, \tilde{h}_{ijkl}^c$	Strain-dependent relative species flux coefficients
I_1, I_2	Strain invariants
$\bar{I}_1, \bar{I}_2$	Deviatoric strain invariants
J	Volume strain
$j_i^{cr}, \tilde{j}_i^{cr}$	Relative species flux
$k_{ij}, \tilde{k}_{ij}$	Hydraulic permeability
K	Stiffness matrix
K'	Drained bulk modulus parameter
$l_{ij}^{\alpha\beta}, \tilde{l}_{ij}^{\alpha\beta}$	Coupling coefficients
n, n_0	Porosity
$\hat{n}_i, \hat{n}_{0i}$	Unit normal vectors
N	Interpolation function
P	Mechanical pressure
p	Vector of nodal unknowns
PHE	Porohyperelastic
PHETS	Porohyperelastic-transport-swelling
PVVs	Principle of virtual velocities
r	Deformed radius; Eulerian radial direction
R	Undeformed radius; Lagrangian radial direction
S_{ij}	Second Piola-Kirchoff stress
t	Time
T_{ij}	First Piola-Kirchoff stress
u_i	Displacement
$\dot{u}_i$	Solid (total) velocity
U^e	Effective strain energy density function
V	Volume
$v_i^{\alpha r}, \tilde{v}_i^{\alpha r}$	Apparent relative velocity of α constituent ($\alpha = f, c$)
v_i^s	Solid velocity ($= \dot{u}_i$)
v_i^α	Velocity of the α constituent ($\alpha = f, c$)
v_B	Bubble velocity
W_{ij}	Vorticity of spin tensor
x_i^η, X_i^η	Position vector of η constituent ($\eta = s, f, c$)
∇	Del operator

Greek

α	Opening angle
δ_{ij}	Kronecker delta
$\delta( )$	Virtual quantity
θ_0	Half polar opening angle

$\lambda_R, \lambda_\Theta, \lambda_Z$ Principle stretch ratios

$\mu^\alpha, \tilde{\mu}^\alpha$ Chemical potential of α constituent ($\alpha = f, c$)

π^f Pore fluid stress

ρ_T^α True density of α constituent (dm^α/dV^α, $\alpha = f, c$)

ρ^α Apparent density of α constituent (dm^α/dV, $\alpha = f, c$)

σ_{ij} Total Cauchy stress

ϕ Quadratic strain parameter

ϕ^α Volume fraction of α constituent

Superscripts

d Drained

e Effective

exp Experimental

f Fluid phase

s Solid phase

η, ξ Constituents, i.e., s, f, c

Subscripts

i, j, k, l, m, n Spatial directions, i.e., 1, 2, 3; R, Θ, Z; etc.

i Internal

e External

INT Intima

MED Media

M, N Node numbers

0   Initial or Reference

References

1. Hayashi, K., Experimental approaches on measuring the mechanical properties and constitutive laws of arterial walls, *J. Biomed. Eng.*, (Special Ed. — 20th Anniversary Biomechanics Symposium) 115, 481, 1993.
2. Humphrey, J. D., Mechanics of the arterial wall: review and directions, *Crit. Rev. in Biomed. Eng.*, 23(1/2), 1, 1996.
3. Huyghe, J. M. R. J., Nonlinear finite element models of the beating left ventricle and the intramyocardial coronary circulation, PhD Thesis, University of Eindhoven, The Netherlands, 1986.
4. Simon, B. R., Wu, J. S. S., Carlton, M. W., Evans, J. H., and Kazarian, L. E., Structural models for human spinal motion segments based on a poroelastic view of the intevertebral disc, *J. Biomech. Eng.*, 107, 293, 1985.
5. Simon, B. R., Wu, J. S. S., Carlton, M. W., Kazarian, L. E., France, E. P., Evans, J. H., and Zienkiwicz, O. C., Poroelastic dynamic structural models of rhesus spinal motion segments, *Spine*, 10, 494, 1985.
6. Mow, V. C., Kuei, S. C., Lai, W. M., and Armstrong, C. G., Biphasic creep and stress relaxation of articular cartilage in compression: theory and experiments, *J. Biomech. Eng,*, 102, 73, 1980.

7. Mow, V. C., Kwan, M. K., Lai, W. M., and Holmes, M. H., A finite deformation theory for nonlinearly permeable soft hydrated biological tissues, in *Frontiers in Biomechanics*, Schmid-Schonbein, G. W., Woo, S. L. Y., and Zweifach, B. W., Eds., Springer-Verlag, New York, 1985, 153.

8. Simon, B. R., Multiphase poroelastic finite element models for soft tissue structures, *Appl. Mech. Rev.*, 45, 191, 1992.

9. Lai, M., Hou, J. S., and Mow, V. C., A triphasic theory for the swelling and deformation behaviors of articular cartilage, *J. Biomech. Engr.*, 113, 245, 1990.

10. Simon, B. R., Kaufmann, M. V., McAfee, M. A., and Baldwin, A. L., Finite element models for soft tissues based on porohyperelastic transport-swelling theory, in *1995 ASME Bioengineering Conference*, Hochmuth, R. M., Langrana, N. A., and Hefzy, M. S., Eds., ASME, New York, 1995, 311.

11. Kaufmann, M. V., Simon, B. R., and Baldwin, A. L., Arterial wall transport simulation — porohyperelastic/transport finite element models, in *1996 Advances in Bioengineering*, BED-Vol. 33, Rastegar, S., Ed., ASME, New York, 1996, 333.

12. Kenyon, D. E., Consolidation in transversely isotropic solids, *J. Appl. Mech.*, 46, 65, 1979.

13. Kenyon, D. E., A mathematical model of water flux through aortic tissue, *Bull. Math. Biol.*, 41, 79, 1979.

14. Jayaraman, G., Water transport in the arterial wall — a theoretical study, *J. Biomech.*, 16, 833, 1983.

15. Chuong, C. J. and Fung, Y. C., Compressibility and constitutive equation of arterial wall in radial compression experiments, *J. Biomech.*, 17, 35, 1984.

16. Kaufmann, M. V., Porohyperelastic analysis of large arteries including species transport swelling effects, Ph.D. Thesis, University of Arizona, Tucson, AZ, 1996.

17. *ABAQUS Theory Manual*, Ver. 5.5, Hibbitt, Karlson and Sorensen, Inc., Pawtucket, RI, 1995.

18. Fung, Y. C., *Foundations of Solid Mechanics*, Prentice-Hall, Englewood Cliffs, NJ, 1965.

19. Malvern, L. E., *Introduction to the Mechanics of a Continuous Medium*, Prentice-Hall, Englewood Cliffs, NJ, 1969.

20. Spilker, R. L., Suh, J. K., Vermilyea, M. E., and Maxian, T. A., Alternate hybrid, mixed, and penalty finite formulations for the biphasic model of soft hydrated tissues, in *Biomechanics of Diarthrodial Joints*, Vol. 1, Mow, V. C., Ratcliffe, A., and Woo, S. L.-Y., Eds., Springer-Verlag, New York, 1990, 400.

21. Hughes, T. J. R., *The Finite Element Method — Linear Static and Dynamic Finite Element Analysis*, Prentice-Hall, Englewood Cliffs, NJ, 1987.

22. Zienkiewicz, O. C., *The Finite Element Method*, McGraw-Hill, New York, 1977.

23. Yu, C.-C. and Heinrich, J. C., Petrov-Galerkin methods for the time-dependent convective transport equation, *Int. J. Num. Meth. Eng.*, 23, 883, 1986.

24. Yu, C.-C. and Heinrich, J. C., Petrov-Galerkin method for multidimensional, time-dependent, convective-diffusion equations, *Int. J. Num. Meth. Eng.*, 24, 2201, 1987.

25. Simon, B. R., Kaufmann, M. V., McAfee, M. A., and Baldwin, A. L., Finite element models for arterial wall mechanics, *J. Biomech. Eng.* (Special Ed. — 20th Anniversary Biomechanics Symposium), 115, 489, 1993.

26. Wiederhielm, C. A., Kobayashi, A. S., Woo, L. Y., and Stromberg, D. D., Structural response of relaxed and constricted arterioles, *J. Biomech.*, 1, 259, 1968.

27. Simon, B. R., Kobayashi, A. S., Strandness, D. E., and Wiederhielm, C. A., Large deformation analysis of the arterial cross section, *Basic Eng., Trans. ASME*, 93D, 138, 1971.

28. Simon, B. R., Kobayashi, A. S., Strandness, D. E., and Wiederhielm, C. A., Reevaluation of arterial constitutive equations — a finite deformation approach, *Circ. Res.*, 30, 491, 1972.

29. Simon, B. R., Kobayashi, A. S., Wiederhielm, C. A., and Strandness, D. E., Deformation of the arterial vasa vasorum at normal and hypertensive arterial pressures, *J. Biomech.*, 6, 349, 1973.

30. Ayorinde, O. A., Kobayashi, A. S., Chen, J. C. H., and Merati, J. K., A finite element large deformation analysis of a tapered aorta, ASME Winter Annual Meeting, Houston, TX, 1975.

31. Ayorinde, O. A., Kobayashi, A. S., and Merati, J. K., Finite elasticity analysis of unanesthetised and anesthetised aorta, *J. Appl. Mech.*, 42, 547, 1975.

32. Vawter, D. L., Influence of lung parenchyma on the boundary conditions for pulmonary hemodynamics, Proceedings of the 1st Mid-Atlantic Conference on Bio-Fluid Mechanics, Virginia Polytechnic Institute and State University, Blacksburg, VA, 1978, 195.

33. Vorp, D. A., Rajagopal, K. R., Smolinski, P. J., and Borovetz, H. S., Identification of elastic properties of homogeneous, orthotropic vascular segments in distension, *J. Biomech.*, 28, 501, 1995.

34. Chuong, C. J. and Fung, Y. C., Residual stress in arteries, in *Frontiers in Biomechanics*, Schmid-Schonbein, G. W., Woo, S. L.-Y., and Zweifach, B. W., Eds., Springer-Verlag, New York, 1986, 117.

35. Liu, S. Q. and Fung, Y. C., Zero stress states of arteries, *J. Biomech. Engr.*, 110, 82, 1988.

36. Fung, Y. C., 1990, *Biomechanics, Motion, Flow, Stress, and Growth*, Springer-Verlag, New York, 1990.

37. Jayaraman, G., Filtration across a porous elastic matrix — an application to the water transport in the artery wall, in *1985 Advances in Bioengineering*, Langrana, N. A., Ed., ASME, New York, 1985, 53.

38. Jain, R. and Jayaraman, G., Theoretical model for water flux through the artery wall, *J. Biomech. Eng.*, 109, 311, 1987.

39. Simon, B. R., Kaufmann, M. V., McAfee, M. A., and Baldwin, A. L., Porohyperelastic finite element analysis of large arteries using ABAQUS, *J. Biomech. Eng.*, submitted.

40. Simon, B. R., Kaufmann, M. V., McAfee, M. A., Baldwin, A. L., and Wilson, L. M., Identification and determination of material properties for porohyperelastic analysis of large arteries, *J. Biomech. Eng.*, submitted.

41. Simon, B. R., Gaballa, M. A., and Short, K., Poroelastic finite element analysis of arterial cross sections, 6th International Congress of Biorheology, Vancouver, B.C., Canada, 1986.

42. Simon, B. R. and Gaballa, M. A., Poroelastic finite element models for large arteries, in *Advances in Bioengineering*, BED Vol. 2, ASME, New York, 1986, 140.

43. Simon, B. R. and Gaballa, M. A., Finite strain, poroelastic finite element models for large arterial cross sections, in *Computational Methods in Bioengineering*, BED Vol. 9, Spilker, R. L. and Simon, B. R., Eds., ASME, New York, 1988, 325.

44. Simon, B. R., Gaballa, M. A., Baldwin, A. L., and Yuan, Y., Porohyperelastic finite element and experimental models for in situ rabbit aortas, 11th U.S. National Congress of Applied Mechanics, Tucson, AZ, 1990.

45. Simon, B. R., Kaufmann, M. V., McAfee, M. A., and Baldwin, A. L., Determination of material properties for soft tissues using a porohyperelastic constitutive law, in *1993 Advances in Bioengineering*, BED Vol. 26, ASME, New York, 1993, 7.

46. Yuan, Y. and Simon, B. R., Constraint relations for orthotropic porohyperelastic constitutive laws and finite element formulations for soft tissues, in *Advances in Bioengineering*, BED Vol. 22, ASME, New York, 1992, 203.

47. Gaballa, M. A., Simon, B. R., Raya, R. E., Yuan, Y., and Goldman, S., Experimental measurement of the mechanical properties of hypertensive rats based on porohyperelasticity, in *Advances in Bioengineering*, BED Vol. 20, ASME, New York, 1991, 621.

48. Gaballa, M. A., Raya, T. E., Simon, B. R., and Goldman, S., Arterial mechanics in spontaneously hypertensive rats: mechanical properties, hydraulic conductance, and finite element models, *Circ. Res.*, 71, 145, 1992.

49. Simon, B. R. and Gaballa, M. A., Poroelastic finite element models for the spinal motion segment including ionic swelling, in *Computational Methods in Bioengineering*, BED Vol. 9, Spilker, R. L. and Simon, B. R., Eds., ASME, New York, 1988, 93.

50. Simon, B. R., Liable, J. P., Pflaster, D. S., Yuan, Y., and Krag, M. H., A poroelastic finite element model formulation including swelling in soft tissue structures, *J. Biomech. Eng.*, 118, 1, 1996.

51. Kaufmann, M. V., Simon, B. R., McAfee, M. A., and Baldwin, A. L., Poroelastic finite element formulation including coupled transport and swelling effects, in *1995 ASME Bioengineering Conference*, BED Vol. 29, Hochmuth, R. M., Langrana, N. A., and Hefzy, M. S., Eds., ASME, 1995, 315.

52. Kaufmann, M. V., Simon B. R., and Baldwin, A. L., Poroelastic-transport finite element formulation including convection and swelling, in *Advances in Bioengineering*, BED Vol. 31, Hull, M. L., Ed., ASME 1995, 329.

53. Simon, B. E., Lui, J., Kaufmann, M. V., and Baldwin, A. L., Data reduction methods for determination of material properties for porohyperelastic-transport-swelling (PHETS) FEMs of large arteries, ASME Summer Bioengineering Conference, Sun River, OR, 1997.

54. Tedgui, A. and Lever, M. J., The interaction of convection and diffusion in the transport of ^{131}I-Albumin within the media of the rabbit thoracic aorta, *Circ. Res.*, 57, 856, 1985.

55. Chandran, K. B., Ray, G., and Ghista, D. N., Clinical implications of pressure-deformation analysis of the aortic arch with local variations in the elastic property, Proceedings of the Applied Mechanics, Fluid Engineering, and Bioengineering Conference, Vol. 32, AMD Symposia Series, ASME, 1979, 125.

56. Fenton, T. R., Taylor, J. R., and Gibson, W. G., Theoretical stress analysis of vasoconstriction at arterial branch points using the finite element technique, Proceedings of the 10th Canadian Medical and Biological Engineering Conference: Biomedical Engineering — The Future of Health Care, Canadian Medical and Biological Engineering Society, Gloucester, 1984, 31.

57. Fenton, T. R., Gibson, W. G., and Taylor, J. R., Stress analysis of vasoconstriction at arterial branch sites, *J. Biomech.*, 19, 501, 1986.

58. Keeny, S. M. and Richardson, P. D., Stress analysis of atherosclerotic arteries, Proceedings of the Ninth Annual Conference of the IEEE Engineering in Medicine and Biology Conference, IEEE, New York, 1987, 1484.

59. MacWilliams, B. A., Savilonis, B. J., and Hoffman, A. H., Finite element stress analysis of stenotic blood vessels, Proceedings of the 15th Northeast Bioengineering Conference, Boston, MA, 1989, 141.

60. MacWilliams, B. A., Savilonis, B. J., and Hoffman, A. H., Modeling lipid transport through the transverse clip stenosis, Proceedings of the 14th Northeast Bioengineering Conference, Boston, MA, 1988, 204.

61. Savilonis, B. J. and Hoffman, A. H., Computer modeling of flow through a two-dimensional arterial stenosis, *Bioengineering*, Proceedings of the Northeast Conference, IEEE, New York, 1985, 135.

62. Vito, R. P., Whang, M. C., Giddens, D. P., Zarins, C. K., and Glagov, S., Stress analysis of the diseased arterial cross-section, in *Advances in Bioengineering*, BED Vol. 17, ASME, 1990, 273.

63. Vito, R. P., Whang, M. C., Glogov, S., and Aoki, T., Distribution of strains and stresses in the arterial cross section, in *Advances in Bioengineering*, BED Vol. 20, ASME, 1991, 639.

64. Thubrikar, M. J., Roskelley, S. K., and Eppink, R. T., Study of stress concentration on the walls of the bovine coronary arterial branch, *J. Biomech.*, 23, 15, 1990.

65. Salzar, R. S., Thubrikar, M. J., and Eppink, R. T., Pressure-induced mechanical stress and artherosclerosis in the carotid artery bifurcation, *Annals Biomed. Eng.*, 19, 587, 1991.

66. Aoki, T. and Ku, D. N., Collapse of diseased arteries with eccentric cross section, in *Advances in Bioengineering*, BED Vol. 20, ASME, 1991, 341.

67. Veress, A. I., Cornhill, J. F., Powell, K. A., Herderick, E. E., and Thomas, J. D., Finite element modeling of atherosclerotic plaque, in *Proceedings of the Computers in Cardiology Conference*, IEEE, Piscataway, NJ, 1993, 791.

68. Hayashi, K. and Imai, Y., FEM analysis of stress in atherosclerotic vascular wall, in *1995 Bioengineering Conference*, BED Vol. 29, Hochmuch, R. M., Langrana, N. A., and Hefzy, M. S., Eds., ASME, 1995, 63.

69. Perktold, K. and Rappitsch, G., Mathematical modeling of arterial blood flow and correlation to atherosclerosis, *Technology and Health Care*, 3, 139, 1995.

70. Perktold, K. and Rappitsch, G., Numerical analysis of arterial wall mechanics and local blood flow phenomena, in *Advances in Bioengineering*, BED Vol. 26, Tarbell, J. M., Ed., ASME, 1993, 127.

71. Rappitsch, C. and Perktold, K., Computer simulation of convective diffusion processes in large arteries, *J. Biomech.*, 29, 207, 1996.

72. Xu, X. Y., Griffith, T. M., Collins, M. W., Jones, C. J. H., and Tardy, Y., Coupled modeling of blood flow and arterial wall interactions by the finite element method, *Proceedings of the Computers in Cardiology Conference*, IEEE, Piscataway, NJ, 1993, 687.

73. Heil, M. and Pedley, T. J., Large Deformations of cylindrical tubes conveying a viscous flow, *1995 Bioengineering Conference*, BED Vol. 29, Hochmuch, R. M., Langrana, N. A., and Hefzy, M. S., Eds., ASME, 1995, 411.

74. Kyriacou, S. K. and Humphrey, J. D., Nonlinear finite element analysis of intracranial saccular aneurysms — a parametric study, in *1995 ASME Bioengineering Conference*, BED Vol. 29, Hochmuch, R. M., Langrana, N. A., and Hefzy, M. S., Eds., ASME, 1995, 67.

75. Oh, S., Kleinberger, M., and McElhaney, J. H., A finite element analysis of balloon angioplasty, *Advances in Bioengineering*, BED Vol. 22, ASME, 1992, 269.

76. Chandran, K. B., Gao, D., Han, G., Baraniewski, H., and Corson, J. D., Finite element analysis of artery-graft anastomosis, *Proceedings of the Third Joint ASCE/ASME Mechanics Conference*, AMD Vol. 98, ASME, San Diego, CA, 1989, 233.

77. Chandran, K. B., Gao, D., Han, G., Baraniewski, H., and Corson, J. D., Finite element analysis of arterial anastomoses with vein, dacron and PTFE grafts, *Med. Bio. Eng. Comp.*, 30, 413, 1992.

78. Ballyk, P. D., Walsh, C., Zhu, S. Y., and Ojha, M., Stress distribution at an end-to-side anastomosis, in *1995 ASME Bioengineering Conference*, BED Vol. 29, Hochmuch, R. M., Langrana, N. A., and Hefzy, M. S., Eds., ASME, 1995, 523.

79. Ballyk, P. D., Walsh, C., Zhu, S. Y., and Ojha, M., Stress distribution at an end-to-side anastomosis, in *Advances in Bioengineering*, BED Vol. 31, Hull, M. L., Ed., ASME, 1995, 327.

80. Borgersen, S. E. and Sakaguchi, R. L., Simulation inflation of a vascular stent using 3-dimensional nonlinear FEA contact analysis, in *Advances in Bioengineering*, BED Vol. 28, Askew, M. J., Ed., ASME, 1994, 129.

81. Gu, W., Lai, W. M., and Mow, V. C., A generalized triphasic theory for multi-electrolyte transport in charged, hydrated soft tissues, in *Advances in Bioengineering*, BED Vol. 28, Askew, M. J., Ed., ASME, 1993, 217.

82. Huang, Y., Rumschitzki, D., Chien, S., and Weinbaum, S., A fiber matrix model for the filtration through ferestral pores in a compressible arterial intima, *Heart Circ. Physiol.*, 41, H2023, 1997.

6

A Three-Dimensional Vascular Model and Its Application to the Determination of the Spatial Variations in the Arterial, Venous, and Tissue Temperature Distribution

6.1   Introduction 6-1
6.2   The Vascular Model 6-3
      Description of the Vascular Model • Geometrical and Vascular
      Parameters • Temperature Profiles due to Different Thermal
      Conditions • Conclusions from the Vascular Model
6.3   The Efficiency Function Approach 6-14
      The Efficiency Function • Validation of the Efficiency Function
      Concept • Conclusions about the Efficiency Function Approach
6.4   The Vascular Model and the Efficiency Function
      Approach in Hyperthermia Treatment 6-18
      Temperature Profiles • Conclusions from Hyperthermia
      Simulations

Jürgen Werner
Ruhr University

Heinrich Brinck
FH Gelsenkirchen

6.1   Introduction

Convective heat transfer via the blood plays an essential role in homeothermy of the human body. Pennes[1] proposed quantifying heat transfer effects in perfused biological tissue by a heat source/sink term. This term is proportional to the perfusion rate and the difference between the tissue temperature and the global arterial blood temperature (bio-heat approach). The underlying assumption was that all heat transfer occurs in the capillaries. Pre-arteriole and post-venule heat transfer was neglected. Weinbaum et al.[2,3] questioned the validity of the bioheat equation that had been used for many years in the analysis of thermal problems in perfused tissue. They proposed a new bioheat transfer equation.

It is the merit of Weinbaum and Jiji[4] that they drew attention to the fact that countercurrent heat exchange in small arteries and veins is of importance under certain conditions. Countercurrent heat transfer helps to conserve heat in conditions of severe cold. Intervessel heat exchange is significant especially during periods of minimal blood flow. A new bioheat transfer equation was presented by Weinbaum and Jiji in 1985[4] and further developed by Zhu et al.[5] and Song et al.[6,7] This model quantifies tissue heat transfer with a single equation that is independent of any blood temperature. To account for the enhancement of heat transfer near vessels not in thermal equilibrium, an enhanced thermal conductivity term is introduced.

In an exacting discussion (Wissler,[8,9] Weinbaum and Jiji[10]), the underlying assumptions in the new model were questioned. In their early work, Weinbaum and Jiji assumed that the mean of arterial (T_a) and venous (Tv) blood temperatures were equal to the mean tissue temperature T_t: (($T_a + T_v$)/2 $\approx T_t$). In later papers (Zhu et al.,[5] Weinbaum and Jiji[11]), a slightly different hypothesis was introduced (see below). They, too, assumed that there is negligible heat transfer between the countercurrent artery–vein pairs and the surrounding tissue. Wissler is convinced that a source/sink term of the form proposed by Pennes belongs to the bioheat equation. In his opinion, the old form of the bioheat equation yields more realistic results than the new form. He presented an alternative model[9] which computes the spatial variations in the arterial, venous, and tissue temperatures. Wissler's model consists of three equations and can be incorporated into finite-difference whole body models. Unfortunately, Wissler did not suggest any algorithm to calculate the heat transfer parameters which are used to quantify the heat flow between the vessels and between the vessels and the tissue.

A model by Baish et al.[12] derives the relationship between the countercurrent and vessel/tissue heat transfer parameters and the morphology of the vascularity. Some crucial simplifications restrict the applicability in whole body models. The model does not consider axial conduction through the tissue cylinder along the vessel axes. Only heat transfer perpendicular to a vessel pair is described.

Charny and Levin[13] incorporated the Baish model into a system of heat balance equations similar to those proposed by Wissler. They called this one-dimensional model the three-equation model and used the predictions of this model to evaluate the models of Pennes and Weinbaum and Jiji. In a later paper, Charny et al.[14] found that the two approximate formulations were each appropriate in different tissue regimes and under certain physiological conditions. They proposed that a hybrid model would be most appropriate for simulations of bioheat transfer in perfused tissue.[14]

The vascular network in the muscle tissue near the skin surface is a region of special interest within the systemic circulation.

Chen and Holmes[15] defined twelve categories of vessels according to their diameter and calculated the equilibration length, i.e., the length over which the temperature difference is reduced by a factor e. This length depends on the blood velocity, which means that a vessel category which is in thermal equilibrium at a basal blood velocity may not be in equilibrium at high blood flow. Nevertheless, from such calculations it became clear that the equilibrium of blood temperature with tissue takes place between the terminal arterial branches and the pre-capillary arterioles, and not in the capillaries, as had been previously assumed. Fig. 6.1 schematically shows the changes of blood temperature as the blood traverses the systemic circulation. Blood at arterial temperature T_0 is distributed to tissues which are warmer (upper branch of the curve in Fig. 6.1) or cooler (lower branch) than T_0. Blood temperature rapidly equilibrates with tissue temperature after leaving the terminal arterial branches and attains the tissue temperature prior to entering the arterioles. Beyond the venules, major changes in temperature are the result of mixing at venous confluences. Mixed vena cava blood returns to the heart at a temperature close to arterial temperature.

The object of our study was to develop a three-dimensional and vascular model[16] in which convective heat exchange between feeder vessels and the tissue is calculated without any assumptions other than values for the Nusselt number. The Nusselt number is a dimensionless parameter to quantify the heat transfer coefficient at the vessel wall. The contributions of these thermally significant vessels are treated individually and not collectively. We examined the way in which the vascular geometry and blood flow rate influence the heat transported in perfused tissue under basal and extreme conditions. A small scale

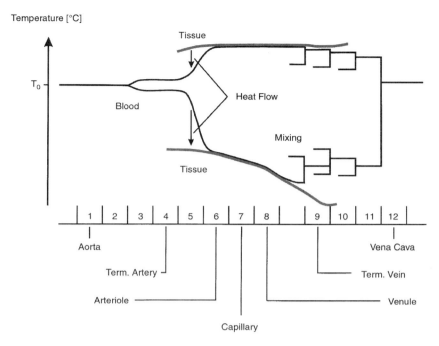

FIGURE 6.1   Schematic representation of the blood temperature profile of the systemic circulation (modified after Chen and Holmes, N.Y. Acad. Sc.,1980). Ta_0 = arterial temperature; numbers 1–12 = category of vessel.

vascular model of this kind can be used for several purposes. For the purpose of studying hyperthermia as a cancer therapy, significant localized variations in temperature and heat flux are of great interest. Since the basic vascular model presented here is calculated without less crucial assumptions, it is possible to test the validity of assumptions inherent in the formulations of other bioheat transfer concepts.

6.2   The Vascular Model

Description of the Vascular Model

A three-dimensional vascular model was constructed to compute the spatial variations in the arterial, venous, and tissue temperatures. This model was developed by considering closely spaced, countercurrent pairs of thermally significant vessels of radii A and V, assumed to run in z-direction (Fig. 6.2). The vessels are surrounded by a tissue cylinder of radius Rt.

The model will be applied to the cross-sectional area of a human extremity. A schematic axial view of a limb is shown in Fig. 6.2, where one can see the idealized three layer organization of the vasculature of the peripheral circulation: the core contains the countercurrent central artery and vein and the surrounding tissue. These vessels are the starting points of a countercurrent arterio-venous network in the muscle layer. The countercurrent pairs of arteries and veins are constantly branching. The counter-current network in the muscle tissue typically consists of eight generations of paired vessels. The muscle layer is surrounded by the skin layer in which the blood is supplied by isolated larger riser vessels. The thickness of each layer is illustrated in the left part of Fig. 6.2.

Core Layer

The core layer in the center of the limb contains the central countercurrent artery and vein. In the calculations presented here, the artery and vein are not incorporated explicitly into the model. Conductive heat transfer through the vessel walls is neglected. A no-flux condition is enforced at the bottom of the

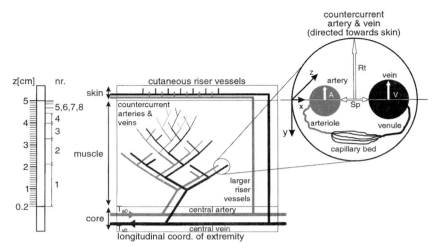

FIGURE 6.2   Schematic view of the three-layer organization of a limb including the vascular of the peripheral circulation. Amplification: Schematic view of a countercurrent vessel pair surrounded by a tissue cylinder of Krogh-type. A = radius of artery; V = radius of vein; Sp = spacing between vessels; Rt = radius of tissue cylinder. The z-coordinate is measured in a radial direction perpendicular to the skin surface which is also the direction of the large scale temperature gradient.

core layer. Blood is supplied in a radial direction through two separate systems of vessels: the muscle circulation and the cutaneous circulation.

Muscle Layer

The total thickness of the muscle layer is $L_m = 0.048$ m. The vasculature is described in detail. It is assumed that the vessels vary in spacing (Sp), radius (A,V), number density (Nd), and flow velocity (U) only as a function of tissue depth z. By cutting the tree in different planes parallel to the surface of the ground one gets eight sections according to the eight vessel generations. Another not-crucial assumption is that the geometrical parameters remain constant along one vessel generation n: Sp(n), A(n), V(n), Nd(n). Convenient functional forms for the variation of the vascular geometry in the muscle layer in the radial direction have been proposed by Song et al.[7] The cross-sectional area of the surrounding tissue cylinder is equal to the inverse of the vessel pair density: $1/Nd(n) = \pi Rt^2(n)$. The Krogh-type tissue cylinder is affected only by the blood vessel pair which is located along its central axis. No-flux boundary conditions are enforced around the tissue cylinder walls in x-direction and y-direction. Although the radius V of a vein is often two or three times larger than the radius A of the corresponding artery, calculation is also done with equal-sized vessel pairs so that the results may be compared to the results of other analytic or simulation models; e.g., the so-called Baish-values[12] can only be calculated under the condition A = V. The model can compute the temperature distribution around vessels of unequal size without changes. In order to simplify the boundary calculations, the circular cross-sections of the blood vessels and the tissue cylinder were transformed to square boundaries of equivalent hydraulic diameter $D_h = 4$ times cross-sectional area/perimeter. The arteriole, venule, and capillary bed heat transfer between blood and tissue was calculated using Pennes' traditional bioheat equation,[1] which is generally accepted.

    Two measures reduce the computational costs: no heat flux occurs across the Krogh-type cylinder walls. Out of each branching generation one vessel pair with its sourrounding tissue cylinder is taken to calculate the temperature profile in three dimensions. Since the cross-sectional area of a tissue cylinder has symmetry in relation to the x-axis, a no-flux condition is assumed across the x-axis to calculate the temperature profile on half of the cross-sectional area.

Skin Layer

Blood is supplied to the skin by isolated larger riser vessels. The arterial blood that enters the cutaneous plexus in the skin is assumed to have the same temperature T_{a0} as the arterial blood in the central artery. Upon entering the small vertical riser vessels, the blood equilibrates immediately to the local tissue temperature. This warm blood perfusion is modeled as a distributed volumetric heat source with the traditional Pennes source term. The skin layer exchanges heat at its surface with its surroundings by radiation, convection, conduction, and evaporation.

Geometrical and Vascular Parameters

The quantities L, Sp, A, Rt, and U/U_0 are functions of the vessel generation number n. These data are summarized in Table 6.1.

TABLE 6.1   Geometrical and Vascular Parameters (mm or N.D.)

n	L	Sp	A = V	Rt	U/U_0	Sp/A
Core	2.00	—	—	12.97	—	—
1	19.13	.1922	.3845	12.97	1.0	.5
2	11.66	.1056	.2112	9.17	.9975	.5
3	7.11	.0611	.1217	6.48	.8968	.5
4	4.33	.0434	.0747	4.59	.6996	.6
5	2.64	.0578	.0496	3.24	.4556	1.2
6	1.61	.1035	.0358	2.29	.2388	2.9
7	.98	.1595	.0281	1.62	.0961	5.7
8	.60	.2074	.0237	1.15	.0257	8.8
Skin	2.00	—	—	1.15	—	—

Variation in angular direction is not considered. Because no heat flux occurs across the Krogh-type cylinder walls, the calculations are not only representative for a longitudinal slice but for the whole extremity in general. The distinction between three basic regions in the limb is typical for surface tissue in general. Only the relative dimension and the layer thickness vary over the body.

Blood flow is considered to be steady and laminar, because in microcirculation, steady or quasi-steady flow conditions dominate. Vessel tapering and wall distensibility are not taken into account. To describe the heat transfer between vessel and tissue in microcirculation, the use of the asymptotic values of the Nusselt number seems to be appropriate. A local Nusselt number is incorporated into the vascular model to account for entrance effects or local marginal layer conditions. In the calculations presented here, however, an overall Nusselt number value of 4.0 is used to describe the heat exchange between vessel and tissue. This value was proposed by Chato.[16]

Equations

The general diffusion equation is solved using implicit finite difference methods with linear and nonlinear boundary conditions. The passive system is described by the diffusion equation in the form:

$$c\rho \cdot \partial T/\partial t = \partial/\partial x\left(\lambda \cdot \partial T/\partial x\right) + \partial/\partial y\left(\lambda \cdot \partial T/\partial y\right) + \partial/\partial z\left(\lambda \cdot \partial T/\partial z\right) + Q \tag{6.1}$$

$$Q = Q_{perf} + Q_{met} \tag{6.2}$$

where c = specific heat [J/kg/°C], ρ = density [kg/m³], λ = thermal conductivity [W/m/°C)], T = tissue temperature °C, t = time [s], Q = heat source [W/m³], Q_{met} = metabolic heat source, Q_{perf} = perfusion heat source, and $x,y,$ and z = coordinates [m].

The heat transfer from the feeder vessels running in z-direction is calculated individually, not collectively in a continuum formulation. Continuum formulations are only used for vessels running perpendicular to the z-direction. Here the arteriole, venule and capillary bed heat transfer between blood and tissue is calculated using the traditional perfusion source term first defined by Pennes. But in this term, the global arterial blood temperature T_{a0} is replaced by the local arterial blood temperature $T_a(z,t)$.

$$Q_{perf}(z,t) = \omega \cdot c_b \rho_b \cdot (T_a - T)$$ (6.3)

where ω = tissue perfusion [m³ blood/s/m³ tissue], c_b = specific heat of blood [J/kg/°C], ρ_b = density of blood [kg/m³], and T_a = arterial blood temperature [°C].

Since the temperature of blood in the vessels changes with time, it is necessary to write two more equations. The arterial and venous temperatures are computed along the length of a one-dimensional countercurrent network in z-direction. Heat transfer at the vessel walls is regarded as laminar heat transfer in a circular tube. The temperatures along the vessels are considered to be quasi steady-state (since the time constant for vessels is essentially the transit time of blood through the vessel[19]). The parameters and coordinates of laminar flow in an artery are demonstrated in Fig. 6.3.

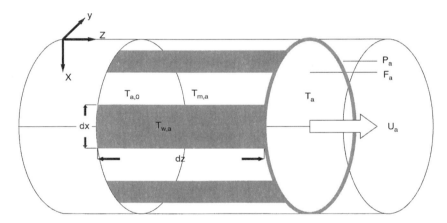

FIGURE 6.3   Three-dimensional view of an artery with parameters and coordinates of laminar flow.

The heat flow through a vessel surface area ($P_a dz$) is equal to the change in energy of the arterial blood along *dz* in steady-state.

$$P_a \cdot dz \cdot \alpha_a (T_{w,a} - T_a) = F_a \cdot U_a dz \cdot c_b \rho_b \cdot dT_a / dz$$ (6.4)

where $T_{w,a}$ = arterial wall temperature [°C], F_a = cross-sectional area of the artery [m²], and U_a = blood velocity in the artery [m/s].

The value α_a is determined by using the local Nusselt number $Nu_a(z)$ to describe this heat exchange. The value

$$St_a = \alpha_a / (U_a c_b \rho_b) = (Nu_a \cdot \lambda / (2 \cdot A)) / (U_a c_b \rho_b)$$ (6.5)

is called the Stanton number. With the relationship

$$k_a = -(P_a / F_a) \cdot St_a = -(Nu_a \cdot \lambda) / (A^2 \cdot U_a \cdot c_b \rho_b)$$ (6.6)

and assuming $\alpha_a = $ const along $[0, \Delta z]$, one obtains by integration

$$T_a(z) = T_{w,a} + \left(T_a(0) - T_{w,a}\right) \cdot \exp(ka \cdot z) \tag{6.7}$$

The mean arterial blood temperature $T_{m,a}$ along Δz can be calculated from

$$T_{m,a} = T_{w,a} + \left(T_a(\Delta z) - T_a(0) / (k_a \cdot \Delta z)\right) \tag{6.8}$$

where $\Delta z = $ nodal spacing in z-direction. The three-dimensional energy balance (Eq. 6.1) is subject to a convective boundary condition along the arterial and venous wall. At the tissue-vessel interface it is:

$$-\lambda \, \partial T / \partial n = \alpha_a \cdot \left(T - T_{m,a}\right) \tag{6.9}$$

where $n = $ the coordinate axis perpendicular to the vessel direction.

T_v is determined the same way as T_a. It is necessary to account for the relative average volumetric blood flow rate g entering the venous flow. Blood is equilibrated within the tissue it perfuses. Upon returning to the vein, the capillary blood will be at the local mean tissue temperature T_t.

$$T_v = \left(1.0 - g\right) \cdot T_v + g \cdot T_t \tag{6.10}$$

The mean tissue temperature $T_t(z, t)$ is calculated using an area-weighted average of the node temperatures $T(x,y,z,t)$. The true venous return temperature depends on the vascular geometry of the venules and small veins re-entering the countercurrent vein. For a basal perfusion rate, the blood in these vessels may re-enter the vein at a temperature near $T_{w,v}$. At high perfusion rates, the venous return temperature assumes a value somewhere between $T_{w,v}$ and T_t. The difficulty is that there is at present no theory for predicting this temperature. As the temperature difference $|T_t(z) - T_{w,v}(z)|$ is comparatively small (Fig. 6.4), it seems to be acceptable to take the T_t limit for the solution procedure. Note that the limit $T_{v,v}$ is not equal to the blood temperature T_v in the countercurrent vein: $T_{w,v}(z) \neq T_v(z)$.

$$\left|T_{w,v}(z) - T_v(z)\right| \approx \left|T_{w,v}(z) - T_t(z)\right| \tag{6.11}$$

At the surface of the skin, the rate at which heat is transported to the surface by conduction is equal to the rate of heat removal from the surface by radiation, conduction, convection, and evaporation.

$$-\lambda \cdot \partial T / \partial n = q_{rad} + q_{cond} + q_{conv} + q_{eva} \tag{6.12}$$

The muscle tissue surrounding the vessel pair was subdivided into blocks. The corresponding three-dimensional nodal network consists of 187,892 nodes.

To get an impression, see the grid pattern for n = 3 with a resolution of 80 × 30 gridpoints (gp). The nodal network together with the absolute and relative dimensions is presented in Fig. 6.5.

A central difference approximation is done using an irregular grid. The nodal spacings in x-direction and y-direction are small around the vessels and in the spacing between the vessels. The nodal spacings at the vessel walls are up to 64 times smaller than the nodal spacings at the tissue cylinder wall. The arterial and venous walls comprise 56, 48, 32, or 24 gridpoints. At the interface in z-direction between two tissue cylinder ends, gridpoints with the same position within the grid are associated with each other. When the three-dimensional tissue temperature profile and the one-dimensional blood temperature profiles are calculated for one cylinder, the neighbouring tissue cylinders provide fixed boundary values.

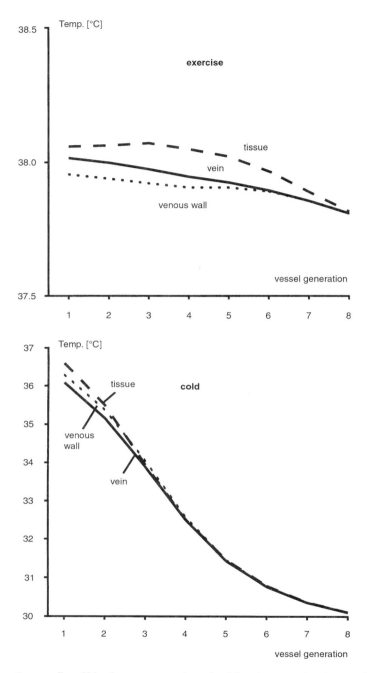

FIGURE 6.4   Mean tissue, wall, and blood temperatures along the eight vein generations in case of exercise and cold.

The correspondent gridpoint temperatures and the temperature of the blood leaving the neighboring tissue cylinder are assumed to be fixed for one time step.

Environmental Conditions

Three different cases denoted R, W, and C were studied. Emphasis was laid on those physiological features which influence the convective heat transfer directly, like the perfusion rate or the blood velocity in the tissue layers. Rather extreme values for these features were chosen to demonstrate the influence of these

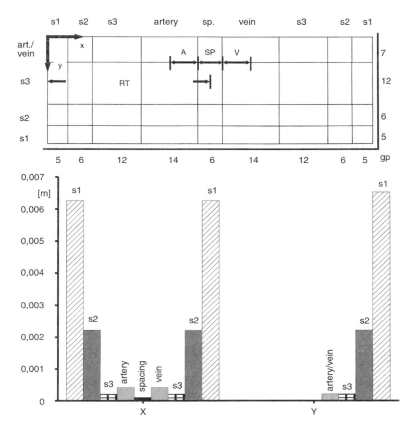

FIGURE 6.5 Schematic view of the irregular grid in x- and y-direction with gridpoint resolution (upper panel) and dimensions within the irregular grid (lower panel). gp = number of gridpoints; s1, s2, and s3 = regions with different nodal distances.

factors. In case R, the steady-state tissue and blood temperature distribution was calculated under resting conditions. In case W, exercise was modeled by increasing the metabolic heat source by a factor of 100.0. The metabolism remained constant at this maximum level until steady-state occurred. This is an idealized assumption because of this high level of exercise. The blood perfusion rate in the muscle was assumed to be elevated to a maximum by a factor of 20.0. In case C, the ambient temperature was decreased to 5.0°C. Subsequently, the skin blood flowrate decreased to the minimum level because of this cold environment. The blood entered the countercurrent network in the muscle tissue, the core, and the skin with a constant temperature T_{a0} throughout the time. The parameters are listed in Table 6.2.

Numerical Methods

The three-dimensional diffusion equation with coefficients as functions of location and time is solved by a finite-difference method, i.e., by a modification of the alternating direction implicit method (ADI), first reported by Douglas and Rachford.[20] The program is processed by a Control Data CYBER 205 vector computer. To achieve increases in computational speed on vector computers, one must use special algorithms for parallel/vector machines. Most of the computational time (T_{CPU}) in the program is used to solve tridiagonal systems of linear equations by the parallel direct solution procedure cyclic reduction (CR). There are great irregularities in the underlying nodal network. The minimum spacing between two nodes is 0.007 mm, and the maximum spacing is 2.1 mm, which is 300 times greater. The explicit computation of the blood temperatures is loosely coupled to the computation of the three-dimensional tissue temperature profiles. For these reasons, one has to choose a very small time step, $\Delta t = 0.1$ s. Each time step requires 1.5 CPU-s. The time consumption for a steady-state computation is about 30.0 CPU-h.

TABLE 6.2   Parameters Used for Three Environmental Conditions

Parameter	Rest	Exercise	Cold	Unit
$Q_{met,core}$	245.0	245.0	245.0	W/m³
$Q_{met,muscle}$	245.0	24500.0	245.0	W/m³
$Q_{met, skin}$	245.0	245.0	245.0	W/m³
ω_{core}	0.0005425	0.0005425	0.0005425	m³/sm³
ω_{muscle}	0.0005425	0.01085	0.0005425	m³/sm³
ω_{skin}	0.0005425	0.0005425	0.00001	m³/sm³
T_u	30.0	30.0	5.0	°C
q_{eva}		10.0		W/m²
v		0.0		m/s
Nu		4.0		N.D.
T_{a0}		37.5		°C
$c_t = c_b$		3800.0		J/kg°C
$\rho_t = \rho_b$		1050.0		kg/m³
$\lambda_t = \lambda_b$		0.465		W/m°C

Temperature Profiles due to Different Thermal Conditions

The vascular model calculates the tissue temperatures $T(x,y,z)$ in three dimensions. Blood temperatures $T_a(z)$, and $T_v(z)$ and heat flows $q_a(z)$, and $q_v(z)$ are calculated in the direction of the vessels. The mean tissue temperature $T_t(z)$ perpendicular to the vessels and the mean temperature $T_{sp}(z)$ in the spacing between the vessels is extracted from $T(x,y,z)$.

Fig. 6.6 shows a two-dimensional tissue temperature profile around a vessel pair embedded in muscle tissue. The profile is located in the middle of the muscle layer. Resting conditions are assumed. Because the underlying grid is not equally spaced, the diagram is not in linear scale. The cross-sectional area around a vessel pair has mirror symmetry in relation to the x-axis.

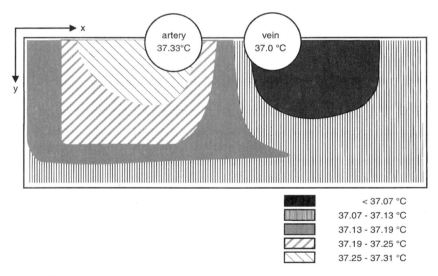

FIGURE 6.6   Two-dimensional temperature profile in the middle of the muscle tissue perpendicular to the vessels under resting conditions.

The tissue temperatures range from temperature $T_a(z)$ of the arterial blood to temperature $T_v(z)$ of the venous blood. The temperature domains are represented by graphic patterns. It can be concluded from the distribution of the patterns that there is a greater temperature gradient around the artery than around the vein, and that the mean tissue temperature $T_t(z)$ is closer to the temperature $T_v(z)$ of the venous blood while the mean tissue temperature of the spacing between the vessel is exactly the arithmetic mean of the blood temperatures: $T_{sp} = (T_a + T_v)/2$.

Characterizing temperatures for this z-location are:

$$T_v = 37.0 < T_{t,min} = 37.02 < T_t = 37.12 < T_{sp} = 37.16 < T_{t,max} = 37.3 < T_a = 37.33$$

Fig. 6.7 demonstrates temperature profiles of the mean tissue temperatures T_t and T_{sp} and the blood temperatures T_a and T_v in the radial direction in cases R, W, and C. The arterial blood temperature T_a increases slightly in case W and decreases in cases R and C along the length of the countercurrent network. In light of this computational result, the assumption of a constant arterial blood temperature throughout the muscle tissue is not correct.[1] In a very cold environment, the arterial blood temperature decreases from 37.5°C to 30°C in the outer section of the muscle tissue.

There must be pre-arteriole heat transfer between the blood and tissue which cannot be neglected, especially in a cold environment and when the perfusion rate is small. As evident from the arterial blood temperatures calculated in case W, the influence of pre-arteriole heat transfer decreases when the perfusion rate is high. At the end of the vascular tree, blood is equilibrated with the surrounding tissue: $T_t = T_a = T_v$.

For equally sized countercurrent vessels, the arithmetic mean of the blood temperatures is equal to the mean tissue temperature of the spacing between the vessels at any given location along the branching network: $T_{sp} \approx (T_a + T_v)/2$. If the metabolic rate and the perfusion rate are high (W), the tissue temperature exceeds the arithmetic mean of the blood temperature by 0.1–0.3°C as well as the blood temperatures. The temperature of small volumes of tissue near the vessel walls is significantly lower than the mean tissue temperature.

In a cold environment, the mean tissue temperature falls below the arithmetic mean of the arterial and venous blood temperatures by about 0.1-0.5°C. The assumption $T_t = (T_a + T_v)/2$ is approximately valid for only a very small region of the muscle tissue near the skin where the blood is nearly equilibrated with the tissue. This, however, is one crucial assumption inherent in the early work of Weinbaum and Jiji.[4] They assumed that for equally sized countercurrent vessels, $(T_a + T_v)/2$ is equal to T_t at any given location along the branching network, which, according to our detailed model, does not seem to be correct, and which was already questioned by Wissler.[8]

Furthermore, results show (Fig. 6.8) that Weinbaum and Jiji's later hypothesis $d(T_a + T_v)/dz \approx dT_t/dz$ is questionable, too. Fig. 6.9 shows the heat flows q_a and q_v W/m from the tissue into the vessels. When the metabolic rate is low, in cases R and C, there is heat loss from the arterial blood and heat gain to the vein. The heat flow out of the artery is slightly greater than the heat flow into the vein. When the metabolic rate is high (W), the direction of the heat flows is reversed. In this case, the amount of heat flow into the artery is twice as great as the heat flow out of the vein. It is interesting that, in spite of the fact that the mean tissue temperature $T_t(z)$ lies above both blood temperatures $T_t(z)$ and $T_v(z)$, there is a heat flow out of the vein. This is a consequence of the tissue cooling by the artery in the neighborhood of the vessel pair: $T_{w,v} \leq T_v \leq T_t$ (see Fig. 6.4). There is a net heat flow from the vessels to the tissue, even though the vessels are closely coupled.

The layer of fat between the muscle and the skin can be treated as a pure conduction layer. This layer is thin compared to the muscle layer. It was shown that the blood in the vessels of the muscle tissue adjacent to the skin is equilibrated with the surrounding tissue. It is not essential to make a distinction between the two layers. The temperature within the vessel is not uniform. Baish et al.[12] note that the thermal resistance within the vessel cannot be neglected when the spacing between the vessels is small. Using a Nusselt number close to 4,[21] allowance is made for the thermal resistance within the vessels. Wissler[18] calculates that the Nusselt number equals 48/11 = 4.364. He notes that this is correct, even when the temperature profile in blood is markedly skewed due to the presence of the other vessel, and that the value does not depend on either intervessel distance or the relative diameters of the two vessels. It is appropriate to use the cup-mixed, mean blood temperatures in computations. Zhu et al.[21] demonstrated

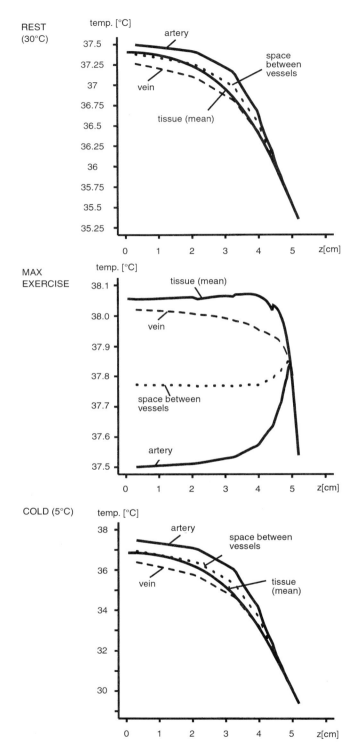

FIGURE 6.7   Radial temperature profiles of tissue temperatures and blood temperatures in cases of rest, exercise, and cold.

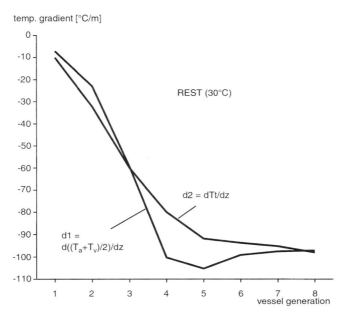

FIGURE 6.8 Tissue temperature gradient dT_t/dz compared with the mean blood temperature gradient $d(T_a + T_v)/2/dz$ under resting conditions in z-direction.

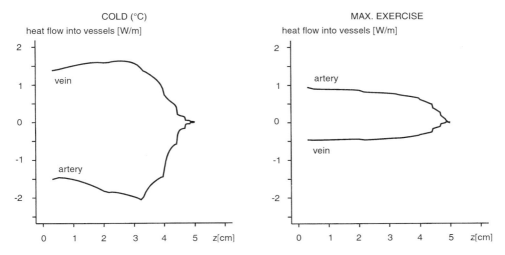

FIGURE 6.9 Heat flow from the tissue into the artery and vein in cases of cold and exercise.

that the result is also true for countercurrent heat exchange in a tissue cylinder in which there is not perfect countercurrent heat exchange. In further computations, the value of 48/11 for the Nusselt number should be used instead of the value 4.

An even more sophisticated model would have to include time varying arterial temperature (T_{a0}) in the central artery and the effects of central and local thermoregulation on blood flow and evaporation rate in the skin layer.

Conclusions from the Vascular Model

The three-dimensional vascular model presented here is unique in its implementation of the vascular system explicitly in terms of the physical details. The pre-arteriole and post-venule vessels of the muscle layer are incorporated into the model. The arterial blood temperature is not constant along the length of the countercurrent network. The deviation from the arterial blood temperature at the beginning of the vascular tree increases when the perfusion rate is small. The temperature equilibrium occurs with the passage of blood through the vascular tree in the muscle tissue near the skin surface. At the end, the blood is equilibrated with the surrounding tissue. Radial temperature profiles of the mean tissue temperature and the mean tissue temperature of the spacing between the vessels were presented for various conditions. Blood temperature profiles and profiles describing heat flow across the vessel walls were calculated.

Assumptions inherent in bioheat formulations of Pennes[1] and Weinbaum and Jiji[4] for the convective heat transfer were questioned.

When the perfusion rate is high and when, at the same time, there is a high metabolic rate, there is a heat flow into the artery and a minor heat flow out of the vein. This is obviously a countercurrent effect. The vascular model presented here treats the effect of blood perfusion on a vessel-by-vessel basis and predicts the temperature in and near individual blood vessels. However, detailed information about vascular geometry is rarely available. Even when it is, the resulting computational task is formidable. Such a model is useful only in modeling small volumes of tissue. Whole body models, however, should be constructed without the physical details of the vascular system.

6.3   The Efficiency Function Approach

The Efficiency Function

Because of its complexity, a vascular model, although delivering valuable results substantially more reliable than those of former nonvascular models, cannot be routinely used as a module for thermoregulatory whole body models, where a simple module is required for the perfusion term (Q_{perf}) in the heat balance differential Eq. 6.1.

The classical bioheat approach by Pennes[1] for perfusion is as follows:

$$Q_{perf} = \omega \cdot c_b \rho_b \left(T_{a0} - T_t \right) \tag{6.13}$$

i.e., the perfusion term is the product of perfusion ω, specific heat c_b of blood, density ρ_b of blood, and difference of arterial blood temperature T_{a0} (assumed to be constant with respect to the local coordinate) and mean local tissue temperature T_t. The results of the vascular model confirmed the assumption that such an approach, although very easy to use, may lead to wrong computations and conclusions, especially at a low perfusion rate or, worse, under non-homogeneous conditions (e.g., local hyperthermia). Therefore, we computed an efficiency function (EF)[23] dependent on perfusion ω and radial coordinate r to compensate for the deficiencies of the bioheat approach:

$$Q_{perf} = EF\left(\omega, r\right) \cdot \omega\left(r\right) c_b \rho_b \left(T_{a0} - T \right) \tag{6.14}$$

Since the function should follow the conditions:

$$
\begin{aligned}
EF\left(\omega, r\right) &\to 0, & \text{if } \omega \to 0, \\
EF\left(\omega, r\right) &\to L\left(r\right), & \text{if } \omega \to \infty, \\
EF\left(\omega, r\right) &\in \left[0, L\left(r\right)\right]
\end{aligned}
\tag{6.15}
$$

we chose the following form:

$$EF(\omega,r)=L(r)/\left[1+F(r)/\omega(r)\right] \qquad (6.16)$$

$L(r)$ and $F(r)$ are parameter functions, the values of which were varied in a way that the computed temperature profiles were in good compatibility with the results of the complex vascular model (see Fig. 6.11).

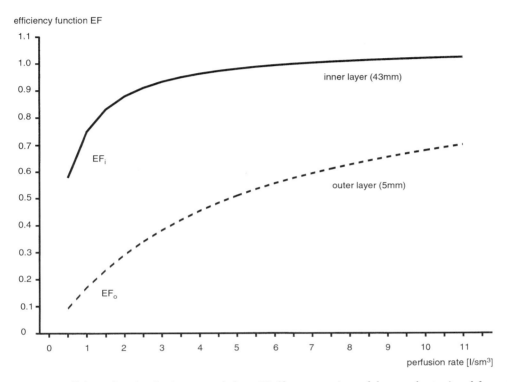

FIGURE 6.10 Efficiency function for inner muscle layer EF_i (four generations of the vascular tree) and for outer muscle layer EF_o (four further generations) as a function of the perfusion rate. $EF = 1$ means no correction of the Pennes bioheat term.

To be able to incorporate this alternative model easily into whole body models,[24–28] the parameter function was simplified by assuming constancy within the generations of the branching vascular tree. Good agreement between the predictions of the vascular model and the predictions of the nonvascular model using the modified perfusion term was achieved, defining the parameter functions L and F for an inner muscle layer (subscript i, four further generations of the vascular tree) and an outer layer (subscript o, four further generations of the vascular tree). The values determined by minimizing both the mean difference of the two predictions and the maximal deviation were $F_i = 0.0004$, $L_i = 1.065$, $F_o = 0.005$, and $L_o = 1.023$. The inner layer typically is 9–10 times thicker than the outer layer. In our standard computations, the thickness of the inner and outer layers were 4.5 cm and 0.5 cm, respectively. Fig. 6.10 shows the EF graph for the inner and outer muscle layers as a function of the perfusion rate. $EF \neq 1$ denotes a correction of the Pennes term. In the perfusion interval, EF_i increases from 0.58 to 1.03 and EF_o from 0.09 to 0.7. This means that the classical Pennes term has to be corrected in the inner muscle layer at low perfusion rates, whereas in the outer layer it also has to be modified at higher perfusion rates. At low rates, the outer layer has, more or less, the properties of a conduction layer ($EF \rightarrow 0$). Outside the muscle tissue, a correction of the Pennes term is not necessary ($EF = 1$).

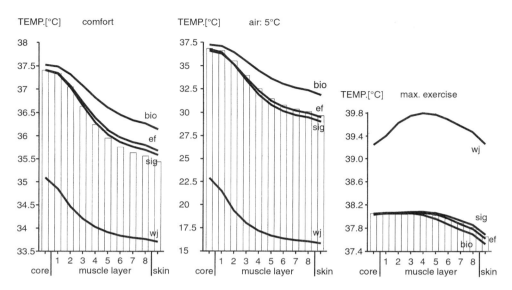

FIGURE 6.11 Predicted temperature profiles for comfort conditions, cold, and maximal exercise. Temperature of central artery of core assumed to be 37.5°. Eight muscle layers (of decreasing thickness) according to vessel generations of countercurrent arteriovenous tree. Bars represent the exact vascular model. Non-vascular approaches are denoted by: bio = Pennes' bio heat; ef = efficiency function; sig = Baish/Wissler/Charny-Levin; and wj = Weinbaum-Jiji.

After determination of EF, the perfusion term Q_{perf} in Eq. 6.14 may be determined and used for the heat balance, Eq. 6.1. Additionally, the heat balance for the blood has to be modified slightly. To calculate mean venous temperature (T_{v0}) of the blood returning from the muscle tissue to the central vein, a local effective temperature ($T_{ef}(r)$ is defined as:

$$T_{ef}\left(r\right) = Ef\left(\omega, r\right) \cdot T\left(r\right) + \left(1 - EF\left(\omega, r\right)\right) \cdot T_{a0} \tag{6.17}$$

T_{v0} is calculated using a perfusion-weighted average of this effective tissue temperature $T_{ef}(r)$:

$$T_{v0} = \int_r \left(\omega\left(r\right) \cdot T_{ef}\left(r\right)\right) dr / \int_r \left(\omega\left(r\right)\right) dr \tag{6.18}$$

Although this efficiency function approach is formally very similar to the bioheat approach, it has essentially different properties as it comprises implicitly pre-arteriole and post-venule heat transfer, as well as countercurrent heat transfer in neighboring artery–vein pairs. The formal simplicity of the essential Eqs. 6.13 and 6.14, however, has not changed.

Validation of the Efficiency Function Concept

First we compared results for a human extremity obtained by the simplified nonvascular models to those computed by the exact vascular model. Again, three situations are examined: (1) resting comfortable conditions: ambient temperature (30°C), basal metabolic rate (245 W/m³), and perfusion ($5.4 \cdot 10^{-4}$ m³/s⁻¹m⁻³); (2) cold stress (5°C) with maximal vasoconstriction in the skin (10^{-5} m³ · s⁻¹ · m⁻³); to create difficult non-homogeneous testing conditions, shivering is not initiated; and (3) maximal exercise at 30°C is simulated by increasing metabolic heat production in the muscles involved by a factor of 100, inducing a maximal perfusion of $\approx 10^{-2}$ m³ · s⁻¹m⁻³. Again, to create extreme testing conditions, sweat production is not initiated. The other physical and physiological parameters used are the same as listed in the above report on the vascular model.

Because we are not dealing with whole body models, the temperature of the central artery in the extremity is chosen arbitrarily (37.5°C in these comparative computations). The comparison is presented in Fig. 6.11. As the x-axis indicates, there is a core layer, eight generations of the vascular tree in the muscle (1–8), and a skin layer; because these 10 layers are of different thickness, there is no linear horizontal scale. If we compare the profiles calculated with the nonvascular models to those gained by the vascular model, it is evident that the bioheat approach delivers satisfactory results in heat stress conditions (which, if the heat is not induced locally, yield a rather homogeneous temperature distribution in the tissue). However, in indifferent and especially cold conditions (where temperatures diverge very much), the bioheat approach delivers temperatures deviating by about 2.5°C from those obtained by the vascular model, an inaccuracy that can be compensated to a large extent by using the EF, the maximal deviations of which are about 0.3°C. As heat transfer from the vessels is neglected in the model of Weinbaum and Jiji (denoted by wj in Fig. 6.11), the tissue profile exceeds or falls below the expected profile. The assumptions made in the wj-model are valid only in a small region of the muscle tissue near skin where the blood is nearly equilibrated with the tissue. The Baish-Wissler/Charny-Levin approach (denoted by sig in Fig. 6.11), mentioned above delivers results comparable to the EF approach, the advantage of the latter being the easy implementation in a whole body model.

In the next step, we tried to validate the computations by available experimental data. This, however, involved new difficulties. Exact temperature profiles within the human body are scarcely available, and if they are, there is a span of experimental values due to intraindividual and interindividual differences. There is not a complete set of necessary parameter values, and there are problems with measuring techniques often implying errors greater than those inherent in recent model computations. Even with modern techniques, exact profiles can hardly be obtained without extensive mathematical corrections.[29,30] The temperature of the tissue surrounding the central artery and vein has not been measured. Moreover, this area, represented in the model as the central core layer, in reality is assymetrically situated within the extremity. Despite of all these problems we used the experimental data by Pennes.[1] Radial temperature profiles in the forearm of nine subjects were measured showing a considerable span of $\approx$ 2°C. Ambient temperature was, on average, 26.6°C. T_{a0} was estimated by Pennes to be 36.25°C. Local metabolic rate, perfusion, and evaporative rate were not recorded. In Fig. 6.12 the averaged experimental profile is compared to the nonvascular model computations (bio, ef, sig, wj) which are, with the exception of the wj-model, within the span and accuracy of the experimentally determined values. As expected, the bioheat approach delivers high temperature values in the muscle tissue near the skin as the heat source term operates in this region with an unrealistically high blood temperature, an error which is avoided in the EF approach.

Conclusions about the Efficiency Function Approach

The main shortcoming of the classical Pennes bioheat approach, which was used for more than 40 years in various applications, is the assumption that convective heat transfer takes place exclusively in the capillaries, thereby neglecting the most essential heat transfer in the arterio-venous networks of medium diameter in muscle tissue. Therefore, the application of the bioheat approach has been seriously questioned, although Pennes himself demonstrated a reasonable compatibility of computed and measured temperature profiles in the human forearm under resting conditions (Fig. 6.12). The vascular model enabled the simulation of vascular architecture with a degree of precision previously unknown, although still idealized, so that its results may be taken as a criterion for the applicability of simple nonvascular models using substitutional processes for the description of the real heat transfer. Their fundamental significance as simple and practicable means for the estimation of heat flow and temperature distributions is not questioned in light of the fact that vascular models are of a complexity that does not allow general application and implementation. We conclude that the classical bioheat approach is, despite its shortcomings, suited to the estimation of temperature distributions of sufficient homogeneity, meaning that its results are good for overall heat stress and may be satisfactory for comfort conditions. Unreliable results must be expected when the body is subjected to cold stress or when local hyperthermia is present,

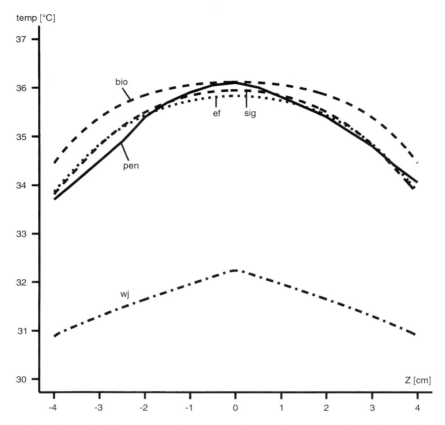

FIGURE 6.12 Experimental validation using Pennes' experimental data (mean values of nine subjects at 26.6°C ambient temperature and an arterial influx temperature of 36.25°C. Radial temperature profile of human forearm compared to the predicted profiles of the nonvascular models. z = coordinate perpendicular to the skin: model specification: bioheat (bio); Weinbaum/Jiji (wj); Charny/Levin (sig); efficiency factor model (ef); and measured profile (pen).

as, for example, applied in clinical radio-frequency or microwave treatments (see below). We offer an improvement of the bioheat approach using the EF, which implicitly takes into account heat transfer in arterio-venous countercurrent networks and is easily applicable, because, in essence, it is a factor with which the bioheat term has to be multiplied. However, the EF has to be used in dependence of the perfusion rate and depth of the tissue. For muscle tissue of the skeleton, the function presented in Fig. 6.10 should be generally applicable, whereas for skin, the bioheat approach does not seem to need substantial correction. Preferably, the outer skin layer should be treated as a pure conduction layer. If significant heat transfer to or from large central vessels is expected, it has to be determined by separate heat balance equations. Estimation of heat transfer in special organs with a vascular architecture significantly different from the arterio-venous tree presented in Fig. 6.2 may need a different EF.

6.4 The Vascular Model and the Efficiency Function Approach in Hyperthermia Treatment

The central aim of loco-regional hyperthermia in conjunction with radio- or chemotherapy is to heat tumor tissue in such a way that the temperature throughout the tumor is above therapeutic level (42–43°C) and to limit increase in the neighboring tissue to avoid any destruction or pain. Tumor regrowth is often observed, possibly because it takes place particularly in the vicinity of vessels[3C] where insufficient heating can be predicted by our computations.

For an estimation of the temperature profiles induced by an external radiofrequency heat source, we used the vascular model and compared the results obtained with those computed by use of non-vascular models.[32] A human extremity under resting conditions (air temperature 29°C, relative humidity 40%, air velocity 0.2 m/s) was submitted to a heat source homogeneously distributing 77,275 W/m³, a value proposed by Charny and Levin.[32] This means that about 73.6 W/kg were applied. The hyperthermia simulation lasted 15 minutes. We used a simple control algorithm of blood flow (suggested by Charny and Levin[32]), assuming an initial perfusion level of 5.4×10^{-4} m³s⁻¹m³, which is augmented by a factor of 10 if tissue temperature exceeds a threshold (42.5 and 43°C were tested). The local heat input temperatures for separate control of core, muscle, and skin were mean temperatures of the core section, the muscles of a Krogh-cylinder, and the corresponding skin section. The inflow temperature of the central artery in the computation was assumed to be constant at 37.5°C. Within the arterio-venous countercurrent branching tree, the diameters of the arterial vessels are supposed to be one-half of those of the venous vessels. Between skin and environment there is a heat exchange by conduction, convection, radiation, and evaporation. The same complete boundary conditions as in whole body models[26–28] were used.

Temperature Profiles

For a first survey, Fig. 6.13 shows an example of the temperature distribution at the end of the hyperthermia simulation computed by the vascular model in a muscle layer perpendicular to the direction of the vessels and containing the vessels of the second generation of the vascular tree. Note again that a highly irregular grid is used in the computation: the outer section comprises the majority of the area and therefore predominantly determines mean tissue temperature. The arterial cup-mixing temperature in this section tends to be about 38°C, the venous temperature is about 42°C, and the tissue temperature varies from about 38°C to about 43°C. From Fig. 6.13 it is obvious that the gradient near the artery is much greater than near the vein. The dynamics of temperature and heat flow in the same muscle layer (second vessel generation) are presented in Fig. 6.14. Mean tissue temperature in that layer starts at 37.6°C and stabilizes, after a strong initial increase, at about 42.4°C. The increase is stopped by the onset of the postulated blood flow regulation (see above). Similar dynamics are observed regarding the temperature of the tissue situated between the vessels (spacing), stabilizing at 39.9°C, and the vessel temperatures, both of which are lower than the mean tissue temperature of the whole muscle section. The arterial temperature is increased several times by about 0.8°C, an effect which is moderately reflected in the venous temperature.

Under resting conditions, we computed an arterial blood velocity in this layer of 0.0296 m/s. Due to the assumed augmented blood flow, this value increases to 0.296 m/s. Therefore, the arterial blood is hardly warmed up on its way through the muscle and its temperature is considerably lower than the mean tissue temperature. The tissue in the vicinity of the artery and the countercurrent blood of the vein are cooled. Consequently, even the venous blood temperature is lower than the mean tissue temperature, although it is warmed up in the capillaries to local tissue temperature. Because of the strong external heat source, the cooling effect of the artery is confined to tissue very near to the vessel. The vessel wall temperature is influenced by both the blood temperature and the temperatures of the surrounding tissue. The controller-induced dynamics are inversely reflected in the heat flows and corresponding blood temperatures.

Fig. 6.15 demonstrates the tissue temperature profiles of non-vascular approaches in comparison with results obtained by the vascular model. The latter were computed for two cases: (1) diameters of the veins equal to those of the arteries; and (2) venous diameters doubled as compared to the arterial ones. From Fig. 6.15 it is obvious that the variation of the proportion of diameters does not influence the results to a significant extent. The results of the wj-approach are not included in this figure, as they are far beyond the range of the other results, if we apply realistic boundary conditions between skin and

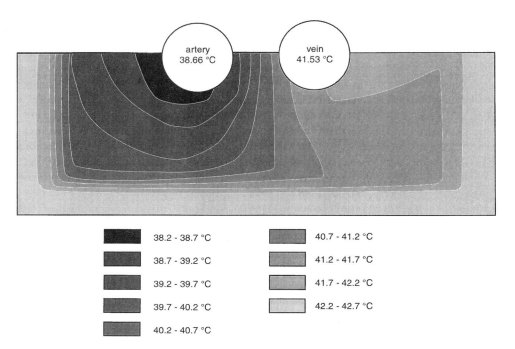

FIGURE 6.13   Vascular model: Hyperthermia simulation, time 15 min. x/y temperature profile of the muscle layer containing the second vessel generation (i.e., the middle of the muscle) of the extremity.

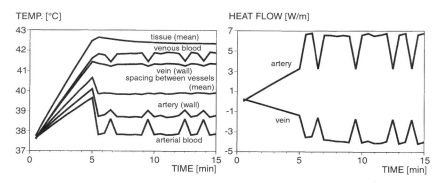

FIGURE 6.14   Vascular model: Hyperthermia simulation. Dynamics of temperatures and heat flows into vessels of the muscle layer of Fig. 6.13.

environment. The profiles would become comparable to the other computation only if arterial influx temperature was changed.

Note that layers 5–8 (Fig. 6.15) constitute only about 10% of the whole muscle thickness. Taking into account that a heat exchange between central vessels and the core would decrease core temperature, the temperature drop from the core to the first muscle section will be smaller in the vascular and EF computations. The distinct maximum in all computations, except in the bioheat approach, is due to the cooling processes within the vascular tree, which decrease with increasing approach to the superficial layers, whereas the cooling power from the environment increases. The bioheat approach exhibits a profile around the chosen threshold temperature (42.5°C). As it does not take into account any local convective processes, the typical course of the profile obtained by the vascular model cannot be obtained. We conclude that the bioheat approach should be considered unsuitable for simulating such local hyperthermia problems. Both the Baish-Wissler/Charny-Levin and the efficiency function approach reflect the

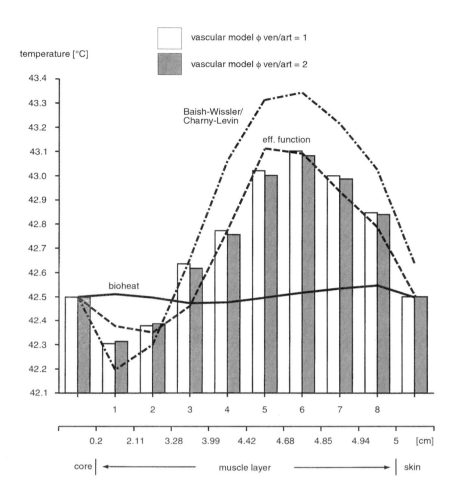

FIGURE 6.15 Radial profiles of spatial mean tissue temperature in an extremity after 15 min. of hyperthermia. Comparison of results from the vascular model (bars) with two different proportions of vessel diameters and from various non-vascular models. Number of muscle layer 1–8 according to vessel generation. Width of muscle layers diminishes dramatically (see non-linear cm-scale and scale in Fig. 6.2).

local temperature pattern. Minimal deviations from the results of the complex vascular model are obtained by the efficiency factor model, although the implementation of this (see above) is very easy, like that of the bioheat approach.

Fig. 6.16 shows arterial and venous temperatures, mean tissue temperature, and the mean temperature in the space between the vessels as a function of depth of tissue at the end of the hyperthermia simulation using the complex vascular model. Blood temperatures are distinctly lower than mean tissue temperatures. The temperature between the vessels is, in spite of the external heat source, determined by the blood temperatures. It is approximately equal to the arithmetic mean of both blood temperatures. This temperature is therefore rather low over a wide area, meaning that there may be insufficient heating at least between the vessels. Looking again at Fig. 6.14, we recognize these cooler areas in the vicinity of the artery. Temperatures lower than 40.7°C comprise layers around the artery of about double vessel radius. However, taking into account the irregular grid of Fig. 6.14, it is computed that more than 90% of the area has a temperature greater than 42.2°C. Cooler areas, although comprising a very small portion of the tissue, are especially obvious around the vessels of the first generation and may be essential for hyperthermia treatment. In the area of the first generation, we computed that 9.6% of the tissue is cooler than 42.0°C. The total area with a temperature that is less than or equal to 42°C is about 6%.

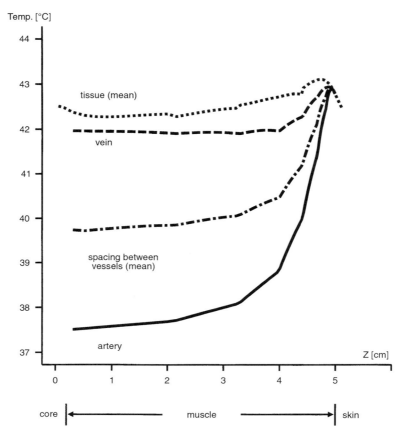

FIGURE 6.16 Various radial temperature profiles after 15 min. of hyperthermia computed by the vascular model. (Taking into account that in contrast to Fig. 6.15 there is a linear abscissa now, the temperature profile of the mean tissue temperature is the same in the two figures).

Conclusions from Hyperthermia Simulations

The strength of the vascular model is that, for the first time, while using the parameter set well established by former investigations,[6,7] heat transfer is taken into account on a vessel-by-vessel basis in a three-dimensional counter current arterio-venous tree, without using a substitutional process. The weak point is that, because of its complexity, it is not generally applicable. That is why we looked, like others before, for an adequate substitutional process that is based on the results of the complex vascular model. By this means, we claim that the Pennes bioheat approach is substantially improved by the EF concept. This was also attempted by the wj-approach which, since its introduction, has undergone many modifications and supplements. We think that its shortcomings in our computations are primarily due to the fact that it does not cope with realistic boundary conditions by conduction, convection, radiation, and evaporation at the skin. Even the modified version from 1992[34] is characterized by the authors as, (Weinbaum et al.,) "of somewhat limited usefulness in whole limb heat transfer." Again a hybrid model, in which either the bioheat or Weinbaum/Jiji terms are applied, is proposed. We argue that in light of the improvement of the bioheat term by the efficiency function, this is an avoidable complexity, particularly when one realizes that the proposed wj-term refers to only 10% of the muscle (layers 5–8). The vascular model may be supplemented in two respects that will, however, not alter the principal results and conclusions. The small amount of heat transfer of central vessels to the core section could be modeled explicitly. Also, the fat layer is not treated explicitly because it can be treated as a pure conduction layer and the blood in the vessels of muscle tissue adjacent to the skin is equilibrated with the surrounding tissue, so it is not essential to make a distinction between this muscle and the fat layer.

We realize that blood flow regulation is far more complicated than in these computing examples. There are both central and local mechanisms that vary throughout the body. Sekins et al.[35] documented the effect of the threshold temperature. According to their results, Charny and Levin[36] suggested the following control mechanism: if tissue temperature reaches a critical level, blood flow increases linearly. If it falls again below this level, blood flow retains the present status. Such dynamics are called type II-properties by Roemer et al.[37] Besides the effect of different blood flow in normal and tumor tissue, we have to take into account that, in practical applications, an attempt is made to focus the external heat on the tumor by using special applicators. It was demonstrated that the bioheat model is not suited to the simulation of local hyperthermia. The assumption of constant arterial temperature throughout the tissue is wrong and causes the flat temperature profiles.

Charny and Levin[35] determined the conditions of a hyperthermia treatment that will lead to a systemic heating, i.e., a warming up of the central blood and, thus, of greater parts of the body. To take into account such effects, whole-body thermal models have to be coupled with vascular models. Consequently, central thermoregulatory control comes into play, and the assumption of a constant central arterial temperature has to be abandoned. This was achieved by Charny and Levin,[38] who presented a three-dimensional thermal and electromagnetic model of whole-limb heating. A three-dimensional finite-element model of a tumor-bearing human lower leg was constructed and mathematically attached to a whole body thermal model. For heat transfer between blood and tissue, the bioheat approach was used. Tests should be made as to whether such a comprehensive and useful model could still profit from the improvement of the bioheat approach, presented here by use of the vascular model and the efficiency function model.

References

1. Pennes, H.H., *J. Appl. Physiol.*, 1, 93, 1948.
2. Weinbaum, S., Jiji, L.M., and Lemons, D.E., *ASME J. Biomech. Eng.*, 106, 321, 1984.
3. Jiji, L.M., Weinbaum, S., and Lemons, D.E., *ASME J. Biomech. Eng.*, 106, 331, 1984.
4. Weinbaum, S. and Jiji, L.M., *ASME J. Biomech. Eng.*, 107, 131, 1985.
5. Zhu, M., Weinbaum, S., and Lemons, D.E., *ASME J. Biomech. Eng.*, 110, 74, 1988.
6. Song, W.J., Weinbaum, S., and Jiji, L.M., *ASME J. Biomech. Eng.*, 109, 72, 1987.
7. Song, W.J., Weinbaum, S., Jiji, L.M., and Lemons, D.E., *ASME J. Biomech. Eng.*, 110, 259, 1988.
8. Wissler, E.H., *ASME J. Biomech. Eng.*, 109, 226, 1987a.
9. Wissler, E.H., *SME J. Biomech. Eng.*, 19, 355, 1987b.
10. Weinbaum, S. and Jiji, L.M., *ASME J. Biomech. Eng.*, 109, 234, 1987.
11. Weinbaum, S. and Jiji, L.M., *ASME J. Biomech. Eng.*, 111, 271, 1989.
12. Baish, J.W., Ayyaswamy, P.S., and Forster, K.R., *ASME J. Biomech. Eng.*, 108, 246, 1986.
13. Charny, C.K. and Levin, R.L., *ASME J. Biomech. Eng.*, 111, 263, 1989.
14. Charny, C.K. and Levin, R.L., *ASME J. Biomech. Eng.*, 112, 80, 1990.
15. Chen, M.M. and Holmes, K.R., *N.Y. Acad. Sci.*, 335, 137, 1980.
16. Brinck, H. and Werner, J., *ASME J. Biomech. Eng.*, 116, 324, 1994a.
17. Chato, J.C., *ASME J. Biomech. Eng.*, 102, 110, 1980.
18. Chen, Z.P. and Roemer, R.B., *ASME J. Biomech. Eng.*, 114, 473, 1992.
19. Wissler, E.H., *ASME J. Biomech. Eng.*, 110, 254, 1988.
20. Douglas, J., Jr. and Rachford, H.H., Jr., *Am. Math. Soc.*, 82, 421, 1956.
21. Eberhart, R.C., *Heat Transfer in Medicine and Biology*, 1st ed., Plenum Press, New York, 1985, 261.
22. Zhu, M., Weinbaum, S., and Jiji, L.M., *Int. J. Heat Mass Transfer*, 33, 2275, 1990.
23. Brinck, H. and Werner, J., *J. Appl. Physiol.*, 77, 1617, 1994b.
24. Wissler, E.H., *J. Appl. Physiol.*, 16, 734, 1961.
25. Stolwijk, J.A.J. and Hardy, J.D., *Pflugers Arch.*, 129, 1966.

26. Werner, J. and Buse, M., *J. Appl. Physiol.*, 65, 1110, 1988.
27. Werner, J., Buse, M., and Foegen, A., *Biol. Cybern.*, 62, 63, 1989.
28. Werner, J. and Webb, P., *Ann. Physiol. Anthropol.*, 12, 123, 1993.
29. Ducharme, M.B., van Helder, W.P., and Radomski, M.W., *J. Appl. Physiol.*, 71, 1973, 1991.
30. Ducharme, M.B. and Tikuisis, P., *Eur. J. Appl. Physiol.*, 64, 395, 1992.
31. Overgaard, J. and Nielsen, O.S., *Ann. N.Y. Acad. Sci.*, 335, 254, 1980.
32. Brinck, H. and Werner, J., *Int. J. Hyperthermia*, 11, 615, 1995.
33. Charny, C.K. and Levin, R.L., *ASME J. Biomech. Eng.*, 110, 277, 1988a.
34. Weinbaum, S., Jiji, L.M., and Lemons, D.E., *ASME J. Biomech. Eng.*, 114, 539, 1992.
35. Sekins, K.M., Lehmann, J.F., Esselmann, P., Dundore, D., Emery, A.F., de Lateur, B.J., and Nelp, W.B., *Arch. Med. Rehabil.*, 65, 1, 1984.
36. Charny, C.K. and Levin, R.L., *IEEE Trans. Biomed. Eng.*, 35, 362, 1988b.
37. Roemer, R.B., Oleson, J.R., and Cetas, T.C., *Am. J. Physiol.*, 249, R153, 1985.
38. Charny, C.K. and Levin, R.L., *IEEE Trans. Biomed. Eng.*, 38, 1030, 1991.

7

Arterial Fluid Dynamics: The Relationship to Atherosclerosis and Application in Diagnostics

James E. Moore Jr.,
Florida International University

Antonio Delfino
Swiss Federal Institute of Technology

Pierre-André Doriot
University Hospital of Geneva

Pierre-André Dorsaz
University Hospital of Geneva

W. Rutishauser
University Hospital of Geneva

7.1   Introduction 7-1
7.2   Basics of Arterial Fluid Mechanics 7-2
7.3   Atherosclerosis and the Involvement of
      Fluid Mechanics 7-5
      Investigations of Arterial Flow Patterns • *In Vitro* Studies of the
      Effects of Flow on the Artery Wall • Additional Studies of
      Arterial Flow Patterns and Atherogenesis
7.4   Application of Arterial Fluid Mechanics in
      Diagnostics 7-11
7.5   Conclusions 7-15

7.1   Introduction

Blood flow patterns have been a topic of interest to engineers and clinicians from the early days of both disciplines. One of the most well-known equations describing fluid flow through a pipe is named for Jean Poiseuille, a physician who derived his equation in an investigation of blood flow in 1843. Beyond the interest in blood flow as a naturally occurring physical phenomenon, much research has been devoted to investigating the possibility that certain blood flow patterns are involved in the formation of diseases. From a clinical point of view, blood flow patterns have been more of a subject of curiosity rather than scientific investigation, although clinicians are increasingly applying knowledge of fluid dynamics to improve diagnostic capabilities.

The common ground that has provoked extensive interaction between engineers and medical scientists in the investigation of blood flow patterns is the study of disease. Atherosclerosis is the most common disease affecting the cardiovascular system and is the leading cause of death and disability in western countries. Atherosclerosis is manifested in the form of narrowing blood vessels called stenoses. If the narrowing becomes severe enough, the supply of blood to the distal tissues may be compromised. The amount of blood that can flow through an artery depends on the pressure gradient, the viscosity of blood, and the diameter. The pressure gradient from the entrance of the aorta to the vena cava is approximately 120 mmHg in systole for healthy individuals. The dynamic viscosity of blood in large and medium size arteries (> 1 mm in diameter) is a constant four centipoise. Changes in diameter are, therefore, the most common causes for reductions in blood supply to a particular region. Basic work in fluid dynamics demonstrates that flow is proportional to diameter to the fourth power. This strongly non-linear relationship means that once a stenosis appears in an artery, further reductions in diameter, even small ones, can create large reductions in flow rate.

Various factors such as high blood pressure, high blood cholesterol levels, cigarette smoking, and a sedentary lifestyle have been identified to help indicate if an individual is at risk of developing atherosclerosis. However, none of these factors can explain why the disease develops only in certain arteries, because their effects are felt throughout the cardiovascular system. The arteries most commonly affected are the carotid, coronary, and femoral arteries, and the abdominal aorta. It is quite rare for other arteries to be affected to a clinically significant degree. The highly localized nature of atherosclerosis has led to the hypothesis that mechanical factors such as blood flow patterns may be involved in its formation. Much research has been devoted to more specifically identify the exact mechanical conditions that lead to atherosclerosis formation. Flow analysis techniques have advanced to the point that the flow in any artery can be accurately modeled with either experimental or numerical techniques. At the same time, vascular biologists and clinical investigators have incorporated the use and understanding of mechanical stimuli such as fluid flow in biological and *in vivo* experiments aimed at better understanding atherosclerosis formation.

7.2   Basics of Arterial Fluid Mechanics

The cardiovascular system presents quite a challenging problem to engineers. Arteries are compliant (flexible) pipes that are continually curving and branching. The heart provides a complex, time-varying output of blood and pulse pressure waves that propagate down the system through the expansion of arteries. Compared to a typical industrial piping system that includes a steady flow producing pump and rigid pipes that remain straight for fairly long lengths, the arterial system is much more complex. However, the tools used to analyze flow through both systems are very much the same. One of the most basic parameters in determining flow patterns in any flow situation is the Reynolds number:

$$\mathrm{Re} = \frac{Vd}{v} \tag{7.1}$$

where V is the average flow velocity, D is a length scale (such as pipe diameter), and v is the kinematic viscosity. The Reynolds number, representing the ratio of inertial to viscous effects, was named after Osborne Reynolds, who performed some of the most important fundamental work on fluid mechanics in the late 1800s. In pipe flow, a Reynolds number of less than approximately 2500 indicates that the flow should be laminar. For laminar flow, the relationship between flow rate and pressure gradient through a long straight pipe (Poiseuille flow) is

$$Q = \frac{\pi D^4}{128 \rho v} \left| \frac{dp}{dz} \right| \tag{7.2}$$

where Q is the volume flow rate, ρ is the fluid density, and dp is the pressure difference along an axial length dz. The velocity profile (fluid velocity plotted vs. radial position) for the Poiseuille flow is parabolic in shape, with the maximum value of velocity occurring at the center of the tube.

At Reynolds numbers of greater than 2500, the flow should begin to become turbulent. Under normal physiological conditions, the largest Reynolds number that would be encountered in the cardiovascular system would be approximately 2100 (for exercise conditions in the ascending aorta). Thus, blood flow in most healthy situations should be laminar, or at least not turbulent. In fact, the flow in most arteries would be most properly classified somewhere in between laminar and turbulent. The presence of curvature and branches can provoke the development of secondary flow patterns such as large-scale vortices. These vortices typically appear as regular, periodic phenomena and thus do not exhibit the chaotic behavior associated with turbulence.

Curved tube flows have long been associated with vortex formation. Dean was one of the first to rigorously study flow in curved tubes.[1,2] He used a perturbation technique to predict the flow field in slightly curved tubes and was able to obtain an analytical first order solution. More recent studies have employed numerical methods that permit the analysis of more strongly curved tubes, as are found in the cardiovascular system.[3] In a curved tube, the fluid swirls from the inner wall of curvature to the outer wall through the middle of the tube, and from the outer wall to the inner wall along the side walls of the tube (Fig. 7.1). Another consequence of curvature is that the velocity profile in the tube is skewed toward the outer wall of curvature (when the flow is fully developed), creating a lower wall shear rate at the inner wall than at the outer wall. Branches may also provoke vortex formation, but the degree to which this occurs depends strongly on the amount of blood that exits the branch. This has been demonstrated most effectively using laboratory models of branching tubes. When the percentage of parent vessel blood flow that exits the branch is greater than 10%, a vortex forms in the parent vessel near the site of the branch.[4] Similar flow patterns occur in bifurcations, where a parent vessel splits into two basically equal daughter branches (Fig. 7.1).

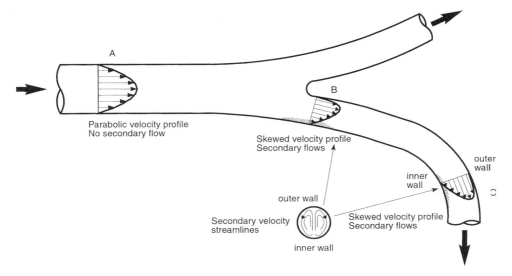

FIGURE 7.1 Illustration of the effects of geometric features on blood flow patterns. For straight tubes (A), the velocity profile is parabolic (in the case of steady flow), with its maximum value located at the center of the tube (Poiseuille flow). There are no secondary (cross-stream) velocities present. In a branching region (B), the velocity profile is skewed toward the flow divider wall. There are secondary velocities present that swirl the fluid from the outer wall to the inner wall along the side walls. In curved sections (C), the velocity profile is skewed toward the outer wall of curvature, and similar secondary flows are present. The shaded areas would be subjected to relatively low shear stresses. For pulsatile flow, these areas may be subjected to oscillating wall shear stress. Such areas of the vasculature are sites of atherosclerosis development.

The pulsatile nature of the cardiac output is also an important factor in determining blood flow patterns. Perhaps because this flow situation rarely occurs in industrial piping systems, not much was known about pulsatile flow phenomena until the 1950s. The most important parameter in describing pulsatile flows is the Womersley parameter:

$$\alpha = \frac{D}{2}\sqrt{\frac{\omega}{v}} \qquad\qquad (7.3)$$

where ω is the radian frequency of the heartbeat. The Womersley parameter represents the ratio of unsteady to viscous effects. This dimensionless parameter arose during an analytical study of oscillatory flow in a straight pipe.[5] In this case, the equations governing fluid flow yield an exact solution. At Womersley parameters of less than approximately 4, the flow is dominated by viscous effects. The velocity profiles tend to be nearly parabolic in shape. At higher Womersley parameter values, unsteady effects become more important, and the velocity profiles may take on complex m-shapes, with a core of nearly inviscid (plug) flow in the middle. The unsteadiness also affects the appearance of turbulence. Experimental studies employing hot film anemometry[6] have demonstrated that higher frequency pulsatile flows are more protected from turbulence because the chaotic velocity fluctuations associated with turbulent flow require a certain amount of time to develop fully. It has also been observed that turbulence is more likely to develop during flow deceleration.[7] Analysis of pulsatile flow in curved tubes has demonstrated that vortex formations similar to those observed for steady flow are present, but the velocity profile may be skewed toward the inner wall of curvature in a purely oscillatory flow.[8] The flow instabilities created by branches are also present with pulsatile flow. However, like turbulent fluctuations, they tend to be less prevalent when the flow is accelerating. The intensity of the vortex formation at a branch increases with increasing values of Re and α.[9]

Blood flow patterns are also affected by the fact that the viscosity of blood is not constant under all conditions. Blood is a non-Newtonian fluid that exhibits shear-thinning behavior. That is, blood becomes more viscous at lower shear rates. This phenomenon occurs due to the formation of chains of red blood cells called rouleaux, and is related to the initiation of the clotting process. Under the flow conditions present in most large and medium-sized arteries, this behavior primarily affects the flow near the middle of the artery where the shear rate is lowest. Thus, the wall shear rates and wall shear stresses are not greatly affected.[10] In most arterial flow studies where the wall shear stress is the primary interest, the non-Newtonian behavior of blood is usually neglected.

Another factor that has a relatively minor effect on overall blood flow patterns is the compliance of the vessel wall. While a reduction in the shear rate amplitude in compliant tubes as opposed to rigid tubes has been noted, the spatial distribution of the wall shear rate in fully three-dimensional arterial models does not change greatly with the addition of wall compliance.[11] As with the non-Newtonian behavior of blood, the importance of vessel wall compliance depends on the type of information desired. If the goal is to determine the overall distribution of wall shear stress in an artery, then the vessel wall compliance may be neglected. However, it is possible with current computer technology to construct a fully three-dimensional model of blood flow in an artery that includes non-Newtonian fluid behavior as well as vessel wall compliance.[11] Similar capabilities exist for constructing laboratory flow models.[12]

One aspect of vessel wall compliance that is important in the overall functioning of the cardiovascular system is the compliance of arteries. The compliance of the aorta is responsible for a phenomenon known as the Windkessel effect. The on-or-off nature of the flow from the heart is smoothed by the expansion of the aorta during systole. Thus, the aorta acts as a capacitor that helps propel blood distally during diastole. Another important consequence of arterial compliance is the propagation of pressure pulse waves along the arterial tree. The contraction of the heart and subsequent ejection of blood into the aorta creates a pressure wave that propagates along the aorta. This propagation does not occur as a sound wave traveling in the blood itself, but rather as a progressive expansion of the arterial wall. However, there are many similarities with the propagation of sound. Thus, the tools for analyzing pressure waves in arteries are often borrowed from acoustics. One factor that is common to both fields is the phenomenon of wave reflections at sites of changing impedance. In the arterial system, the most significant pressure wave reflection sites are branches. The speed of propagation of pressure waves in arteries may be approximated with the Moens-Korteweg equation:

$$c = \sqrt{\frac{Eh}{\rho D}} \tag{7.4}$$

where E is the elastic modulus of the wall material, h is the wall thickness, and ρ is the density of blood. The propagation speed in the aorta would be expected to be approximately 10 m/s (as compared to the maximum blood flow velocity of approximately 1 m/s). Thus, wave propagation phenomena in arteries are often addressed with the assumption that the wavelengths are much larger than the vessel diameters. This simplifying assumption has led to much progress in understanding wave propagation in arteries.[13]

7.3   Atherosclerosis and the Involvement of Fluid Mechanics

Investigations of Arterial Flow Patterns

Interest in the role of blood flow in the development of arterial diseases predates much of the work in basic fluid mechanics. One of the early pioneers to study the cellular aspects of disease formation was Ludwig Virchow, whose first publication concerning arterial disease appeared in 1847.[14] Virchow readily subscribed to the theory that an inflammation of the inner arterial surface was the starting point for atherosclerosis. He proposed that mechanical or chemical factors led to inflammatory reactions in the arterial wall. Virchow described the histological structure of atherosclerotic plaques with great detail and clarity and correctly recognized that they formed beneath the intima. His numerous histological observations led him to contend that atherosclerosis did not develop due to clotting. Furthermore, he showed considerable insight by distinguishing between atherosclerotic plaques and fatty streaks, which he stated were superficial collections of fat and had very little to do with even initiating atherosclerosis.

Much of the work that followed Virchow's mechanical irritant hypothesis was directed more toward the solid mechanics of the artery wall. Beginning with the work of Roy,[15] several other investigators[16–18] studied the mechanics of the artery wall in an attempt to explain the localization of atherosclerosis. The shift toward the study of blood flow patterns and their possible involvement in atherogenesis began in the 1950s with the work of Gofman and Texon. (In fact, Virchow's original hypothesis did not exclude blood flow patterns, which certainly qualify as mechanical. It just appears that subsequent researchers interpreted it as solid mechanical.) Gofman appears to have been the first to mention fluid mechanics specifically as a localizing factor,[19] but he did not pursue his hypothesis with any scientific study. Texon[20] followed shortly thereafter with a more specific hypothesis of how hemodynamic factors can account for the distribution of atherosclerosis. Noting the predilection of the disease to the inner walls of curved arteries, Texon applied the knowledge gained from Dean's studies of flow in curved tubes[1,2] to propose the first specific relationship between flow and atherogenesis. Texon proposed that because the pressure at the inner wall of a curved tube is lower than that at the outer wall, the endothelium at the inner wall may be denuded by a suction effect. Of course, the static arterial pressure is much greater than the local reduction in pressure due to curvature, so negative pressures would never be reached in normal physiological flow situations. Nevertheless, this theory opened the door to more detailed studies of arterial flow patterns and their involvement in atherogenesis.

Fox and Hugh[21] were the first to perform blood flow simulations aimed at a better understanding of atherogenesis. Noting that atherosclerosis generally forms in the inner walls of curved vessels and along the lateral, or outer walls of branched vessels, they proposed that boundary layer separation occurs at these locations and leads to the deposition of blood-borne platelets and cholesterol. In this context, boundary layer separation refers to the phenomenon of retrograde velocities that may occur near the outer walls of branching arteries. These are typically zones of relatively low flow velocity, i.e., zones of low wall shear stress (Fig. 7.1). Fox and Hugh constructed two-dimensional channel flow models of arteries and scattered aluminum powder over the surface to visualize the flow patterns. Their experiments clearly demonstrated flow separation along the inner walls of curved channels and along the lateral walls of branching channels, but they employed only steady flow conditions. In pulsatile flow, the definition of boundary layer separation is less clear than for steady flow. Pulsatile flow in straight tubes can effectively separate because flow deceleration often leads to negative near-wall velocities, meaning that the entire tube circumference would be subjected to boundary layer separation for at least some portion of the cardiac cycle.

The potentially damaging effects of flow on the cells that line the insides of arteries were first demonstrated by Fry.[22] His experiments consisted of inserting a cylinder with an eccentric passageway into the aortas of dogs so that the section of arterial wall still exposed to flow would be subjected to high shear stresses. He demonstrated that shear stresses in excess of 380 dynes/cm^2 damage the cells that line the inner surfaces of arteries (endothelial cells). He proposed that the endothelium in sections of the arterial system that are subjected to such high shear stresses would be damaged and develop atherosclerosis. However, shear stresses in the normal, non-diseased cardiovascular system would never be expected to be so high, even under exercise conditions. Nevertheless, the debate over the merits of the high shear stress and low shear stress (i.e., boundary layer separation) theories lasted well into the 1980s.

Subsequent flow studies in more realistic arterial models tended to support the low wall shear stress theory. Caro et al.[23] performed steady flow simulations in a cast of a human aorta and noted that locations of early atheroma coincided with locations of relatively low near-wall flow velocity. They proposed that the mechanism for atherosclerosis development may be flow-mediated mass transfer of blood-borne substances into and out of the artery wall. In one of the first detailed, quantitative studies of blood flow, Friedman et al.[24] modeled blood flow in the aortic bifurcation and found that both high and low wall shear stresses occurred where lesions develop. The highest wall shear stresses that were predicted were approximately 30 dynes/cm^2, an order of magnitude below Fry's damage threshold. It was therefore concluded that low wall shear stresses were likely to be more consistent with plaque development. The flow model in this case was a two-dimensional numerical model. While the use of a two-dimensional model limited the applicability of the results, the use of a numerical technique in the mid–1970s predated the appearance of user-friendly computational fluid dynamics (CFD) software packages. Later work by the same group[25] made use of a transparent cast of an actual human aorta obtained at autopsy. This cast was mounted in a pulsatile flow system where three-dimensional laser Doppler anemometer (LDA) measurements were performed. Correlations with intimal thickening measurements from the same aorta demonstrated that mean and pulse (maximum–minimum) wall shear rates correlated inversely with intimal thickening (e.g., a low mean wall shear rate coincided with high intimal thickening). The aorta model used in this study did not include any of the major branches that appear just distal to the diaphragm, nor the curvature of the aorta as it follows the spine in the lower back. Nevertheless, the use of an LDA to perform accurate measurements of arterial flow patterns represents a significant advancement in understanding their role in atherogenesis.

Numerous arterial model flow studies, performed in the 1980s, attempted to elucidate further the relationship between shear stress and intimal thickening. Friedman's group continued to employ computational methods and the LDA to explore the effects of individual variations in geometry on the flow patterns that were previously associated with intimal thickening.[26,27] They concluded that naturally occurring individual variations in factors such as branching angle, branch asymmetry, and flow divider curvature are large enough to create or eliminate the low wall shear stresses associated with atherogenesis. This may help explain why some people are more susceptible to plaque development than others. Their studies were mainly concerned with the aortic bifurcation, but the hemodynamic phenomena observed in their studies should be applicable to some extent to all branching vessels.

In a more extensive study of aortic bifurcation flow, Friedman et al.[28] performed LDA measurements in casts of ten different human aortas and correlated the results with intimal thickness data from the same aortas. They noted that early intimal thickening was greatest in locations of relatively high shear stress, but the thickness in these areas appeared to plateau over time. The intimal thickness in areas of low shear stress, initially lower than that found in high shear stress areas, gradually increased to higher values. Thus, in the long-term, low wall shear stress appeared to be more consistent with clinically significant disease formation.

While Friedman's work concentrated on individual vessel geometries, there was other ongoing research at that time that analyzed the flow patterns in vessel geometries obtained by averaging geometric data from a larger population. Steady flow visualization[29] and LDA measurements[30] were performed in a carotid artery model constructed with the average geometry of 124 angiograms. Those studies identified the lateral wall of the carotid sinus as a location of low wall shear stress. The stagnation of flow in this region was further confirmed using steady and pulsatile flow visualization.[31] Quantitative LDA measurements under pulsatile flow conditions confirmed that the mean wall shear stress was relatively low at the lateral wall of the carotid sinus, and clearly demonstrated the oscillatory nature of the shear stress in this region. In one of the most often cited studies in this field, Ku et al.[32] correlated specific shear stress indices with intimal thickness data obtained from victims of non-atherosclerosis related deaths at autopsy. It was found that mean shear stress correlated negatively with intimal thickness, as did maximum shear stress. Additionally, a new index of shear stress oscillation was established, the oscillatory shear index (OSI),

$$\mathrm{OSI} = \frac{\int_0^T |\tau_w^*| \, dt}{\int_0^T |\tau_w| \, dt} \qquad (7.5)$$

where τ_w is the wall shear stress, τ_w^* is the component of wall shear stress acting in the direction opposite to the mean flow, and T is the period of the cardiac cycle. There was a significant positive correlation noted between OSI and intimal thickness, indicating that atherosclerosis is more likely to develop in locations where the shear stress oscillates between positive and negative values within the cardiac cycle.

Investigations of flow patterns in other commonly diseased vessels have resulted in similar conclusions. In the left main coronary artery bifurcation, Friedman et al.[33] noted negative correlations between mean shear stress and intimal thickness, and maximum shear stress and intimal thickness. The shear stresses were evaluated using LDA measurements in a cast model, as was done earlier by the same authors for the aortic bifurcation. No shear stress oscillation was noted in their results. He and Ku[34] constructed a numerical model of the same left coronary artery bifurcation based on average measurements compiled from different sources. They correlated their shear stress measurements with the distribution of atherosclerotic lesions as demonstrated in a previous study. They reported an inverse relationship between mean wall shear stress and lesion formation, as well as between maximum wall shear stress and lesion formation. Their results also demonstrated very little shear stress oscillation. However, both of these coronary artery studies featured a static geometry. As may be ascertained from coronary angiograms, the geometry of the coronary arteries is dynamic in nature due to the contraction of the myocardium. The movement and deformation of the coronary arteries occur most swiftly in systole and can have significant effects on the flow patterns. Moore et al.[35] studied the effects of physiological axial movement on pulsatile flow in a straight tube model of the left coronary artery and concluded that taking the movement into account doubled the OSI predicted when movement was neglected. The presence of oscillating shear without vessel movement (not noted in the study of Friedman et al.) was due to the fact that a slightly different volume flow waveform was used. The effects of movement and deformation need to be explored in more detail in order to conclusively establish the relationship between wall shear stress and intimal thickness in the coronary arteries.

Flow studies in a model of the abdominal aorta based on an average geometry demonstrated correlations between mean wall shear stress and intimal thickness and between OSI and intimal thickness, but not between maximum shear stress and intimal thickness.[36] The glass tube aorta model used in this study was based on the average measurements of 55 bi-planar angiograms and included seven major branches as well as the curvature of the aorta. Volume flow waveforms in the suprarenal and infrarenal aorta were carefully controlled to exhibit the bi-phasic and tri-phasic shapes that have been clearly demonstrated *in vivo*.[37] Pulsatile flow visualization revealed significant vortex formation beginning at the level of the renal arteries, and a stagnation zone along the posterior wall of the infrarenal that split and continued along the lateral-posterior walls of the aortic bifurcation.[38] Subsequent pulsatile velocity measurements using magnetic resonance imaging velocimetry (MRIV) confirmed this general flow behavior and provided shear stress data for correlation with intimal thickness measurements. As with the carotid and coronary arteries, a negative correlation between mean shear stress and intimal thickness was noted. The OSI was found to positively correlate with intimal thickness. In contrast to the studies of the coronary and carotid artery, no significant correlation with maximum shear stress or its inverse could be found. In this study, the spatial variations in maximum shear stress were very different from the spatial distribution of mean wall shear stress. For the cases of the carotid and coronary arteries, these shear stress indices had similar distributions, resulting in similar correlations with intimal thickness. Thus, the unique aspects of flow in the abdominal aorta provided delineation between the correlations with mean and maximum shear stresses that could not have been detected in the carotid and coronary artery studies.

In practically all of these studies that correlated wall shear stress and atherosclerosis development, rigid wall models and Newtonian fluids were employed. As mentioned earlier, these factors have a minor effect on the overall flow patterns. The more specific question of whether or not these factors affect the correlation between wall shear stress and atherogenesis was addressed by Friedman et al.[39] They concluded that using rigid wall models and Newtonian fluids had no effect on the slopes of the correlations between intimal thickness and any normalized shear rate measure.

Advances in CFD capabilities in recent years facilitated the return to the question of flow-mediated mass transfer. The analogy between heat and mass transfer meant that CFD techniques that couple flow solutions with heat transfer analysis could be exploited to produce information on the mass transfer of blood-borne solutes into and out of the artery wall. At first, these techniques were applied to the study of flow in simplified arterial geometries such as axisymmetric narrowings.[40–42] In general, it was found that solute concentration and wall flux are reduced in the separation zones distal to the narrowings. The effects of flow on mass transfer were even more significant when the permeability of the wall was specified as a function of the shear rate.[11]

In Vitro Studies of the Effects of Flow on the Artery Wall

Research on the effects of fluid shear stress on vascular endothelial cells has also advanced since Fry's experiments. Aided by advances in cell culturing techniques and the biological sciences in general, devices to subject cells to physiologically realistic mechanical conditions *in vitro* were developed with the goal of providing information on the interaction of fluid mechanics and biological responses. Such information would be complementary to the flow studies outlined above in determining if and why certain flow patterns lead to atherosclerosis formation. There were two devices developed that were capable of subjecting cultured endothelial cells to shear stress. The first was the parallel plate flow chamber, in which a nourishing culture medium was pumped through a chamber with a rectangular cross-section. The bottom plate of the chamber was coated with endothelial cells that could be viewed directly through the transparent walls of the chamber using phase-contrast microscopy. The length and width of the chamber were much larger than the height, providing Poiseuille flow conditions over the vast majority of the surface. Knowledge of the height of the gap between the upper and lower plates, the flow rate and fluid properties allowed the calculation of the shear stress in the chamber. This device was first employed to study the effects of flow over bovine kidney cells.[43] The other device, similar to a cone-and-plate viscometer, features a spinning cone that is placed a short distance above a flat plate onto which cells have been cultured.[44] The small gap and slight angle of the cone provide a nearly uniform shear stress field on the plate.

The results of studies that employed these devices demonstrate a very strong relationship between flow and endothelial cell behavior. At first it was observed that the morphology of endothelial cells changes with the presence of flow-induced shear stress. In static culture, endothelial cells have a relatively round, cobble-stone appearance. Within 24–48 hours following the application of flow, the cells take on a more elongated shape and align with the direction of flow.[44] Later experiments with a parallel plate flow chamber confirmed this behavior.[45] The effects of pulsatile shear stress on endothelial cell morphology depend on whether or not the shear stress and sheer stress amplitude reverse during the cardiac cycle. Helmlinger et al.[46] used a parallel plate flow chamber to subject endothelial cells to varying degrees of oscillating shear stress. The imposition of a low amplitude oscillating component on a steady mean shear stress caused the cells to be less aligned than those subjected to the same level of steady shear stress. With high amplitude oscillating shear stress, the cells detached completely within 24 hours.

Two separate studies of vascular endothelial cell morphology obtained from autopsy specimens supported these *in vitro* results. Examination of fetal aortas demonstrated that endothelial cells in relatively straight regions of the aorta appeared uniformly aligned with the flow direction.[47] In regions near branches, the cells' alignment was along the direction of streamlines that one would expect for flow near a branch. On the surface of the aortic valve sinus, where randomly oriented turbulent-like flow patterns were noted, the cells were irregular in shape, size, and orientation. A study of the endothelial cell structure of rabbit aortas also reported alignment of cells in the expected flow directions in straight and branched segments.[48]

In addition to morphological effects, other aspects of endothelial cell behavior are also affected by shear stress. Advances in biological assay capabilities and molecular biology techniques have allowed researchers to investigate many aspects of endothelial cell behavior in the presence of flow. It is known that shear stress affects the production of vasoactive substances such as endothelin[49] and prostacyclin.[50] The ability of these substances to provoke the contraction or relaxation of the medial vascular smooth muscle suggests that endothelial cells play a role in determining the local resistance to flow. The level of production of these substances depends on the type of shear stress applied (steady, pulsatile, reversing, or non-reversing), suggesting a highly sophisticated flow-sensing mechanism within the cell. Other substances whose production by endothelial cells has been shown to be affected by flow include the attachment molecules ICAM–1, VCAM–1, and E-selectin.[51] These substances may play an important role in atherogenesis because of their role in the recruitment of monocytes into the arterial wall. Excessive monocyte recruitment is considered a key event in atherosclerosis development. Other aspects of endothelial cell behavior that have been shown to be affected by flow include cytoskeletal structure, cell stiffness, proliferation rates, and endocytosis.[52]

All of these aspects of endothelial cell behavior that are affected by flow are expected to play at least a minor role in the development and/or progression of atherosclerosis. The morphology of the cells is important because the presence of gaps between cells can lead to increases in endothelial layer permeability.[53] Additionally, if endothelial cells detach in response to highly oscillating shear stress *in vivo*, as was observed *in vitro*,[46] then the permeability in regions exposed to oscillating shear stresses may be greatly increased. The production of endothelin was found to be maximal at very low shear stresses.[49] In addition to its vasoactive properties, endothelin is also a simulator of smooth muscle cell proliferation, another key event in atherosclerosis formation. It has also been observed that the expression of attachment molecules increases with oscillating shear stress.[54] Thus, in separate experiments (often performed with different types of cells and different apparati), endothelial cell activities that would seem to lead to atherosclerosis formation increase in the presence of low and/or oscillating shear stress. These results tend to support the hypothesis put forth by the arterial model flow studies that low and oscillating wall shear stresses lead to atherosclerosis formation.

Additional Studies of Arterial Flow Patterns and Atherogenesis

There have also been several flow-related *in vivo* studies that have contributed to the knowledge of the potential role of fluid mechanics in atherogenesis. The inherent difficulty of performing *in vivo* experiments has been a limiting factor, as has the fact that atherosclerosis forms over decades (much longer than a typical research grant). Nevertheless, the information gained appears to support the results obtained with *in vitro* flow models and cell culture studies. Zarins et al.[55] surgically created local coarctations in the aortas of cynomolgus monkeys that were subsequently fed an atherogenic diet. It was found that the region directly within the coarctation was spared of lesions. Distal to the coarctations, significantly more lesions formed and appeared to be localized to the regions where one would expect flow separation. Later work by the same group showed that increasing the flow rate through an artery by surgically creating an arterio-venous fistula provoked diameter changes that maintained the same level of shear stress as sham operated controls.[56] Other *in vivo* studies have demonstrated that endothelial permeability varies along the arterial tree, with areas near branches showing higher degrees of permeability.[57]

The flow modeling techniques applied in the study of flow in healthy vessels have also been extended to the analysis of flow in vessels where disease is or has been developing. Much of the basis for this work extends from the hypotheses put forth by Friedman.[58] He proposed that different thickening histories at sites exposed to high and low shear rates were due to the presence of multiple, competing shear-dependent processes in the vessel wall, and he proposed a model to describe this behavior. Numerical flow simulations coupled with a similar wall shear stress–plaque formation law were used to simulate the formation of atherosclerosis in the aortic bifurcation.[59,60] They concluded that low wall shear stresses were most consistent with the onset of the disease, but that areas of both low and high wall shear stress showed accelerated plaque growth during subsequent disease formation. A similar flow simulation in the carotid bifurcation also produced lesions similar to those found *in vivo*.[61] In that study, it was also noted that the shear stress field created by the presence of slight intimal thickening was more homogeneous than with no intimal thickening. This indicates that intimal thickening may be a means by which the artery attempts to even out the wall shear stress distribution. The development of clinically significant atherosclerosis may be the result of this process becoming unstable.

The presence of stenosis within a blood vessel can have a variety of effects on the blood flow patterns. Clinically, the most important effect is the increase in flow resistance that, when the body fails to properly compensate, can lead to tissue death. The pressure drop across a region of stenosis depends most strongly on the percent reduction in diameter (or area), which produces a large amount of pressure head loss from flow separation and turbulence.[62] The actual viscous pressure head loss is much smaller under most physiological blood flow conditions. As will be discussed below, the occurrence of turbulence near the stenosis in and of itself is an important and useful clinical phenomenon. Stenoses that reduce the cross-sectional area of the artery by more than 89% can initiate the onset of turbulent flow distal to the stenosis, even at Reynolds numbers as low as 200.[62]

Another potentially important clinical phenomenon associated with stenoses is the collapse of the artery. While arteries typically become stiffer with age, atherosclerotic stenoses may still be compliant. As blood flows through these constrictions, the static pressure within the lumen falls because of the Bernoulli effect. It has been hypothesized by several authors that such high-grade stenoses may collapse under physiological conditions.[63,64] If this is the case, the flow through the artery may be severely limited due to the even greater reduction in cross-sectional area. The subsequent collapse of the tube can have additional effects on the flow rate. Flow through elastic tubes can be limited (choked) under certain circumstances, as described by Conrad.[65] Flow choking is an important physiological consequence which could limit blood flow during a high-demand situation where the distal resistance is dramatically lowered. Shapiro and others have developed a one-dimensional fluid mechanic model of tube collapse and flow choking which uses an analogy to transonic, compressible flows and supercritical flows in channels.[66,67] Ku et al.[68] have applied this theory to estimate the flow conditions needed to induce collapse in a high-grade arterial stenosis. Collapse was predicted for a wide variety of physiological arterial stiffnesses, flows, and pressures.

7.4 Application of Arterial Fluid Mechanics in Diagnostics

When the presence of diseases such as atherosclerosis restricts the supply of blood to certain tissues, there is obvious clinical interest in the important variables such as pressure and flow. The measurement of blood pressure by direct or indirect methods is a very common diagnostic procedure. The capability to measure flow is more recent but is gaining in importance. Advances in diagnostic capabilities have provoked clinicians to learn more about pressure, flow, and other important flow variables. This additional knowledge is also applied in developing and evaluating new treatment procedures.

The concept that blood pressure, particularly pulse pressure, might be indirectly related to disease far predates all studies of any direct relationship. One can indeed imagine that an abnormally weak pressure pulse at palpation was recognized as a negative symptom thousands of years ago. Nowadays, we know quite a lot about pressure pulses, and palpation yields much more differentiated diagnostic information than it did 100 years ago. For example, diseases of the heart valves can be identified more or less precisely in certain cases.[69] Palpation of arterial pulses is, of course, limited to the few sites where a greater artery is relatively close to the skin (e.g., the carotid arteries). Palpation of the femoral arteries is possible only in certain cases. In 1971, the American Heart Association found it useful to publish a monograph that lists publications issued on cardiovascular auscultation and photocardiography from 1820 to 1966.[70]

The stethoscope allows one to hear, rather than merely feel, arterial pulses under certain conditions. The microphone (which was the high-tech version of the stethoscope thirty years ago) set the stage for developments in the field of mechanography (phonocardiogram, apexcardiogram, carotidogram, etc.).[71] With apexcardiograms and carotidograms, the low frequency pressure fluctuations felt in palpation are transduced into audible frequencies. While stethoscopes and microphones may yield a rather deformed image of pressure pulses, they usually result in a better representation of flow pulses. To record apexcardiograms or carotidograms, one uses, instead, special transducers with a rather large time constant (on the order of seconds).

The development of more powerful techniques in recent decades has, regrettably, reduced somewhat the importance of empirical knowledge of pressure and flow pulses. Doppler ultrasound introduced a more quantitative means to hear flow pulses than either the stethoscope or microphone. It also extended the subcutaneous depth to which one could evaluate flow. Furthermore, tight arterial stenoses can be localized on the basis of the higher audio frequencies (higher Doppler shift) associated with increased flow velocity. The spectral broadening in the Doppler signal may also be used as a diagnostic parameter.[72] In some cases, one may even determine if an artery that cannot be directly visualized is severely obstructed by the use of a deductive diagnostic scheme. For example, one can determine which one of the two posterior cerebral arteries is impaired by a severe stenosis by performing different Doppler measurements on the anterior arteries.[73]

In the 1970s, ultrasound imaging modalities became commercially available which were soon completed by Doppler capabilities. Today, one can hear and visualize blood flows (in color) in their anatomical context (Fig. 7.2).[74] This is, of course, superior to what can be achieved by palpation or with a microphone. In many cases, Doppler ultrasound can be used to assess instantaneous blood flow semi-quantitatively (for example, the flows across the cardiac valves) or even quantitatively (in some larger arteries). In the last few years, methods for use in nuclear magnetic resonance (NMR or MRI) examinations were also developed.[75,76] For example, one may obtain detailed information on flow in the larger vessels such as the aorta[77] or the carotid artery.[78] The usefulness of MRI angiography remains limited due mainly to spatial resolution problems. The applicability to coronary arteries in particular is further hindered by the movement of these arteries during the cardiac cycle and by the movement of the chest cavity as a whole due to breathing. A further problem that arises is the signal loss that occurs due to turbulence distal to stenoses.[79] Thus, there remain several significant obstacles preventing the use of MRI for the evaluation of coronary artery disease.

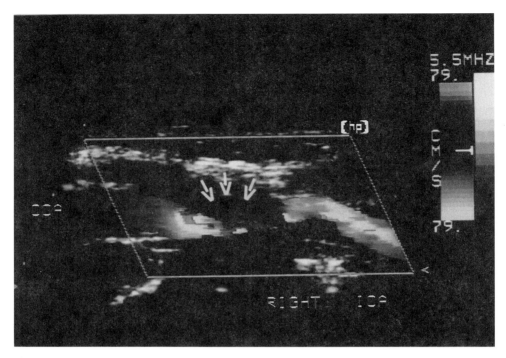

FIGURE 7.2 Color Doppler ultrasound image of a stenosis on the right internal carotid artery (ICA). Where the vessel caliber is reduced by the atheromatous plaques (indicated by arrows), the flow velocity is higher (blue instead of yellow or red).

Due to their non-invasiveness, the techniques mentioned so far provide only limited information. Furthermore, the quality of the information can vary greatly. Therefore, non-invasive techniques have not been able to totally replace invasive ones. Among the invasive techniques, (x-ray) angiography is probably the most powerful because of its unsurpassed spatial and temporal resolutions.[80] It is indispensable in the examination of the coronary and cerebral arteries. Although angiography is one of the most powerful tools used to diagnose atherosclerosis, there are significant drawbacks. It requires invasive catheterization and the injection of radiopaque dye into the patient (Fig. 7.3). Thus, angiography is a potentially expensive and at least mildly risky procedure. Other disadvantages include the lack of information regarding flow reserve, the vessel wall, and the exact nature of the plaque.

Angiography may also provide hemodynamic information. For example, a record of the instantaneous left ventricular pressure is an important, basic tool for evaluating all types of heart disease.[81] In valvular diseases, additional pressure recordings in other cardiac chambers and in the great cardiac vessels are usually required. For better accuracy, one can use strain-gauge pressure transducers mounted on the tips of catheters or guide wires. Pressure pulses can then be recorded without distortions, which is not the case with fluid-filled catheters.[82] Radiopaque contrast agents also provide some valuable hemodynamic information. For instance, observation of the spatial-temporal propagation of the contrast agent provides a subjective impression of the flow pulse.[83] Quantitative assessments of flows or flow pulses require, however, mathematical-physical models and sophisticated computations that involve the gray levels of the image pixels.[84]

A further reason why angiography still plays a dominant role is that it is the only imaging modality that simultaneously allows for intravascular interventions. Techniques such as balloon angioplasty are gaining popularity due to the fact that the trauma and expense associated with surgery can be avoided in many cases. Intravascular interventions are now routinely performed directly following angiography. In neuroradiology, their importance is growing rapidly. As will be illustrated later, recent developments in these fields could make hemodynamic aspects become increasingly important.

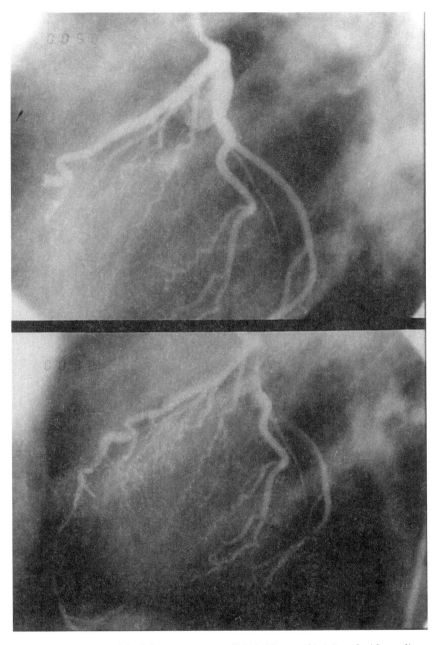

FIGURE 7.3   Angiographic view of the left coronary artery (LCA). The vessel is injected with a radiopaque solution for temporary opacification. Upper panel: As clearly visible at left, one main branch of the LCA, the left anterior descending coronary artery (LAD) ends abruptly because it is occluded. Lower panel: Same view after successful (artificial) recanalization.

Because the heart is a vital organ, cardiac diseases can have very severe or life-threatening consequences. Furthermore, heart failure eventually affects other organs (the lungs, for example).[85] The development of stenoses in the coronary arteries (commonly referred to as coronary artery disease or CAD) is, therefore, one of the most important clinical manifestations of this disease.

Even though the limitation of blood flow is the most important aspect of CAD, the diagnosis of CAD does not require the measurements of flow or flow pulses. The occurrence of chest pain and dyspnea usually yields a reliable indication, which is often reinforced by a positive ECG exercise test.[86] The initial diagnosis is then confirmed usually by cardiac catheterization, which basically includes a recording of the instantaneous left ventricular pressure, a left ventricular angiogram, and coronary angiography. These evaluations provide reliable information on the morphologic state of the coronary arteries as well as a subjective evaluation of the flow conditions. There are also rare cases in which these arteries seem completely normal. The diagnosis may then be "Syndrome X." Various hypotheses have been proposed to explain this phenomenon: disease of the micro-vessels, steal effects, regional spasmic constriction of the arterioles, etc. Pressure recordings in the cardiac chambers and in the aorta provide quite a lot of information about the state of the myocardium and the valves.[87] As a rule, cardiac output (mean forward flow per minute) is only calculated in the case of valvular disease.[87] The cardiac flow pulse could be determined by computing the temporal evolution of the left ventricular volume, but this is practically never done. (The left ventricular volume can itself be calculated by geometric or densitometric methods. As a rule, it is determined only for end diastole and/or end systole and used to calculate left ventricular ejection fraction.)

The therapeutic goal in the case of CAD is simple: to restore as best as possible the equilibrium between the oxygen demand of the myocardium and its supply by the coronary arteries, at least for the situations in daily life and, hopefully, for several years. In some cases (e.g., stenosis of the left main coronary artery), coronary artery bypass graft (CABG) surgery is the preferred method.[88] In other patients, vessel patency can be achieved by drug therapy (e.g., reduction of the oxygen demand by decreasing the resistance of the peripheral vessels or of the coronary arteries and/or the arterioles, etc.). Another technique that is gaining in popularity is percutaneous transluminal coronary angioplasty (PTCA).[89]

The drawback to PTCA is that it may be only moderately effective, despite a high rate of initial reopening (about 90% for skilled operators).[89] The primary reason is that the dilated lumen may close again after some minutes, hours, or days (known as sudden occlusion). This occurs when the stenotic plaque has not been sufficiently deformed by balloon inflation. Plaques may be quite elastic, and may require enormous deformation to be plastically deformed. The second reason is that restenosis occurs in a significant percentage of patients (15 to 50%, depending on the patient population, the kind of lesions, etc.)[90] In an attempt to solve these problems, many clinicians implant stents, but the success is only moderate, at least at present.[90]

Because the problem of restenosis remains, fluid dynamics could become attractive for the following reasons. First of all, PTCA is relatively expensive, especially if it is supplemented by stenting or if it must be repeated. Furthermore, some patients will require CABG surgery sooner or later.[91] Consequently, it is important to restrict PTCA to cases where it is most likely to be beneficial. PTCA also has other negative aspects. PTCA partially removes the obstacle to flow, but the lesion is merely pushed back into the vessel wall. Stenting may help by holding the lesion away from the lumen, but the lesion will still be present. It may therefore be preferable not to dilate stenoses which do not limit flow under resting conditions.[88] If this is the case, the risk remains that the plaque will continue to develop into a more serious blockage. Unfortunately, there are no reliable means of predicting the course of development. PTCA also assumes that the patient accepts living with some restrictions. Assuming that all these aspects will be clarified one day, one will have to determine, in the catheterization laboratory, to what extent a stenosis limits the flow.

Methods to determine flow limitations as a result of stenosis are already available. One can, for instance, assess quantitatively the coronary flow reserve distal to a stenosis by the use of angiographic flow measurement techniques.[92] Alternatively, a method based on pressure measurements has been proposed.[93] These techniques are, however, not yet widely used. When they are used, it is often to assess the success of PTCA and stenting, rather than in the initial treatment decision process.

A third possibility is the recently developed FloWire Doppler device (Cardiometrics, Mountain View, CA).[94] It is essentially an ultrasonic transducer mounted at the tip of a catheter guide wire (with a diameter of 0.46 or 0.36 mm) connected to an apparatus which computes and displays the instantaneous maximal velocity in the region of the sample volume. With this device, the time evolution of the flow velocity during the cardiac cycle can be continuously recorded at any (accessible) arterial site. Normal and abnormal flow pulse patterns in the left and right coronary arteries have also been assessed, as well as the ranges of maximal and average flow velocity in the different coronary arteries.[95] One can also compare the averaged maximal velocities proximal and distal to a stenosis. According to some authors, the ratio of these two averaged velocities provides an indication of the functional severity of a stenosis.[95]

In summary, pressures, pressure pulses, flows, and flow pulses provide diagnostic information, but this information is seldom sufficient for the final diagnosis in any situation when assessed non-invasively. This is not really surprising because these four parameters are only symptoms and not causes. The availability of powerful imaging modalities means that someday one will be able to see why pressure or flow pulses are abnormal instead of merely documenting that they are abnormal. For example, with Doppler echography once can see blood regurgitates through a diseased cardiac valve, but one would like to see the valve itself to determine if valve replacement is indicated. For the same reasons, broad-band screenings of populations at risk of CAD are not highly attractive. However, in the near future, hemodynamic aspects could be more and more involved in the decision makings in PTCA.

7.5 Conclusions

There is much evidence that blood flow patterns are involved in the development and progression of atherosclerosis. Unfortunately, all of this evidence is indirect. Atherosclerosis develops over decades, and so it is difficult to track the development of the disease in detail. It is even more difficult to monitor arterial flow patterns over such a time period, especially considering all of the changes that occur in the factors that determine blood flow patterns. Arterial geometry changes may occur over periods of months to years. The mean flow rates through individual arteries and the pulse rate may change from minute to minute, depending on the level of activity. These factors will be extremely difficult, if not impossible, to overcome. Advancement in the knowledge of the role of hemodynamics in atherogenesis will come from designing scientific studies that more closely approximate the *in vivo* situation.

There are several researchers who are making efforts to combine studies of the solid mechanics of the artery wall with the fluid dynamics studies outlined above. Knowledge of the solid wall stresses in the artery wall is limited due to the extremely complex nature of the problem. This complexity is being somewhat alleviated by the application of state-of-the-art finite element techniques that can more closely replicate the *in vivo* geometry and complex mechanical behavior.[11,97,98] There are also a few groups that are studying the effects of combined fluid shear stress and cyclic substrate stretch on endothelial cells.[99,100] The morphological effects of combined shear stress and cyclic stretch include an enhanced elongation and alignment with the flow direction than that which is found for cells subjected to the same levels of only one mechanical stimulus or the other.[101] All of these studies indicate that the inclusion of solid and fluid mechanics will be important in future investigations of atherogenesis.

The further application of fluid mechanics in diagnostic medicine will enhance the prevention and treatment of the deadly effects of atherosclerosis. As with all new medical technologies, it will take time for fluid mechanics to be fully accepted as a standard diagnostic tool.

Acknowledgments

The authors would like to acknowledge the support of the following organizations: The Swiss National Science Foundation, The Whitaker Foundation, The (USA) National Science Foundation, The Burroughs Wellcome Trust, The Florida International University Foundation, Oak Ridge Associated Universities, the National Institutes of Health, and the American Heart Association, Florida Affiliate.

References

1. Dean, W.R., *Philos. Mag.*, 4, 208, 1927.
2. Dean, W.R., *Philos. Mag.*, 5, 673, 1928.

3. Tada, S., Oshima, S., and Yamane, R., *ASME J. Biomech. Eng.*, 118, 311, 1996.
4. Pinchak, A.C. and Ostrach S., *J. Appl. Physiol.*, 41, 646, 1976.
5. Hale, J.F., McDonald, D.A., and Womersley, J.R., *J. Physiol.*, 128, 629, 1955.
6. Nerem, R.M., Rumberger, J.A., Gross, D.R., Muir, W.W., and Geiger, G.L., *Cardiovasc. Res.*, 10, 301, 1976.
7. Seed, M.A. and Wood, R., *Cardiovasc. Res.*, 5, 319, 1971.
8. Pedley, T.J., *The Fluid Mechanics of Large Blood Vessels*, Cambridge University Press, Cambridge, 1980.
9. Fukushima, T., Homma, T., Harakawa, T., Sakata, N., and Azuma, T., *ASME J. Biomech. Eng.*, 110, 166, 1988.
10. Dutta, A. and Tarbell, J.M., *ASME J. Biomech. Eng.*, 118, 111, 1996.
11. Perktold, K. and Rappitsch, G., *J. Biomech.*, 28, 845, 1995.
12. Ku, D.N. and Liepsch, D., *Biorheology*, 23, 359, 1986.
13. Nichols, W.W. and O'Rourke, M.F., *McDonald's Blood flow in Arteries*, Lea and Febiger, Philadelphia, PA, 1990.
14. Rather, L.J., *A Commentary on the Medical Writings of Rudolf Virchow*, Norman Publishing, San Francisco, CA, 1990.
15. Roy, C.S., *J. Physiol.*, 3, 125, 1880.
16. Krafka, F., *Arch. Pathol.*, 23 1, 1937.
17. Hirsch, S., *Cardiologia*, 5, 122, 1941.
18. Leary, T., *Arch. Pathol.*, 32, 507, 1941.
19. Gofman, J.W., *Ann. N.Y. Acad. Sci.*, 64, 590, 1956.
20. Texon, M., *Arch. Intern. Med.*, 99, 418, 1957.
21. Fox, J.A. and Hugh, A.E., *Br. Heart J.*, 26, 388, 1966.
22. Fry, D.L., *Circ. Res.*, 22, 165, 1968.
23. Caro, C.G., Fitzgerald, J.M., and Schroter, R.C., *Proc. R. Soc. London B*, 177, 109, 1971.
24. Friedman, M.H., O'Brien, V., and Ehrlich, L.W., *Circ. Res.*, 36, 277, 1975.
25. Friedman, M.H., Hutchins, G.M., and Bargeron, C.B., *Atherosclerosis*, 39, 425, 191.
26. Friedman, M.H., Deters, O.J., Mark, F.F., Bargeron, C.B., and Hutchins, G.M., *Atherosclerosis*, 46, 225, 1983.
27. Friedman, M.H. and Ehrlich, L.W., *J. Biomech.*, 17, 881, 1984.
28. Friedman, M.H., Deters, O.J., Bargeron, C.B., Hutchins, G.M., and Mark, F.F., *Atherosclerosis*, 60, 161, 1986.
29. Bharadvaj, B.K., Mabon, R.F., and Giddens, D.P., *J. Biomech.*, 15, 349, 1982.
30. Bharadvaj, B.K., Mabon, R.F., and Giddens, D.P., *J. Biomech.*, 15, 363, 1982.
31. Logerfo, F.W., Nowak, M.D., and Quist, W.C., *J. Vasc. Surg.*, 2, 263, 1985.
32. Ku, D.N., Giddens, D.P., Zarins, C.K., and Glagov, S., *Arteriosclerosis*, 5, 293, 1985.
33. Friedman, M.H., Bargeron, C.B., Deters, O.J., Hutchins, G.M., and Mark, F.F., *Atherosclerosis*, 68, 2, 1987.
34. He, X. and Ku, D.N., *ASME J. Biomech. Eng.*, 118, 74, 1996.
35. Moore, J.E., Jr. et al., *ASME J. Biomech. Eng.*, 116, 302, 1994.
36. Moore, J.E., Jr., Xu, C., Glagov, S., Zarins, C.K., and Ku, D.N., *Atherosclerosis*, 110, 225, 1994.
37. Holenstein, R. and Ku, D.N., *Biorheology*, 25, 835, 1988.
38. Moore, J.E., Jr., Ku, D.N., Zarins, C.K., and Glagov, S., *ASME J. Biomech. Eng.*, 114, 391, 1992.
39. Friedman, M.H., Bargeron, C.B., Duncan, D.D., Hutchins, G.M., and Mark, F.F., *ASME J. Biomech. Eng.*, 114, 317, 1992.
40. Ma, P., Li, X., and Ku, D.N. *Int. J. Heat Mass Transfer*, 37, 2723, 1994.
41. Rappitsch, G. and Perktold, K., *ASME J. Biomech. Eng.*, 29, 207, 1996.
42. Rappitsch, G. and Perktold, K., *ASME J. Biomech. Eng.*, 118, 511, 1996.
43. Krueger, J.W., Young, D.F., and Cholvin, N.R., *J. Biomech.*, 4, 31, 1971.

44. Dewey, C.F., Bussolari, S.R., Gimbrone, M.A., and Davies, P.F., *ASME J. Biomech. Eng.*, 103, 177, 1981.

45. Levesque, M.J. and Nerem, R.M., *ASME J. Biomech. Eng.*, 107, 341, 1985.

46. Helmlinger, G., Geiger, R.V., Schreck, S., and Nerem, R.M., *ASME J. Biomech. Eng.*, 113, 123, 1991.

47. Gau, G.S., Ryder, T.A., and Mackenzie, M.L., *J. Pathol.*, 131, 55, 1980.

48. Nerem, R.M., Levesque, M.J., and Cornhill, J.F., *ASME J. Biomech. Eng.*, 103, 172, 1981.

49. Sharefkin, J.B., Diamond, S.L., Eskin, S.G., McIntire, L.V., and Dieffenbach, C.W., *J. Vasc. Surg.*, 14, 1, 1991.

50. Frangos, J.A., Eskin, S.G., McIntire, L.V., and Ives, C.L., *Science*, 227, 1477, 1985.

51. Sampath, R., Kukielka, G.L., Smith, C.W., Eskin, S.G., and McIntire, L.V., *Ann. Biomed. Eng.*, 23, 247, 1995.

52. Nerem, R.M., *ASME J. Biomech. Eng.*, 115, 510, 1993.

53. Weinbaum, S. and Chien, S., *ASME J. Biomech. Eng.*, 115, 602, 1993.

54. Varner, S.E., Chappell, D.C., Medford, R.M., Alexander, R.W., and Nerem, R.M., Flow and the associated shear stress regulates VCAM-1 gene expression and transcription in human vascular endothelial cells, *Bioeng. Conf.*, BED Vol. 29, ASME, Beaver Creek, 1995, 517.

55. Zarins, C.K., Bomberger, R., and Glagov, S., *Circulation*, 64, 221, 1981.

56. Zarins, C.K., Zatina, M.A., Giddens, D.P., Ku, D.N., and Glagov, S., *J. Vasc. Surg.*, 5, 413, 1987.

57. Hermann, R.A., Malinauskas, R.A., and Truskey, G.A., *Arteriosclerosis and Thrombosis*, 14, 313, 1994.

58. Friedman, M.H., *Arteriosclerosis*, 9, 511, 1989.

59. Nazemi, M. and Kleinstreuer, C., *ASME J. Biomech. Eng.*, 111, 311, 1989.

60. Nazemi, M. and Kleinstreuer, C., *ASME J. Biomech. Eng.*, 111, 316, 1989.

61. Lee, D. and Chiu, J.J., *J. Biomech.*, 29, 1, 1996.

62. Young, D.F. and Tsai, F.Y., *J. Biomech.*, 6, 395, 1973.

63. Logan, S.E., *IEEE Biomed. Eng.*, BME 22, 327, 1975.

64. Binns, R.L. and Ku, D.N., *Arteriosclerosis*, 9, 842, 1989.

65. Conrad, W.A., *IEEE Trans. Biomed. Eng.*, BME 16, 284, 1969.

66. Shapiro, A.H., *ASME J. Biomech. Eng.*, 99, 126, 1977.

67. Elad, D., Sahar, M., Avidor, J.M., and Einav, S., *ASME J. Biomech. Eng.*, 114, 84, 1992.

68. Ku, D.N., Ziegler, M.N., and Downing, J.M., *ASME J. Biomech. Eng.*, 112, 444, 1990.

69. Hurst, W.J. et al., Eds., *The Heart, Arteries and Veins*, 7th ed., McGraw-Hill, New York, 1990, chaps. 11, 40-45.

70. Ravin, A. and Frame, F.K., Eds. *International Bibliography of Cardiovascular Auscultation and Phonocardiography*, Monogr. 31, The American Heart Association, Inc., New York, 1971.

71. Tavel, M.E., Ed., *Clinical Phonocardiography and External Pulse Recording*, 4th ed., Year Book Medical Publishers, Chicago, IL, 1985.

72. Giddens, D.P., Zarins, C.K., and Glagov, S., *ASME J. Biomech. Eng.*, 115, 588, 1993.

73. Keller, H.M., Schubiger, O., Krayenbuehl, C., and Zumstein, B., Cerebrovascular Doppler examination and cerebral angiography-alternative or complementary, *Neuroradiology*, 16, 140, 1978.

74. Polak, F.J., Ed., *Peripheral Vascular Sonography: A Practical Guide*, Williams and Wilkins, Baltimore, MD, 1992.

75. Nayler, G.L., Firmin, D.N., and Longmore, D.B., Blood flow imaging by cinemagnetic resonance, *J. Comp. Assisted Tomography*, 10, 715, 1986.

76. Potchen, E.J., Haacke, E.M., and Siebert, J.E., Eds., Magnetic resonance angiography, *Concepts and Applications*, Mosby, St. Louis, MO, 1930.

77. Moore, J.E., Jr., Maier, S.E., Ku, D.N., and Boesiger, P., *J. Appl. Physiol.*, 76, 1520, 1994.

78. Caro, C.G., Dumoulin, C.L., Graham, J.M., Parker, K.H., and Souza, S.P., *ASME J. Biomech. Eng.*, 114, 147, 1992.

79. Siegel, J.M., Oshinski, J.N., Pettigrew, R.I., and Ku, D.N., *J. Biomech.*, 29, 1665, 1995.

80. Skorton, D.J., Brundage, B.H., Schelbert, H.R., and Wolf, G.L., Relative merits of imaging techniques, in *Heart Disease*, 5th ed., Braunwald, E., Ed., W.B. Saunders, Philadelphia, PA, 1997, chap. 11.

81. Davidson, C.J., Fishman, R.F., and Bonow, R.O., Cardiac catheterization, in *Heart Disease*, 5th ed., Braunwald, E., Ed., W.B. Saunders, Philadelphia, PA, 1991, chap. 6.

82. The design of manometers, in *Blood Flow in Arteries*, 2nd ed., McDonald, D.A., Ed., Edward Arnold Publishers, Ltd., London, 1974, chap. 8.

83. Rutishauser, W., Simon, H., Stucky, J.P., Schaad, N., Noseda, G., and Wellauer, J., Evaluation of roentgen cinedensitometry for flow measurement in models and in the intact circulation, *Circulation*, 36, 951, 1967.

84. Eigler, N.L., Pfaff, J.M., Zeiher, A., Whiting, J.S., and Forrester, J.S., Digital angiographic impulse response analysis of regional myocardial perfusion: linearity, reproducibility, accuracy, and comparison with conventional indicator dilution curve parameters in phantom and canine models, *Circ. Res.*, 64, 853, 1989.

85. Braunwald, E., Heart failure, in *Harrison's Principles of Internal Medicine, Part 7, Section 1, Contribution 195*, 13th ed., Isselbacher, K.J., Braunwald, E., Wilson, J.D., Martin, J.B., Fauci, A.S., and Kasper, D.L., Eds., McGraw-Hill, New York, 1994, 998.

86. Hurst, J.W., Atherosclerotic coronary heart disease: historical benchmarks, methods of study and clinical features, differential diagnosis, and clinical spectrum, in *The Heart, Arteries and Veins*, 7th ed., Hurst, W.J. et al., Eds., McGraw-Hill, New York, 1990, chap. 50.

87. Davidson, C.J., Fishman, R.F., and Bonow, R.O., Cardiac catheterization, in *Heart Disease*, 5th ed., Braunwald, E., Ed., W.B. Saunders, Philadelphia, PA, 1997, chap. 6.

88. Marco, J. et al., Eds., Seventh complex coronary angioplasty course, Unité de Cardiologie Interventionnelle, Toulouse, 1996, 161 and 475.

89. Topol, E.J., *Textbook of Interventional Cardiology*, Saunders, Philadelphia, PA, 1999.

90. Ruygrok, P. and Sprenger de Rover, C., Eds., *Second Thoraxcenter Course on Coronary Stenting*, Barjesteh, Meeuwes & Co., Rotterdam, 1995, 200.

91. Marco, J. et al., Eds., Seventh complex coronary angioplasty course, Unité de Cardiologie Interventionnelle, Toulouse, 1996, 177.

92. Pijls, N.H.J., Uijen, G.J.H., Aengevaeren, W.R.M., Hoevelaken, A., Pijnenburg, T., Van Leeuven, K., and van der Werf, T., Concept of maximal flow ratio for immediate evaluation of percutaneous transluminal coronary angiography result by videodensitometry, *Circulation*, 83, 865, 1991.

93. De Bruyne, B., Baudhuin, T., Melin, J.A., Pijls, N.H.J., Sys, S.U., Bol, A., Paulus, W.J., Heyndricks, G.R., and Wijns, W., Coronary flow reserve calculated from pressure measurements in humans. Validation with positron emission tomography, *Circulation*, 89, 1013, 1994.

94. Doucette, J.W., Corl, P.D., Payne, H.M., Flynn, A.E., Goto, M., Nassi, M., and Segal, J., Validation of a Doppler guide wire for intravascular measurement of coronary artery flow velocity, *Circulation*, 85(5), 1899, 1992.

95. Ofili, E.O., Labovitz, A.J., and St. Vrain, J.A. et al., Analysis of coronary blood flow dynamics in angiographically normal and stenosed arteries before and after endoluminal enlargement by angioplasty, *J. Am. Coll. Cardiol.*, 21, 308, 1993.

96. Donohue, T.J., Kern, M.J., and Aguirre, F.V. et al., Assessing the hemodynamic significance of coronary artery stenoses: analysis of translesional pressure-flow relationships in patients, *J. Am. Coll. Cardiol.*, 22, 449, 1993.

97. Salzar, R.S., Thubrikar, M.J., and Eppink, R.T., *J. Biomech.*, 28, 1333, 1995.

98. Delfino, A., Moore, J.E., Jr., Stergiopulos, N., and Meister, J.J., *J. Biomech.*, in press.

99. Benbrahim, A. et al., *J. Vasc. Surg.*, 20, 184, 1994.

100. Moore, J.E., Jr. et al., *Ann. Biomed. Eng.*, 22, 416, 1994.

101. Zhao, S. et al., *Arteriosclerosis, Thrombosis Vasc. Biol.*, 15, 1781, 1995.

8

Computational Fluid Dynamics Modeling Techniques Using Finite Element Methods to Predict Arterial Blood Flow

8.1 Introduction 8-1
8.2 Background and Literature Review................................8-2
8.3 Uncertainties in the Proposed Studies............................8-4
   Experimental Uncertainties • Numerical Uncertainties
8.4 Experimental Method..8-4
   Technical Details, Instrumentation and Data Acquisition
   System • Arterial Geometry • Formulation • Boundary
   Conditions • Non-Newtonian Blood Viscosity
8.5 Results 8-13
   Animal Hemodynamic Data • Flow Calculation Description
8.6 Discussion 8-22
   Discussion for Study One • Discussion for Study Two
8.7 Summary and Conclusions 8-23
   Experimental Study • Numerical Study

R. K. Banerjee
Kettering University

L. H. Back
Jet Propulsion Lab

Y. I. Cho
Drexel University

8.1 Introduction

In the past, the study of hemodynamics has been approached from many different directions, by researchers from disciplines as diverse as medicine and engineering. In attempting to match the knowledge of fluid dynamics to the clinical reports of the patterns of arterial disease, researchers have discovered insufficiencies on both sides. This study identifies a suitable methodology to address some unanswered questions. The broad objective of this hemodynamic study is to develop the capability for numerical prediction.

This investigation presents *in vivo* measurements of flow quantities and subsequent numerical flow and pressure calculation and validation, using measured flow parameters, vessel dimensions, and blood viscosity in a mildly tapered femoral artery of a living dog. While such a combined experimental–calculation approach may appear straightforward in principle, it is fraught with concerns about the reliability of *in vivo* measurements and the appropriateness of the method of flow calculation with regard to the non-Newtonian viscosity of blood, wall and inlet boundary conditions, and handling of the non-linear convective acceleration terms and coupling of the pressure and velocity fields during the cardiac cycle. Moreover, giving statistical significance to such an approach would generally require numerous experiments and theoretical comparisons — a procedure which is beyond the scope of most detailed flow investigations such as this one.

Our study is modest and more along the lines of a bioengineering approach. We posed the question: What is the most accurate procedure for numerically calculating and validating arterial flows using *in vivo* measurements? In order to answer the question realistically, we identified two possible approaches and posed subsequent questions:

- What level of approximation is required to adequately describe the measured variation of pressure change with time, $\Delta p(t)$ and time-averaged pressure drop Δp_a, across a mildly tapered femoral artery segment of a dog given the measured velocity waveform $u(t)$?

- What level of approximation is required to adequately describe the measured variation of velocity with time, $u(t)$, and time-averaged velocity, u_a, given the measured pressure change with time, $\Delta p(t)$?

Clearly, the two approaches above are diametrically opposed, but the answer to either question would lead to the objective of this study. This question requires accurate measurements of $\Delta p(t)$, $u(t)$, and, thus, Δp vs. u, which is of interest in and of itself for physiological flow and is rarely measured or reported in the literature. In the flow calculations, blood viscosity was assumed to be dependent upon shear rate as prescribed through the second invariant of the rate of strain tensor which, in turn, results in relatively unimportant elastic and time relaxation effects. This is believed to be a reasonable approximation for blood flow through arteries the size of the femoral artery in a dog.

On the other hand, since the arterial wall is, in general, very complex, being anisotropic (elastin, collagen, smooth muscle), viscoelastic (creep, stress relaxation, hysteresis), and under a mean and fluctuating state of three-dimensional stresses, it is very difficult to specify its dynamic behavior with certainty in computational methods. Clearly, the simplest assumption from a calculation point of view is to neglect wall interaction and motion, which is known to be relatively small, carry out the flow calculation for an assumed rigid wall, and then compare the calculations to *in vivo* flow data to appraise the extent to which the flow and pressure fields can be described. This is the approach used in this investigation. It is believed to provide some insight into arterial hemodynamics and flow calculation assumptions, albeit for a single vessel of specific size, shape, segment length, and particular pressure and velocity waveforms. Also, some implications with regard to *in vitro* arterial vessel model flow studies are inferred from this study since it provides an *in vivo* reference datum for construction of models with similar Δp and u characteristics during the cardiac cycle.

8.2 Background and Literature Review

In order to achieve the above-mentioned objectives, various issues must first be addressed. For any study in arterial blood flow dynamics, the following important issues require review:

1. The behavior of blood, the principle fluid;
2. The physiological flow regime, expressed in terms of the Reynolds number, the Womersley number, and the Dean number (for flow in curved geometry); and
3. The characteristics of blood ejection from the heart, which is pulsatile (unsteady) in nature.

First, we must consider the nature of the principal fluid, blood, itself. In large arteries (and in smaller ones like the femoral artery) which originate from the heart, blood, for most practical purposes, behaves as a homogeneous and continuous fluid medium, with characteristics similar to a Newtonian fluid, that is, a fluid in which the shear stress is linearly proportional to the rate of deformation. Further down the circulatory tree this assumption becomes increasingly less satisfactory, and by the time blood reaches the arterioles that feed the capillary bed, its rheology is complex. In flow through the capillaries, the concept of a continuous fluid medium loses meaning, and the motion must instead be treated in terms of individual deformable erythrocytes (red blood cells) being transported by the surrounding plasma through narrow passageways sometimes smaller in diameter than the diameter of the erythrocyte itself in its relaxed state.

Second, when dealing with physiological applications, we must closely study the flow regimes, which are at least as complex and varied as the nature of the fluid itself. The character of a flow regime may be expressed, in most cases, in terms of its Reynolds number, a dimensionless parameter which measures the ratio of inertial force to viscous force acting on a fluid. In flows with a high Reynolds number, fluid inertia is the dominant force; in flows with a low Reynolds number, a viscous effect predominates. In the heart and aorta, the Reynolds number is high, approaching values for the transition (Re = 2000 ~ 3000) from a laminar to a turbulent flow. The Reynolds number decreases as the blood proceeds down the arterial tree; in the capillaries, the flow is in the so-called Stokes regime, where viscous forces overwhelm the inertial forces. Given its Reynolds number, the flow through most arteries such as the femoral artery is considered to be laminar.

A third factor requiring close attention is the pulsatile (unsteady) nature of blood flow from the heart. This adds to the complexity of the problem. Arterial vessels are often curved and branched, producing complicated secondary motions of blood and entry phenomena which are far more difficult to analyze than the fully developed flow in a straight tube of infinite length. A further complication is the distensibility of arterial vessels, which deform in a viscoelastic fashion under pressure variations associated with the pulsatility of the blood flow.

This study focuses only on flow behavior. However, many other arguments based on chemical processes, mass transfer, and various other unexplained, naturally occurring phenomena have been made. Although various authors have reported experimental and numerical procedures for obtaining the pressure–flow relationship, the numerical prediction of an *in vivo* pressure drop using the Navier–Stokes equations with the convective terms in an arterial geometry with physiological shape has not been documented in full.

Flow reversal in a dog's artery and phase lag between the pressure gradient and the flow have been reported by McDonald and his co-workers[17] and have also been numerically studied by Womersley.[21,22] A factor of only ± 1.8% radial dilitation $\Delta r_w / \bar{r}_w$ is reported for the resting iliac artery of a dog.[17] Compliance of the resting femoral artery of a dog will be even smaller than the iliac artery. Ling et al.[15] have reported that, for a dog aorta, flow profiles developed locally and transiently during the passage of the pressure-gradient wave. *In vivo* velocity and shear rate data have indicated relatively flat velocity profiles. Back et al.[2] have measured *in vitro* pulsatile velocity and pressure in a human femoral artery model with a reverse lumen curvature. Talbot and Berger[19] have illustrated the relationship of the acceleration and deceleration of velocity to the pressure distribution. For a steady flow analysis, Back et al.[1] have measured flow resistance in models of a curved femoral artery of a human. They have predicted that in order to have a fully developed secondary flow, a relatively high Dean number and curved segment angle are required.

The measured non-Newtonian properties of blood, i.e., shear thinning and viscoelasticity, have been reported in the literature.[14,16] The importance of the non-Newtonian properties of blood in flow simulations have been demonstrated.[7,10] Flow measurements in an atherosclerotic, curved, tapered femoral artery model of a human have been performed by Back et al.[3] For a straight tapered model of the smooth femoral artery, the mean flow measurements have shown significant pressure drop which can be estimated from momentum considerations using the assumption of a local Poiseuille flow. For a mildly tapered femoral artery of a dog, the numerical study of Banerjee et al.[7] has predicted oscillatory shear stress and a phase lag between pulsatile pressure drop and velocity.

Based on the questions posed earlier and discussion of possible approaches, this study was divided into two parts. Study 1 attempted to propose a method to calculate pressure changes in an artery of a dog using an arterial angiogram, the non-Newtonian viscosity of blood, and an instantaneous velocity measurement by non-invasive means. Study 2 attempted to propose an alternate method to calculate velocities using instantaneous pressure gradient measurements. Clearly, either of the two studies validated experimental results with numerical methods.

8.3   Uncertainties in the Proposed Studies

Experimental Uncertainties

Rather than using *in vitro* measurements, these studies aimed to compare and validate the numerical calculations with *in vivo* measurements. It is to be noted that, in the past, *in vitro* measurements were generally found to be easier to compare with numerical calculations since it is comparatively simpler to control any *in vitro* experiment[8] than to control an *in vivo* experiment. In view of this, Studies 1 and 2 were difficult exercises since they involved the greater uncertainties that accompany *in vivo* measurements.

Also, it is practically impossible to measure, and hence numerically verify, *in vivo* velocity profiles at any instant of time near the wall region. At any instant of time, the Doppler flow cuff measures a spatial average velocity only in the core region and excludes the near-wall region of an artery. However, through calibration, the spatial average velocity across the artery is determined. Nevertheless, this poses a difficulty in specifying temporal and spatial dependent velocity boundary conditions, as may be observed in Study 1. In contrast, the measurement of pressure drop is usually more accurate and reliable; hence, such a pressure boundary condition is easy to apply, as may be seen in Study 2. Clearly, from a calculational point of view, Study 1 has greater uncertainties than Study 2.

Numerical Uncertainties

It should be noted that the heart rate of a dog (128 beats/min) is one-and-a-half times faster than the heart rate of a human (75 beats/min), and, therefore, the rate of change of an instantaneous boundary condition over a pulse cycle is higher for a dog than for a human. In general, a numerical calculation for a dog artery has a more stringent condition than for a human artery.

For a numerical simulation of flow, particularly pulsatile flow, calculation of pressure is inherently difficult. Since pressure calculation involves second-order spatial derivatives and a first-order temporal derivative of velocity, any sharp variation in the inlet boundary conditions, as in the case of pulsatile flow, significantly affects the pressure calculation.

Since velocity is used as an inlet boundary condition, as in Study 1, an inherent mass balance is automatically achieved. Only the momentum balance needs to be performed in the flow domain. In contrast, when a pressure boundary condition is applied, as in Study 2, both mass and momentum balances need to be achieved. Clearly, from a numerical point of view, Study 2 is more difficult than Study 1.

8.4   Experimental Method

The animal experiments were performed in Dr. Crawford's laboratory at the USC School of Medicine using instrumentation transported from the Jet Propulsion Laboratory (JPL) to USC with subsequent data reduction at JPL. The animals used were mongrel dogs which were unclaimed at local government pounds. They were certified by USC veterinarians for single experiment utilization, and the experimental protocol was approved by the USC Animal Experimental Board.

Methods for measuring *in vivo* static pressure changes along a segment of the femoral artery without inserting a catheter into the lumen, which can disturb the flow, were investigated initially. One way to take this measurement is by inserting a small plastic tube with a flared end through an incision in the artery wall. By suturing the artery around the tube with the flared end flush with the inner arterial wall surface, we are able to replicate a sidewall static pressure tap. This technique, however, has not proved successful in measuring the *in vivo* static pressure change along a length of an artery because the procedure causes local vaso-constriction and some thrombosis at the injured sites. Consequently, the narrowing of the arterial cross-sectional area in the vicinity of the plastic tubes precludes obtaining meaningful measurements.

Another way to measure static pressure changes is to use small arterial branches as pressure taps. This involves severing a branch some distance from the main lumen, inserting a small plastic tube, and tying the tube externally to prevent blood leakage. Care must be exercised during the experiment to prevent clotting in the plastic tube by frequent flushing with heparin. This technique is used in the animal experiments, and details are reported here.

Animals were lightly anesthetized with ketamine–xylazine given intramuscularly, and a transcutaneous catheter was placed into a dorsal foreleg vein, maintained open with a heparin lock, for control in case vascular support or euthanasia became necessary. Under this light anesthesia, transoral tracheal intubation was performed, and anesthesia was controlled thereafter using inhalation nitrous oxide/methoxyflurane with 20% oxygen and assisted respiration. An angiography catheter was then introduced into a femoral artery and advanced to the aortic bifurcation for the injection of standard angiographic contrast material.

The opposite femoral artery was surgically exposed from 1 cm beyond the inguinal ligament to the adductor canal, and side (muscular) branches were identified. Most of these were ligated, but at appropriate distances some are cannulated for lateral wall pressures. An L and M Doppler flow meter is positioned just after the proximal pressure tap. Flow measurements, absolute and differential pressures, and angiograms are taken during resting and vasodilated flow (adenosine or the radiographic contrast agent renografin). Correlation of angiograms and physiological data is made by use of a radiation sensitive voltage producing device in the radiographic field. Just after euthanasia by a large injection of pentobarbital while the animals are still anesthetized, post-mortem casts are made of the vascular segment using a silicone rubber bolus.[11] Care is taken to obtain mean *in vivo* arterial reference diameters by adjusting the perfusion pressure. Details on the measurements, instrumentation, and data acquisition system are given below.

Technical Details, Instrumentation and Data Acquisition System

For studying the femoral artery, contralateral femoral artery catheters were used to measure iliac bifurcation pressures with the tip positioned with a fluoroscope. The tip of the probe was closed and faced downstream. The static pressure holes were on the side. The transducer control instrumentation was part of the EforM in the USC laboratory. This transducer was calibrated with a mercury manometer.

Based on a pre-test angiogram, a section of the femoral artery was selected between two branches. Any intervening branches were tied off. The two end branches were isolated, cannulated with a static pressure tube, and each branch was sutured tightly around its tube. A fluoroscope was used to position the pressure tap tube normal to the artery with the tip at the wall location. The tube was then secured in position to the surrounding tissue. The animal's leg was positioned so that the arterial section was level. The pressure tap tubes were connected to the two sides of the Validyne differential pressure transducer (model number DP103-20), which has a frequency response of at least 1000 Hz for incompressible fluid such as blood, and appropriate valving to permit frequent flushing and zero checking (Fig. 8.1). The transducer was filled with heparin solution used in frequent flushing to prevent clotting at the tips. 100 cm of water at the full scale diaphragm plate were used in the transducer to minimize the time delay of the response due to fluid movement as the transducer diaphragm deflected. A 200 Hz low-pass internal filter was used. The excitation voltage and calibration for the differential pressure transducer were determined at JPL under static and steady flow conditions with a water manometer. The instantaneous *in vivo* pressure difference signals were quite smooth.

The velocity measurement was made with an L and M Doppler flowmeter. Based on the pre-test angiogram, the position for the flow cuff in a relatively straight section of the artery was selected. The appropriate sized flow cuff was placed around the vessel. The skin incision flaps were pulled up and secured, forming a wall to hold the saline solution in which the flow cuff was immersed. For blood flow, an 8.2 MHz excitation signal provided a good output. Both the time-mean and instantaneous fluctuating

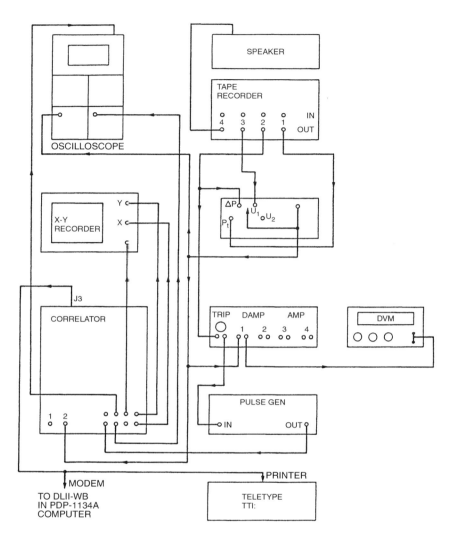

FIGURE 8.1   Instrumentation schematic for acquiring the fluctuating pressure, pressure drop, and velocity measurements during the *in vivo* animal experiments at USC for display on the EforM (channels 1, 2, 4, and 6), evaluation of the time-averaged readings during the test, and recording on the cassette tape recorder for subsequent analysis.

signals were obtained with the unit. The fluctuating signal had a high noise level even with a steady flow in bench experiments at JPL. A 30 Hz internal low pass filter was used to obtain a localized 0.03 s average throughout the approximate 0.47 s heartbeat of a dog. Filtering below 30 Hz has been observed to reduce the level of the peaks. The instrument was calibrated at JPL for steady flow of blood through silicon tubing. These calibrations have been shown to be good for the mean of a fluctuating flow in bench tests at JPL.

Additional instrumentation (Fig. 8.1) from JPL provided for the amplification/attenuation and/or time averaging of the three fluctuating signals and their recording on a magnetic tape cassette in the ± 1 volt range. A fourth channel was used for voice documentation during the tests.

Several angiograms of the arterial section were obtained during the experiments. The renografin solution injection for these angiograms caused vasodilation. Data was normally obtained during this process.

At a later date, the recorded data was reduced at JPL. The schematic for the instrumentation used to retrieve the tape recorded signals is given in Fig. 8.2. The smoothly varying differential pressure signal was used to provide a marker via a trip circuit and a pulse generator. This marker was used to excite the correlator and initiate a 256 point additive accumulation during the next window, which was typically 0.4 s. Generally, 80 to 160 such sums were averaged at each time t in the cardiac cycle. The resulting distribution was displayed on the oscilloscope, could be plotted on the X-Y plotter, and was sent in digital form at 110 baud to a computer for further processing.

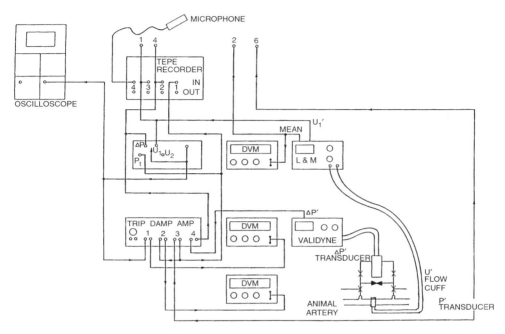

FIGURE 8.2 Instrumentation schematic for the analysis of the tape recorded fluctuating signals at JPL. Averaged cyclic distributions synchronized using the pressure drop signal were obtained on the correlator for subsequent processing on the computer. The time average of the signals was also evaluated.

Correlator settings and reference information such as the mean readings into and out of the tape recorder were also keyed into the computer via a teletype. Each signal on a given segment of tape was thus processed individually, but always relative to the timing of the differential pressure signal. The computer corrected for the calibrations of the various instruments used to process the signal and provided a plot of the data.

Interpretation of these results required knowledge of the physical topography of the vessel section and the blood properties. The former was obtained using the densitometric image scanning methods developed by the JPL Biomedical Image Processing Laboratory. Therefore, the lateral extent and the cross-sectional area along the vessel section could be obtained from the angiograms. Background interference could be corrected for by using a pre-angio x-ray. Direct visual comparisons could also be made with the silicone cast of the section.

The viscosity of the blood drawn from the animal was measured with a Brookfield cone and plate viscometer, which measures the shear stress at shear rates up to 450/s. These measurements were made at several temperatures including the *in vivo* temperature with a chemically stabilized sample stored under refrigeration. The density was obtained by weighing 10 cc of this blood sample.

Arterial Geometry

A tracing of the x-ray of a portion of the femoral artery of a dog where the measurements are taken is shown in Fig. 8.3. The pressure drop across the segment is measured by using two small branch arteries which are ligated and connected by tubing to a Validyne transducer. The cuff for the Doppler flowmeter is located near the first branch, as shown in Fig. 8.3.

For Studies 1 and 2, the vessel segment (Fig. 8.3) is simplified and kept relatively straight with mild taper. The vessel diameter at the first branch tap (referred to as port 1) is $d_1 = 3.8$ mm, and at the second branch tap (referred to as port 2) is $d_2 = 3.6$ mm. The axial distance between the branch pressure taps is 52 mm, so that the ratio of axial distance and diameter d_1 of the vessel segment is 13.7.

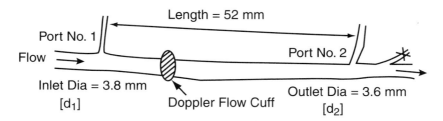

FIGURE 8.3 X-ray tracing of a portion of the femoral artery of a living dog. Ligated small branch arteries are marked port 1 and 2.

Formulation

The objective of this numerical study was to obtain the time-dependent solution of an incompressible, non-Newtonian fluid for the selected geometry (Fig. 8.3). The flow is described by the conservation equations of fluid mass and momentum. A finite element method (FEM) was used to solve the two conservation equations and, therefore, to obtain the velocity, wall shear stress, and pressure distributions. These two equations are presented as follows:

$$u_{j,j} = 0 \tag{8.1}$$

and

$$\rho\left[\frac{\partial u_i}{\partial t} + u_j u_{i,j}\right] = \sigma_{ij,j} + \rho f_i \tag{8.2}$$

where $i, j = 1, 2$ for axisymmetric flows, u_i is the ith component of the velocity vector, ρ is density, σ_{ij} is stress tensor, and f_i is the body force. Furthermore,

$$\sigma_{ij} = -p\,\delta_{ij} + \tau_{ij} \tag{8.3}$$

$$\tau_{ij} = 2\eta_{ij}\varepsilon_{ij} \tag{8.4}$$

$$\varepsilon_{ij} = 0.5\left(u_{i,j} + u_{j,i}\right) \tag{8.5}$$

Here, p is the pressure, τ_{ij} is the deviatoric stress tensor, ε_{ij} is the shear rate tensor, η_{ij} is the tensor viscosity, and δ_{ij} is the Kronecker delta. This study was conducted using the axisymmetric coordinate system with symmetry about the z-axis, i.e., all field quantities were independent of θ, and the circumferential component of velocity u_θ was zero. Axial components of flow were reported as a function of r along the z-direction.

The stress vector s_i at a point on the boundary of a fluid element is defined by

$$s_i = \sigma_{ij} n_j \tag{8.6}$$

For a known element and the solution field, the stress component s_i on the boundary at the Gaussian integration points was evaluated. Subsequently, the normal and tangential components of stress vectors were obtained after applying the appropriate transformations.

The Galerkin formulation[5] using finite elements was applied here in order to discretize the above continuity and momentum equations; this resulted in a set of non-linear algebraic equations of the form

$$M\dot{V} + K(U)V = F(U), \tag{8.7}$$

where $U = (u_1\ u_2)$, $V = (u_1\ u_2\ P)$, $K(U)$ is the global system matrix developed from the momentum balance, M is the mass matric, u_i is the velocity unknown, and F is the forcing function (including body forces and boundary conditions). Four nodal quadrilateral elements were considered for this study. The mesh plot for the artery is shown in Fig. 8.4.

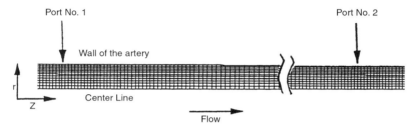

FIGURE 8.4 The mesh plot of the femoral artery shown in Fig. 8.3. For Studies 1 and 2 the vessel segment is simplified and kept relatively straight with mild taper.

In order to numerically solve the momentum and continuity equations for velocity and pressure fields, various solution techniques were used. In a FEM, solution techniques include mixed and penalty formulations. For the mixed formulation, both velocity and pressure in the momentum and continuity equations are independent variables and were solved simultaneously, whereas for the penalty formulation, only velocity is an independent variable, and pressure was approximated by the dilation rate of liquid. In other words, the FEM was not directly applied to the system of equations but rather to a perturbed system of equations in which the continuity requirement was weakened by a penalty parameter, ε. The pressure was approximated as follows:

$$u_{i,i} = -\varepsilon p, \quad \text{where } \varepsilon = 1 \times 10^{-9} \tag{8.8}$$

For high aspect ratios of the elements, a small penalty parameter is recommended. Physically, this can be equated to simulating a flow that has an insignificant compressibility effect. This approach has the advantage of eliminating one of the dependent variables p^*, which is then recovered by post-processing from the velocity field by

$$p^* = -u_{i,i}^* / \varepsilon \tag{8.9}$$

Clearly, both the pressure and velocity fields must be determined in the calculation method. Considering the merits and demerits of numerical schemes and associated inlet boundary conditions, a penalty formulation was used for Study 1, whereas a mixed formulation was chosen for Study 2.

The matrix equation (8.7), representing a discrete analog of the original equations for an individual fluid element, was constructed, assembled, and solved. For spatial integration, the number of iteration steps was limited to ten at each time step with a combination of the successive substitution and quasi-Newtonian scheme.

The numerical simulation of a pulsatile flow required a time integration method. The implicit time integration scheme used in this study was the second-order trapezoidal method with a variable time step, which is dependent on the magnitude of temporal inlet velocity and its gradient change. Depending on the physiological velocity pulse shape, the time steps were varied between 1×10^{-4} to 1×10^{-5} s. The finite-element computer code[13] was used to formulate and solve this matrix equation. For the present study, a Sun Ultrasparc 2 with a speed of 200 MHz, 256 MB RAM, and a 4 GB disk was used and the post-processed results are down loaded to an IBM-PC computer for plotting the results.

In comparison to the core elements of the artery, the element sizes near the wall were kept small in order to achieve accuracy for flow parameters. The aspect ratio of the elements was chosen to be less than 10. For validation of the numerical computation, two separate computer modeling runs were performed at peak systolic flow with different convergence criteria as follows: both the relative velocity error with respect to the previous step and the relative residue error compared to the initial value were set to be 2% and 1%.[9] Furthermore, the overall convergence was confirmed by increasing the total number of meshes by 20% over that of the previous run, and the two results were compared to check for accuracy. When the improvement with 20% more meshes was less than 1% in velocity vectors, wall shear stress, and pressure, the computation was considered to be accurate. In addition, the available experimental data and the numerical prediction of pressure drop were compared. The analysis of results is from the computation with the least CPU time, i.e., with less than 2% for both relative velocity error and relative residue error. The CPU time for each time step is approximately 4.38 s for Study 1 and 2.86 s for Study 2.

Boundary Conditions

Study 1 was conducted with a pulsed velocity inlet condition whereas a pressure gradient was used in Study 2. The inlet conditions used in both studies were obtained from *in vivo* experimental data. The outflow boundary conditions did not need to be specified at the exit since its values were effectively determined by extrapolation similar to the finite difference schemes. A no-slip boundary condition was specified on the rigid arterial wall. Since the artery was relatively straight, the flow domain was modeled as axi-symmetric. Specific details of boundary conditions for each study are provided below.

Boundary Conditions for Study 1

Obtaining an accurate *in vivo* instantaneous velocity inlet profile with a non-invasive method is difficult; such an instantaneous velocity profile could have a spatial velocity distribution that is either parabolic, uniform, or any combination of the two. Given this experimental difficulty and uncertainty, the measured core velocity is the most accurate and must be considered. Since the femoral segment is deep within the circulation and is branch-free upstream, the numerical calculations were performed for the instantaneous velocity and a parabolic spatial inlet flow condition for which u_{cl} is twice the calibrated spatial mean velocity u_c. Fig. 8.5a shows the calibrated *in vivo* spatial mean velocity, as measured by an ultrasound Doppler flow cuff. Further details are provided in the results section below.

The inlet core velocity u_{cl} obtained by curve fitting is shown in Fig. 8.6a, and used as an input for the present numerical simulation. The calculations are started at an axial distance of 2 cm upstream of port 1 for calculational stability purposes. The selected time steps for flow analysis are marked from 1 to 8 on the pulse.

Boundary Conditions for Study 2

Measured pulsed pressure drop was applied at the inlet section where port 1 is located (Fig. 8.6b). Since there was hardly any arterial curvature, the pressure variation in the radial direction was kept constant. Based on the applied pressure drop at the inlet, the flow field developed in the flow domain. For Study 2, since pressure drop was the boundary condition, the flow domain was truncated at port 1 and port 1 locations.

Determination of multi-dimensional arterial flow through viscoelastic vessels is beyond the scope of current computational capability. In the literature, physiological wall motion data for a dog is available[4] which may be useful in developing simpler interactive computer codes.

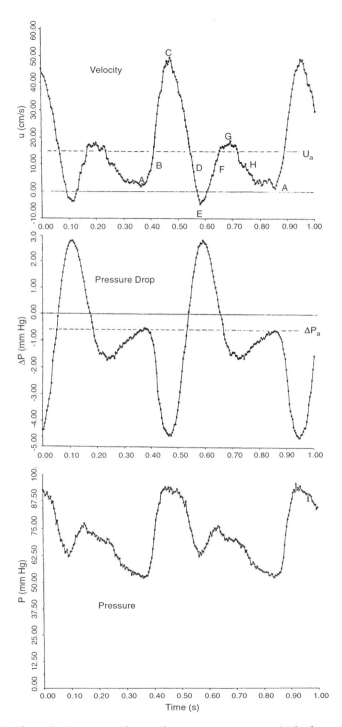

FIGURE 8.5 Doppler flow velocity, pressure drop, and pressure measurements in the femoral artery of a dog. The dashed curves are mean values. The *in vivo* spatial mean velocity u_c in the femoral artery along a pulse cycle was obtained by a calibrated Doppler flow cuff (Fig. 8.5a). Arterial pressure drop and pressure are shown in Fig. 8.5b and Fig. 8.5c, respectively.

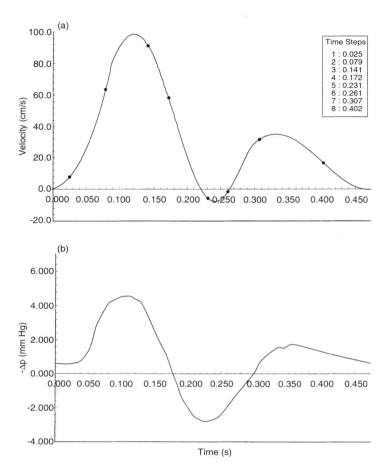

FIGURE 8.6 *In vivo* core velocity u_{cl} pulse along the time steps for which flow data are reported in this chapter (Fig. 8.6a). *In vivo* pressure drop between ports 1 and 2 is shown in Fig. 8.6b. The pulse rate is 128 beats/min.

Non-Newtonian Blood Viscosity

In order to calculate the shear rate-dependent non-Newtonian viscosity in the flow field, the local shear rate, $\dot{\gamma}$, was calculated from the velocity gradient through the second invariant of the rate of strain tensor, $II\dot{\gamma}$, as follows:

$$\dot{\gamma} = \sqrt{\frac{1}{2} II} = \sqrt{\frac{1}{2}\left[\sum_i \sum_j \dot{\gamma}_{ij}\dot{\gamma}_{ji}\right]} \qquad (8.10)$$

After the local viscosity is determined by the blood model, Eq. 8.11, the local shear stress, τ ($= \eta\dot{\gamma}$), is calculated. Schneck's best three variable model (B3VM)[20] was used in our earlier calculations. However, during the course of Studies 1 and 2, a more accurate blood model, namely the Carreau model,[10] was used.

Non-Newtonian Blood Viscosity for Studies 1 and 2

The Carreau model was used to represent the shear rate-dependent non-Newtonian blood viscosity whose model constants were obtained by a curve fitting of available shear rate-dependent blood viscosity data in the literature.[10]

$$\eta = \eta_{\infty} + \left(\eta_0 - \eta_{\infty}\right)\left[1 + \left(\lambda\dot{\gamma}\right)^2\right]^{\frac{n-1}{2}} \qquad (5.11)$$

where λ (characteristics time) = 3.313 s, n = 0.3568, η_O = 0.56 poise, and η_∞ = 0.0345 poise. The dimensionless frequency parameter (α = 0.5 d $\sqrt{(\omega/\nu)}$) calculated based on infinite shear rate viscosity for a dog was 3.7.

8.5 Results

Animal Hemodynamic Data

The Doppler blood flow velocity measurements and pressure drop measurements are shown in Fig. 8.5 as a function of time in the same plot. From these two measurements, Δp vs. u is plotted to show hysteresis effects (Fig. 8.7).

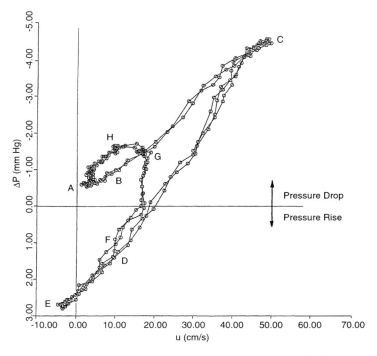

FIGURE 8.7 Hysteresis plot of pressure change Δp vs. u_c in the femoral artery of a dog. Points A through H are identified in Fig. 8.5a along the velocity pulse for a cardiac cycle.

The flow signal was tri-phasic with a brief period of reverse flow during the early part of diastole. At peak systole, the largest instantaneous pressure drop of −4.5 mm Hg was measured. The largest instantaneous pressure rise of +2.7 mm Hg occurred during flow reversal. During both of these peak conditions, there was little phase lag as is evident in Fig. 8.5.

To help identify the loop type hysteresis curve (Fig. 8.7), points labeled A through H are shown along the velocity wave form in Fig. 8.5 for a cardiac cycle beginning with the systolic phase (point A). During blood flow acceleration to peak flow, points A, B, and C lay along the upper branch of the hysteresis curve (Fig. 8.7), while during the flow deceleration phase from peak to reverse flow, points C, D, and E lay along the lower branch. For the second flow acceleration phase during diastole, points E, F, and G transize from the lower branch of the hysteresis curve to the upper branch vicinity. During the latter part of diastole where flow deceleration occurs again, points G, H, and A loop above the upper branch of the hysteresis curve to begin again the next cardiac cycle which is also shown in Fig. 8.7.

In addition to the pulsatile flow data, there is also interest in the mean flow data. The resting heart rate of a dog is 128 beats/min and therefore, the period, T, of a heart beat is 0.47 s. Using a digital voltmeter as shown in the previous section, the time-averaged blood flow velocity, u_a = 15.1 cm/s, and time-averaged pressure drop, Δp_a = –0.59 mm Hg, were obtained and are shown in Fig. 8.5 by the dashed curves. In this case, the ratio of peak pressure drop at peak velocity to the mean value, $\Delta p_p/\Delta p_a$, was 7.6, thus indicating the relatively large variations in instantaneous pressure change compared to the mean value, including during the flow reversal phase. The corresponding value of the ratio of peak flow velocity to the mean value, u_p/u_a, was 3.3, a value which is less than half of the pressure drop ratio, $\Delta p_p/\Delta p_a$. A more complicated pressure change–flow relationship relative to the mean flow value was evident during diastole.

To place the hemodynamic data in perspective, some comments are in order. Measurement of the viscosity of the dog's blood at the *in vivo* temperature gave η = 0.037 poise, and the measured blood density, ρ, was equal to 1.04 g/cm³. The time-averaged flow rate Q_a (= $u_a A_1$) based on the upstream vessel diameter d_1 (3.8 mm) was 102 ml/min. Since the mean flow Reynolds number Re_a (= $4 Q_a/\pi \nu d_1$) was only 161, the flow was expected to be laminar. Considering the mean diameter ($\bar{d}$ = 3.7 mm), the time-averaged value of the pressure drop Δp_a using the Poiseuille relation

$$\Delta p_a = -128\eta l Q_a/\pi \bar{d}^{\,4} \qquad (8.12)$$

was –0.54 mm Hg. A further correction for mildly uniform taper was made from the momentum consideration;[2] this then gave a value of Δp_a= –0.57 mm Hg. Since the measured value of Δp_a was only about 3% higher, this afforded a reasonable estimate of the mean pressure drop and instilled confidence in the accuracy of the measurement technique. Also considering the mean diameter ($\bar{d}$ = 3.7 mm), the time-averaged value of the wall shear stress $(\tau_w)_a$ using the Poiseuille relation

$$\left(\tau_w\right)_a = 32\eta Q_a/\pi \bar{d}^{\,3} \qquad (8.13)$$

was 12.7 dynes/cm².

Finally, although the dimensionless frequency parameter, (α = 0.5 $d_1\sqrt{(\omega/\nu)}$) was 3.7 and hysteresis effects were evident during the flow acceleration and deceleration phases of the cardiac cycle, phase variations between the velocity and pressure difference signals were relatively small at peak systole.

With the experimental data at hand, we turned to the task of numerically calculating the flow field and pressure differences by solving the momentum conservation equations, including the shear rate-dependent viscosity of blood.

Flow Calculation Description

This section has been divided in two parts: the results of Study 1 are followed by the results of Study 2. This section clearly demonstrates the agreement and disagreement between experimental measurement and numerical calculation. The calculations in Study 1 for a parabolic inlet profile were consistent with the Doppler flowmeter calibration where core flow velocities were about twice the spatial mean velocities shown in Fig. 8.5, unlike the earlier calculations[7] which corresponded to assumed lower centerline velocities, which were incorrectly interpreted from the Doppler calibration, by a factor of about two. Also, these calculations used a more accurate Carreau model for shear rate-dependent non-Newtonian blood viscosity (η) than the Wilburn and Schneck model[20] used in the earlier calculations.[7]

Flow Calculation for Study 1

Temporal and spatial variations of the velocity profile were obtained along the radial directions of ports 1 and 2, whereas shear rate and non-Newtonian viscosity were reported at each port location. Shear stress was calculated along the arterial wall between ports 1 and 2. *In vivo* pressure drop measurements across the ligated arteries (ports 1 and 2), which were connected to the tubing leading to a pressure transducer, were reported along with the validation of numerical pressure drops between ports 1 and 2.

This section presents detailed flow analysis for parabolic inlet velocity profiles. Results for the systolic acceleration (Fig. 8.6a) are represented by time steps 1 and 2, whereas results for systolic deceleration are represented by time steps 3 and 4. Time steps 5 and 6 represent an accelerating negative flow and a decelerating negative flow for early diastole, respectively. Time steps 7 and 8 represent an accelerating and decelerating positive flow for late diastole, respectively. Fig. 8.6a indicates the exact time that corresponds to the number of the time step; the figure shows that the velocity at the inlet is positive for time steps 1-4, 7, and 8, and negative for 5 and 6.

Velocity Distribution for Parabolic Inlet Velocity Profile
Fig. 8.8a shows the spatial distribution of the axial velocity profile for different time steps at port 1. The instantaneous velocity profiles of curves 1–4 are positive, with a centerline velocity that rises to 93 cm/s near peak systole; for curve 5, a flow reversal with velocity –7.2 cm/s is obtained near the wall, although the core velocity is still positive with a value of 9.7 cm/s. In contrast, the velocity for curve 6 is less negative with a maximum value of –2 cm/s near a distance halfway along the radius. For late diastolic flow, curves 7 and 8 show similar velocity profiles with a lower magnitude than that observed during systolic flow. Flow reversal near the wall during the late decelerating flow with a positive core velocity is an interesting phenomenon. Since the calculations were initiated upstream of port 1, some flow development occurred before port 1 during the cardiac cycle. For example, during flow acceleration phases (curves 2 and 7) the core velocity profiles became flatter than the inlet profiles, and during the reverse flow phase (curves 5 and 6) the velocity profiles became M-shaped as noted.

Similar trends were observed at port 2 (Fig. 8.8b) with variations of velocity caused by the mild taper. Due to the gradual taper of the artery, a higher core velocity of 51 cm/s was obtained at port 2 (compared to 45 cm/s at port 1) for curve 2 during the accelerating phase of the systolic flow. This trend was partially reversed during the decelerating phase of the systolic flow. At port 2, the centerline velocity for curves 3 and 4 were 90 cm/s and 70 cm/s, respectively, whereas at port 1 the values were 93 cm/s and 69 cm/s, respectively. Furthermore, in comparison to port 1, curve 5 at port 2 showed a slightly higher value for the reversed peak flow with a value of –7.6 cm/s. The velocity for curve 6 had a maximum value of –2.1 cm/s.

Shear Rate for Parabolic Inlet Velocity Profile
For a single cardiac pulse with a parabolic velocity profile, an oscillating wall shear may be observed in Figs. 8.9a and b. Spatial variations in the shear rate for selected time steps 1-8 (Fig. 8.6a) are shown in Fig. 8.9a (port 1) and 8.9b (port 2) with arrow marks indicating the oscillations at the wall regions. The magnitude of the wall shear rate at each time step is larger at port 2 than at port No. 1, indicating the effect of the taper.

Fig. 8.9c shows the complete time history for the entire pulse cycle. At port 1, the shear rate varied from zero to a maximum value of 1350 s^{-1} at t = 0.089 s (before peak systole, between time steps 2 and 3 of Fig. 8.9) and attained a zero shear rate during the decelerating phase of late systole (between time steps 4 and 5). It then reached a maximum negative shear rate of –310 s^{-1} at t = 0.225 s (early diastole, near the peak negative velocity after time step 5) which was followed by a shear rate of 494 s^{-1} at t = 0.296 s (before the late diastolic peak positive flow, after time step 7). At port 2 (Fig. 8.9c) the shear rate varied from zero to a maximum positive value of 1555 s^{-1} at t = 0.089 s and attained a zero shear rate value during the decelerating phase of late systole, before reaching a minimum shear rate of –340 s^{-1} at t = 0.225 s, which was followed by a shear rate of 565 s^{-1} at t = 0.296 s before the peak diastolic positive flow.

The steeper slope in the curve of the wall shear rate vs. time at port 2 compared to port 1 (Fig. 8.9c) was due to the taper during both acceleration and deceleration. This in turn affected the non-Newtonian viscosity, the shear stress, and the pressure drop. The systolic decelerating slope of the shear rate at port 2 was greater than at port 1, causing the shear rate lines (Fig. 8.9c) to intersect after time step 4 at

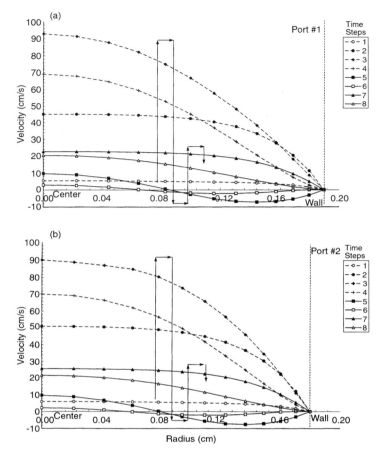

FIGURE 8.8 Radial velocity distribution, with a parabolic spatial inlet condition, for different time steps along a pulse cycle at port 1 (Fig. 8.8a) and port 2 (Fig. 8.8b).

t = 0.205 s. At port 2, the magnitude of the extremum values of the wall shear rate was larger than that for port 1. This can be attributed to the arterial taper.

Non-Newtonian Viscosity of Blood

The non-Newtonian viscosity of blood was modeled using whole blood. The non-Newtonian blood viscosity at the wall is presented in Fig. 8.10 for the parabolic inlet flow. It is apparent that for high shear, the non-Newtonian viscosity is low and vice-versa. In Fig. 8.9c, the shear rate at port 2 is higher than that at port 1, resulting in a lower non-Newtonian viscosity. The viscosity during systole, in general, is smaller than that during diastole. At peak systole in Fig. 8.10, the non-Newtonian viscosity is 0.037 poise (1 poise = 0.1 Pa/s) which is the infinite shear rate viscosity measured using dog blood in a cone and plate viscometer. Near the zero shear rate, the non-Newtonian viscosity is 0.16 poise, which is approximately four times the infinite shear rate viscosity. The non-Newtonian viscosity oscillates during the decelerating part of systole which is due to the oscillation in shear rate between positive and negative values.

Due to the higher systolic decelerating slope for the parabolic inlet condition (Fig. 8.9c), the shear rate at port 2 at the end of systole and at the beginning of diastole was lower (larger negative value) than at port 1, which caused a locally lower value of the non-Newtonian viscosity (Fig. 8.10).

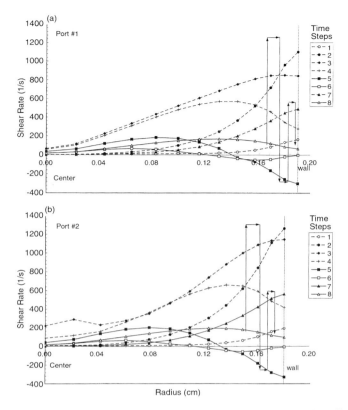

FIGURE 8.9a-b Radial shear rate distribution, with a parabolic spatial inlet condition, for different time steps along a pulse cycle at port 1 (Fig. 8.9a) and port 2 (Fig. 8.9b).

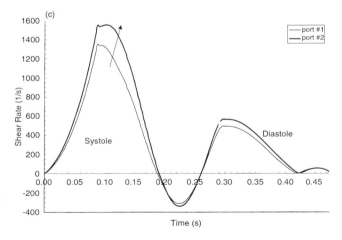

FIGURE 8.9c Temporal variation of wall shear rate distribution at port 1 and port 2 for a parabolic spatial inlet condition.

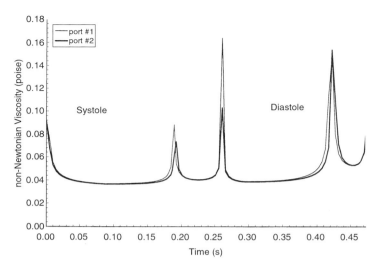

FIGURE 8.10   Temporal variation of non-Newtonian viscosity at the wall for port 1 and port 2 for parabolic spatial inlet conditions.

Shear Stress

Fig. 8.11 shows the instantaneous shear stress along the arterial wall for parabolic inlet conditions. The shear stress at the wall illustrates the combined effect of instantaneous wall shear rate and local non-Newtonian viscosity.

During the systolic acceleration at port 2, the wall shear stress shown in Fig. 8.11 increases from zero to more than 47 dynes/cm^2, followed by a drop to a value of less than -13 dynes/cm^2 during early diastole (time step 5). Subsequently, during the late diastolic phase (time step 7), the shear stress rises to a positive value of more than 22 dynes/cm^2 followed by a reduction to zero at the end of a pulse cycle. the maximum value of the wall shear stress is 57.0 dyn/cm^2 at t = 0.089 s (before peak systole between time steps 2 and 3). As observed in Fig. 8.11, the effect of the taper is indicated by a gradual rise in the magnitude of shear stress with axial distance for all the time steps.

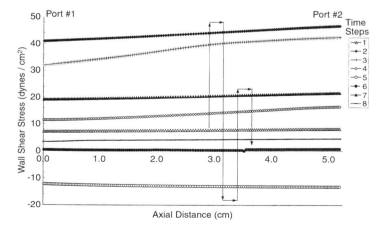

FIGURE 8.11   Wall shear stress distribution for different time steps along a pulse cycle between ports 1 and port 2 for parabolic spatial inlet conditions.

Pressure Drop

Since Banerjee et al.[6] reported pressure drop results calculated for different discontinuous pressure formulations (e.g., bilinear, linear with local basis function, and linear with global basis function), only the results obtained using the discontinuous bilinear pressure formulation[5] are presented here for the parabolic inlet condition. The pressure drop calculated for different pressure formulations is presented in Banerjee et al.[6] Curve 1 in Fig. 8.12 represents the pressure drop for the parabolic inlet condition. Curve 2 represents the experimentally obtained *in vivo* pressure drop measurements.

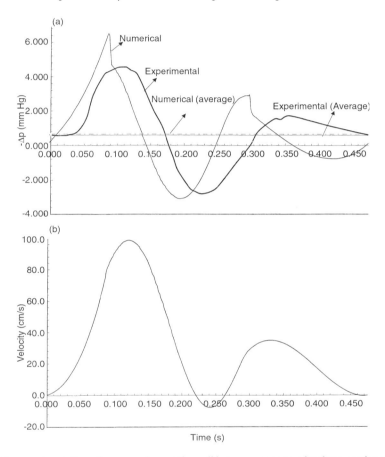

FIGURE 8.12 Temporal variation of pressure drop at the wall between ports 1 and 2 along a pulse cycle (Fig. 8.12a). Instantaneous pressure drops for parabolic spatial conditions are plotted along with *in vivo* data obtained in a living dog. Also plotted is the time-averaged pressure. Fig. 8.12b is the measured inlet core velocity u_{cl} (same plot as in Fig. 8.6a). Figs. 8.12a and b should be read in conjunction to calculate the phase angle.

For the parabolic inlet condition, the calculated pressure drop has a maximum value of −6.7 mm Hg at t = 0.087 s (represented by the upward spike) during peak systole and reaches 3.1 mm Hg at t = 0.192 s during early diastole (represented by the downward spike). Experimentally measured peak pressure drops are −4.5 mm Hg at t = 0.110 s during systole and 2.7 mm of Hg at t = 0.225 s during the early diastolic phase of flow. For mid-diastole, the second peak pressure drop is also overestimated (−3.0 compared to −1.7 mm Hg) and in late diastole, a pressure rise was calculated but a pressure drop was measured. Numerical calculation of the average pressure drop is −0.645 mm Hg for the parabolic inlet condition, whereas experimental data show a value of −0.59 mm Hg. The 10% difference between the calculated and measured average pressure drops, shown in Fig. 8.12 by the horizontal lines, is relatively small compared to the Δp oscillations during the cardiac cycle.

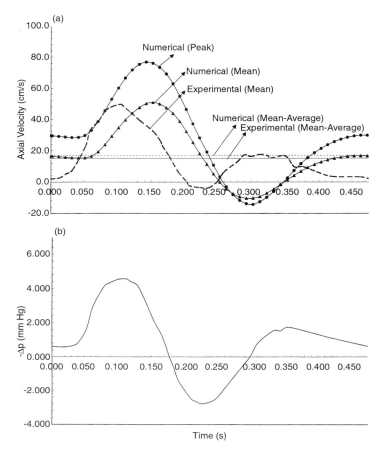

FIGURE 8.13 Temporal variation of velocity at cuff location along a pulse cycle (Fig. 8.13a). Instantaneous velocities (spatial average and peak) are plotted along with the *in vivo* spatial mean u_c data obtained in a living dog. Also plotted are the time-averaged velocities. Fig. 8.13b is the measured $-\Delta p$ in mm Hg (same plot as in Fig. 8.6b). Figs. 8.13a and b should be read in conjunction to calculate the phase angle.

During the acceleration part of the flow, the predicted peak pressure drop $-\Delta p_p$ is obtained at t = 0.087 s for the parabolic inlet conditions (Fig. 8.12a). From Fig. 8.12b, a peak Doppler velocity $u_{c\text{-}p}$ is obtained at t = 0.12 s. This clearly indicates a phase lag between predicted Δp_p and measured $u_{c\text{-}p}$ that amounts to $\theta = -25°$, since $u_{c\text{-}p}$ lags predicted $-\Delta p_p$. At the beginning of diastole, flow reversal occurred ($u_{cl} = -7.7$ cm/s at t = 0.24 s). The peak in the predicted adverse pressure gradient $+\Delta p_p$ occurred at an earlier time so that the phase angle was $-37°$. For the second smaller peak flow, $u_{c\text{-}p}$ at t = 0.33 s, the phase angle was about $-27°$. These calculated phase angles are within the range of $0°$ to $-60°$ known for physiological flows, and indicate a hysteresis effect between predicted Δp peaks and measured u_c. Moreover, the predicted Δp peaks appear to precede the measured Δp peaks by phase angles of $-18°$ to $-46°$, progressively increasing during the cardiac cycle.

Flow Calculation for Study 2

Results of the numerical calculations where the measured pressure gradient $\Delta p(t)$ was used to calculate the flow field are shown in Figs. 8.13a and b. Fig. 8.13a shows the pulsatile Doppler blood flow mean velocity $u_c(t)$ and the spatial average instantaneous mean velocity, defined as

$$\bar{u}(t) = \frac{1}{A} \int_A u \, dA \tag{8.14}$$

and centerline velocity $u_{cl}(t)$ obtained from the calculations. Also shown are the time-averaged velocities of the pulse cycle. Plotted below these results is the measured pressure drop $\Delta p(t)$ (Fig. 8.13b). The phase angle θ between the measured peak Δp and peak u_{c-p} signals was calculated as before. That is, if the peak Δp signal occurs at ωt_1, and the peak u_{c-p} signal occurs at $\omega t_2 = \omega t_1 - \theta$, then since the circular frequency ω is equal to πf ($= 2\pi/\tau$), the phase angle θ becomes $360 (t_2 - t_1)/T°$.

Fig. 8.13a shows that during systole where flow acceleration occurs, at peak flow the mean $\bar{u}_p$ is 50.8 cm/s at t = 0.152 s, which is very near the Doppler velocity u_{c-p} (49.4 cm/s at t = 0.107 s), and the phase angle θ between Δp_p and $\bar{u}_p$ is about $-35°$ since $\bar{u}_p$ lags $-\Delta p_p$. At the beginning of diastole, the adverse pressure gradient is peak $+\Delta p_p$, flow reversal occurs, and the peak mean value $-\bar{u}_p$ is -10.3 cm/s at t = 0.293 s, which is about 2.9 times larger than $-u_{c-p}$ (-3.58 cm/s at t = 0.236 s). The $+\Delta p_p$ $-\bar{u}_p$ phase angle increased to about $-50°$. For the second smaller peak flow during mid-diastole where flow acceleration occurred again, $\bar{u}_p \sim u_{c-p}$, similar to the value during systole. The $-\Delta p_p$, $-\bar{u}_p$ phase angle further increased to about $-75°$. The next cardiac cycle began before the mean $\bar{u}$ decreased much at the end of diastole during the deceleration phase, presumably due to the large phase lag. Fig. 8.13a also shows that the ratio of predicted center line to mean velocity $u_{cl-p}/\bar{u}_p$ is less than the Poiseuille value of 2, being 1.5 at peak systole, 1.4 for peak reverse flow, and 1.8 for the second smaller peak flow.

In the physiological case, phase angles are near zero for peak flow (systole) and reverse flow, and positive for the second peak flow (mid-diastole). Values of θ are positive because u_{c-p} occurred before Δp_p.

Calculated phase angles are associated with the pressure gradient

$$-\frac{\partial p(t)}{\partial z}$$

accelerating or decelerating the mean flow, i.e., $Q(t) = \bar{u}(t) A$ including convective acceleration terms. Radial variation in pressure across the flow is negligible. The force balance on an element volume $A dz$ in the axial direction z is

$$-A dp - 2\pi r_w \tau_w dz = \rho d \int_A u^2 dA + \rho \frac{\partial Q(t)}{\partial t} dz \tag{8.15}$$

Thus, the pressure force must overcome the wall shear force to provide the net force necessary to provide convective and unsteady flow accelerations and decelerations. The calculated phase angles between measured Δp_p and predicted $\bar{u}_p$ are nearly within the range of $0°$ to $-60°$ known for physiological flows. However, using the measured Δp as input, the calculated velocities, particularly during diastole, are out of phase and consequently not in agreement with the measured mean flow velocities. Flow reversal in femoral artery of a dog and phase lag between pressure gradient and flow have been reported by McDonald and his co-workers.[17] In their case however, the agreement is much better between observed (high speed cinematography) and calculated velocities from a derived pressure gradient

$$\frac{\partial p}{\partial z} = \frac{1}{c} \frac{\partial p_a}{\partial t} \tag{8.16}$$

where

$$\frac{\partial \overline{p_a}}{\partial t}$$

is the time derivative of arterial pressure, and c is the recorded value for the peak to peak wave speed.

Fig. 8.13a also shows the time-averaged mean velocity $\overline{u}_a$ obtained by integration over the cardiac cycle along with the measured time integrated mean flow velocity u_{c-a}. The value of $\overline{u}_a = 16.9$ cm/s is about 12% larger than the Doppler value $\overline{u}_{c-a} = 15.1$ cm/s, but the difference is relatively small compared to the velocity variations during the cardiac cycle.

In the calculations, the femoral artery of the dog was assumed to be rigid. Whereas the dynamic distensibility D or compliance $c(= D/2)$ of the vessel wall was not measured, the measured arterial blood pressure $P_a(t)$ at the iliac bifurcation was used to estimate the magnitude of wall movement during the cardiac cycle from

$$\frac{dr_w}{r_w} = \frac{1}{2}Dd\left(p_a - \overline{p_a}\right) \tag{8.17}$$

The mean blood pressure $\overline{P_a} = 72$ mm Hg and the peak systolic and minimum diastolic pressures were 92 and 52 mm Hg, respectively (Fig. 8.5). By using a value of $D = 0.0013$ 1/mm Hg for the normal dog femoral artery,[18] the systolic and diastolic excursions from the mean wall radius were estimated to be about the same, and the total excursion

$$\frac{\Delta r_{ws} + \Delta r_{wd}}{\overline{r}_w}$$

was 2.6%. These estimated changes in lumen size during a heartbeat are not believed to significantly alter the findings of this investigation in which wall motion was neglected during systole where axial flow velocities are high. However, during early diastole, where flow velocities are low and flow reversal occurs, radial recoil of the vessel wall may induce axial velocities no longer negligible. In this case, the coupled equations of blood flow and wall motion would need to be considered.

8.6   Discussion

The numerical estimation of an instantaneous pressure drop for a pulsatile velocity or vice-versa is a challenging task. For blood flow, the non-Newtonian nature of blood viscosity poses an additional challenge in the calculation of the flow field. In the present case, temporal and spatial variations of shear rate and velocity add further complexity to the problem. Sharp changes in the instantaneous inlet boundary condition create numerical oscillations in the calculated data. In order to minimize these oscillations, the time-dependent inlet boundary condition is smoothened, as shown in Fig. 8.6a.

In the present analysis, numerical calculations are conducted for two consecutive pulse cycles in order to compare them and to obtain accurate results; the results for the second pulse cycle are calculated in continuation of the first one. Complete numerical data are reported only for the second cycle. Some important observations are noted as follows:

Discussion for Study 1

- Near peak systole, where sharp changes in velocity occur, less oscillation of pressure data are observed for the second cycle than the first cycle.

- The numerical calculation begins during the late accelerating phase of the systole, i.e., at about 0.05 s prior to the peak systolic velocity. Hence, the peak systolic pressure drop for the first cycle is meaningless and is ignored. When the peak pressure drop is compared between the second and third cycles, only a 0.13% change is observed.
- Determining the *in vivo* velocity pulse by non-invasive means such as the Doppler flow cuff allows the physiological pressure drop to be calculated. The *in vivo* prediction of pressure drop using an angiogram alone has a potential application from a medical diagnostic point of view, because it allows quantification of the physiological flow parameters in arterial blockages or stenoses.

Discussion For Study 2

- The problem is solved for three consecutive pulse cycles. It was observed that the results of the second and third cycles are within 1% of each other. It is evident that a repetitive solution is obtained from the second cycle and, therefore, the results for the second cycle are presented.
- For a practical scenario, the *in vivo* pressure measurements can be obtained using invasive means such as catheters. However, the pressure measured by a catheter needs to be modified in order to avoid the errors caused by the change in the pressure field due to insertion of a catheter.

8.7   Summary and Conclusions

The observations made in the present study may be summarized as follows.

Experimental Study

The velocity waveform for the femoral artery of a dog was tri-phasic with relatively small phase variations between the measured velocity and pressure drop at peak forward and reverse flow conditions. Hysteresis effects were observed during the flow acceleration and deceleration phases of the cardiac cycle.

Numerical Study

Summary and Conclusions for Study 1

1. Determining the *in vivo* velocity pulse by non-invasive means such as the Doppler flow cuff allows the physiological pressure drop to be calculated.
2. The calculation of instantaneous pressure drop across a segment of the femoral artery of a dog was found to be in reasonable agreement with the measured *in vivo* pressure drop during the cardiac cycle. The shape of the inlet velocity profile was assumed to be parabolic since the femoral segment was deep in the circulation and branch-free upstream (pressures were measured in ligated branches). Numerical calculation of the time-averaged pressure drop over the cardiac cycle was about 10% higher than the measured *in vivo* value, which could also be estimated by using the mean flow Poiseuille relation with a momentum correction for the vessel taper.
3. A phase lag between predicted peak Δp_p and measured peak Doppler velocity u_{c-p} at systolic peak flow, early in diastole, and in mid-diastole, was calculated in the range of phase angles θ of $-25°$ to $-37°$. The predicted Δp peaks preceded the measured Δp peaks at peak systolic flow, early in diastole, and in mid-diastole by phase angles of $-18°$ to $-46°$.
4. For the parabolic inlet condition, the mild taper of the artery caused some changes in both shear rate and stress; specifically, a higher magnitude of oscillating shear rate at the wall during both systole and diastole was observed downstream. Furthermore, the steeper gradient of wall shear rate due to its taper affected not only the non-Newtonian viscosity of blood and the shear stress, but also the pressure.

Summary and Conclusions for Study 2

1. A progressive increase in phase lag between the inlet pressure gradient and calculated mean flow was predicted. The phase lag increased from a few degrees during early systole to about $-75°$ during late diastole. A similar result has been reported by McDonald and his co-workers.[17] Their results show an increase of phase lag from zero degrees during early systole to about $-58°$ during late diastole.
2. The time-averaged mean velocity $\bar{u}_a$ obtained by integration over the cardiac cycle is found to be about 12% larger than the measured time-integrated Doppler flow velocity $u_{a\text{-}c}$.
3. The magnitude of the measured peak velocities u_c is found to be in fair agreement with the calculated mean velocities $\bar{u}_a$. This fact is clearly evident when the systolic part of the pulse cycle and the time-averaged values are compared. This observation can also be made with caution in the diastolic region. In the diastolic region, the calculated phase difference is comparatively higher than the experimental observations, resulting in apparent disagreement in the instantaneous velocities. This needs to be investigated further from the perspective of viscoelastic wall effect.

From the comparisons of numerical Studies 1 and 2, we conclude that to estimate the pressure drop–flow rate relationship in the mildly compliant femoral artery wall of the dog by using rigid wall theory, it may be better to use the measured velocity than the measured pressure gradient as input into the calculation. Hence, it is essential to use an accurate temporal and spatial distribution of velocity inlet profile for the calculation in order to obtain an accurate pressure distribution. An earlier study[12] has suggested this approach, i.e., used the measured flow rate to estimate the mean wall shear stress in an elastic artery model. Our study determined that by using the measured Doppler flow velocity as input, the calculated Δp is in fairly good agreement with the measured Δp, but is still somewhat out of phase. On the other hand, using the measured Δp as input, the calculated velocities, particularly during diastole, are more out of phase and not in agreement with the magnitude of the measured flow velocities.

It is believed that the calibrated Doppler flowmeter provides reliable flow velocity measurements. The very high frequency excitation signal provides nearly instantaneous Doppler frequency shift data even with an internal low pass filter as discussed in the Experimental Method section above. The validyne differential pressure transducer has a high frequency response, and, with calibration, also accurately measures the instantaneous Δp. Therefore, the physiological phase difference between the Δp and u signals are shown in Fig. 8.5 and in the Δp-u hysteresis curve in Fig. 8.7. Reproduction of physiological measurements such as these is a challenging chore for computational fluid mechanics.

In vivo measurements in the laboratory have repeatedly shown that the velocity pulse and the pressure drop are not only dependent on the physiological location of the artery, but also vary from one individual dog to another. The results reported in earlier work[6] are for a femoral artery of a different dog with a similar trend of measured instantaneous velocity but with a different magnitude. Hence, a unique set of results for any dog, or, in other words, a generalization of results, is practically impossible. This study has attempted to propose a method to calculate pressure and velocity changes in an artery using an arterial angiogram, the non-Newtonian viscosity of blood, and an instantaneous velocity or pressure pulse. In conclusion, the hemodynamic measurements reported herein are believed to provide a set of consistent data that may be useful in subsequent studies of arterial flows and numerical flow calculation methods. Clearly, more work is needed to acquire a better understanding of arterial flow and pressure variation along vessels during the cardiac cycle.

Acknowledgments

The animal experiments were carried out at the USC School of Medicine by the late Dr. D. W. Crawford with the assistance of Dr. R. F. Cuffel from the Jet Propulsion Laboratory. This chapter presents the experimental results of one phase of research carried out by the Jet Propulsion Laboratory at the California Institute of Technology under a contract with the National Aeronautics and Space Administration. The authors are indebted to Mr. Jack Blaylock, Dr. Art Gooding, Dr. Ram Ramanan, and Dr.

Vahe Haroutunian of Fluent Inc. for their valuable advice and sustained cooperation in conducting the numerical calculations.

Nomenclature

A	=	cross-sectional area
D	=	distensibility
c	=	compliance
d	=	lumen diameter
l	=	length of the vessel segment
n_j	=	normal vector
p	=	pressure
p_1	=	wall pressure at reference location (initial node)
p_i	=	wall pressure at any downstream nodes
Q	=	flow rate
r_w	=	lumen radius
s_i	=	stress vector
t	=	time
T	=	period of heart beat
u_i	=	velocity vector
z	=	distance along the main lumen
α	=	dimensionless frequency parameter, or Womersley number
$\dot{\gamma}$	=	shear rate
ε	=	penalty parameter
η	=	viscosity
θ	=	phase angle between Δp and u
ν	=	kinematic viscosity ($= \eta/\rho$)
ρ	=	density
σ_{ij}	=	stress tensor
ω	=	circular frequency, $2\pi/T$
τ_w	=	wall shear stress

Subscripts

a	=	time-averaged
c	=	Doppler measurement
cl	=	centerline condition
p	=	peak value
Δ	=	difference
$\star$	=	numerically computed value

Superscripts

$(\bar{\ })$	=	spatial average

References

1. Back, L.H., Cho, Y.I., Crawford, D.W., and Blankenhorn, D.H., Flow resistance in curved femoral artery flow models of man for steady flow, *ASME J. Biomech. Eng.*, 109, 90, 1987.
2. Back, L.H., Back, M.R., Kwack, E.Y., and Crawford, D.W., Flow measurements in a human femoral artery model with reverse lumen curvature, *ASME J. Biomech. Eng.*, 110, 300, 1988.
3. Back, L.H., Kwack, E.Y., and Crawford, D.W., Flow measurements in an atherosclerotic curved, tapered femoral artery model of man, *ASME J. Biomech. Eng.*, 110, 310, 1988a.
4. Back, M.R., Kopchok, G.E., Mueller, M., Cavaye, D.M., Donayre, C.A., and White, R.A., Changes in arterial wall compliance after endovascular stenting, *J. Vasc. Surg.*, 19, 905, 1994.
5. Baker, A.J., *Finite Element Computational Fluid Mechanics*, Hemisphere, Washington, D.C., 1983, 153.
6. Banerjee, R.K., Cho, Y.I., and Back, L.H., Pressure drop in a tapered femoral artery of a dog: Pulsatile flow, *Adv. Bioeng.*, BED 22, 285, 1992.
7. Banerjee, R.K., Cho, Y.I., and Back, L.H., Pressure drop measurement and calculation in a tapered femoral artery of a dog, *Biorheology*, 32(6), 655, 1995.
8. Banerjee, R.K., Cho, Y.I., and Back, L.H., Numerical studies of 3-D arterial flows in reverse curvature geometry: part I, peak flow. *ASME J. Biomech. Eng.*, 115, 316, 1992.
9. Banerjee, R.K., Cho, Y.I., and Kensey, K.R., A study of local hemodynamics in a 90 degree branch vessel with extreme pulsatile flows, *Int. J. Computational Fluid Dynamics*, 9, 23, 1997.
10. Cho, Y.I. and Kensey, K.R., Effects of the non-Newtonian viscosity of blood on flows in a diseased arterial vessel: part 1, steady flows, *Biorheology*, 28, 241, 1991.
11. Crawford, D.W., Barndt, R., Jr., and Back, L.H., Surface characteristics of normal atherosclerotic human arteries, including observations suggesting interaction between flow and intimal morphology, *Lab. Invest.*, 34, 463, 1976.
12. Dutta, A., Wang, D.M., and Tarbell, J.M., Numerical analysis of flow in an elastic artery model, *ASME J. Biomech. Eng.*, 114, 26, 1992.
13. FIDAP Manual, Fluent Incorporated, Lebanon, NH, 1996.
14. LIEPSCH, D., Thurston, G., and Lee, M., Studies of fluids simulating blood-like rheological properties and application in models of arterial branches, *Biorheology*, 28, 39, 1991.
15. Ling, S.C., Atabek, H.B., and Carmody, J.J., Pulsatile flow in arteries, *Proceedings of the 12th International Congress of Applied Mechanics*, Springer-Verlag, New York, 1968, 277.
16. Mann, D.E. and Tarbell, J.M., Flow of non-Newtonian blood analog fluid in rigid curved and straight artery models, *Biorheology*, 27, 711, 1990.
17. Nichols, W.W. and O'Rourke, M.F., *McDonald's Blood Flow in Arteries*, 3rd ed., Edward Arnold, London, 1990, 12.
18. Pedley, T.J., The Fluid Mechanics of Large Blood Vessels, Cambridge University Press, 1980, 1.
19. Talbot, L. and Berger, S.A., Fluid-mechanical aspects of the human circulation, *Am. Scientist*, 62, 671, 1974.
20. Wilburn, F.J. and Schneck, D.J., A constitutive equation for a whole blood model, *Biorheology*, 21, 201, 1976.
21. Womersley, J.R., Method for the calculation of velocity, rate of flow and viscous drag in arteries when the pressure gradient is known, *J. Physiol.*, 127, 553, 1955.
22. Womersley, J.R., Oscillatory motion of a viscous liquid in a thin-walled elastic tube-I: The linear approximation for long waves, *Philos. Mag.*, 46, 199, 1955a.

9

Numerical Simulation Techniques and Their Application to the Human Vascular System

Ding-Yu Fei
Virginia Commonwealth University

Stanley E. Rittgers
University of Akron

Don Fei
Forest Hills, New York

Sunil Acharya
University of Akron

9.1   Introduction **9**-1
9.2   Numerical Methods **9**-3
       Governing Equations for Blood Flow • Finite Element
       Formulation • Shape Function and Local and Global
       Coordinates • Considerations for Pulsatile Flow
9.3   Applications **9**-11
       Numerical Study of Flow in End-to-Side Distal Anastomosis:
       Three-Dimensional Models under Steady Flow Conditions •
       Numerical Study of Flow in Y-Shaped Bifurcations: Three-
       Dimensional Models under Steady Flow Conditions •
       Numerical Study of Flow through Stenosed Coronary Arteries:
       Two-Dimensional Models under Pulsatile Flow Conditions •
       The Effects of Non-Newtonian Flow and Vessel Distensibility
9.4   Summary **9**-31

9.1   Introduction

Numerical or computational model studies have been widely accepted as extremely valuable tools to study and predict the behavior of biomechanical systems. When a numerical model is correctly established and tested, it can be used to interpret experimental results, predict changes in outcome parameters with respect to variations of other variables, evaluate existing and new theories, and help design new biomechanical systems such as artificial organs. Along with theoretical analysis and experimental methods, numerical simulation has become one of the three major approaches for biomechanical studies.

Generally speaking, theoretical studies can provide fundamental information and governing equations for the behavior of mechanical systems. For most biomechanical systems, the geometrical structures are complex and vary individually within a relatively large range. Therefore, theoretical studies of those systems usually involve more complex equations and boundary conditions than those of simple mechanical systems. The equations involved in biomechanical systems are often partial differential equations. Except for some simple geometries with simple physical conditions, no exact solutions may be obtained for these equations. Therefore, while theoretical studies often provide the governing equations for the behavior of biomechanical systems, the solutions for particular systems may not be available.

Normally, experimental studies are capable of providing information most closely resembling real situations. However, suitable equipment and proper experimental settings are required. For biomechanical systems, these experiments may include either *in vitro* or *in vivo* models. Although experimental studies are widely used to investigate biomechanical systems, they can suffer from practical difficulties and high operating costs. Due to the limitations of current technologies, some physical parameters in biomechanical systems are difficult to measure. For example, wall shear rate measurements, which are of prime interest in hemodynamic studies in the vascular system, are difficult to obtain experimentally. For *in vitro* experiments, such measurements are typically taken using vessel models with transparent walls and refractive index-matched fluids using laser Doppler anemometry at numerous axial sites. Obtaining accurate velocity recordings near the wall is difficult, however, due to the relatively small velocities present and lower particle density. In addition, because of the complexity and variation of geometries in biomechanical systems and the sophisticated nature of the measurement instrumentation, the operating costs may be very high. For example, an attempt to study a physical phenomenon as a function of several parameters in the system requires a series of models. Due to the complexity of geometries required in these models, the experiments would be very expensive as well as time consuming.

Numerical model studies are generally capable of solving problems with complex geometries and physical conditions. Many problems which cannot be solved by theoretical studies may be solved numerically. Numerical model studies usually include several variables as input parameters. With proper modifications of a numerical model, it is not difficult to study the changes in one or several parameters as functions of other parameters. For this reason, numerical model studies are usually more flexible than the corresponding experimental studies. Another advantage of the numerical method over the experimental method is that some physical parameters which are difficult to measure experimentally are readily obtained with numerical simulations. For example, shear rates along the wall at any site in a flow system can be calculated easily from the numerical results whereas they are difficult to obtain using experimental methods as mentioned previously. The major limitation of the numerical model study is that it is unable to solve problems if they are governed by phenomena which cannot be described in mathematical terms. In addition, the accuracy of some numerical model studies may be limited by computer execution time and/or memory space when complex equations and boundary conditions are involved. With improvement of computer and modeling technologies, more complex structures and more input parameters may possibly be included in numerical simulations of biomechanical systems. It is expected that numerical models will play a more important role in the investigation of biomechanical systems in the future.

It should be noted that, although a numerical model study can sometimes replace an experimental study, numerical simulations and experiments are often complementary. In practice, experiments are often preceded by numerical model studies, and numerical simulations are validated by experiments. Sometimes, the input parameters for a numerical study may be obtained directly from an experimental study. The combination of an analytical study, a numerical simulation, and experimental observation provides the most powerful approach to comprehensive analysis of biomechanical systems.

The numerical method applies a variety of technologies including mathematics, physics, computer science, and engineering. Finite element technologies are usually employed in these studies. Applications of numerical model studies in biomechanical systems include a variety of fields, such as studies of the hemodynamics of the cardiovascular system, analyses of the mechanics of major human structures such as the head, heart, lung, bones, and joints, analyses of soft tissue mechanics, and mechanics of artificial organs. Many textbooks have been published to introduce the methodologies of numerical methods and their applications in many different fields.[1,18,57,60,62] The goal of this chapter is to introduce techniques involved in a numerical model study of hemodynamics in the vascular system with the emphasis on large and middle size vessels. Particular examples of this application include the fluid flow fields in three-dimensional steady flow models of an end-to-side graft anastomosis and the Y-branched bifurcation, and two-dimensional pulsatile flow models of a stenosed coronary artery. The reader is assumed to have a fundamental knowledge of fluid dynamics and finite element techniques.

9.2   Numerical Methods

Governing Equations for Blood Flow

Conservation of linear momentum and conservation of mass are the primary laws governing blood flow in humans. The equations of conservation of linear momentum, or Navier-Stokes equations in a Cartesian coordinate system for an incompressible and Newtonian fluid, are:

$$
\begin{aligned}
\frac{\partial u_b}{\partial t_b} + u_b \frac{\partial u_b}{\partial x_b} + v_b \frac{\partial u_b}{\partial y_b} + w_b \frac{\partial u_b}{\partial z_b} + \frac{1}{\rho} \frac{\partial p_b}{\partial x_b} + f_{xb} &= \frac{\mu}{\rho} \left(\frac{\partial^2 u_b}{\partial x_b^2} + \frac{\partial^2 u_b}{\partial y_b^2} + \frac{\partial^2 u_b}{\partial z_b^2} \right) \\
\frac{\partial v_b}{\partial t_b} + u_b \frac{\partial v_b}{\partial x_b} + v_b \frac{\partial v_b}{\partial y_b} + w_b \frac{\partial v_b}{\partial z_b} + \frac{1}{\rho} \frac{\partial p_b}{\partial y_b} + f_{yb} &= \frac{\mu}{\rho} \left(\frac{\partial^2 v_b}{\partial x_b^2} + \frac{\partial^2 v_b}{\partial y_b^2} + \frac{\partial^2 v_b}{\partial z_b^2} \right) \\
\frac{\partial w_b}{\partial t_b} + u_b \frac{\partial w_b}{\partial x_b} + v_b \frac{\partial w_b}{\partial y_b} + w_b \frac{\partial w_b}{\partial z_b} + \frac{1}{\rho} \frac{\partial p_b}{\partial z_b} + f_{zb} &= \frac{\mu}{\rho} \left(\frac{\partial^2 w_b}{\partial x_b^2} + \frac{\partial^2 w_b}{\partial y_b^2} + \frac{\partial^2 w_b}{\partial z_b^2} \right)
\end{aligned}
\tag{9.1}
$$

The continuity equation is:

$$
\frac{\partial u_b}{\partial x_b} + \frac{\partial v_b}{\partial y_b} + \frac{\partial w_b}{\partial z_b} = 0
\tag{9.2}
$$

where the subscript b denotes the biological or physical system; u_b, v_b, w_b are the velocity components, and f_{xb}, f_{yb}, and f_{zb} are the components of body force per unit of volume along the x_b, y_b, and z_b directions, respectively; p_b is the pressure, and ρ and μ are the density and dynamic viscosity of the fluid, respectively. Here we assume that the numerical models use a fluid with the same properties as the biological system, i.e., ρ and μ are the same for the physical system and the model.

To simplify the expressions, the above equations can be expressed in vector form as:

$$
\frac{\partial \overline{u_b}}{\partial t_b} + \left(\overline{u_b} \cdot \nabla \right) \overline{u_b} + \frac{1}{\rho} \nabla p_b + \overline{f_b} = \frac{\mu}{\rho} \nabla^2 \overline{u_b}
\tag{9.3}
$$

$$
\nabla \cdot \overline{u_b} = 0
\tag{9.4}
$$

where $\overline{u_b} = \overline{i} u_b + \overline{j} v_b + \overline{k} w_b$ is the velocity vector, and $\overline{f_b}$ is the vector of body force per unit of volume. The effect of body force in the above equations is generally not significant and can be ignored for the study of blood flow in vessels.

When performing a numerical study, we can use non-dimensional parameters to simplify the simulation. In these cases, reference values are selected for normalization purposes. When different references are selected, the appearance of the equations may be different. Because blood flows in vessels are generally considered as either steady or pulsatile, we can select an l_0 as the reference or characteristic length, U_0 as the reference velocity, and $\omega_0 = 2\pi/T_0$ as the reference angular frequency, where T_0 is the reference period. Usually, the vessel diameter, the average velocity along a cross section of the vessel, and the angular frequency of the base harmonic of the pulsatile flow are selected as the reference length, velocity, and angular frequency, respectively. The normalized non-dimensional length, time, velocity, and pressure are expressed as:

$$x = \frac{x_b}{L_0}$$

$$t = t_b \omega_0 = t_b \frac{2\pi}{T_0}$$

$$\bar{u} = \frac{\bar{u}_b}{U_0}$$

$$p = \frac{p_b}{\rho U_0} \tag{9.5}$$

Substituting the above relationships into Eqs. 9.3 and 9.4 and ignoring the terms of body force, the modified form of the Navier-Stokes equation for a pulsatile, incompressible, and Newtonian fluid can be expressed as:

$$\frac{4\alpha^2}{R_e} \frac{\partial \bar{u}}{\partial t} + (\bar{u} \cdot \nabla)\bar{u} + \nabla p = \frac{1}{R_e} \nabla^2 \bar{u} \tag{9.6}$$

The continuity equation is

$$\nabla \cdot \bar{u} = 0 \tag{9.7}$$

where $\bar{u} = \bar{i}u + \bar{j}v + \bar{k}w$ is the normalized non-dimensional flow velocity vector, and R_e is the Reynolds number:

$$R_e = \frac{L_0 U_0}{\nu} \tag{9.8}$$

where $\nu = \mu/\rho$ is the kinematic viscosity and α is the Womersley number:

$$\alpha = \frac{L_0}{2} \sqrt{\frac{\omega_0}{\nu}} \tag{9.9}$$

To study blood flow in the vascular system, the initial and boundary conditions can be described as:

$$\bar{u}_{in} = \bar{h}(t) \tag{9.10}$$

$$\bar{u}_{wall} = 0 \tag{9.11}$$

where $\bar{u}_{in}$ and $\bar{u}_{wall}$ are the velocities at the inlet boundary and at the wall, respectively; $\bar{h}(t)$ is a function of the inlet velocity. The outflow is assumed traction free and the surface traction free condition can be described as

$$-p\bar{n} + \frac{1}{R_e} \frac{\partial \bar{u}}{\partial \bar{n}} = 0 \tag{9.12}$$

where $\bar{n}$ is the normal unit vector at the outflow boundary. The shear stress at a vessel wall is

$$\tau_w = \mu\gamma_w = -\mu\left.\frac{\partial u_t}{\partial\bar{n}}\right|wall \tag{9.13}$$

where γ_w is the shear rate; u_t is the tangential velocity and $\bar{n}$ is the normal unit vector at the wall.

Finite Element Formulation

From the above discussion, it is apparent that for a three-dimensional flow field, we need to solve a set of coupled equations consisting of the three momentum equations and the continuity equation to obtain the three velocity components and the pressure in the field. As a general consideration, one could put the four variables together to solve the equations. However, since the continuity equation does not include the pressure variable, many methods can be used to simplify the calculation and save computational time and computer memory. To simplify the description of the derivation, we first put the four variables together to solve the equations. The methods used to simplify the calculation in the realistic development of algorithms will then be introduced.

To solve the equations, let us first define two operators $L(\bar{u},p)$ and $W(\bar{u})$ as:

$$L\left(\bar{u},p\right) = \frac{4\alpha^2}{R_e}\frac{\partial\bar{u}}{\partial t} + \left(\bar{u}\cdot\nabla\right)\bar{u} + \nabla p - \frac{1}{R_e}\nabla^2\bar{u} \tag{9.14}$$

and

$$W\left(\bar{u}\right) = \nabla\cdot\bar{u} \tag{9.15}$$

For the differential equations defined in domain Ω, we need to find the solutions of:

$$L\left(\bar{u},p\right) = 0 \text{ and } W\left(\bar{u}\right) = 0 \tag{9.16}$$

We seek the approximate solutions for the equations as:

$$u = \sum_{k=1}^{m} u_k N_k, \quad v = \sum_{k=1}^{m} v_k N_k, \quad w = \sum_{k=1}^{m} w_k N_k, \text{ and } p = \sum_{k=1}^{m} p_k N_k$$

and simplify the expression as:

$$\bar{u} = \sum_{k=1}^{m}\bar{u}_k N_k \text{ and } p = \sum_{k=1}^{m} p_k N_k \tag{9.17}$$

where $\bar{u}k = \bar{i}u_k + \bar{j}v_k + \bar{k}w_k$, and p_k are the $\bar{u}$ and p values on node k, N_k is called the shape function, m is the number of nodes for an element, and $k = 1, 2, 3, \ldots, m$.

The errors ϵ_l and ϵ_w for these approximations are

$$\epsilon_l = L\left(\sum_{k=1}^{m}\bar{u}_k N_k, \sum_{k=1}^{m} p_k N_k\right) \neq 0 \tag{9.18}$$

$$\epsilon_w = W\left(\sum_{k=1}^m \overline{u}_k N_k\right) \neq 0 \tag{9.19}$$

Let us first consider ϵ_l. We can use a weighted residual method which requires that the weighted error, or residual, goes to zero in the domain Ω. That is,

$$\epsilon_l = \int_\Omega \epsilon_l \, \Psi_n d\Omega = 0 \tag{9.20}$$

where Ψ_n represents arbitrary weighted functions. The introduction of the weighted functions will allow us to obtain the necessary number of linearly independent algebraic equations to match the unknown parameters. As derived in Appendix A at the end of this chapter, three sets of algebraic matrix equations can be obtained for each element as:

$$\sum_{n=1}^m \left(\sum_{k=1}^m \frac{\partial \overline{u}_k}{\partial t} D_{kn} + \sum_{k=1}^m \overline{u}_k \left(u_k + v_k + w_k\right) E_{kn} + \sum_{k=1}^m p_k H_{kn} - \sum_{k=1}^m \overline{u}_k F_{kn} + \sum_{k=1}^m \overline{u}_k G_{kn}\right) = 0 \tag{9.21}$$

where the expressions of the parameters D_{kn}, E_{kn}, H_{kn}, F_{kn}, and G_{kn} are shown in Appendix A.

Similarly, when we consider the error ϵ_w, a fourth set of algebraic matrix equations can be obtained as:

$$\sum_{n=1}^m \sum_{k=1}^m \overline{u}_k \cdot \overline{I_{kn}} = 0 \tag{9.22}$$

where the expression of $\overline{I_{kn}}$ is shown in Appendix A.

The four sets of algebraic matrix equations for each element with the boundary conditions are the algebraic equations we need to solve to obtain the approximate solutions for the governing equations.

As mentioned previously, since the parameter of pressure is not included in the continuity equation, several methods can be used to simplify the calculation. One of them is to use a lower order shape function for the pressure variable or to set it as a constant value for each element. Because pressure does not appear in the continuity equation, it is not the primary parameter and does not need to be continuous across the inter-element boundaries. To simplify the simulation, we can replace Eq. 9.17 with:

$$\overline{u} = \sum_{k=1}^m \overline{u}_k N_k \quad \text{and} \quad p = \sum_{k=1}^e p_k J_k \tag{9.23}$$

where N_k and J_k are shape functions of different orders with $e < m$.[57] Following a similar procedure to that described above, we can obtain the approximate solutions. The minimum continuity requirements for each element for this method are: (u, v, w) are linear in x, y, and z, and p is a constant.

Another method, the penalty function method,[22,66,71] considers the continuity equation as a constraint to the Navier-Stokes equation. In this method, a term involving the pressure is added to the continuity equation as:

$$\nabla \cdot \overline{u} + \xi p = 0 \tag{9.24}$$

where ξ is a small parameter of order 10^{-6}. This modification allows the momentum and continuity equations to be decoupled but introduces a small error in the order of $\leq \xi^{1/2}$.[55,56] Cuvelier et al.[22] indicate that the penalty function method yields a 10-fold decrease in computational requirements.

A different and popular approach to solving the governing equations is to use the stream function Λ and the vorticity ζ as the primary variables and indirectly calculate the velocity and pressure in the flow field.[1,57]

Let us consider two-dimensional situations. The stream function and vorticity in the flow field can be expressed as:

$$\frac{\partial \Lambda}{\partial y} = u, \quad \frac{\partial \Lambda}{\partial x} = -v, \quad \zeta = \frac{\partial u}{y} - \frac{\partial v}{\partial x} = \nabla^2 \Lambda \tag{9.25}$$

Because the continuity equation is identically satisfied by the stream function, the three coupling equations for a two-dimensional flow field can be reduced to two second-order equations. Furthermore, if only the stream function Λ is introduced, the three coupling equations for a two-dimensional flow field can be reduced to a single fourth-order equation. Since the number of coupled equations is reduced, the computation is greatly simplified.

Shape Function and Local and Global Coordinates

The determination of the shape or interpretation function is an important step in the numerical simulation. Let us consider two-dimensional situations and define N_k as a shape function which is a component in a linear equation

$$U(x, y) = \sum_{k=1}^{m} U_k N_k \tag{9.26}$$

where $U(x,y)$ is a potential function within an element and U_k is the potential value at node k ($k = 1, 2, \ldots m$). Fig. 9.1 shows samples of an element having m nodes (Fig. 9.1a) and 3 nodes (Fig. 9.1b) in a global coordinate system. As an example, the shape functions derived for the triangle element shown in Fig. 9.1b are shown in Appendix B at the end of this chapter.

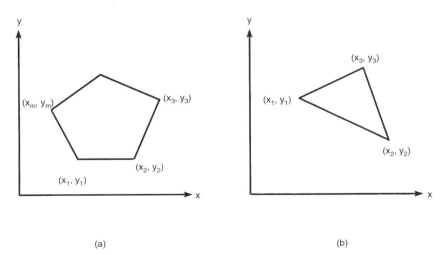

FIGURE 9.1 A two-dimensional element having (a) m nodes and (b) three nodes in a global coordinate system.

For an element having a more complex shape function and/or a large number of nodes, derivation of the shape function is more complex. It can be greatly simplified if local or element coordinates are assigned for each element. For example, a square element with the origin of its local coordinates attached to its geometric center (Fig. 9.2a) can be used for a two-dimensional, 4-node element (Fig. 9.2b) in the global coordinates. Also, a cube element with the origin of its local coordinates attached to the center of the cube (Fig. 9.3a) can be used for a three-dimensional, 8-node element (Fig. 9.3b). The procedure to derive the shape functions in the local coordinates is similar to that in global coordinates. The results for the elements shown in Figs. 9.2a and 9.3a are listed in Appendix C at the end of this chapter. The shape functions obtained in local coordinates need to be transferred to corresponding global coordinates. The formula for the transformation between local and global coordinates in two-dimensional situations is also provided in Appendix C.

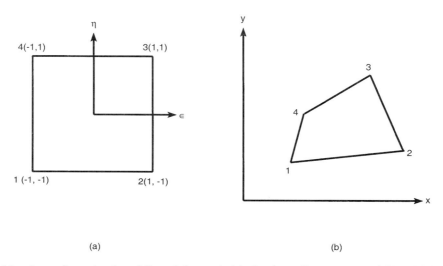

(a)   (b)

FIGURE 9.2   A two-dimensional quadrilateral element in (a) a local coordinate system and (b) a global coordinate system.

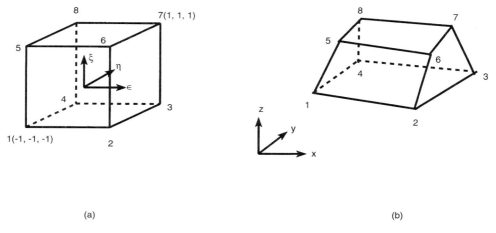

(a)   (b)

FIGURE 9.3   A three-dimensional hexahedral element in (a) a local coordinate system and (b) a global coordinate system.

Considerations for Pulsatile Flow

For simulating blood flow, we can use either a steady flow or pulsatile flow model. For steady flow, the time-dependent term in the Navier-Stokes equation, i.e., $\partial \bar{u}/\partial t$, can be ignored and the calculation greatly simplified. When a pulsatile flow model is used, the problem naturally becomes time-dependent. The finite element formulation of time-dependent problems involves two steps: spatial approximation and temporal approximation. Derivation of formulations presented in the previous sections deal with spatial approximation. The time-dependent term has not yet been considered. The step of time approximation often uses finite difference formulae for the time derivatives. This step converts the differential equations into a set of algebraic equations among the velocity values $\bar{u}^s$ at time $t_s = s\Delta t$, where Δt is the time increment between two simulations of the spatial approximation and s is an integer. All time approximation schemes seek to find $\bar{u}^s$ at time t_s using the known value(s) of $\bar{u}$ from a previous time or times, i.e., $\bar{u}^{s-1}$, $\bar{u}^{s-2}$, etc. The choice of using one or more previous value(s) is dependent on the requirement of accuracy and the relative slope of the change of velocity to the time interval selected. Because those approaches are discussed in detail by many textbooks, we will not include those formulae here.

In the temporal approximation, the velocity profiles and/or pressure gradient in the inlet of the vessel at different times or phases within the cardiac cycle are usually needed to serve as the boundary conditions. To simulate real vessels or experimental models, we can measure the flow rate or the pressure gradient waveform experimentally in the vessel or model to provide the boundary conditions. Womersley[73] and Nichols and O'Rourke[47] described a method to predict the velocity profiles from the pressure gradient in an axisymmetrical rigid pipe. Very often, the flow rate waveform may be easier to measure because currently available pressure transducers normally need access holes on the pipe wall while many transducers for flow rate measurements do not have such a requirement. By modifying Womersley's formulae,[27] we can also use the measured flow rate to predict the velocity profiles at different phases within a pulsatile cycle.

Using a Fourier series, any shape of a pulsatile flow rate can be expressed as:

$$q_b(t_b) = Q_{0b} + \sum_{n=1}^{\infty} Q_{nb} \sin(n\omega_b t_b - \theta_{nb}) \tag{9.27}$$

where the subscript b again represents the biological or physical system, q_b is the measured flow rate waveform, Q_{0b} is the steady flow component, n is an integer, Q_{nb} is the nth harmonic of the flow rate, ω_b and t_b are angular frequency and time, respectively, and θ_{nb} is the angular phase of the nth harmonic. The velocity profiles in a straight rigid pipe at any phase can be calculated from the superimposition of the velocity profiles generated from each component of flow rate at the same phase. For a fully developed flow, the velocity profile generated by the steady flow term Q_{0b} is a parabolic distribution, i.e.,

$$v(r_b/R_b) = V_{mb}\left[1 - \left(\frac{r_b}{R_b}\right)^2\right] = \frac{2Q_{0b}}{\pi R_b^2}\left[1 - \left(\frac{r_b}{R_b}\right)^2\right] \tag{9.28}$$

where r_b is the distance between the observation point to the central axis of the pipe, R_b is the radius of the pipe, and V_{mb} is the peak velocity at the central axis of the pipe.

For an arbitrary sinusoidal flow rate with an angular frequency $n\omega_b$,

$$q_b(t_b) = Q_b \sin(n\omega_b t_b - \theta_{nb}) \tag{9.29}$$

the pressure gradient that drives the flow rate along the tube is

$$\Delta p_b = \Delta P_b \cos(n\omega_b t_b - \Phi_{nb}) \tag{9.30}$$

where Δp_b is the magnitude of the pressure gradient with the relationship of

$$\Delta p_b = \frac{\mu}{\pi R_b^4 \left(M_{10}' / \alpha_n \right)} Q_b \tag{9.31}$$

and

$$\Phi_{nb} = \epsilon_{10} + \theta_{nb} \tag{9.32}$$

where μ is the viscosity and α_n is the Womersley number for that frequency with

$$\alpha_n = R_b \sqrt{\frac{n \omega_b}{v}} \tag{9.33}$$

M_{10}' and ϵ_{10} are two parameters related to α_n, which can be found from the tabulated data by Womersley[73] and McDonald.[42] When $\alpha > 10$,

$$M_{10}' = 1 - \frac{\sqrt{2}}{\alpha_n} + \frac{1}{\alpha_n^2}$$

$$\epsilon_{10} = \frac{\sqrt{2}}{\alpha_n} + \frac{1}{\alpha_n^2} + \frac{19}{24 \left(\sqrt{2} \right) \alpha_n^3}$$

When the pressure gradient is determined, we can follow the procedure described by Womersley[47,75] to calculate the velocity profiles at different phases for this frequency. In the same way, we can calculate the velocity profiles for different harmonics. As mentioned previously, the velocity profiles at different phases for a pulsatile flow can be obtained by superimposing the velocity profiles calculated from the steady flow rate plus a series of harmonics. The number of harmonic terms included in the simulation will depend on the accuracy required.

The Reynolds number and Womersley number are two important parameters for characterizing the flow features. In order to keep dynamic similarity, when we use numerical methods to study steady flow, we should use the same Reynolds number (Eq. 9.8) for both the biological system and the numerical model, i.e.,

$$\frac{D_b U_b}{v} = \frac{D U_0}{v} \tag{9.34}$$

where D_b is the diameter of the vessel, U_b is the average velocity in the vessel, and v is the kinematic viscosity of the fluid. This is called Reynolds number modeling. When performing a numerical study under pulsatile flow conditions, both the Reynolds number and the Womersley number (Eq. 9.9) should be kept the same for the biological system and the model. In these cases, we should also have

$$\frac{D_b}{2} \sqrt{\frac{\omega_b}{v}} = \frac{D}{2} \sqrt{\frac{\omega}{v}} \tag{9.35}$$

where ω_b and ω are the angular frequencies of the basic harmonic for the biological system and model, respectively.

9.3   Applications

Extensive investigations have been performed to study the hemodynamic features of the vascular system using numerical simulations. Emphasis has been placed on the hemodynamics of vessels with complex geometry such as branched bifurcations and graft anastomoses, and diseased vessels with stenoses and aneurysms. The basic information sought in these studies includes (1) the fundamental hemodynamic characteristics of different types of vessels; (2) the variations in those hemodynamic features with respect to changes in vessel geometry and flow conditions; and (3) the relationships between the hemodynamic features and vessel diseases. The primary goal of these studies is to use the findings from these observations to improve the diagnosis and/or treatment of those diseases.

In numerical simulation, the basic hemodynamic features studied in vessels are the velocity vector fields in different planes and the velocity profiles along different cross-sections. The overall flow pattern, which may include flow skewing, reversed flow, flow separation, and stagnation, and secondary flow are revealed using graphical representations of those data. Various other parameters such as shear stress, streamline contours, vorticity contours, and particle paths can be derived based on the velocity and pressure data.

Among the derived parameters, the shear stress along the vessel wall is particularly important because there is increasing evidence that shear stress plays an important role in the formation and progress of some vascular diseases, particularly atherogenesis. Currently, two theories based on low shear and high shear exist to explain the role of shear stress in atherogenesis. Caro et al.[11] reported that low shear stress is associated with arterial atherogenesis. This may be caused by the reduced mass transfer rate in the low shear stress area of the arterial wall. Here, nutrients are prevented from reaching the vessel wall[23] and lipids are easily stagnated along the wall.[10] Many other studies[6,7,24,38,45,58,68,77] also support this view of low shear effects. On the other hand, Fry[32] reported that high shear stress on the wall may damage the endothelium of the vessel wall. His experiments showed that the walls of vessels obtained from dogs became degraded after being exposed to a shear stress of 380 dyne/cm^2 for one hour. This finding is supported by the clinical study of femoral atherogenesis by Cho et al.[16] and Back et al.[2] Because both low and high shear stresses may be causal agents, it is necessary to analyze the effects of wall shear stress according to the particular conditions of the vessel.

In this section, we present three typical applications of numerical simulation in the vascular system. First, the results of three-dimensional steady flow models of end-to-side distal graft anastomoses with different angles and flow rates are presented. The emphasis in this application is on the variation of velocity vector plots and shear rates in the central plane of the vessels. This application demonstrates the utility of numerical simulation in helping surgeons make decisions about revascularizing diseased or injured arteries. Second, numerical results of three-dimensional steady flow in Y-shaped bifurcation models are presented. Here, the emphasis is placed on the velocity distributions within different cross-sectional planes. This application demonstrates the utility of numerical methods for evaluating the feasibility and/or accuracy of new experimental techniques. Third, numerical results of two-dimensional stenosed coronary artery models under pulsatile flow conditions are presented. In addition to the velocity distribution and shear rate, the results of particle tracking and pressure contours are also used to assess hemodynamic features. This application demonstrates the use of numerical methods to study changes in hemodynamic features associated with the severity of disease.

The results presented for each application are based on vascular models studied by the authors' group. Results from similar studies performed by other groups are presented for comparison. Since this chapter deals mainly with numerical simulation, detailed descriptions of other experimental methods such as magnetic resonance imaging and ultrasound Doppler color imaging will not be included. Readers may refer to the cited papers and other textbooks to familiarize themselves with those techniques.

Numerical Study of Flow in End-to-Side Distal Anastomosis: Three-Dimensional Models under Steady Flow Conditions

Numerical studies of the hemodynamics of end-to-side anastomosis have been performed by many groups. Pietrabissa et al.[54] reported results under two-dimensional steady flow conditions. Fei et al.[28] presented results in three-dimensional models under steady flow conditions; this work included investigations of the effect of different angles between the graft and the host artery and different flow rates. Steinman et al.[66] presented results of a two-dimensional model under pulsatile flow conditions. The effect of wall distensibility with a two-dimensional model under pulsatile flow conditions was studied by Steinman and Ethier,[65] and the effect of non-Newtonian fluid was reported by Ballyk et al.[3] As mentioned previously, the basic results to be presented in this section were obtained by the authors' group.[28]

The Models and Results of Numerical Simulation

Fig. 9.4 shows a typical three-dimensional mesh of the distal anastomosis models generated for the study. Fluid enters at the upper left vessel (referred to as the graft) and exits from the right end of the horizontal vessel (referred to as the artery). The left end of the artery is blocked to simulate an occluded vessel. While arterial occlusion may or may not be present clinically, incorporation of this feature would require specification of boundary conditions (i.e., magnitude of retrograde flow, proximal artery pressure, etc.) not readily known. The length of the graft is 20 cm and the length of the artery is 28 cm. Both vessels have a radius of 1 cm. A zero reference position within the model was defined as the intersection of the geometric center lines of the graft and artery branches. A total of 6272 brick elements consisting of eight nodes each were generated throughout the entire region with the mesh density being gradually increased in the region close to the anastomosis. Using fluid properties typical of those for human blood ($\mu = 0.035$ poise, $\rho = 1$ gm/cm^3, and $\nu = 0.035$ cm/s), steady, parabolic velocity profiles at Reynolds numbers of 100 and 205 were specified as entrance boundary conditions. These Reynolds number values are typical of iliofemoral bypass grafts with marginal and adequate clinical flow rates, respectively, which may vary with the patency of the patient's distal runoff vessels.[4,5,78] Zero velocities were assigned to all vessel walls and the occluded end of the artery as boundary conditions. A total of seven geometric models were examined by varying the angle of distal anastomosis over a range of 20, 30, 40, 45, 50, 60, and 70 degrees. Approximately 40 velocity profiles were taken along the artery in each model using a cross-section perpendicular to the flow axis. The component of the velocity vector parallel to the vessel walls was used to calculate the shear rate at each wall. This shear rate was then converted to a normalized shear rate, NSR, representing the ratio of the local shear rate at the wall to the corresponding value at the entrance of the graft.

The results are shown as follows.

Velocity Vector Plots
Fully developed parabolic flow was obtained at the graft inlet and at the artery outlet (L/D = 10) of all models with Re = 100, while the profile was skewed slightly toward the outer wall at the artery outlet for all models with Re = 205. Enlarged views of the velocity vector plots in the anastomotic region are shown in Figs. 9.5a and 9.5b for the case of a 45° distal anastomotic angle with Re = 100 and 205, respectively. Within the anastomosis, the velocity profile is skewed toward the outer wall with a stagnation point occurring almost directly opposite the midpoint of the graft lumen along the outer wall. A weak vortex is formed within the entire diameter of the closed end of the artery just upstream of the stagnation point. A very small separated flow region is seen along the inner wall of the artery just downstream of the anastomosis for Re = 205. These characteristic features are generally seen in all modeled cases with the exception that the inner wall separation zone does not occur in cases with Re = 100 for anastomotic angles less than 60° and in cases with Re = 205 for anastomotic angles less than 45°. Furthermore, it was observed that the site of the stagnation point consistently moved further downstream relative to the toe of the graft/artery anastomosis with decreasing anastomotic angles. Velocity vectors within cross-sectional planes perpendicular to the arterial axis of 30° and 45° models (Figs. 9.6a and 9.6b) show that symmetric vortices are formed upstream in the graft/artery anastomosis. These vortices are initially centered toward

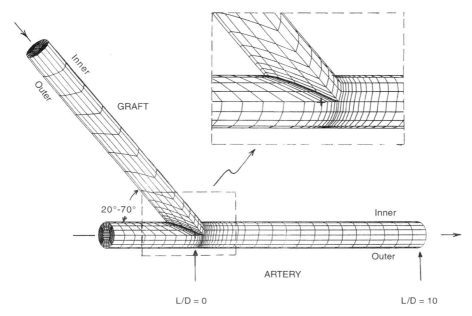

FIGURE 9.4   Three-dimensional mesh used in numerical simulation of arterial bypass graft distal anastomosis with angles of 20°–70°. Insert shows mesh detail at anastomotic junction.

the outer and lateral walls of the artery and gradually shift toward the vessel center line with increasing distance downstream. The cross-sectional velocity magnitudes for Re = 205 are two times those for Re = 100 and are seen to rapidly dissipate with axial distance. The flow patterns are very similar for the anastomotic angles of 30° and 45°, with the maximum velocities being reduced by approximately 35% in the 30° angle model from those in the 45° angle at corresponding Reynolds numbers.

Wall Shear Rates on the Graft

For Re = 100, the normalized shear rates (NSR) along the inner wall were all larger than or equal to 1.0 for angles greater than 45°. In the case of the 45° model, for example, NSR = 1.0 over a distance from the entrance of the graft (L/D = –10) to a position at L/D = –1.2 and then increased to approximately 2.2 at the vessel wall junction. For angles less than 40°, the inner wall NSR was less than 1.0 along a distance near the junction (–2 ≤ L/D ≤ –0.8, –2.8 ≤ L/D ≤ –0.5, and –3.8 ≤ L/D ≤ –0.4 for angles of 40°, 30°, and 20°, respectively) with the lowest values being 0.95, 0.85, and 0.63 for the 40°, 30°, and 20° angles, respectively. Along the outer wall, NSR increased from 1.0 at the graft entrance to values up to 1.68 (range: 1.68 to 1.08) at the wall junction for all angles from 20° to 50°. For the 60° and 70° angle cases, NSR dropped below 1.0 between –1.5 ≤ L/D ≤ –1.0 and continued to decrease to 0.95 and 0.78, respectively at the wall junction.

For Re = 205, the normalized shear rates along the inner wall were all larger than or equal to 1.0 for angles greater than 40°, increasing sharply near the junction (range: 1.58 to 2.9 for angles 40° to 70°). For the 30° and 20° cases, NSR fell to 0.94 and 0.70, respectively within –1 ≤ L/D ≤ 0. Along the outer wall, NSR increased from 1.0 at the graft entrance to values up to 1.50 (range: 1.50 to 1.06) at the wall junction for all angles from 20° to 45°. For the 50°, 60°, and 70° angle cases, NSR dropped below 1.0 between –1.5 ≤ L/D –0.5, and continued to decrease to 0.99, 0.85, and 0.68, respectively at the wall junction.

Wall Shear Rates on the Artery

Normalized shear rates at Re = 100 for anastomotic angles of 20° to 70° are displayed along the artery inner and outer walls in Figs. 9.7 and 9.8, respectively. Inner wall NSR (Fig. 9.7) upstream of the anastomosis sharply fell to zero and then became slightly negative compared to relatively high graft inlet

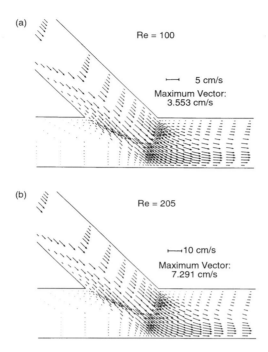

FIGURE 9.5   Velocity vectors in the plane of symmetry for a 45° anastomotic angle at (a) Re = 100 and (b) Re = 205. (Note: Velocity vectors are plotted at mesh nodes and do not lie along tube diameters.)

values with the overall shear rates being consistently lower for larger angles. Inner wall NSR downstream of the anastomosis likewise decreased rapidly from graft inlet values and were negative for angles greater than 60°. For all angle cases, however, NSR returned to inlet values within 10 diameters downstream. Outer wall NSR (Fig. 9.8) were negative upstream of the anastomosis and decreased to zero within three diameters for all angles with the maximum negative values increasing with larger angles. A stagnation point was also seen, upstream of the anastomosis, whose location approached the axial reference with increasing angles. Shear rates increased downstream of the stagnation point with the magnitude and location of the peak increasing with the degree of the anastomotic angle. Again, NSR returned to inlet values within 10 diameters downstream in all angle cases.

NSR at Re = 205 for anastomotic angles of 20° to 70° are displayed along the artery inner and outer walls in Figs. 9.9 and 9.10, respectively. The overall findings for models with Re = 205 corresponded to those for Re = 100 with the primary exception that inner wall shear rates (Fig. 9.9) were considerably lower just downstream of the anastomosis and became negative for angles greater than 45°. Additionally, NSR did not return to inlet values within 10 diameters downstream for any angle models. Along the outer wall (Fig. 9.10), the same general trends were seen as with Re = 100 except that the NSR were consistently more negative upstream of the anastomosis and more positive downstream of the anastomosis than corresponding values observed at the lower flow rate. Again, shear rates did not return to inlet values within 10 diameters downstream for any angle models.

Utility of the Computational Simulation: Helping Surgeons Make Decisions in Vascular Surgery

The findings from the above study clearly document the potent influences of graft angle and flow rate on flow patterns and wall shear rates within bypass graft end-to-side distal anastomoses. Characteristic of all distal end-to-side anastomoses is a skewed velocity profile which produces low, and often reversed, velocities along the wall in the proximal artery and in regions of the anastomotic toe and heel, while producing elevated forward velocities along the distal outer wall of the artery. In particular, for a proximally occluded artery, a graft distal anastomosis with a 20° angle produced the least amount of flow

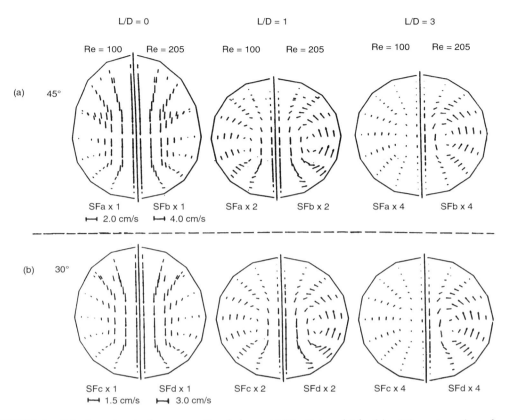

FIGURE 9.6 Velocity vectors in the cross-sectional planes at L/D = 0, 1, and 3 for (a) a 45* anastomotic angle and (b) a 30° anastomotic angle. The cases of Re = 100 (left) and Re = 205 (right) along with maximum in-plane velocities, VM, and relative graphical velocity scale factors (SF) are shown in each panel.

separation and shear rate extremes along the artery walls as compared to anastomoses with angles up to 70°. This finding is primarily related to the occurrence of velocity skewing toward the outer wall as flow enters from the graft into the artery. The resultant high velocities along the outer wall are offset by lower, potentially reversed, velocities which maintain a net constant flow rate at all cross-sections. With a low angle of entry, the fluid inertial forces undergo less redirection in the anastomosis, and, thus, experience reduced skewing effects. For a given anastomotic angle, an increase in Reynolds number from 100 to 205 decreased the normalized wall shear rates along the inner arterial wall while increasing the magnitude of the normalized wall shear rates along the outer arterial wall. The absolute values of the wall shear rate for all arterial wall sites, however, were generally higher at Re = 205.

The implications of the presence of well defined vortices within the artery cross-section would be expected to produce a flushing effect especially along the inner artery wall. Presence of these secondary motions does not affect shear rates along the plane of symmetry but would increase shear rates along the artery wall away from the plane of symmetry in a circumferential direction. Since the strength of these secondary motions and, therefore, the convection of high axial momentum from the outer wall toward the inner wall, increased with the degree of the anastomotic angle, it is concluded that these effects account for the shorter length of the corresponding separation zones seen near the toe region (Figs 9.7 and 9.9). This effect appears especially obvious for angles greater than or equal to 50° for Re = 100 and greater than or equal to 40° for Re = 205.

The above numerical results correlate well with several *in vitro* experiments.[21,36,48] In particular, the model used by Keynton et al.[36] exactly matched conditions of the above study for distal anastomotic angles of 30°, 45°, and 60°. Those results documented flow patterns similar to the above study, where

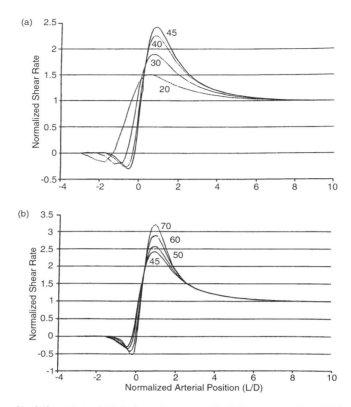

FIGURE 9.7   Normalized Shear Rates (NSR) along the inner wall of the artery at Re = 100 for anastomotic angles of (a) 20° to 45° and (b) 45° to 70°.

peak shear rates were found to be located at similar axial positions but with generally higher values. The discrepancy in magnitudes may be due to measurement noise artifacts in the experimental study and larger radial intervals in the numerical study. Keynton et al.[36] also observed flow reversals in a separation zone at the toe region near the anastomosis but only for the 60° angle of distal anastomosis at a Reynolds number of 205. This differs somewhat with the above results where flow reversals were seen at Re = 100 for graft angles of 60° and 70° as well as at Re = 205 for graft angles greater than or equal to 45°, and may be due to the limited ability of the laser Doppler anemometer to access the entire toe region. Based on the relative magnitude of lowest shear rates, those investigators recommended an angle of 45° or possibly 30° for the anastomosis and concluded that 60° would be undesirable.

While the effects of pulsatile flow were not incorporated into the above simulations, a further analysis was performed in the same distal anastomotic models under pulsatile conditions simulating those in the iliofemoral region ($Re_{mean} = 459$, $Re_{peak} = 459$, and $\alpha = 3.56$) using ultrasonic Doppler color flow imaging.[59] While it was not possible to obtain the same degree of measurement resolution as in the above numerical model, the findings from that study showed that there was wide variation in velocities over time. However, the areas of low velocity and flow separation along the inner arterial wall and within the entire anastomosis were minimized for the 30° angle at each of three critical time intervals during the cardiac cycle. Flow visualization images obtained by Bassiouny et al.[5] within models made from transparent silicon casts of actual anastomotic junctions also revealed a stagnation point along the arterial floor together with low and oscillating shear. High shear occurred in the hooded graft while low shear regions formed along the lateral walls and the heel of the anastomoses.

The above comparisons with experimental results are encouraging and give credibility to the numerical techniques employed in the above study. Steinman et al.[65] used a numerical simulation to obtain flow patterns and wall shear stress maps under pulsatile flow conditions in a two-dimensional rigid walled

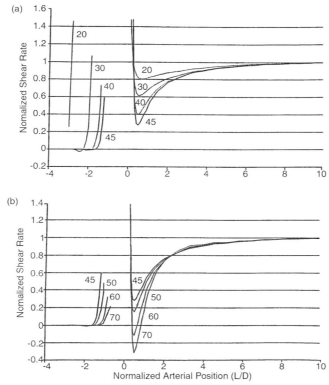

FIGURE 9.8   Normalized Shear Rates (NSR) along the outer wall of the artery at Re = 100 for anastomotic angles of (a) 20°–45° and (b) 45°–70°.

model of a 45° distal anastomosis. They found a mobile recirculating region along the anastomotic toe with an initially elevated wall shear stress which sharply dropped to negative values over the next one diameter downstream along the artery. Narrowly elevated wall shear stresses were also seen at the heel region which rapidly fell to zero further upstream. A transition in wall shear stress was seen along the bed of the artery going from a negative value upstream of the anastomosis, through a stagnation point, to positive downstream of the anastomosis. The authors concluded that sites of intimal hyperplasia correlated with large values of wall shear stress and/or large temporal variations in wall shear stress. While these results are consistent with the above basic results, emphasis on high wall shear stresses along relatively narrow distances at the graft/artery junctions rather than on the nearby low wall shear stresses acting over larger distances results in an apparently conflicting conclusion regarding the nature of hemodynamic effects upon intimal hyperplasia. From chronic canine studies using vein and polytetra-fluoroethylene grafts,[6,64] it has been shown that intimal hyperplasia at the distal anastomosis occurs primarily in graft regions just proximal to the anastomosis and along the toe, heel, and floor of the host artery. Furthermore, inverse correlations between wall shear rate and thickness of intimal hyperplasia have been consistently reported by other studies.[5,7,24,38,45] It is our conclusion, therefore, that it is important to find ways to minimize the extent of low shear rate regions in the bypass graft distal anastomosis.

The results obtained from the above numerical simulation would suggest the following recommendations for the region of a bypass graft distal anastomosis:

1. Grafts should be placed with a minimal distal anastomotic angle (preferably 20° or less).
2. Grafts should be exposed to high flow as opposed to low flow conditions.

The preceding application demonstrates that a numerical study may assist surgeons in making decisions about graft surgery since graft angle is a factor which can be directly controlled by the surgeon

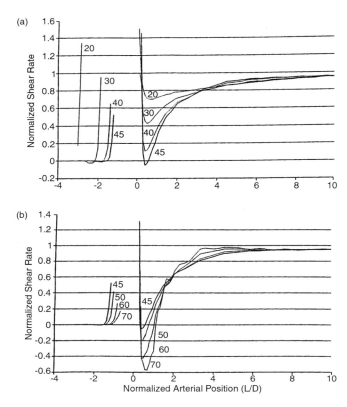

FIGURE 9.9 Normalized Shear Rates (NSR) along the inner wall of the artery at Re = 205 for anastomotic angles of (a) 20°–45° and (b) 45°–70°.

through operative procedures. The requirement for high flow conditions might be implemented by pre-operative selection of vessels with good distal run-off and/or use of a post-operative exercise regimen.

Numerical Study of Flow in Y-Shaped Bifurcations: Three-Dimensional Models under Steady Flow Conditions

The hemodynamic features in the aortic (Y-shaped) bifurcation have been studied by many groups using numerical simulations. Friedman et al.[30] (two-dimensional pulsatile flow), Friedman and Ehrlich[31] (two-dimensional steady flow), Einav and Stolero[25] (two-dimensional pulsatile flow), Yung et al.[76] (three-dimensional steady flow), Thiriet et al.[69] (three-dimensional steady flow), and Fei et al.[29] (three-dimensional steady flow) reported the results of numerical simulations under different geometry and flow rate conditions with rigid wall and Newtonian flow models. Lou and Yang studied the effect of non-Newtonian blood flow[40] and flexible walls[39] under two-dimensional pulsatile flow conditions. Again, the example presented in this section is based on results obtained by the authors' group.[29]

The Model and the Results of Numerical Simulation

Fig. 9.11 shows the generated mesh of the model used in this study. The diameters (D) of the three branches were all 2.54 cm and the bifurcation angle was 60°. A reference position (A in Fig. 9.11) was defined as the intersection of the central axes of the inlet vessel and the two outlet branches. The length of the inlet vessel was 2 D and the lengths of the two outlets were 5.5 D from the reference position. A total of 13,140 brick elements consisting of eight nodes each were generated throughout the entire model, with the mesh density being gradually increased near the branch point. In Cartesian coordinates, the mid-plane of the bifurcation (passing through the three central axes of the tubes) was oriented along the

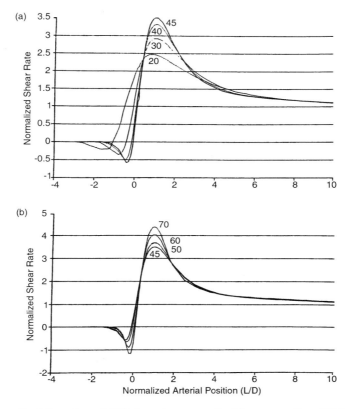

FIGURE 9.10 Normalized Shear Rates (NSR) along the outer wall of the artery at Re = 205 for anastomotic angles of (a) 20°–45° and (b) 45°–70°.

XZ plane, with the inlet branch along the Z axis. Again, the fluid viscosity and density were set to 0.035 poise and 1 mg/cm^3, respectively. A rigid vessel wall and an incompressible, Newtonian fluid were assumed. From the limited data available in man, the peak Reynolds number for the abdominal aorta proximal to the bifurcation is between 400 and 1100, and for the common iliac artery it is between 390 and 620.[47] To study the hemodynamic features in reasonably physiological flow conditions, Reynolds numbers of 250, 500, and 750 were selected to perform the simulation. For each Re number, corresponding parabolic velocity profile was assigned on the inlet surface of the inlet vessel. Again, zero velocities were assigned at all vessel walls as boundary conditions.

To demonstrate the results, several planes are pre-defined. The mid-plane of the bifurcation mentioned above is denoted L1 (for longitudinal plane 1). Another plane L2 is defined to be parallel to L1 but off-center by 0.2 D; the offset direction is arbitrary, due to symmetry. The orientations of other particular planes selected for examination are shown in Fig. 9.12. Five cross-sectional planes (normal to both L1 and the vessel axis) present transverse views of the fluid motion. One (C0) is on the inlet side, 1 D upstream from the reference point. Four others (C1, C2, C3, and C4) are on the rightward outlet branch at distances of 1.5 D, 2.5 D, 3.5 D, and 4.5 D downstream from the reference point, respectively. Three planes normal to L1, positioned along the right outlet branch direction are also defined as normal plane N2 (passing through the central axis) and N1 and N3, which are parallel to N2 but 0.4 D off the central axis toward the inner and outer walls, respectively.

Fig. 9.12 shows the velocity vector plot at Re = 500 in plane L1 (the mid-plane) of the bifurcation. It can be seen that the velocity profiles are laminar and nearly parabolic in the inlet vessel and skew toward the dividing (inner) wall in the outlet branches. Fig. 9.12 also shows large asymmetric velocity peaks near the divider walls at the inlets of the daughter branches, which then diminish and migrate toward the

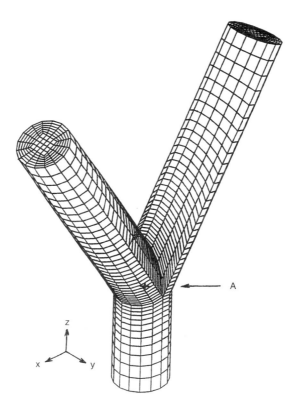

FIGURE 9.11   Three-dimensional mesh used in numerical simulation of Y-shaped bifurcation with a branch angle of 60°. Point A indicates the intersection of the three vessel axes.

center line as the fluid proceeds downstream. The peak velocities obtained at different Re at the intersections of plane L1 with planes C0 and C1 and the ratios of peak velocities at C1 and C0 (peak ratio) are listed in Table 9.1. It can be seen that with higher Re the ratio increases, indicating that the peak velocity within each outlet branch approaches that of the inlet vessel.

Fig. 9.13 displays the superimposed velocity profiles at four sites from longitudinal planes L2 (off-center) and L1 (mid-plane) at Re = 750. The profiles originate from the intersections of planes L2 and C1–C4 (Fig. 9.13a) and the intersections of planes L1 and C1–C4 (Fig. 9.13b), as indicated in the figure legends. In the off-center velocity profiles (Fig. 9.13a), a distinct secondary velocity maximum is observed near the outer wall in the outlet branch, causing the profile to assume a skewed M-shape. Conversely, in the mid-plane velocity profiles (Fig. 9.13b) the secondary peaks are much less apparent. In these figures (and subsequent velocity profile plots), r/R is the normalized radial distance, where r is the distance from the observed point to the central axis of the tube and R is the tube radius. Negative and positive values denote proximity to the inner and outer walls, respectively.

Flow profiles in the three normal planes (N1–N3) of the rightward outlet branch were also examined. Fig. 9.14 presents velocity profiles from the inner wall (a), central axis (b), and outer wall planes (c) at Re = 250. As before, the four superimposed velocity profiles represent the four axial positions C1–C4. The origin of each velocity profile in Fig. 9.14 is again defined by the intersections of planes N1, N2, and N3 with C1–C4, as indicated in the figure legends. As expected, all velocity profiles in these planar orientations exhibit axial symmetry, but the flow patterns vary significantly among the three (N1–N3) planes. The N1 plane (Fig. 9.14a) reveals a relatively high flow velocity near the branch point, but reduced velocity distally. In the N2 plane (Fig. 9.14b), the velocity profile exhibits a central minimum proximal to the branch point, but gradually approaches a parabolic shape further downstream. Contrary to the results from plane N1, the flow velocity within plane N3 (Fig. 9.14c) is virtually zero near the bifurcating

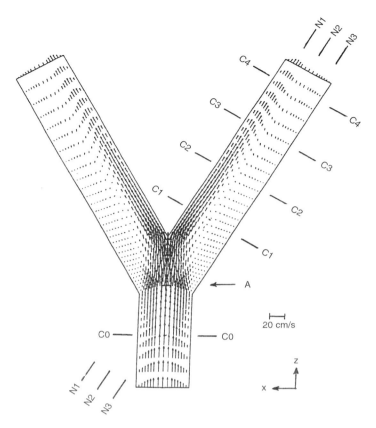

FIGURE 9.12 Numerical velocity vector plot in mid-plane (L1) at Re = 500, showing planes selected for study.

TABLE 9.1 Peak Velocity Measured on the Mid-Plane L1 at the Cross Section Line C0 on the Inlet Vessel and at C1 on the Right Outlet Vessel Using Numerical Simulation (NS) and MRI

Re	Peak at C0 NS	(cm/s) MRI	Peak at C1 NS	(cm/s) MRI	Peak Ratio NS	(%) MRI
250	6.9	7.3	5.0	5.3	72.5	72.6
500	13.8	14.4	12.9	13.5	93.5	93.8
750	20.7	21.8	19.8	20.7	95.7	95.0

position and accelerates only moderately thereafter, as seen from the velocity scale. This is expected, because plane N3 passes through the zone of stasis in the outlet branch. The results also indicate the appearance of a slight central velocity minimum at the most proximal position, with subsequent restoration of a parabolic profile.

Finally, cross-sectional views along the outlet branch of the model at positions C1 through C4 were also examined (Fig. 9.15) at Re = 250. The view orientation in these figures is such that the opposite outlet branch is located to the left. From Fig. 9.15a it can be seen that the radial flow patterns exist as two mirror-image vortices separated by a horizontal plane of symmetry, where the highest radial components of velocity are located near the flow divider wall. Further downstream (Fig. 9.15b–d) the radial flow intensity gradually subsides. Although not shown, similar results were observed at higher Re, with proportional increases in radial flow velocities, and increased compression of the highest radial velocity components along the divider wall. Furthermore, within the inlet vessel (plane C0; results not shown), the numerical results revealed a complete absence of radial motion at any Re.

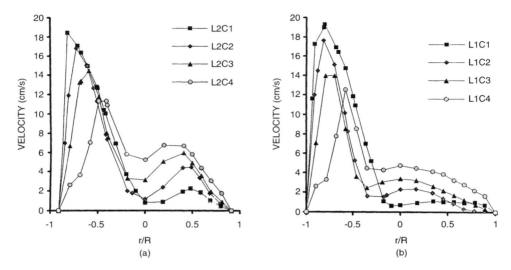

FIGURE 9.13   Numerical velocity profiles at Re = 750 originating at the intersections of planes C1-C4 with the (a) off-center plane (L2) and (b) mid-plane (L1) of the bifurcation.

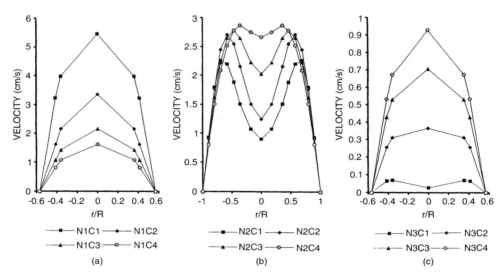

FIGURE 9.14   Numerical velocity profiles at Re = 250 originating at the intersections of planes C1-C4 with the normal planes (a) N1, (b) N2, and (c) N3.

Utility of the Computational Simulation: Evaluating a New Experimental Method or Procedure

The results of the numerical simulations in the above study agree very well with others previously reported[69,76] in terms of both axial and radial flow patterns. In addition to the observation of the flow patterns at different cross-sections, the results of the above simulation were also used to evaluate several MRI bolus-tracking sequences for visualizing the flow field. In these cases, the numerical results served as standard references to compare with the results obtained from the MRI study. The MR images were acquired from an *in vitro* experimental model having the same geometry and flow conditions as those in the numerical simulation. A detailed description of the experimental model and the MRI method have been presented previously.[29] Conceptually, the MRI system used several modified pulse sequences to excite several bands or tags almost simultaneously on a selected slice, which is the image plane.

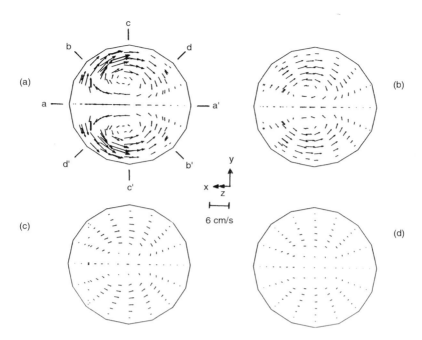

FIGURE 9.15   Numerical radial velocity vector plots at Re = 250 in cross-sectional planes (a) C1, (b) C2, (c) C3, and (d) C4.

After an adjustable time-of-flight (TOF) delay, the position information of the tags was acquired. Upon reconstruction, any flow-induced motion was revealed as a displacement of the tags in the image. This procedure permits the simultaneous observation of axial or radial flow profiles at multiple sites. By altering the orientation and position of the imaging plane, the flow profiles were examined in various planar views through the model bifurcation.

Fig. 9.16 shows the MR image in the mid-plane (L1) of the bifurcation at Re = 500. The flow conditions are the same as those shown in Fig. 9.12. Velocity profiles along horizontal lines at intervals of 0.21 D are displayed in the image. The TOF for this image is 100 ms. The original tagged positions are the parallel dark bands evident in the static fluid outside the bifurcation. Within the flow vessel, each tag has been swept forward by fluid motion over the TOF interval. It is clear that the velocity profiles are laminar and nearly parabolic in the inlet vessel and skew toward the dividing (inner) wall in the outlet branches as can be seen in the numerical results (Fig. 9.12). Very good agreement was observed in the flow pattern from these two figures. Very good agreements were also observed in the flow patterns in MR images (not shown) on the longitudinal planes L2 (off-center) and L1 (mid-plane) at Re 750 (compare to Fig. 9.13) and on the three normal planes (N1–N3) of the rightward outlet branch at Re 250 (compare to Fig. 9.14). Fig. 9.17 shows the MRI velocity profiles in the cross-section planes C1–C4. Radial motion in Fig. 9.17 was investigated using four narrow bands intersecting at the center of the tube; the TOF was 200 ms. These results can be compared with the numerical velocity vector plots shown in Fig. 9.15. For easy comparison, the corresponding tag positions from the MR images are indicated in Fig. 9.15a as aa', bb', cc', and dd'. Again, two symmetrical vortices with velocity distributions in accord with the numerical results are evident, as is the corresponding reduction of vortical motion downstream.

To make quantitative comparisons with the numerical results, the peak velocities at the intersections of plane L1 with planes C0 and C1 as well as the ratio of peak velocities at C1 and C0 (peak ratio) were calculated from the MR images and listed in Table 9.1. The agreement was found to be very good, with a maximum discrepancy of 6% (referenced to the numerical results) in the peak velocities. The discrepancy in peak ratios between the two methods was also very small (less than 1%). The above results demonstrated that numerical simulation can be used to evaluate the feasibility and accuracy of a developed experimental method.

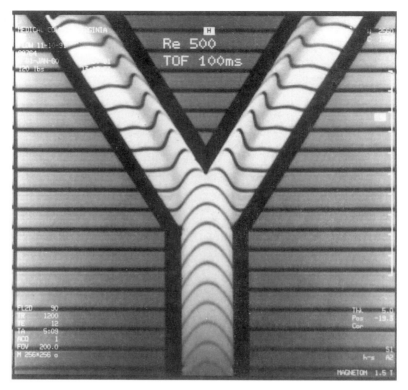

FIGURE 9.16   MR image using time-of-flight (TOF) of 100 ms, showing velocity profiles in mid-plane (L1) at Re = 500.

FIGURE 9.17   MR image using TOF = 200 ms, showing radial velocity profiles at Re = 250 in cross-sectional planes (a) C1, (b) C2, (c) C3, and (d) C4.

Numerical Study of Flow through Stenosed Coronary Arteries: Two-Dimensional Models under Pulsatile Flow Conditions

Many studies have been carried out to observe the flow patterns in stenosed coronary arteries using numerical simulations.[37,46,70,74] The results presented in this section are based on a study from our group which made detailed investigations of the flow patterns through moderate (75% area reduction) and severe (95% area reduction) asymmetric stenoses of the coronary artery. The purpose of the study was to determine if the extreme hemodynamic conditions produced were sufficient to stimulate clotting responses in the blood which would lead to a rapid thrombus formation. The existence of such a sequence of events would provide a better understanding of the underlying initiating conditions as well as insight into possible preventative treatments.

The Models and the Results of Numerical Simulation

Fig. 9.18 shows the two-dimensional mesh of one of the two stenosed arterial models (95% area reduction case) used in the study. Since stenoses are predominantly asymmetric and varied in shape, a Gaussian geometry with arbitrary standard deviation was used along the mid-plane. The height of the Gaussian distribution was suitably adjusted to obtain two different degrees of stenoses: 75% and 95% area reductions.

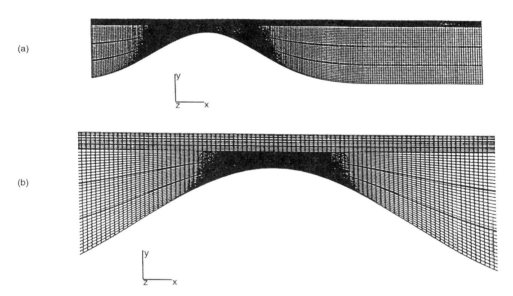

(a)

(b)

FIGURE 9.18   (a) Full two-dimensional mesh of axisymmetric 95% area reduction stenosis. (b) Zoom of stenotic throat region showing fine mesh density near crown of stenosis and disease-free wall.

In order that the simulation approximate actual coronary arterial conditions, the initial and boundary conditions of the flow were matched to those of patients under resting conditions. These were first simulated in a scaled-up *in vitro* experimental model[9] and then the inlet pressure waveform required to generate the desired flow waveform was used as the driving boundary condition for the computational model. The differential pressure waveforms obtained from *in vitro* models are shown in Fig. 9.19. The average and first 10 harmonic terms for each waveform were used for the simulation. The average Reynolds numbers at the inlet of the model and the peak Reynolds numbers at the throat of the stenoses were used as matching parameters. The arterial model simulated was considered to be rigid, and a Newtonian laminar flow was assumed in order to simplify the problem.

Velocity Data

Overall observations of the flow fields for the two degrees of stenoses (Figs. 9.20 and 9.21) showed a strong jetting effect at the stenosis through the majority of the cardiac cycle. Also, there was a recirculation zone downstream, whose length was determined by the jet reattachment length. Velocity fields for the 75% area reduction stenosis are presented in Fig. 9.20 where the velocity plot is color coded with red corresponding to the extreme high velocities and blue corresponding to the extreme low velocities. Velocities in early diastole were small (5 cm/s) and there was no jet formation. The jet velocity accelerated, however, with the up-slope of the input pressure waveform, reaching its maximum (~1 m/s) just after the peak inlet diastolic pressure was achieved (Figs. 9.20a and c). As the jet became stronger in late diastole, the reattachment point moved away from the stenosed wall and elongated the recirculation zone up to 5 diameters. This effect was reversed in systole when the jet weakened, causing shrinkage of the recirculation zone and strong flow reversal.

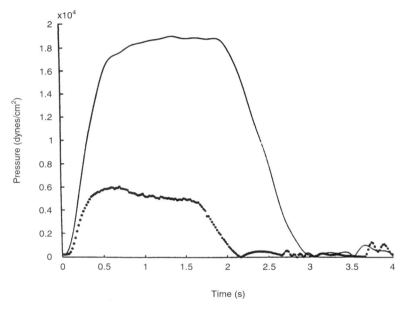

FIGURE 9.19   Differential pressure wave forms obtained from *in vitro* models and employed as input driving conditions for the numerical simulations. Broken line = 75% area reduction; solid line = 95% area reduction.

The velocity field for the 95% area reduction stenosis (Fig. 9.21) showed a trend qualitatively similar to that of the 75% area reduction stenosis. Flow acceleration was higher compared to the 75% case, since the time required to reach peak pressure was approximately the same in both cases and the peak pressure for the 95% case was much higher than that for the 75% case. The greater area reduction for the 95% case caused the jet to become very narrow (Fig. 9.21a) in the throat region compared to the 75% case.

However, the stronger jet created much stronger vortices alongside the jet. As peak flow was attained, the jet reached a maximum velocity of 1.57 m/s with a maximum reattachment length of about 4 diameters. The recirculation zone, just downstream of the stenosis, had a larger height compared to that of the 75% area reduction case and split into three predominant vortices at peak flow conditions. As the diastolic phase ended, the flow decreased and caused strong reversals. This effect decreased as the upstream low pressure front traveled downstream. The velocities then reached their lowest values in late systole (~ 8 cm/s).

Shear Data

The velocity fields described above were used to directly compute shear rates present in the flow field at corresponding times. Figs. 9.20 and 9.21 compare the velocity profile at the time interval of peak flow

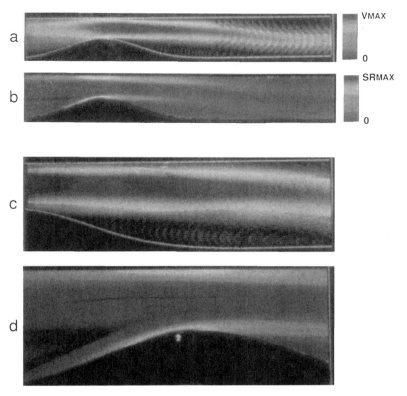

FIGURE 9.20   Solutions for 75% area reduction stenosis at peak flow: (a) velocity field, (b) shear rate field, (c) zoom of velocity field at stenotic throat, and (d) zoom of shear rate field at stenotic throat. Color scale for velocity: $V_{MAX} = 89$ cm/s. Color scale for shear rate: $SR_{MAX} = 488$ 1/s.

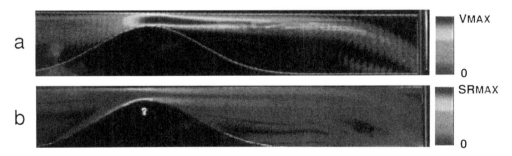

FIGURE 9.21   Solutions for 95% area reduction stenosis at peak flow: (a) velocity field and (b) shear rate field. Color scale for velocity: $V_{MAX} = 139$ cm/s. Color scale for shear rate: $SR_{MAX} = 1630$ 1/s.

with the corresponding shear contours for the 75 and 95% area reduction cases, respectively. The highest shear zones were observed just upstream of the stenosis crown where the jet began separating from the stenotic wall (Figs. 9.20b and d and 9.21b) with the shear rates on the jet boundary near the undiseased wall being nearly as elevated. Other high shear zones were concentrated along the jet boundary and at the moving reattachment point downstream. The elevated shear rates alongside the jet decreased in magnitude with distance downstream. Furthermore, when flow reversal was strongest in late diastole, the shear rate at the reattachment point became comparable to those at the crown of the stenosis. The maximum shear rate for the 75% area reduction case was of the order of 700 1/s (Fig. 9.20b), while the corresponding shear rate for the 95% area reduction stenosis was of the order of 1900 1/s (Fig. 9.21b). A comparison of shear and pressure contours at the peak velocity interval for the 95% area reduction case is shown in Fig. 9.22. It is clear that an extreme pressure buildup (Fig. 9.22b) occurs just upstream of the stenosis with low negative pressures at the crown of stenosis. Thus, the maximum pressure gradient occurs in a region very close to where the highest shear stresses exist, compounding the potential for mechanical disruption of the plaque.

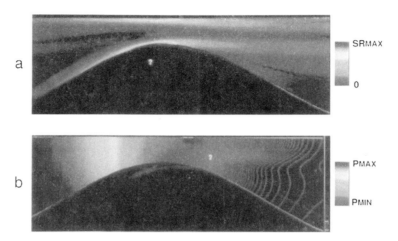

FIGURE 9.22   Solutions for 95% area reduction stenosis at peak flow: (a) shear rate field and (b) pressure field zoomed at stenosis crown. Color scale for shear rate: $SR_{MAX} = 1630$ 1/s. Color code range for pressure: $P_{MIN} = -2850$ dynes/cm²; $P_{MAX} = 8750$ dynes/cm².

Particle Tracking

Particle tracking data was obtained for several particles "seeded" at specific locations using the velocity and pressure data for each time step stored in memory. To evaluate the potential for particles to be stimulated by elevated hemodynamic shear stress,[34] a platelet stimulation function along a given path was defined as

$$PSF = \tau \times \left(T_{\exp} \right)^{0.45} \tag{9.36}$$

where τ is the shear stress and T_{exp} is the exposure time. Fig. 9.23a shows various particle paths through the 75% area reduction stenosis. These particles were chosen to illustrate movement along (1) the disease-free wall; (2) the mid-stream; and (3) the diseased wall and, thus, provide a broad range of PSF values. Particles at similar locations in the 95% area reduction simulation were also observed (Fig. 9.23b). Shear stresses and PSF values were computed along each particle path over a cardiac cycle. Of particular interest were the results for particles labeled #1 in the stenoses simulations (Figure 9.23b) as these were the cases which experienced the maximum shear and PSF values, generally occurring near the crown of the stenosis. A shear history along the path of particle #1 is shown in Fig. 9.24. Here, stresses are seen to gradually increase upon approach to the stenosis and then suddenly rise 1–2 orders of magnitude near the crown, falling sharply upon exit. Similarly, the PSF is seen to abruptly rise from a near-zero level to an extreme value at the stenosis crown before sharply falling to unstenosed values just downstream. At no point, however, did these values reach the minimum necessary to elicit serotonin release from platelets under steady flow conditions.[34]

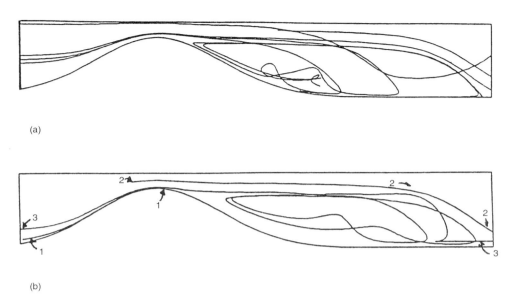

(a)

(b)

FIGURE 9.23   Particle paths through 95% area reduction stenosis: (a) arbitrary particles entering along vessel wall and in mid-stream, and (b) selected particles exhibiting maximum shear stresses (#1), travel outside the separation zone (#2), and complex path within the separation zone (#3).

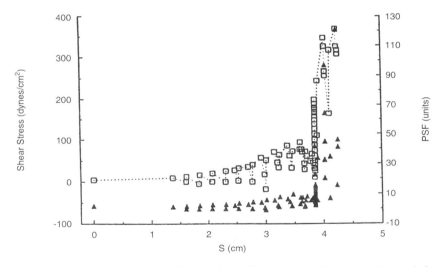

FIGURE 9.24   Shear stress (squares) and platelet stimulation function (triangles) values along path for particle #1 (Fig. 9.23b) from entrance to crown of stenosis.

Utility of the Computational Simulation: Hemodynamic Changes Associated with Disease Severity

The above basic results are in good agreement with other computational[70] and experimental[8,17,48] studies. Numerical simulation of stenosed arterial flow by Tutty[70] showed qualitatively similar features. In particular, the changes in vortex structure, reattachment length, and the location of the flow separation zone are qualitatively in agreement with the above basic results. The variation of shear stress along the diseased wall also showed similar patterns with the peak shear stress occurring at the flow separation point upstream of the stenosis crown. A quantitative comparison is not possible as the input flow conditions and degree of stenoses used were different from the basic simulation.

The time variation of the velocity field throughout the entire cardiac cycle confirms the ultrasonic Doppler Color Flow studies by Boreda et al.[8] The variation of velocity at the inlet and throat of the stenosis was qualitatively similar to the input pressure variation during a cardiac cycle. However, the velocity at the throat of the stenosis reaches its peak at an instant following the time at which the pressure at the inlet is maximum. This time lag was smaller for the 95% area reduction stenosis compared to the 75% area reduction stenosis as the initial acceleration (in early diastole) for the 95% stenotic case was much higher.

The magnitude of shear rates obtained in the basic results are of a comparable order to those obtained by Markou et al.[41] and Sakariassen and Barstad.[61] However, the peak wall shear rates (at a given Reynolds number) obtained in the above results are underestimates of corresponding values presented by Cho and Kensey.[17] This was primarily attributed to the particular geometry of the stenosis used in the present study. The pressure distribution obtained at each time of the simulation agreed closely with the study by Tutty.[70] Cho and Rittgers[79] have shown that the maximum pressure drop is increased at higher Reynolds numbers and that the variation of pressure drop along the axis is decreased at higher Reynolds numbers. These results were in agreement with observations from the results of our study. Even though pressure variations by themselves were not the primary objectives of the study, their validity also implies the correctness of a pressure driven simulation.

An experimental investigation of coronary flow through arteries with similar degrees of stenosis by Boreda et al.[8] showed shear stresses of the same magnitude as the present simulation. This is in agreement with studies by Markou et al.[41] and Sakariassen and Barstad[61] which showed increased platelet deposition with increased degrees of stenoses under controlled biochemical conditions.

The above results demonstrated that numerical simulation is a suitable means for observing hemodynamic changes associated with different disease severity in the coronary artery.

The Effects of Non-Newtonian Flow and Vessel Distensibility

Most of the numerical simulations of blood flow in the large and middle size vessels consider blood as a Newtonian fluid and use rigid vessel walls to simplify the computation. In general, the non-Newtonian effect of flow is small in those vessels because the shear rate is high and the viscosity of fluid can be considered constant. However, low shear exists in some regions such as near bifurcations, graft anastomoses, stenoses, and aneurysms. Blood in those regions appears to have non-Newtonian properties. The effects of a non-Newtonian fluid on flow patterns were studied by several groups.

When considering non-Newtonian effects, the viscosity term in Eq. 9.1 or Eq. 9.6 will not be a constant but rather a function of shear rate. Generally speaking, the Casson model[12] is known to adequately represent blood rheology. For the Casson model, an effective or apparent viscosity is defined[40] as:

$$\gamma' = \frac{\tau}{\gamma} = \frac{\left(\sqrt{\eta}\sqrt{\gamma} + \sqrt{\tau_y}\right)^2}{\gamma} \tag{9.37}$$

where τ is shear stress, γ is shear rate, τ_y is the yield shear stress, and η is the Casson viscosity or limiting apparent viscosity given by

$$\eta = \rho\nu$$

where ρ is the density and ν is the kinematic viscosity. The value of the yield shear stress for blood can be found in the literature.[40]

The Casson equation is valid for blood viscosity over a wide range of cell concentrations and shear rates of 0.1 to 20 1/s[19,43] or 1 to 100,000 1/s.[13,14] At shear rates over 50 1/s[15] or 100 1/s,[35] the blood behaves like a Newtonian fluid.

To approximate the complex rheological behavior of blood in a pulsatile flow model, the apparent viscosity used by Perktold and Rappitch[53] was

$$\mu(\gamma) = \mu_\infty + (\mu_0 - \mu_\infty) \frac{1}{(1 + \lambda_\gamma)^b} \tag{9.38}$$

The parameters in this equation (μ_∞, μ_0, λ_γ, and b) were obtained from experimental data using a curve-fitting procedure.

Lou and Yang[39] studied the non-Newtonian behavior of blood under pulsatile flow conditions at the aortic bifurcation. Their results showed that the non-Newtonian property of blood did not drastically change the flow patterns, but caused an appreciable increase in the shear stresses and a slightly higher resistance to both flow separations and phase shifts between flow layers.

In stenosed arterial flow, the introduction of a non-Newtonian viscosity model has been shown to affect the reattachment length.[17] This is primarily due to the low shear rate present in the jet and recirculation zones. The results by Nakamura and Tadashi[46] showed that the non-Newtonian property of blood reduced the distortion of the flow pattern, as well as the pressure and shear stress at the portion of the wall associated with the stenosis.

Perktold et al. studied the pulsatile non-Newtonian flow characteristics in a bifurcation with an aneurysm[50] and in a three-dimensional carotid bifurcation model.[51,52] Lou and Yang[39] and Perktold and Rappitch[53] reported results on non-Newtonian effects in a bifurcation flow field. All comparisons between Newtonian and non-Newtonian conditions showed no significant changes in general flow patterns and rather minor differences in the basic flow characteristics.

The effects of vessel distensibility have also been studied. In those studies, the vessel walls were modeled as thin-walled vessels. Steinman and Ethier[67] reported results in a distensible end-to-side anastomosis model. The comparison between distensible-walled and rigid-walled simulations showed moderate changes in wall shear stress at isolated locations, primarily the bed, toe, and heel. Lou and Yang[39] performed computer simulations of blood flow at the aortic bifurcation assuming flexible walls. Their results showed that wall expansion reduced flow reversals or eddies during the decelerating systole while wall contraction restricted them during diastole. The shear stresses at a flexible bifurcation were about 10% lower than those at a rigid one. The effect of wall distensibility in a stenosed model was studied by Siebes[63] who showed that the wall motion downstream from a coronary stenosis was greater than that in a normal artery and increased the possibility of post-stenotic dilation. The shear stress was not observed to change dramatically. Results from the carotid bifurcation model[53] showed that flow separation and recirculation decreased slightly in the sinus and increased somewhat in the bifurcation region; the wall shear stress decreased by 25% in the distensible model.

9.4 Summary

Numerical model studies in biomechanical systems encompass very broad and different fields. We have presented a few examples of the application of numerical simulation techniques to the vascular system. Typical flow features in bypass graft end-to-side distal anastomoses, Y-shaped arterial bifurcations, and coronary arterial stenoses have been presented. It has been shown that numerical methods can be used to (1) study the major flow features in the vascular system; (2) observe the relationship between hemodynamic features and vessel disease; and (3) potentially assist in the treatment of some of those diseases. By benefiting from further developments in computer technology and modeling techniques, it is expected that numerical model studies will play an even more important role in cardiovascular biomechanics in the future.

Acknowledgments

The authors thank Dr. Kenneth A. Kraft, Dr. James Thomas, and Dr. Danhua Liu for their work in the collection of the numerical and experimental data presented in this chapter. Thanks also to Niley Mukhjee for his comments about the manuscript.

References

1. Anderson, D.A., Tannehill, J.C., and Pletcher, R.H., *Computational Fluid Mechanics and Heat Transfer*, McGraw-Hill, New York, 1984.
2. Back, M.R., Cho, Y.I., and Back, L.H., Fluid dynamics study in a femoral artery branch casting of man with upstream main lumen curvature for steady flow, *J. Biomechanical Eng.*, 107, 240, 1985.
3. Ballyk, P.D., Steinman, D.A., and Ethier, C.R., Simulation of non-Newtonian blood flow in an end-to-side anastomosis, *Biorheology*, 31, 565, 1994.
4. Bandyk, D.F., Seabrook, G.R., Moldenhauer, P. et al., Hemodynamics of vein graft stenosis, *J. Vasc. Surg.*, 8, 688, 1988.
5. Bassiouny, H.S., Leiber, B.B. et al., Quantitative inverse correlation of wall shear stress with experimental intimal thickening, *Surgical Forum: Cong. Am. Coll. Surg.*, 328, 1988.
6. Bassiouny, H.S., White, S., Glagov, S. et al., Anastomotic intimal hyperplasia: mechanical injury or flow induced, *J. Vasc. Surg.*, 15, 708, 1992.
7. Berguer, R., Higgins, R.F., and Reddy, D.J., Intimal hyperplasia: an experimental study, *Arch. Surg.*, 115, 332, 1980.
8. Boreda, R., Fatemi R.S., and Rittgers, S.E., Potential for platelet stimulation in critically stenosed carotid and coronary arteries, *J. Vasc. Invest.*, 1, 26, 1995.
9. Cao, J. and Rittgers, S.E., Particle motion within *in vitro* models of stenosed internal carotid and left anterior descending coronary arteries, *Ann. Biomed. Eng.*, in press.
10. Caro, C.C., Fitz-Gerald, J.M., and Schroter, R.C., Arterial wall shear and distribution of early atheroma in man, *Nature*, 223, 1159, 1969.
11. Caro, C.G., Fitz-Gerald, J.M., and Schroter, R.C., Atheroma and arterial wall shear observation correlation and proposal of a shear dependent mass transfer mechanism for atherogenesis, *Proc. R. Soc. London B*, 177, 109-159; 1971.
12. Casson, N., A flow equation for the pigment oil suspensions of the printing ink type, in *Rheology of Dispersed Systems*, Mill, D.C., Ed., Pergamon Press, Oxford, 1959, 84.
13. Charm, S.E. and Kurland, G.S., The flow behavior and shear-stress shear rate characteristic of canine blood, *Am. J. Physiol.*, 203, 417, 1962.
14. Charm, S.E. and Kurland, G.S., Viscometery of human blood for shear rates of 0 to 100,000 1/sec, *Nature*, 206, 617, 1965.
15. Charm, S.E. and Kurland, G.S., *Blood Rheology in Cardiovascular Fluid Dynamics*, Vol. 2, Bergel, D.H., Ed., Academic Press, London, 1972, chap. 15.
16. Cho, Y.I., Back, L.H., and Crawford, D.W., Experimental investigation of branch flow ratio, angle, and Reynolds number effects on the pressure and flow fields in arterial branch model, *J. Biomechanical Eng.*, 107, 257, 1985.
17. Sho, Y.I. and Kensey, K.R., Effects of the non-Newtonian viscosity of blood on flows in a diseased arterial vessel. Part I: Steady flows, *Biorheology.* 28, 241, 1991.
18. Chow, C.Y., *An Introduction to Computational Fluid Mechanics*, Seminole Publishing Company, 1983.
19. Cokelet, G.R., Merrill, E.W., Gililand, E.R., Shin, H., Britter, A., and Wells, R.E., Rheology of human blood: measurement near and at zero rate, *Trans. Soc. Rheol.*, 7, 303, 1963.

20. Collins, M.W. and Jones, C.J., Flow studies in canine artery bifurcations using a numerical simulation method, *J. Biomechanical Eng.*, 114, 504, 1992.

21. Crawshaw, H.M., Quist, W.C., Serrallach, E. et al., Flow disturbance at the distal end-to-side anastomosis, *Arch. Surg.*, 115, 1280, 1980.

22. Cuvelier, I.G., Segal, A., and Van Steenhoven, A.A., *Finite Element Methods and Navier-Stokes Equations*, D. Reidel, Dordrecht, 1986.

23. Dintenfass, L., Thixotropy of blood and proneness to thrombus formation, *Circ. Res.*, 11, 233, 1962.

24. Dobrin, P.B., Littoony, F.N., and Endean, E.D., Mechanical factors predisposing to intimal hyperplasia and medial thickening in autogenous vein grafts, *Surgery*, 105, 393, 1989.

25. Einav, S. and Stolero, D., Pulsatile blood flow in an arterial bifurcation: numerical solution, *Med. Biol. Eng. Comput.*, 25, 12, 1987.

26. Fei, D., Kuo, J.T., and Teng, Y.C., Waveform inversion and multi-layer neural network, *J. Comput. Acoust.*, 3, 9, 1995.

27. Fei, D.Y., Kraft, K.A., and Fatouros, P.P., Model studies of nonsteady flow using magnetic resonance imaging, *J. Biomechanical Eng.*, 112, 93, 1990.

28. Fei, D.Y., Thomas, J.D., and Rittgers, S.E., The effect of angle and flow rate upon hemodynamics in distal vascular graft anastomoses: a numerical model study, *J. Biomechanical Eng.*, 116, 331, 1994.

29. Fei, D.Y., Kraft, K.A., Rittgers, S.E., and Liu, D.D., Velocity profiles in a bifurcation model: a correlation study of magnetic resonance imaging and numerical simulation, *J. Vasc. Invest.*, 2, 87, 1996.

30. Friedman, M.H., O'Brien, V., and Ehrlich, L.W., Calculations of pulsatile flow through a branch: implications for the hemodynamics of atherogenesis, *Circ. Res.*, 36, 277, 1975.

31. Friedman, M.H. and Ehrlich, L.W., Numerical simulation of aortic bifurcation flows: the effect of flow divider curvature, *J. Biomechanics*, 17, 881, 1984.

32. Fry, D.L., Acute vascular endothelial changes associated with increased blood velocity gradients, *Circ. Res.*, 22, 165, 1968.

33. Gallagher, R.H., Simon, B.R., Johnson, P.C., and Gross, J.F., *Finite Elements in Biomechanics*, Vol. 3, John Wiley & Sons, New York, 1982.

34. Hellums, J.D., Peterson, D.M., Stathopoulos, N.A., Moake, J.L., and Giorgio, T.D., *Studies on the Mechanisms of Shear Induced Platelet Activation, Cerebral Ischemia and Hemorheology*, Hartmann, A. and Kuschinsky, W., Springer-Verlag, Berlin, 1987, 480.

35. Hussian, A.K., Mechanics of pulsatile flows of relevance to the cardiovascular system, in *Cardiovascular Fluid Dynamics and Measurements*, Hwang, N.H.C. and Normann, N.A., Eds., University Park Press, Baltimore, 1977, chap. 15.

36. Keynton, R.S., Shu, C.S., and Rittgers, S.E., The effect of angle and flow rate upon hemodynamics in distal vascular graft anastomoses: an *in vitro* model study, *J. Biomechanical Eng.*, 113, 458, 1991.

37. Lee, D. and Chiu, J.J., A numerical simulation of intimal thickening under shear in arteries, *Biorheology*, 29, 337, 1992.

38. LoGerfo, F., Quist, W., Nowak, M. et al., Downstream anastomotic hyperplasia: amechanism of failure in dacron grafts, *Ann. Surg.*, 4, 479, 1982.

39. Lou, Z. and Yang, W.J., A computer simulation of the non-Newtonian blood flow at the aortic bifurcation, *J. Biomechanics*, 26, 37, 1993a.

40. Lou, Z. and Yang, W.J., A computer simulation of the blood flow at the aortic bifurcation with flexible walls, *J. Biomechanical Eng.*, 115, 306, 1993b.

41. Markou, G.P. et al., Effects of blocking the platelet GP IB interaction with Von Willebrand factor (VWF)under a range of shearing forces, in BMES Abstracts, Pergamon Press, New York, 1993.

42. McDonald, D.A., *Blood Flow in Arteries*, 2nd ed., Williams & Wilkins, Baltimore, 1974.

43. Merrill, E.W., Cokelet, G.R., Britter, A., and Wells, R.E., Rheology of human blood and the red cell plasma membranes, *Bibl. Anat.*, 4, 51, 1964.

44. Misra, J.C., Patra, M.K., and Misra, S.C., A non-Newtonian fluid model for blood flow through arteries under stenotic conditions, *J. Biomechanics*, 26, 1129, 1993.

45. Moringa, K., Okadome, K., Kuroki, M. et al., Effect of wall shear stress on intimal thickening of arterially transplanted autogenous veins in dogs, *J. Vasc. Surg.*, 2, 430, 1985.
46. Nakamura, M. and Tadashi, S., Numerical study on the flow of a non-Newtonian fluid through axisymetric stenosis, *J. Biomechanical Eng.*, 110, 137, 1988.
47. Nichols, W.W. and O'Rourke, M.F., *McDonald's Blood Flow in Arteries: Theoretic, Experimental and Clinical Principles*, 3rd ed., Lea & Febiger, Philadelphia, 1990, 12.
48. Ojha, M., Ethier, C.R., Johnston, K.W., and Cobbold, R.S.C., Steady and pulsatile flow fields in an end-to-side arterial anastomosis model, *J. Vasc. Surg.*, 12, 747, 1990.
49. Ojha, M., Cobbold, R.S.C., Johnston, K.W., and Hummel, R.L., Pulsatile flow through constricted tubes — An experimental investigation using photochromatic tracer methods, *J. Fluid. Mech.*, 203, 173, 1989.
50. Perktold, K., Peter, R., and Florian, H., Pulsatile non-Newtonian blood flow simulation through a bifurcation with aneurysm, *Biorheology*, 26, 1011, 1989.
51. Perktold, K., Peter, R., and Florian, H., Pulsatile non-Newtonian flow characteristics in a three-dimentional human carotid artery bifurcation, *J. Biomed. Eng.*, 113, 464, 1991a.
52. Perktold, K., Peter, R.O., Resch, M., and Langs, G., Pulsatile non-Newtonian blood flow in three-dimensional carotid bifurcation models: a numerical study of flow phenomena under different bifurcation angles, *J. Biomed. Eng.*, 13, 507, 1991b.
53. Perktold, K. and Rappitsch, G., Computer simulation of local blood flow and vessel mechanics in a compliant carotid artery bifurcation model, *J. Biomechanics*, 28, 845, 1995.
54. Pietrabissa, R., Inzoli, F., and Fumero, R., Simulation study of the fluid dynamics of aorto-coronary bypass, *J. Biomed. Eng.*, 12, 419, 1990.
55. Reddy, J.N., Penalty function analysis of 3D Navier-Stokes equations, *Comp. Meth. Appl. Mech. Eng.*, 35, 87, 1982a.
56. Reddy, J.N., On penalty function methods in finite element analysis of flow problems, *Int. J. Numer. Meth. Fluids*, 2, 151, 1982b.
57. Reddy, J.N., *An Introduction to the Finite Element Method*, 2nd ed., McGraw-Hill, New York, 1993.
58. Rittgers, S.E., Karayannacos, P.E., Guy, J.F., Nerem, R.M., Shaw, G.M., Hostetler, J.R., and Vasko, J.S., Velocity distribution and intimal proliferation in autologous vein grafts in dogs, *Circ. Res.*, 42, 792, 1978.
59. Rittgers, S.E. and Bhambhani, G.H., Doppler color flow images of iliofemoral graft end-to-side distal anastomoses models, *Ultrasound Med. Biol.*, 19, 257, 1993.
60. Rockey, K.C., Evans, H.R., Griffiths, D.W., and Nethercot, D.A., *The Finite Element Method*, John Wiley & Sons, New York, 1975.
61. Sakariassen, K.S. and Barstad, R.M., Model systems of thrombus formation in flowing human blood at the apex of eccentric stenoses, in *BMES Abstracts*, Pergamon Press, New York, 1993.
62. Shaw, C.T., *Using Computational Fluid Dynamics*, Prentice Hall, New York, 1992.
63. Siebes, M., Effect of nonlinear wall mechanics on compliant coronary stenoses: a flow simulation study, *Adv. Bioeng.*, 113, 330, 1991.
64. Sottiurai, V.S., Yao, J.S.T., Batson, R.C. et al., Distal anastomotic intimal hyperplasia: histopathologic character and biogenesis, *Ann. Vasc. Surg.*, 1, 26, 1989.
65. Steinman, D.A., Ethier, C.R., and Johnston, K.W., Further studies on flow in an end-to-side anastomosis, *Adv. Bioeng.*, 20, 383, 1991.
66. Steinman, D.A., Vinh, B., Ethier, C.R., Ojha, M., Cobbold, R.S., and Johnston, K.W., A numerical simulation of flow in a two-dimensional end-to-side anastomosis model, *J. Biomechanical Eng.*, 115, 112, 1993.
67. Steinman, D.A. and Ethier, C.R., The effect of wall distensibility on flow in a two-dimensional end-to-side anastomosis, *J. Biomechanical Eng.*, 116, 294, 1994.
68. Svindland, A. and Walloe, L., Distribution pattern of sudanophilic plaques in the descending thoracic and proximal abdominal human aorta, *Atherosclerosis*, 57, 219, 1985.

69. Thiriet, M., Pares, C., Saltel, E., and Hecht, F., Numerical simulation of steady flow in a model of the aortic bifurcation, *J. Biomechanical Eng.*, 114, 40, 1992.

70. Tutty, O.R., Pulsatile flow in a constricted channel, *J. Biomechanical Eng.*, 114, 50, 1992.

71. Van de Vosse, F.N., Segal, A., Van Steenhoven, A.A., and Janssen, J.D., A finite element approximation of the unsteady two-dimensional Navier-Stokes equations, *Int. J. Numerical Methods Fluids Eng.*, 6, 427, 1986.

72. Wille, S.O., Numerical simulations of steady flow inside a three dimensional aortic bifurcation model, *J. Biomed. Eng.*, 6, 49, 1984.

73. Womersley, J.R., Method for the calculation of velocity, rate of flow and viscous drag in arteries when the pressure gradient is known, *J. Physiol.*, 127, 553, 1955.

74. Wong Paul, C.K. et. al., Computer simulation of blood flow patterns in arteries of various geometries, *J. Vasc. Surg.*, 14, 658, 1991.

75. Yeager, R.A., Hobson, R.W., Lynch, T.G. et al., Analysis of factors influencing patency of polytetrafluoroethylene prostheses for limb salvage, *J. Surg. Res.*, 32, 499, 1982.

76. Yung, C.N., De Witt, K.J., and Keith, T.G., Jr., Three-dimensional steady flow through a bifurcation, *J. Biomechanical Eng.*, 112, 189, 1990.

77. Zarins, C.K., Giddens, D.P., Bharadvaj, B.K., Sottiurai, V.S., Mabon, R.F., and Glagov, S., Carotid bifurcation atherosclerosis: quantitative correlation of plaque localization with flow velocity profile and wall shear stress, *Circ. Res.*, 5, 293, 1983.

78. Prendiville et al., *J. Vasc. Surg.*, 1, 1517, 1990.

Appendix A. Derivation of the Finite Element Formulation

Using the Galerkin method, we choose $\Psi_n = N_n$ for Eq. 9.20 in the text. The following Galerkin formulae can be obtained:

$$\in_l = \int_\Omega \in_l N_n d\Omega = 0$$

That is

$$\in_l = \int_\Omega L\left(\sum_{k=1}^m \bar{u}_k N_k, \sum_{k=1}^m p_k N_k\right) N_n d\Omega = 0$$

$$\in_l = \int_\Omega \frac{4\alpha^2}{R_e} \frac{\partial\left(\sum_{k=1}^m \bar{u}_k N_k\right)}{\partial t} N_n d\Omega + \int_\Omega \left(\left(\sum_{k=1}^m \bar{u}_k N_k\right)\cdot\nabla\right)\left(\sum_{k=1}^m \bar{u}_k N_k\right) N_n d\Omega$$

$$+ \int_\Omega \nabla\left(\sum_{k=1}^m p_k N_k\right) N_n d\Omega - \int_\Omega \frac{1}{R_e}\nabla^2\left(\sum_{k=1}^m \bar{u}_k N_k\right) N_n d\Omega = 0.$$

We define

$$L_1 = \int_\Omega \frac{4\alpha^2}{R_e} \frac{\partial \left(\sum_{k=1}^m \overline{u}_k N_k \right)}{\partial t} N_n d\Omega$$

$$L_2 = \int_\Omega \left(\left(\sum_{k=1}^m \overline{u}_k N_k \right) \cdot \nabla \right) \left(\sum_{k=1}^m \overline{u}_k N_k \right) N_n d\Omega$$

$$L_3 = \int_\Omega \nabla \left(\sum_{k=1}^m p_k N_k \right) N_n d\Omega$$

$$L_4 = \int_\Omega \frac{1}{R_e} \nabla^2 \left(\sum_{k=1}^m \overline{u}_k N_k \right) N_n d\Omega$$

such that

$$\in_l = L_1 + L_2 + L_3 - L_4 = 0 \qquad (A1)$$

Because $\overline{u}_k$ are constants for the calculation at a fixed time, when we derive L_1, the term $\partial \overline{u}_k / \partial t$ can be taken out of the integration and

$$L_1 = \sum_{k=1}^m \frac{\partial \overline{u}_k}{\partial t} \int_\Omega \frac{4\alpha^2}{R_e} N_k N_n d\Omega = \sum_{k=1}^m \frac{\partial \overline{u}_k}{\partial t} D_{kn} \qquad (A2)$$

where

$$D_{kn} = \int_\Omega \frac{4\alpha^2}{R_e} N_k N_n d\Omega \qquad (A3)$$

This can be derived as

$$L_2 = \int_\Omega \left(\left(\sum_{k=1}^m \overline{u}_k N_k \right) \cdot \nabla \right) \left(\sum_{k=1}^m \overline{u}_k N_k \right) N_n d\Omega$$

$$= \int_\Omega \left(\overline{u} \cdot \nabla \right) \overline{u} N_n d\Omega$$

$$= \int_\Omega \left\{ \left[\left(u \frac{\partial u}{\partial x} + v \frac{\partial u}{\partial y} + w \frac{\partial u}{\partial z} \right) N_n \right] \overline{i} + \left[\left(u \frac{\partial v}{\partial x} + v \frac{\partial v}{\partial y} + w \frac{\partial v}{\partial z} \right) N_n \right] \overline{j} + \left[\left(u \frac{\partial w}{\partial x} + v \frac{\partial w}{\partial y} + w \frac{\partial w}{\partial z} \right) N_n \right] \overline{k} \right\} d\Omega$$

$$= \int_\Omega \left\{ \left[\left(\left(\sum_{k=1}^m u_k N_k \right) \frac{\partial \left(\sum_{k=1}^m u_k N_k \right)}{\partial x} + \left(\sum_{k=1}^m v_k N_k \right) \frac{\partial \left(\sum_{k=1}^m u_k N_k \right)}{\partial y} + \left(\sum_{k=1}^m w_k N_k \right) \frac{\partial \left(\sum_{k=1}^m u_k N_k \right)}{\partial z} \right) N_n \right] \overline{i} \right.$$

$$+ [..] \overline{j} + [..] \overline{k} \right\} d\Omega$$

$$= \sum_{k=1}^{m} u_k \left(u_k + v_k + w_k \right) \int_{\Omega} \left[\left(\frac{\partial \left(N_k \right)}{\partial x} + \frac{\partial \left(N_k \right)}{\partial y} + \frac{\partial \left(N_k \right)}{\partial z} \right) N_k N_n \right] \bar{i} \, d\Omega$$

$$+ \sum_{k=1}^{m} v_k \left(u_k + v_k + w_k \right) \int_{\Omega} \left[\left(\frac{\partial \left(N_k \right)}{\partial x} + \frac{\partial \left(N_k \right)}{\partial y} + \frac{\partial \left(N_k \right)}{\partial z} \right) N_k N_n \right] \bar{j} \, d\Omega$$

$$+ \sum_{k=1}^{m} w_k \left(u_k + v_k + w_k \right) \int_{\Omega} \left[\left(\frac{\partial \left(N_k \right)}{\partial x} + \frac{\partial \left(N_k \right)}{\partial y} + \frac{\partial \left(N_k \right)}{\partial z} \right) N_k N_n \right] \bar{k} \, d\Omega$$

$$= \sum_{k=1}^{m} \bar{u}_k \left(u_k + v_k + w_k \right) \int_{\Omega} \left[\left(\frac{\partial \left(N_k \right)}{\partial x} + \frac{\partial \left(N_k \right)}{\partial y} + \frac{\partial \left(N_k \right)}{\partial z} \right) N_k N_n \right] d\Omega$$

This can be expressed as

$$L_2 = \sum_{k=1}^{m} \bar{u}_k \left(u_k + v_k + w_k \right) E_{kn} \tag{A4}$$

where

$$E_{kn} = \int_{\Omega} \left[\left(\frac{\partial \left(N_k \right)}{\partial x} + \frac{\partial \left(N_k \right)}{\partial y} + \frac{\partial \left(N_k \right)}{\partial z} \right) N_k N_n \right] d\Omega \tag{A5}$$

L_3 can be derived as

$$L_3 \int_{\Omega} \nabla \left(\sum_{k=1}^{m} p_k N_k \right) N_n d\Omega = \sum_{k=1}^{m} p_k \int_{\Omega} \nabla \left(N_k \right) N_n d\Omega = \sum_{k=1}^{m} p_k H_{kn} \tag{A6}$$

where

$$H_{kn} = \int_{\Omega} \nabla \left(N_k \right) N_n d\Omega \tag{A7}$$

L_4 can be derived as

$$L_4 = \int_{\Omega} \frac{1}{R_e} \nabla^2 \left(\sum_{k=1}^{m} \bar{u}_k N_k \right) N_n d\Omega$$

$$= \int_{\Omega} \frac{1}{R_e} \nabla^2 \bar{u} N_n d\Omega$$

$$= \int_{\Omega} \frac{1}{R_e} \left(\frac{\partial^2 \bar{u}}{\partial x^2} + \frac{\partial^2 \bar{u}}{\partial y^2} + \frac{\partial^2 \bar{u}}{\partial z^2} \right) N_n d\Omega$$

In order to simplify the computation, we can use the following formulae from vector analysis. That is[26]

$$\int_\Omega a\nabla^2\varphi d\Omega = \int_s a\varphi d\bar{s} - \int_\Omega \nabla\varphi \cdot \nabla a d\Omega$$

where $\bar{s}$ is the outward normal unit vector. Therefore

$$L_4 = \int_s \frac{N_n}{R_e}\bar{u}d\bar{s} - \int_\Omega \frac{1}{R_e}\left(\frac{\partial\bar{u}}{\partial x}\frac{\partial N_n}{\partial x} + \frac{\partial\bar{u}}{\partial y}\frac{\partial N_n}{\partial y} + \frac{\partial\bar{u}}{\partial z}\frac{\partial N_n}{\partial z}\right)d\Omega$$

$$= \int_s \frac{N_n}{R_e}\left(\sum_{k=1}^m \bar{u}_k N_k\right)d\bar{s} - \int_\Omega \frac{1}{R_e}\left(\frac{\partial\left(\sum_{k=1}^m \bar{u}_k N_k\right)}{\partial x}\frac{\partial N_n}{\partial x} + \frac{\partial\left(\sum_{k=1}^m \bar{u}_k N_k\right)}{\partial y}\frac{\partial N_n}{\partial y} + \frac{\partial\left(\sum_{k=1}^m \bar{u}_k N_k\right)}{\partial z}\frac{\partial N_n}{\partial z}\right)d\Omega$$

$$= \sum_{k=1}^m \bar{u}_k \int_s \frac{N_n}{R_e}N_k d\bar{s} - \sum_{k=1}^m \bar{u}_k \int_\Omega \frac{1}{R_e}\left(\frac{\partial\left(N_k\right)}{\partial x}\frac{\partial N_n}{\partial x} + \frac{\partial\left(N_k\right)}{\partial y}\frac{\partial N_n}{\partial y} + \frac{\partial\left(N_k\right)}{\partial z}\frac{\partial N_n}{\partial z}\right)d\Omega$$

This can be written as

$$L_4 = \sum_{k=1}^m \bar{u}_k F_{kn} - \sum_{k=1}^m \bar{u}_k G_{kn} \tag{A8}$$

where

$$F_{kn} = \int_s \frac{N_n}{R_e}N_k d\bar{s} \tag{A9}$$

$$G_{kn} = \int_\Omega \frac{1}{R_e}\left(\frac{\partial\left(N_k\right)}{\partial x}\frac{\partial N_n}{\partial x} + \frac{\partial\left(N_k\right)}{\partial y}\frac{\partial N_n}{\partial y} + \frac{\partial\left(N_k\right)}{\partial z}\frac{\partial N_n}{\partial z}\right)d\Omega \tag{A10}$$

The results can be summarized as

$$\sum_{k=1}^m \frac{\partial\bar{u}_k}{\partial t}D_{kn} + \sum_{k=1}^m \bar{u}_k\left(u_k + v_k + w_k\right)E_{kn} + \sum_{k=1}^m p_k H_{kn} - \sum_{k=1}^m \bar{u}_k F_{kn} + \sum_{k=1}^m \bar{u}_k G_{kn} = 0 \tag{A11}$$

Considering n from 1 to m, we obtain three sets of m^2 matrix equations for the element:

$$\sum_{n=1}^m\left(\sum_{k=1}^m \frac{\partial\bar{u}_k}{\partial t}D_{kn} + \sum_{k=1}^m \bar{u}_k\left(u_k + v_k + w_k\right)E_{kn} + \sum_{k=1}^m p_k H_{kn} - \sum_{k=1}^m \bar{u}_k F_{kn} + \sum_{k=1}^m \bar{u}_k G_{kn}\right) = 0 \tag{A12}$$

Using the similar derivation for error ϵ_w (Eq. 9.19 in the text), we have

$$\epsilon_w = \int_\Omega \left[\frac{\partial}{\partial x}\left(\sum_{k=1}^m u_k N_k\right)N_n + \frac{\partial}{\partial y}\left(\sum_{k=1}^m v_k N_k\right)N_n + \frac{\partial}{\partial z}\left(\sum_{k=1}^m w_k N_k\right)N_n \right]d\Omega = 0$$

$$\sum_{k=1}^m\left[u_k \int_\Omega \left(\frac{\partial(N_k)}{\partial x}\right)N_n d\Omega + v_k\int_\Omega \frac{\partial(N_k)}{\partial y}N_n d\Omega + w_k\int_\Omega \frac{\partial(N_k)}{\partial z}N_n d\Omega \right] = 0$$

These equations can be expressed in a vector form as

$$\sum_{k=1}^m \overline{u_k}\cdot\overline{I_{kn}} = 0 \qquad\qquad (A13)$$

where

$$\overline{I_{kn}} = \bar{i}\int_\Omega\left(\frac{\partial}{\partial x}(N_k)\right)N_n d\Omega + \bar{j}\int_\Omega\frac{\partial}{\partial y}(N_k)N_n d\Omega + \bar{k}\int_\Omega\frac{\partial}{\partial z}(N_k)N_n d\Omega \qquad (A14)$$

Considering n from 1 to m, we obtain the fourth set of m^2 matrix equations for the element:

$$\sum_{n=1}^m\sum_{k=1}^m \overline{u_k}\cdot\overline{I_{kn}} = 0 \qquad\qquad (A15)$$

Appendix B. Derivation of the Shape Function for a Triangle Element

Assume $U(x,y)$ varies within the triangle element shown in Fig. 9.1b as

$$U(x,y) = a + bx + cy$$

then,

$$U_1 = a + bx_1 + cy_1$$
$$U_2 = a + bx_2 + cy_2$$
$$U_3 = a + bx_3 + cy_3$$

The solution for this system equation provides a, b, c as a function of x_1, x_2, and x_3, y_1, y_2, and y_3, and U_1, U_2, and U_3. Using Cramer's rule:

$$a = \frac{1}{D} \det \begin{vmatrix} U_1 & x_1 & y_1 \\ U_2 & x_2 & y_2 \\ U_3 & x_3 & y_3 \end{vmatrix}$$

$$b = \frac{1}{D} \det \begin{vmatrix} 1 & U_1 & y_1 \\ 1 & U_2 & y_2 \\ 1 & U_3 & y_3 \end{vmatrix}$$

$$c = \frac{1}{D} \det \begin{vmatrix} 1 & x_1 & U_1 \\ 1 & x_2 & U_2 \\ 1 & x_3 & U_3 \end{vmatrix}$$

$$D = \begin{vmatrix} 1 & x_1 & y_1 \\ 1 & x_2 & y_2 \\ 1 & x_3 & y_3 \end{vmatrix}$$

After simple algebra derivation, we obtain

$$U(x,y) = \sum_{l=1,2,3} \left(a_l + b_l + c_l \right) U_l$$

where the coefficients of the linear equation are

$$a_1 = x_2 y_3 - x_3 y_2$$
$$b_1 = y_2 - y_3$$
$$c_1 = x_3 - x_2$$

and the terms a_2, b_2, and c_2 and a_3, b_3, and c_3 can be derived from the cyclic permutations of indices. Let A = the area of the element, then

$$A = \frac{1}{2} D$$

$$N(x,y)_1 = \frac{1}{2A} \left(a_1 + b_1 x + c_1 y \right)$$

$$N(x,y)_2 = \frac{1}{2A} \left(a_2 + b_2 x + c_2 y \right)$$

$$N(x,y)_3 = \frac{1}{2A} \left(a_3 + b_3 x + c_3 y \right)$$

The derivatives of the shape function are

$$\frac{\partial N_1}{\partial x} = \frac{b_1}{2A}$$

$$\frac{\partial N_2}{\partial x} = \frac{b_2}{2A}$$

$$\frac{\partial N_3}{\partial x} = \frac{b_3}{2A}$$

and

$$\frac{\partial N_1}{\partial y} = \frac{c_1}{2A}$$

$$\frac{\partial N_2}{\partial y} = \frac{c_2}{2A}$$

$$\frac{\partial N_3}{\partial y} = \frac{c_3}{2A}$$

The element matrices for the linear element can be easily derived using formulae from a math handbook; that is,

$$\int_A \left(N_1\right)^a \left(N_2\right)^b \left(N_3\right)^c dA = \frac{a!b!c!}{(a+b+c+2)!} 2A$$

where a, b, and c are the exponents of the shape function N_1, N_2, N_3. For example,

$$\int_A N_1 N_2 dA = \frac{1!1!0!}{(1+1+0+2)} 2A = \frac{A}{2}$$

and

$$\int_A N_1 \frac{\partial N_1}{\partial x} dA = \int_A N_1 \frac{b_1}{2A} dA = \frac{b_1}{2A} \int_A N_1 dA = \frac{b_1}{3}$$

Appendix C. Shape Functions for a Square and Cubic Element in Local Coordinates, and the Transformation between Local and Global Coordinates

For a square element in local coordinates shown in Fig. 9.2a, the derivation of the shape function and its derivatives are very similar to the procedure described in Appendix B. The results are

$$N_i = \frac{1}{4}\left(1 + \epsilon_i \epsilon\right)\left(1 + \eta_i \eta\right)$$

$$\frac{\partial N_i}{\partial \epsilon} = \frac{1}{4}\epsilon_i\left(1 + \eta_i \eta\right)$$

$$\frac{\partial N_i}{\partial \eta} = \frac{1}{4}\eta_i\left(1 + \epsilon_i \epsilon\right)$$

where i = 1, 2, 3, 4; ε_i = −1, 1, 1, −1; and η_i = −1, −1, 1, 1.

For a cubic element (Fig. 9.3), the shape function and its derivatives are

$$N_i = \frac{1}{8}\left(1 + \epsilon_i \epsilon\right)\left(1 + \eta_i \eta\right)\left(1 + \zeta_i \zeta\right)$$

$$\frac{\partial N_i}{\partial \epsilon} = \frac{1}{8}\epsilon_i\left(1 + \eta_i \eta\right)\left(1 + \zeta_i \zeta\right)$$

$$\frac{\partial N_i}{\partial \eta} = \frac{1}{8}\eta_i\left(1 + \epsilon_i \epsilon\right)\left(1 + \zeta_i \zeta\right)$$

$$\frac{\partial N_i}{\partial \zeta} = \frac{1}{8}\zeta_i\left(1 + \epsilon_i \epsilon\right)\left(1 + \eta_i \eta\right)$$

where $i = 1, 2, 3, 4, 5, 6, 7, 8$; $\epsilon_i = -1, 1, 1, -1, -1, 1, 1, -1$; $\eta_i = -1, -1, 1, 1, -1, -1, 1, 1$; and $\zeta_i = -1, -1, -1, -1, 1, 1, 1, 1$.

The relationship between two-dimensional and global coordinates is described as follows. Suppose that the global coordinate and the local coordinate have a transformation

$$x = x\left(\epsilon, \eta\right)$$
$$y = y\left(\epsilon, \eta\right)$$

We have

$$\begin{vmatrix} \dfrac{\partial W}{\partial \epsilon} \\ \dfrac{\partial W}{\partial \eta} \end{vmatrix} = J \begin{vmatrix} \dfrac{\partial W}{\partial x} \\ \dfrac{\partial W}{\partial y} \end{vmatrix}$$

where W is an arbitrary variable and J is Jacobi's matrix as

$$\det J = \begin{vmatrix} \dfrac{\partial x}{\partial \epsilon} & \dfrac{\partial y}{\partial \epsilon} \\ \dfrac{\partial x}{\partial \eta} & \dfrac{\partial y}{\partial \eta} \end{vmatrix}$$

$$\begin{vmatrix} \dfrac{\partial W}{\partial x} \\ \dfrac{\partial W}{\partial y} \end{vmatrix} = J^{-1} \begin{vmatrix} \dfrac{\partial W}{\partial \epsilon} \\ \dfrac{\partial W}{\partial \eta} \end{vmatrix},$$

where

$$J^{-1} = \frac{1}{|J|} = \begin{vmatrix} \dfrac{\partial y}{\partial \eta} & -\dfrac{\partial y}{\partial \epsilon} \\ -\dfrac{\partial x}{\partial \eta} & \dfrac{\partial x}{\partial \epsilon} \end{vmatrix}$$

and

$$\iint f(x,y)dxdy = \iint f\big(x(\in,\eta), y(\in,\eta)\big)\|J|d\in d\eta.$$

These formulas provide the transformation between local and global coordinates.

10

Flow in Thin-Walled Collapsible Tubes

10.1 Introduction..10-1
      Context • Concepts
10.2 A Cursory Literature Review..10-3
      The Starling Resistor • The Pressure–Area Relationship • Flow
      Experiments • Forced Flow Experiments
10.3 Tube Law 10-6
      Unstressed Configuration Dependent Collapse • Description of
      the Tube Law • The Opposite Wall-Edge Contact •
      Normalization • Normalized Similarity Law • Analytical
      Expressions of the Tube Law • Positive Transmural Pressure
10.4 Tube Law Dependent Data...10-14
      Wave Speed and Critical Flow Rate • The Reynolds Number •
      Head Losses and Shape Factor • Wall Shear Stresses in
      Uniformly Collapsed Tubes • Kinetic Energy and Momentum
      Coefficients
10.5 The One-Dimensional Model 10-23
      Similarity Conditions
10.6 Collapsed Tube Flow Experiments............................10-27
      Steady Flow Experiments • Unsteady Flow Experiments
10.7 Physiological Applications...10-36
10.8 Conclusion 10-40

M. Thiriet
INRCA

S. Naili
University of Paris

A. Langlet
L'Universite d'Orleans-Bourges

C. Ribreau
LGMPB-IUT

10.1 Introduction

Context

The importance of collapsible tube flow emerges from cardiovascular, pulmonary, and urinary physiological or pathophysiological problems and related diagnostic (e.g., forced expiration in lung function testing) or therapeutic (e.g., external cuff) techniques. Studies were aimed at understanding the flow behavior in the deformable conduit.[1-3] Physiological fluids (blood, air, and urine) are indeed conveyed in compliant vessels which undergo deformations under varying transmural pressures p, defined as the difference between the internal and the external pressures $p = p_i - p_e$.

Arteries dilate while pressure waves travel through the vascular bed. The cross-sectional luminal area A_i varies whereas p evolves in a range of positive values. The cross-sectional shape may be affected due to the non-uniform distribution of p over the entire tube perimeter and to the neighboring structures. Veins, respiratory conduits, and urethrae may experience changes both in cross-sectional area and shape when they are subjected to negative transmural pressures during natural or functional testing maneuvers.

0-8493-9049-4/01/$0.00+$.50
© 2001 by CRC Press LLC

The collapsing process is characterized by large variations in A_i associated with small variations of p when p is slightly negative.

The collapsing process of a straight flexible pipe is illustrated in Fig. 10.1. The tube loading is such that contact occurs between the opposite walls. This phenomenon is emphasized throughout this chapter.

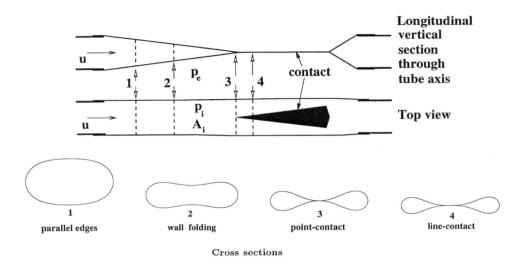

FIGURE 10.1 Stationary collapse of a horizontal straight flexible tube attached at both ends to rigid circular ducts. The thin-walled tube has an unstressed elliptical cross section. The cross-sectional area and shape depend on the transmural pressure distribution, i.e., from the flow-dependent internal pressure field when the external pressure is uniform. Contact occurs between opposite walls; the tube lumen is then composed of two lobes in the present example. Top: pipe axial configuration in the vertical centerplane. Mid: upper tube face with the contact region (dark area). Bottom: cross-sectional shape, before and after contact, are given for some labeled stations (1 to 4).

Concepts

The dynamics of collapsible pipes depend upon the coupling between the fluid and the flexible wall via the non-linear relationship, the tube law, which relates p to A_i. The tube law is described in section 10.3, after a brief literature review (section 10.2).

The pressure waves of small amplitudes propagate with a speed c, which depends on the fluid inertia and the wall compliance (transverse propagation mode). This physical quantity can thus be derived from the tube law. The critical flow-rate q^* (superscript $*$ stands for critical) is related to the luminal area by $q^* = A_i c$. For a given specific energy $p + \rho U^2/2$ in the cross-section (where U is the cross-sectional average fluid velocity), it is the optimal flux of a fluid that can be conveyed by the local adaptation of the tube cross-section.[4] The relationship between q^* and p is given in section 10.3.

The hydraulic quantities of interest, in particular head loss and wall shear stress, are determined in specific configurations corresponding to a uniform longitudinal cross-section along the entire tube length ($dA_i/dx = 0$, $\forall x$; x: axial direction, see section 10.4). These configurations may be demonstrated by flows with combined gravity and friction for particular angles of tube inclination at given transmural pressures.[5,6]

The one-dimensional study of collapsible tube flow has been developed by different research groups. The transmural pressure is supposed to be uniformly distributed in every tube section and transversal bending effects are assumed to be predominant. The one-dimensional model takes into account the wave propagation; it shows, thus, the possible occurrence of critical conditions. A significant dimensionless parameter is indeed associated with this theory:[1] the speed index $S = U/c$, which plays the role of the Froude number in open-channel flows or the Mach number in isentropic gas flow, due to resemblances between these three types of flows. For slightly negative transmural pressures, the tube collapse induces a fluid acceleration while the high compliance entails a low wave speed; the flow may thus become critical

($S = 1$). In other words, critical conditions are reached when the volume flow rate q becomes equal to the critical flow rate q^*. When $q = q^*$, the local specific energy reaches a minimum. The cross-section where $S = 1$ is the critical section. In the standard steady one-dimensional model, the occurrence of a critical condition may predict flow limitation (the outlet pressure information cannot migrate upstream from the critical section). Possible transitions from subcritical to supercritical flows (or conversely) may occur. In particular, a transition to a supercritical flow may be induced by a forced tube throat (the narrowest cross section).[1] An elastic jump, where $S = 1$, may appear between the supercritical and subcritical flow segments, where locally $dA_i/dx \to \infty$; the elastic jump is characterized by a strong flow deceleration.[7–9] Choking may arise when the longitudinal space gradient of the cross-sectional area $dA_i/dx \to -\infty$.

Collapsible duct flows exhibit several physical phenomena of interest, such as the occurrence of (1) critical flows ($S = 1$ in a critical section); (2) transcritical flows from subcritical ($S < 1$) to supercritical ($S > 1$) flows and conversely; (3) transitions between laminar and turbulent flow regimes; and (4) oscillations of various modes under either previously steady or unsteady conditions (forced flow or forced wall motions).

The critical phenomenon is evoked as one of the possible causes for the breakdown of steadiness. Other mechanical factors may induce flow unsteadiness, like the flow separation in the divergent downstream from the throat of the collapsed tube.[2,10–12]

The classical equations of fluid dynamics are provided in two different applications of the one-dimensional model (section 10.5). Typical experiments on flow through collapsed tubes are given in section 10.6.

10.2   A Cursory Literature Review

Some early theoretical models were based on the assumption of a continuous variation in cross-sectional area without any change in circular shape, the effective cross-sectional area being equal to the collapsed tube at the same pressure.[13,14] The mechanical features of collapsible tubes as related to the contact between opposite walls were omitted. Both tube shape and cross-sectional area are actually quite significant. The lumped parameter model is based on the assumption that the geometry of the flexible pipe is specified by the cross-sectional area A_i at the throat of the collapsed tube, which depends on the throat transmural pressure.[2,10,15] Wave propagation and its related phenomena are neglected.

In the most developed one-dimensional models, axial wall bending is taken into account.[11,12,16,17] In the one-dimensional theory, the homogeneous incompressible fluid is supposed either inviscid or viscous, with or without spatial distribution of wall stiffness and the unstressed cross-sectional area, whether body forces exist or not.

Theoretical approaches have been, until recently, solely devoted to limited dynamical aspects due to the complex behavior of the compliant tube-fluid couple. Different classes of models may be defined: (1) models describing the steady flow before the occurrence of any steadiness breakdown; (2) models aimed at stability analysis; and (3) models intended for oscillation analysis. Improved numerical simulations are now emerging. For instance, the oscillation modes have been shown to be strongly affected by the motion of the separation point behind the downstream throat.[18] A three-dimensional tube deformation has been very recently investigated at a very low Reynolds number.[19]

Most of the above-mentioned collapsible tube flow properties arise from experimental observations. The earliest experiments were performed on a very simple test section, the so-called Starling resistor.[20–24]

The Starling Resistor

The Starling resistor is composed of a thin-walled compliant straight pipe of a given length L, mounted on rigid tubings at its ends and enclosed in a rigid transparent chamber (Fig. 10.2). The chamber pressure p_e surrounding the flexible pipe can be adjusted. The upstream end is connected to a constant-head supply reservoir via an upstream valve. The downstream end drains the fluid out of the test segment via

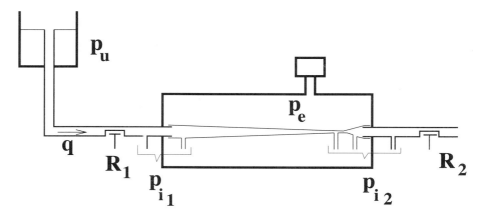

FIGURE 10.2 The Starling resistance. Straight thin-walled flexible pipe mounted on rigid tubes at its ends and enclosed in a rigid transparent chamber. The external pressure p_e is adjustable. The flow rate q is caused by the upstream pressure p_u. The internal pressures p_i are measured at two stations: (1) either in the downstream segment of the entry rigid tube or in the entrance region of the floppy tube and (2) either more or less upstream from the compliant tube outlet or downstream from it. The results are strongly affected by the location of the pressure measurement sites. Local adjustable constrictions are sometimes added to the flow circuit upstream R_1 and downstream R_2 from the compliant tube.

a possible downstream adjustable constriction. Both the inlet p_{i1} and the outlet p_{i2} internal pressures can be set at fixed values.

Among the input data to be measured prior to any investigation, some can be defined as reference quantities, because they either affect the tube mechanics or are scaling factors. The tube deformation depends strongly on the tube ellipticity k_0.[25] The ellipticity k_0[†] is the ratio between the major semi-axis a_0 and the minor semi-axis of the elliptical mid-line in the unstressed (subscript 0) tube configuration (obviously, k_0 is equal to unity when the unstressed cross-section is circular). The description of the collapse of a flexible pipe shows that the characteristic pressure K is proportional to the flexural rigidity D. The scale of K depends upon geometrical and mechanical constants in the unstressed configuration, such as unstressed wall thickness h_0, the Young modulus E, and the Poisson coefficient v:

$$K = Eh_0^3/\left(6\left(1-v^2\right)a_0^3\right).$$

The tube deformation law must be known to interpret the experimental results observed in a simple physical model like the Starling resistor.

The Pressure–Area Relationship

Mechanical models of the wall deformation of a collapsible tube of infinite length have been created in order to derive the tube law as applied to homogeneous material.[25–28] The relationship between the transmural pressure and the luminal area of a collapsible tube exhibits a sigmoidal shape with a high tube compliance in the range of slightly negative transmural pressures, i.e., before any wall contact. Different experimental techniques have been developed to validate the mechanical models and to measure the cross-sectional area during experiments. The easiest method is based on volume change measurements for given transmural pressure variations in tubes of two different lengths in order to eliminate end effects.[29] The luminal shape and area can be investigated from collapsed tube casts[5,30] by the endoscopic technique or by cross beaming on the collapsed pipe with a laser sheet.[31] The optical stereoscopic

---

[†]The geometrical parameter k_0 is named the ellipticity so that it cannot be confused with the eccentricity $(1 - k_0^2)^{1/2}$ or the width-to-height ratio of the collapsed cross section defined also as the eccentricity.[19]

technique monitors the pipe shape and computes the outer cross-sectional area along the whole tube length.[32,33] This method can be used under either static conditions or during flow; the luminal area is obtained from the external tube geometry, assuming a constant wall thickness during the deformation. Impedance methods sense the local axial electrical field produced by an impressed current, using catheters inside the tube lumen (Mc Clurken method).[7] However, the luminal area is measured independently from the pipe shape. Besides, such catheters disturb both the flow and the tube collapse. The transformer principle method[34] uses a single-turn sensing coil embedded in the tube wall, which is very fragile and which cannot be used in unsteady experiments. Ultrasound imaging has been used as the reference method, but it necessitates a careful positioning of the probes. This method proposes to compute the bending stiffness and the derived mechanical quantities.[35]

Flow Experiments

In flow experiments (set up under steady conditions but with a possible unsteadiness during manipulations), different kinds of relationships between the pressure drop $\Delta p = p_1 - p_2$ and the volume flow rate q are obtained, depending on whether the external pressure p_e, the downstream transmural pressure $p_2 = p_{i2} - p_e$, or the upstream transmural pressure $p_1 = p_{i1} - p_e$ are held constant[30] (Fig. 10.3).

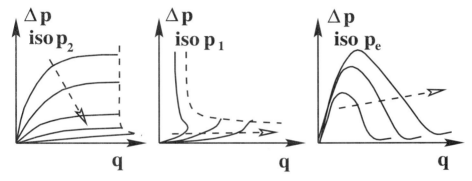

FIGURE 10.3   Relationships between the pressure drop Δp and the volume flow rate q for a given constant outlet transmural pressure p_2 (left), inlet transmural pressure p_1 (mid), or external pressure p_e (right) in the Starling resistor. Dashed-line arrows indicate an increase in constant-set algebraic pressure and dashed curves indicate the critical conditions. $\Delta p - q$ curves in iso-p_2 conditions exhibit pressure drop limitation, flow limitation in iso-p_1 conditions, and negative resistance in iso-p_e conditions.[23,38] (*Source:* Bonis, M. and Ribreau, C., *ASME, J. Biomech. Eng.*, 103, 27–31, 1981. With permission.)

In an initial set of experiments,[23] p_e remains constant whereas q changes. The iso-p_e $q - \Delta p$ curve, obtained for a constant constriction R_2 downstream from the collapsible tube, reaches a peak at the end of the fully collapsed tube stage, when the internal pressure measured upstream from the collapsible tube p_{i1} is close to p_e.[23] Larger flow rates induce a progressive opening of the tube from the tube entry and cause the tube hydraulic resistance to be negative (during the partially collapsed tube phase, the pressure drop vs. the flow rate relationship exhibits a negative slope). Oscillations can be observed in the pressure–flow domain. At high flow rates, the pressure drop in the distended tube is approximately the same as that of a liquid flow in a rigid tube of the same shape (open tube phase).

In a second set of experiments, the pressure p_e is adjusted in order to maintain p_2 constant for the whole range of investigated values of the flow rate q (iso-p_2 curves). At rest in a horizontal tube, the internal pressure is uniform and the tube is entirely collapsed if p_2 is set at a sufficiently low value. With increasing p_{i1} or decreasing p_{i2}, while the upstream region of the tube gradually opens, the flow rate increases up to a plateau for a particular level of pressure drop: a pressure drop limitation (pressure drop independent of the flow rate) is observed.[37–42] Unsteadiness may arise in the plateau region.[38,43]

In a third set of experiments, a flow rate limitation (flow rate independent of the pressure drop) occurs when the transmural pressure p_1 is kept constant (iso-p_1 curves).[44] For a particular value of the pressure drop, q reaches its maximum. Unsteadiness occurs after this maximum.

Instabilities or/and oscillations are observed in the plateau region of the iso-p_2 curves in the neighborhood of the maximal q of the iso-p_1 curves and in the region of negative resistance of Conrad's experiment. In particular, self-sustained oscillations may arise.[23,38,43,45,46] The modes of oscillation of specific wave shape and frequency can be mapped on a plane where coordinates stand for the upstream head propelling the fluid and the downstream transmural pressure.[47]

Forced Flow Experiments

Experiments with either periodic input of flow or with imposed wall motions can be carried out. In the 1960s, the first forced flow experiments were performed in collapsible tubes.[23] They led to the forced Van der Pol's equation. Iso-$\bar{p}_1$ pressure–flow curves in pulsatile flow (non-zero mean $\bar{p}$ sinusoidal pressure) exhibit flow limitation similar to the steady situation previously reviewed. However, the maximum flow rate depends on both the amplitude and the frequency of the input flow oscillations.[48] Such studies may be closely in keeping with medical applications; for instance, a multicompartment lumped-parameter model (each vascular element is composed of a capacitance, a resistance, and an inertial inductance) can be aimed at studying regional blood flow under cardiopulmonary resuscitation by intrathoracic and abdominal pressure variations.[49] The filling process was shown to depend on both the circumferential and axial stiffness. Collapsible tube flow experiments were performed to study the sliding massage effects on venous flow;[50] an additional flow due to the pulling motion of a roller on the floppy tube was shown to depend on the initial flow rate and transmural pressure, degree of obstruction, roller speed, tube properties, and possible flow via an in-parallel tube modeling a venous anastomosis. In collapsible tubes subjected to pressure ramps, like the blood vessels in limbs swathed in cuffs, a supercritical flow stage can be observed after the peak flow in the downstream segment or in the entire tube for a sufficient pressurization rate.[51] In the context of leg pressurization–depressurization cycles, time-dependent filling of collapsed vessels was investigated theoretically and experimentally.[52] Imposed wall motions, more precisely the sinusoidal motion of one compliant wall of a rigid channel, were also used to study the unsteady flow separation and eddy development which are involved in collapsible tube flow behavior.[53]

10.3   Tube Law

Unstressed Configuration Dependent Collapse

The reference pipe configuration is supposed to be stress-free (in particular, $p = 0$). When the unstressed cross-section is circular and $p < 0$, the compliant tube keeps a circular cross-section down to the buckling pressure p_b. Moreover, from this mechanical state, a small decrease in transmural pressure p produces a large change in cross-sectional shape and area A_i.

Different modes of collapse can be observed according to the number N of lobes, lobes being the open parts of the collapsed tube lumen which are associated with symmetry axes. Using computational thin-shell models of deformation in flexible tubes of infinite length and with a purely elastic wall, the buckling pressure is shown to be proportional to $n^2 - 1$.[26,54] Experimental evidence of such collapsing modes was obtained, the cross-section shape displaying usually four, three, or two lobes either in tubes subjected to longitudinal bending effects[30] or in short and thin-walled pipes.[31]

However, physiological vessels and their polymeric models more often present a non-circular unstressed cross section, the shape of which is commonly assumed to be elliptical. In that case, the tube collapses with two-fold symmetry at a distance from the tube ends.

Description of the Tube Law

The presentation of the tube law is limited to the collapse of tubes of elliptical unstressed cross-sections (ellipticity k_0, Fig. 10.4). The origin of the bidimensional Cartesian system (y, z) is located at the ellipse center. The tube center line is the x-axis with unit vector i. Let s be the curvilinear abscissa along the mid-line of any cross-section of the tube wall, with the origin at point M_0 of minimal curvature in the unstressed configuration,[†] where the opposite wall distance becomes minimal during the tube deflection. At each point M of the mid line, unit tangent t, unit normal n, and i define the local directed coordinate frame. Let θ be the oriented angle between the local horizontal axis and the tangent.

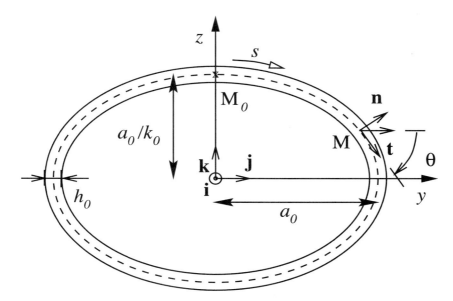

FIGURE 10.4 Cartesian frame (y, z) of the cross-section with unit vector set $\{j, k\}$. The tube center line is the x-axis with unit vector i. The dashed line represents the elliptical mid-line (ellipticity k_0) of the wall of thickness h_0 and the local oriented coordinate system (n, t), where n is the unit outward normal and t the tangent on the curvilinear abscissa s ($s = 0$ at M_0).

The infinitely long tube is supposed to be straight, and the wall thin, homogeneous, and purely elastic. Both the geometry and the mechanical properties are uniform. The flexible pipe is subjected to a uniform transmural pressure p. The wall thickness, much less than the lowest curvature radius of the wall mid-line ($h_0(d\theta/ds) \ll 1$), is assumed to remain constant during the collapse. Moreover, the mid-surface is deformed without extension.

Three characteristic transmural pressures are of interest in the collapse: (1) the ovalisation pressure p_p, for which the curvature at M_0 is locally equal to zero (an oval-shaped cross-section with parallel opposite edges); (2) the point-contact pressure p_c, at which the opposite sides touch for the first time; and (3) the line-contact pressure p_l, when the curvature at the contact point is equal to zero for the second time. Three modes of collapse are thus defined using the distinguishing pressures 0, p_c, and p_l (Fig. 10.5):

1. Mode 1 corresponds to the collapse before contact ($p_c < p \le 0$), characterized by a high tube compliance. The transversal density of the distributed external force f induced by the transmural pressure load is given by $\mathbf{f} = p\mathbf{n}$. The stress resultant acting from one part of the wall to the other $\mathbf{T}(s)$ is continuous everywhere.

[†]At the material point M_0 $(0, a_0/k_0)$, $s = 0$.

2. Mode 2 is characterized by a contact at a single point ($p_l \leq p \leq p_c$). The curvature at the contact point decreases from a finite value down to zero. A contact reaction appears (see below), and the stress resultant is discontinuous at the contact point ($s = 0$).
3. Mode 3 is defined by a contact on a line segment ($p < p_l$). The contact segment lengthens while the transmural pressure decreases from p_l. Just below p_l, the contact reaction splits into two concentrated reactions applied at both ends of the contact segment $\pm s_c$ where $\mathbf{T}$ is continuous. The transversal density $\mathbf{f}$ is thus given by either $\mathbf{f} = p\mathbf{n}$ along the open part of the cross-section or by $\mathbf{f} = (r(s) + p)\mathbf{n}$ on the contact line ($r(s)$: normal reaction distribution).

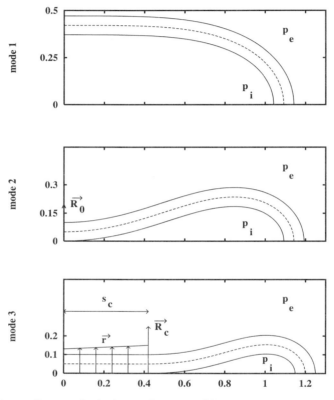

FIGURE 10.5   The three collapse modes in the case $k_0 = 1.6$ and $h_0/a_0 = 0.1$: top $p_c < p \leq 0$, mid $p_l \leq p \leq p_c$, and bottom $p < p_l$. Reaction loading at contact between opposite walls of the flexible pipe:

$$R_0(p_c) \to R_0(p_l) \xrightarrow{\ p<p_l\ } \begin{cases} R_c(p) & s = s_c \\ r(s,p) & 0 \leq s < s_c \end{cases}.$$

The stress resultant $\mathbf{T}$ and the bending moment $\mathbf{M}$ verify the equilibrium equations of the elastic wall on each side of any concentrated force application point:

$$d\mathbf{T}/ds + \mathbf{f} = 0 \tag{10.1}$$

$$d\mathbf{M}/ds + t \times \mathbf{T} = 0 \tag{10.2}$$

The loading singularity is added to the set of equations via a discontinuity condition. Integration of the equilibrium equations gives **T** and **M**.[25,28] The constitutive law of the elastic material states that **M** is linearly proportional to the curvature change in any point of the wall cross-section mid-line:

$$M = d\left(d\theta / ds - d\theta_0 / ds\right) \qquad (10.3)$$

where $D = Eh_0^3 / (12(1-v^2))$ is the flexural rigidity. Note that when the mid-line is assumed to be superimposed to the wetted perimeter χ_i, as soon as the contact occurs the boundary conditions become $z(0) = 0$ rather than $z(0) = h_0/2$ in mode 2 and $z(s_c) = 0$ rather than $z(s_c = h_0/2$ in mode 3. The merging of the mid-line and the luminal line is pure mathematical abstraction and the computed cross-sectional area is approximated (Fig. 10.6). The tube law is highly non-linear: as soon as p becomes slightly negative, the tube transverse configuration can undergo huge changes.

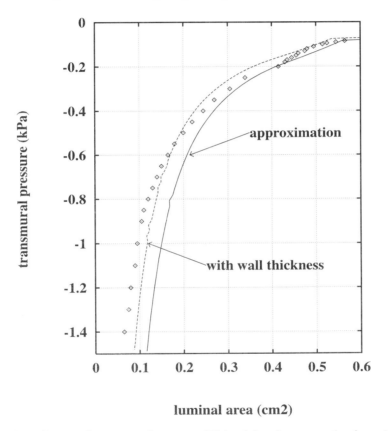

luminal area (cm2)

FIGURE 10.6   Relation between the transmural pressure p (*kPa*) and the tube cross-sectional area A_i (cm²), focused on collapse modes 2 and 3. Comparison between simulations, using either the condition $z(0) = 0$ (continuous line) or the condition $z(0) = h_0/2$ (dashed line) at contact, and experimental results (◊) on a flexible pipe with the following properties:

a_0 (mm)	k_0	h_0 (mm)	n	$K (N/m^{-2})$
9.52	1.337	0.43	1.60	24.47

The Opposite Wall-Edge Contact

The contact generates a local reaction R_0 at the contact point ($s = 0$), which increases when p decreases from p_c ($R_0(p_c)$) down to p_l ($R_0(p_l)$). As soon as p undergoes an infinitesimal decrease, say $p = p_{l-}$, the reaction initiates its splitting into two components R_c (p_{l-}) = $(1/2)$ R_0 p_l). When $p < p_l$, the reaction is distributed along the contact segment of length 2 s_c, with maxima $R_c(p)$ located at both ends of the contact segment ($s = \pm s_c$). These points, associated with a concentrated force, migrate laterally when p continues to decrease, whereas the reaction amplitude exerted on the line of contact, which spreads out, increases.

The contact reactions induce discontinuities in the first derivative dA_i/dp at p_c and in the second one $d^2 A_i/dp^2$ at p_l.[27,55] Such discontinuities probably affect the mechanical behavior of the fluid–tube couple. The discontinuities in the first derivative at p_c and in the second derivative at p_l are exhibited by a break in the slope and by a bending of the tube law respectively (Fig. 10.8). The discontinuities were not always taken into account, and continuously derivable analytic expressions of the $A_i(p)$ relationship were proposed over the whole range $p \leq 0$.[1]

Normalization

Dimensionless variables are defined as:[†]

$$\tilde{y} = y/a_0, \tilde{z} = z/a_0, \tilde{A}_i = A_i/A_{i_0}, \tilde{\chi} = \chi/a_0.$$

The pressure scale is given by $K = Eh_0^3/(6(1-\nu^2)a_0^3)$. The quantity $(K/\rho)^{1/2}$ is taken as the speed scale[††] (ρ = fluid density). Consequently, the flow rate scale is $A_{i0}(K/\rho)^{1/2}$.

The tube deformation, from rest to the line-contact pressure p_l, is illustrated in Fig. 10.7, using normalized quantities $\tilde{p}$ and $\tilde{A}_i$. A two-dimensional bending model of a homogeneous tube shows that the tube law depends upon the ellipticity k_0 (Fig. 10.8). With the following kinematic boundary conditions $\tilde{z}(0) = \tilde{h}_0/2$ in mode 2 and $\tilde{z}(\tilde{s}_c) = \tilde{h}_0 2$ in mode 3, both $\tilde{p}_c$ and $\tilde{p}_l$ are affected by k_0 and $\tilde{h}_0$ (Fig. 10.9).[28†††]

Normalized Similarity Law ($\tilde{p} \leq \tilde{p}_l$)

The pressure–area relationship for $\tilde{p} \leq \tilde{p}_l$ is expressed by the following equation:

$$\tilde{p} = -B\left(k_0, \tilde{h}_0\right)\tilde{A}_i^{-n\left(k_0, \tilde{h}_0\right)} \tag{10.4}$$

[†]The unstressed cross-sectional area A_{i0} is the characteristic area used in general. However, due to the elliptical configuration and to the twin lobe shape, the square of the unstressed major semi-axis (the transverse length scale) of the tube cross-section is a possible area scale.[56]

[††]The usual characteristic velocity $(K/\rho)^{1/2}$ is a propagation-like scale in elastic medium, but other scales can be used.

[†††]The iso-$\tilde{h}_0$ $\tilde{p}_c(k_0)$ and iso-$\tilde{h}_0$ $\tilde{p}_l(k_0)$ relationships depend on the way the normalization is defined. One usually uses the unstressed major semi-axis a_0 in the expression of the pressure scale K. For a given geometry k_0 of the midline, an increase in $\tilde{h}_0$ implies an algebraic increase in both $\tilde{p}_c$ and $\tilde{p}_l$. Depending on k_0 $\tilde{h}_0$, minimal values for both distinguishing contact pressures are observed for $\sim 1.4 < k_0 < \sim 2$. This apparent paradox is attributed to the length scale. The normalization with scale a_0 is associated with different unstressed perimeters χ_0 for various k_0 (a larger ellipticity is associated with a smaller perimeter and hence with less material over which the external forces are distributed). The quantity $\chi_0/4$ can be used as transverse length scale; it gives various normalized shapes with constant perimeter whatever k_0. The pressure scale becomes $K = Eh_0^3/[6(1-\nu^2)(\chi_0/4^3]$ and the characteristic flow rate is $\chi_0^2(K/\rho)^{1/2}/16$.

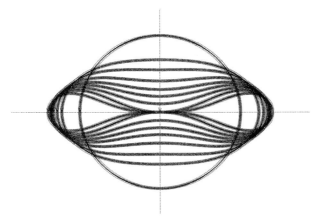

FIGURE 10.7 Deformation of a collapsible tube obtained from a computation of the bending without extension of a thin homogeneous elastic shell subjected to a uniform loading. Unstressed elliptical cross section (ellipticity of 1.005) with the following boundary conditions: $\tilde{z}(0) = 0$ in contact-point mode and $\tilde{z}(\tilde{s}_c) = 0$ in contact-line mode. In a bilobal collapsing mode, the outer tube edges bulge out whereas the faces collapse toward the cross-section center. Wall positions are plotted during the collapse from rest ($\tilde{p} = 0$) down to the contact point ($\tilde{p}_c = -2.54$) and to the contact line ($\tilde{p}_l = -5.20$) with the following set of dimensionless transmural pressures $\tilde{p} = p/K(K = Eh_0^3/[6(1-v^2)a_0^3])$ and cross-sectional areas $\tilde{A}_i = A_i/A_{i_0}$:

$\tilde{p}$	0	−1.60	−1.70	−1.85	−2.00	−2.15	−2.35	−2.64	−5.20
A_i	1	0.89	0.79	0.67	0.57	0.48	0.38	0.27	0.21

where $B = -\tilde{p}_l \tilde{A}_{il}^n$, with $\tilde{A}_{il} = \tilde{A}_i(\tilde{p}_l)$ (Fig. 10.10).[28] This equation, inferred from numerical data, means that the tube shape remains similar to the twin-lobed distinguishing configuration, i.e., to the line-contact shape at $\tilde{p}_l$, when $\tilde{p}$ is further decreasing.[†]

The similarity tube law is classically written $\tilde{p}/\tilde{p}_l = (\tilde{A}_i/\tilde{A}_{il})^{-n}$; $n = 3/2$ when the unstressed cross-section is circular and $\tilde{h}_o = 0$.[26,57] In the latter case, the similarity solution is exact. The similarity law exponent n increases with k_0 for a given h_0. For instance for $h_0 = 0$ (when the mid-line is merged with the wetted perimeter), n rises from 3/2 for $k_0 = 1.005$ to 2 for $k_0 = 2.8$ and up to 9/4 for $k_0 = 10$.

Analytical Expressions of the Tube Law

Analytical expressions are very helpful in collapsible tube flow computations or experimental post-processings. Three types of formulae can be defined: (1) expressions derived from the similarity law; (2) fit polynomials; and (3) equations based on integration of the critical flow curve.[††]

Several expressions are proposed by different groups of investigators (Table 10.1). These expressions do not take into account the derivative discontinuities of the tube law and its related physical phenomena.

[†]When $\tilde{p} < \tilde{p}_l$, a specific mapping can always be determined between any collapsed cross-section of the investigated pipe of ellipticity k_0, and a distinguishing collapsed cross-section of an equivalent tube of unstressed ellipticity k_{00}, having a similar shape at its own line-contact transmural pressure, say $\tilde{p}_{ll}(k_{00})$.[56] The $\tilde{p}_{ll}$ cross-section can be obtained from the $\tilde{p}_l$ cross-section by an affine transformation in the $\tilde{z}$-direction with the ratio $G_2 = k_0/k_{00} = (\tilde{p}/\tilde{p}_l)^{(2ln-3)/5n} = (\tilde{A}_i/\tilde{A}_{il})^{(3-2n)/5}$; the similar open region of the collapsed cross-section is computed with the ratio $G_1 = \chi \times \chi_0 = (\tilde{p}/\tilde{p}_l)^{-(1+n)/5n} = (\tilde{A}_i/\tilde{A}_{il})^{(1+n)/5}$. The lower limit of the validity range of G_1 and G_2 is at least equal to about 30 $\tilde{p}_l$.

[††]The algebraic description of the tube law in each defined pressure range must be consistent with the one-dimensional theory since both $\tilde{A}$ vs. $\tilde{p}$ and $\tilde{A}$ vs. $\tilde{q}^*$ relationships are relevant (see section 10.5) The analytical equations of the tube law insure the discontinuity of the functions $\tilde{c}(\tilde{p})$ and $\tilde{q}^*(\tilde{p})$ at the contact condition $\tilde{p} = \tilde{p}_c$. The $\tilde{q}^*(\tilde{p})$ curve is fitted piecewise from numerical data.

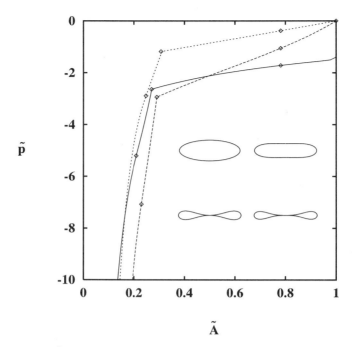

FIGURE 10.8 Numerical $\tilde{p}(\tilde{A}_i)$ laws for three ellipticities: $k_0 = 1.005$ (continuous line), 2.8 (dashed line), and 10 (dotted line), with the characteristic values (◊) corresponding to the displayed characteristic shapes (unstressed elliptical, oval-shaped, point-, and line-contact).

TABLE 10.1 Some Analytical Tube-Law Expressions of the Literature

References	Analytical Expressions of the Tube Law	Validity Range
1, 10	$\tilde{p} = (a - \tilde{A}_i^{-3/2})$	$\tilde{A}_i < 1$
10	$\tilde{p} = 100(\tilde{A}_i - 1)$	$\tilde{A}_i \geq 1$
11	$\tilde{p} = -(\kappa_1 + \tilde{A}_i^{-3/2})$	$\tilde{A}_i \leq 0.95$
	$\tilde{p} = 100(\tilde{A}_i - \kappa_2 + \kappa_3 \exp\{\kappa_4(0.95 - \tilde{A}_i)\})$	$\tilde{A}_i > 0.95$
58	$\tilde{p} = (\tilde{A}_i^{20} - \tilde{A}_i^{-3/2})$	
59	$\tilde{p} = (\tilde{A}_i^{4} - \tilde{A}_i^{-3/2})$	

Algebraic expressions of type (3) are more recent,[56] using nine coefficients, which are plotted in Figs. 10.11 to 10.13 for $k_0 = 1.005$ and for the ellipticity range.[1,2,10] The relationships $\tilde{p}(\tilde{A}_i)$ have been given for the description of the tube law in the three pressure intervals:

$$\tilde{p} = \alpha_{lc}^2 \ln \tilde{A}_i - 2\alpha_{lc}\beta_{lc}^2 / \left(2\tilde{A}_i^2\right) + \gamma_{lc} \qquad \left(\tilde{p}_l \leq \tilde{p} \leq \tilde{p}_c\right)$$

$$\tilde{p} = \exp\left\{2\alpha_{cp}\right\}\tilde{A}_i^{2(\beta_{cp}-1)} / \left[2(\beta_{cp}-1)\right] + \gamma_{cp} \quad \left(\tilde{p}_c \leq \tilde{p} \leq \tilde{p}_p\right) \qquad (10.5)$$

$$\tilde{p} = \exp\left\{2\alpha_{po}\right\}\tilde{A}_i^{2(\beta_{po}-1)} / \left[2(\beta_{cp}-1)\right] + \gamma_{po} \quad \left(\tilde{p}_p \leq \tilde{p} \leq 0\right)$$

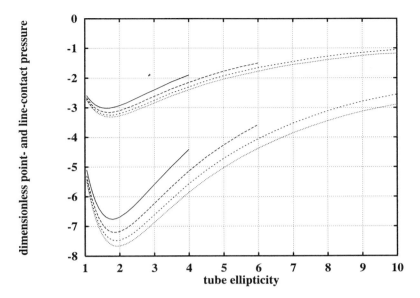

FIGURE 10.9 Dimensionless point-contact and line-contact pressure changes with the tube ellipticity k_0 for four values of the dimensionless wall thickness $\bar{h}_0$ (0.1, 0.05, 0.02, and 0 from top to bottom).

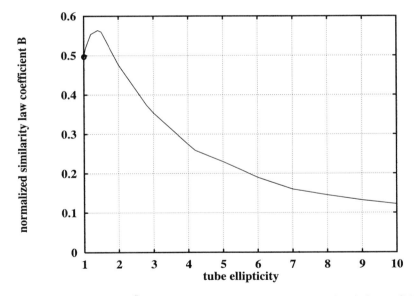

FIGURE 10.10 Coefficient $B(k_0) = -\tilde{p}_l \bar{A}_l^n$ of the similarity law (Eq. 10.4), computed with the condition $z(0) = 0$ at contact. The symbol o is used for the tube ellipticity $k_0 = 1.005$.

Positive Transmural Pressure

The cross-section is circular with unstressed radius a_0; the pressure forces are balanced by wall tangential stresses. The deformed cross-section remains circular when $p > 0$. Let $\{\lambda_k\}_{k=1}^3$ be the tension coefficients associated with the principal axes of the tube $(\lambda_1 = L/L_0, \lambda_2 = a/a_0, \lambda_3 = h/h_0)$.

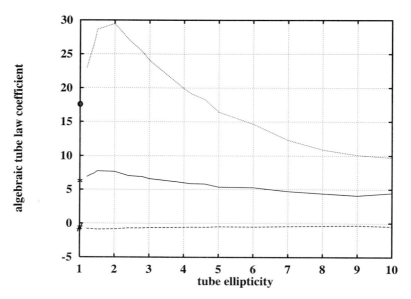

FIGURE 10.11   Variations of the coefficients α_{lc} (mid-continuous line), β_{lc} (bottom-dashed line), γ_{lc} (top-dotted line) of the analytical generalized tube law (Eq. 10.5) with the tube ellipticity for the pressure range $[\bar{p}_i, \bar{p}_c]$ (collapse mode 2). Symbols are used for the peculiar ellipticity $k_0 = 1.005$: * for α_{lc}, # for β_{lc}, and o for γ_{lc}.

The non-linear elastic behavior of the tube can be given by the following expression:[29]

$$p = \left(\xi / \lambda_1^3\right)\left(h_0/a_0\right)\left(\lambda_1^2 - \left(A_{i_0}/A_i\right)^2\right)$$
(10.6)

where ξ is an empirical coefficient taking into account, among other factors, the non-isotropy of the tube material. The coefficient ξ is close to the proposed correction factor:

$$G(1+h_0/2a_0)/(1+h_0/a_0)^2, \text{ with } G = E/2(1+\nu).$$

10.4   Tube Law Dependent Data

The tube law dependent hydraulic quantities are determined in uniform configurations of the collapsed tube, i.e., of a constant cross-section along the entire tube length. Velocity fields are computed assuming a fully developed, steady, incompressible laminar flow through a smooth-walled collapsed pipe.

Wave Speed and Critical Flow Rate

The speed c of small pressure wave propagation in a long straight collapsible tube containing an incompressible fluid is given by the following equation:

$$c^2 = \left(A_i/\rho\right)\left(dp/dA_i\right)$$
(10.7)

This relationship assumes (1) a purely elastic thin wall and a constant geometry along the whole tube length in the reference configuration; (2) a negligible wall inertia; and (3) a one-dimensional fluid motion. The speed of the elasto-hydrodynamics coupling wave is thus computed from the tube law. Experimental observations have shown that the wave speed reaches its minimum when contact between the opposite walls occurs.[55]

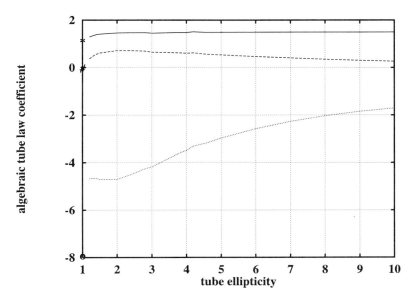

FIGURE 10.12 Variations of the coefficients α_{cp} (top-continuous line), β_{cp} (mid-dashed line), γ_{cp} (bottom-dotted line) of the analytical generalized tube law (Eq. 10.5) with the tube ellipticity for the pressure range $[\tilde{p}_c, \tilde{p}_p]$ (collapse mode 1). Symbols are used for the peculiar ellipticity $k_0 = 1.005$: $\star$ for α_{cp}, # for β_{cp}, and o for γ_{cp}.

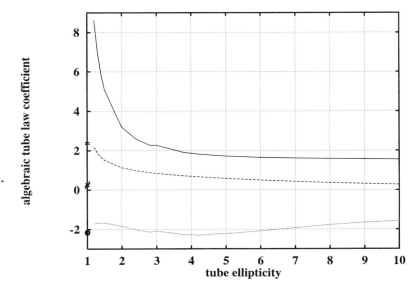

FIGURE 10.13 Variations of the coefficients α_{po} (top-continuous line), β_{po} (mid-dashed line), γ_{po} (bottcm-dotted line) of the analytical generalized tube law (Eq. 10.5) with the tube ellipticity for the pressure range $[\tilde{p}_p, 0]$ (collapse mode 1). Symbols are used for the peculiar ellipticity $k_0 = 1.005$: $\star$ for α_{po}, # for β_{po}, and o for γ_{po}.

A critical flow rate $q^* = A_i c$ can be computed for each area of the cross-section. Due to the discontinuity induced by the wall contact, the critical conditions are different on the right (q^*_{c+}) and left (q^*_{c-}) sides of the discontinuity (Fig. 10.14). The slope of the q^* vs. A_i curve depends on the tube law parameter $\mathcal{M}$:

$$dq^*/dA_i = (\mathcal{M}/2)(q^*/A_i) \tag{10.8}$$

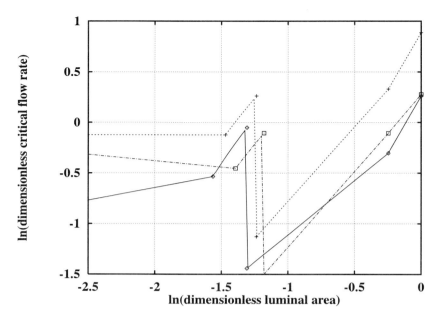

FIGURE 10.14 Relationship between the natural logarithm of the dimensionless cross-sectional area $\bar{A}_i$ and the natural logarithm of the normalized critical flow rate $\tilde{q}^*$ for different tube ellipticities: $k_0 = 1.005$ (continuous line), 2.8 (dotted line), and 10 (dashes line), using the condition $z(0) = 0$ at contact. Between the characteristic values ($\bar{A}_{i0}$, $\bar{A}_{ip}$, $\bar{A}_{ic}$, $\bar{A}_{il}$), the lines correspond to affine functions which fit the non-displayed computed points. When $p \leq p_l$, both the value and the sign of the slope of the relationship are affected by the tube ellipticity.

with $\mathcal{M} = 3 + \bar{A}_i(d^2\bar{p}/d\bar{A}_i^2)/(d\bar{p}/d\bar{A}_i)$. The sign of parameter $\mathcal{M}$ influences the pressure wave propagation.[61] When $\mathcal{M} > 0$, a compressive wave steepens and a rarefaction wave broadens, and the converse is true when $\mathcal{M} < 0$.[1,9]

In the pressure range $p \leq p_l$, the parameter $\mathcal{M}$ is given for thin-walled collapsible tubes of unstressed elliptical cross-sections by $\mathcal{M}[n(k_0, \bar{h}_0)] = 2 - n$. It takes the following values for given (k_0, $\bar{h}_0$) couples: $\mathcal{M}[n(1, 0)] = 1/2$, $\mathcal{M}[n(2.8, 0)] = 0$, and $\mathcal{M}[n(10, 0)] = -1/4$. The slope of the relationship between q^* and A_i is positive when $\mathcal{M} > 0$ and negative when $\mathcal{M} < 0$.[28]

As will be seen in section 10.5, the flow depends not only on the sign of $(1 - S^2)$ but also on the sign of the wave number $\mathcal{M}$.[1,62] Both the wave speed c and the wave number $\mathcal{M}$ are very important properties of the tube law.

In the pressure range $p \leq p_l$, the similarity law gives:

$$\left(\tilde{c}/\tilde{c}_l\right)^2 = \left(\tilde{A}_i/\tilde{A}_{il}\right)^{-n}$$
$$\left(\tilde{q}^*/\tilde{q}_l^*\right)^2 = \left(\tilde{A}_i/\tilde{A}_{il}\right)^{\mathcal{M}} \tag{10.9}$$

The Reynolds Number

The Reynolds number Re can be expressed in dimensionless form[†] by:

$$Re = 4\pi\gamma k_0^{-1}\tilde{\chi}_i^{-1}\tilde{q} \tag{10.10}$$

[†]The Reynolds number is expressed by $Re = 4\gamma\tilde{q}/\tilde{\chi}_i$ when $\chi_0/4$ is used as the transverse length scale, and $\gamma = \chi_0(\rho K)^{1/2}/4\mu$.

where $\gamma = a_0(\rho K)^{1/2}/\mu$ is a viscous effect parameter,[†] and μ is the fluid dynamic viscosity.[††]

In the range $p \le p_l$, the fluid domain may be decomposed into two fluid regions: an area of very small velocity magnitude near the contact point, where the viscous effects are predominant, and an outer zone. One may suppose that the flow regime is mainly affected by this second region. Consequently, the usual cross-length scale, i.e., the hydraulic perimeter $d_h = 4A_i/\chi_i$, and the inertial effects might be underestimated. The supposed laminar-turbulent transition may be computed using the equivalent-tube model (Fig. 10.15). Each lobe of the line-contact cross-section can be modeled by a fluid domain of the same area and the same wetted perimeter, which includes an inner rectangular zone of large aspect ratio, and an outer circular region. The radius of the latter is calculated such that the head loss through the equivalent cross-section is equal to the head loss through the actual one. The ratio between the Reynolds numbers in the actual and circular domains is found to be proportional to the ratio between the model radius and the actual wetted perimeter. Assuming a critical Reynolds number of 2000 in the circular duct, a critical value in the collapsed tube is proposed to be equal to 1200 for the whole range $p \le p_l$ when $k_0 = 1.005$.[63] Note that only experiments should determine the transition regime, hence the critical Reynolds number.[†††]

Head Losses and Shape Factor

The viscous head losses per unit of length $f_v(q)$ in smooth-walled uniformly collapsed tubes are given by[††††]

$$f_v = \frac{\Lambda}{d_h} \frac{\rho U^2}{2} \tag{10.11}$$

where Λ is the friction head loss coefficient: $\Lambda = f_L/Re$ and $\Lambda = f_T/Re^{1/4}$ (f_L and f_T are laminar and turbulent shape factors).[†††††] The shape factors depend on the transmural pressure p and on the ellipticity k_0.

Laminar Pattern

The analytical solution is known for $\tilde{p} = 0$:

$$f_{L_0} = 128\pi^2 \left(1 + k_0^{-2}\right)\tilde{\chi}_0^{-2} \tag{10.12}$$

Additionally, the laminar shape factor f_L is computed for the following characteristic values of the transmural pressures $\tilde{p}_p$, $\tilde{p}_c$, and $\tilde{p}_l$, (Table 10.2).[63]

In the range $\tilde{p} \le \tilde{p}_l$,

[†] Note that γ is the ratio between the square root of the product of the inertia and the elastic forces and viscous forces.

[††] Note that in a given tube conveying a given fluid at a given flow rate, when the tube wetted perimeter remains constant, i.e., for $p \ge p_l$, Re is constant along the whole collapsed tube length. But when the opposite walls are in contact over a line segment, the hydraulic perimeter and consequently Re vary along the tube length.

[†††] The critical Reynolds number for laminar-turbulent transition was found experimentally to be equal to about 2800 for the approximately point-contact shaped duct.[63]

[††††] Another usage gives the friction coefficient C_f, related to the wall shear stress τ_w by the formula $\tau_w = C_f \rho U^2/2$. Here, $f_v = (\chi_i/A_i)\tau_w$, which means that the dimensionless hydraulic loss factor $\Lambda = 4 C_f$ for uniformly collapsed conduits.

[†††††] It is well known that the stronger the flow resistance induced by a given cross-section shape, the higher the shape factor: f_L increases with the width to height ratio for a rigid duct of rectangular cross-section, with the ellipticity of rigid pipes of elliptical cross-section, and with the internal to external diameter ratio of annuli between two rigid concentric cylinders.

TABLE 10.2   Values of the Laminar Shape Factor in the Characteristic
Tube Shapes for Different Ellipticities

k_0	1	2	3	10
f_{L0}	64	67	71	77
f_{Lp}	71	75	79	86
f_{Lc}	50	49	48	47
f_{Ll}	37	36	35	33

$$f_L = \mathcal{F}_l\left(k_0 / \mathcal{G}_2\right) \tag{10.13}$$

where $\mathcal{G}_2$ is

$$k_0 / k_{00} = \left(\tilde{p} / \hat{p}\right)^{(2n-3)/5n} = \left(\tilde{A}_i / \tilde{A}_{il}\right)^{(3-2n)/5}$$

and

$$\mathcal{F}_l\left(k_0 / \mathcal{G}_2\right) \text{ is } 39.0731 - 2.0929\, k_0 / \mathcal{G}_2 + 0.2484\left(k_0 / \mathcal{G}_2\right)^2 - 0.0101\left(k_0 / \mathcal{G}_2\right)^3.$$

The theoretical values are close to experimental data in a tube of ellipticity $k_0 = 1.6$ in the range $[\overline{p}_p,\ 0]$ (difference of about 2%); they are slightly overestimated in the range $[\tilde{p}_l, \tilde{p}_p]$ (difference of about 10%).

Turbulent Pattern

The value of the turbulent shape factor $f_T = 0.26$ in the transmural pressure range $\tilde{p} \leq \tilde{p}_c$ was deduced from experimental data, specifically in an inclined collapsed tube in uniform state.[56] During collapse mode 1, the classical value $f_T = 0.316$ is used.

Wall Shear Stresses in Uniformly Collapsed Tubes

The Navier-Stokes equation, with the previous hypotheses, can be written as:

$$\mu \Delta u\left(y, z\right) = -f_\upsilon \tag{10.14}$$

The normalization of the fluid velocity may be chosen in order to keep constant either f_υ, whatever A_i, or q consistently with the mass conservation when A_i is continuously varying.[†] The velocity scale is given either by $(a_0^2 f_\upsilon / \mu)$ or by q/a_0^2 in the first and second kind of normalization respectively. Let $\upsilon = \mu u / a_0^2 f_\upsilon$.

[†]The velocity scale can be derived from the normalization of Eq. 10.14. When the velocity scale $a_0^2 f_\upsilon / \mu$ is chosen, the pressure drop per unit length $f_\upsilon = \Delta p_g / L$ is constant, whatever A_i ($\Delta p_g = \Delta p$ in a horizontal tube). For a given cross-section, q and f_υ are proportional; the proportionality factor is the so-called conductivity $\sigma(A_i)$:[26] $q = \sigma f_\upsilon$. When the flow rate q is kept constant,[19] the velocity scale is defined by the ratio between the flow rate and the suitable area scale a_0^2. In both cases, the velocity scale is different from the previously defined scale. The ratio of the constant pressure drop normalized fluid velocity to constant flow rate normalized fluid velocity is given by the ratio of the two scales, say $\mu q / (a_0^4 f_\upsilon) = \mu \sigma / a_0^4$.

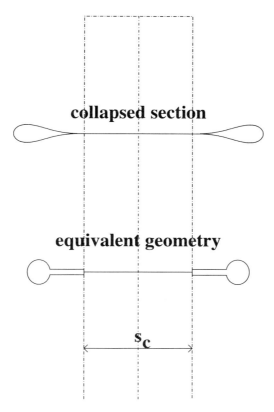

FIGURE 10.15   Actual line-contact cross-section $(k_0 = 4, \tilde{p} = 9.84\,\tilde{p}_l)$ and its model for Reynolds number computations. The vertical lines define the position of the contact point $(\pm\ \tilde{s}_c)$ and the origin of the wall curvilinear abscissa.

The Poisson equation (10.14) in the fluid domain $\tilde{\Omega}$, which here is the cross-section of the tube lumen, with the no-slip conditions on the cross-section boundary, becomes:

$$\Delta \upsilon = -1 \text{ in } \tilde{\Omega} \tag{10.15}$$

Examples of velocity distributions are illustrated for three different transmural pressures in Figs. 10.16 to 10.18. The fluid moves in parallel layers whose boundaries are shaped by the tube wall.

The axial component of the shear stress[†] is given by:

$$\tilde{\tau}_w = \frac{\partial v}{\partial \tilde{y}}\,n_y + \frac{\partial v}{\partial \tilde{z}}\,n_z, \tag{10.16}$$

where n_y and n_z are the components of the outer normal unit vector **n** along the cross-section wall.

The distribution of the wall shear stresses along the lumen perimeter $\tilde{\chi}_i$ is affected by the tube cross-sectional shape. Figs. 10.19 and 10.20 show the variations of the wall shear stress and of the wall curvature over one-fourth of the perimeter for $\tilde{p}_p$ and $\tilde{p}_c$. The ratio between the two normalized shear stresses is given by the inverse of the ratio of the velocity scales, for example $\mu s/a_0^4$. The conductivity vs. cross-sectional area relationship is displayed in Fig. 10.21.

[†]The shear stress scale is given by $a_0 f_\upsilon$ and $\mu q/a_0^3$ in first and second kinds of normalization respectively.

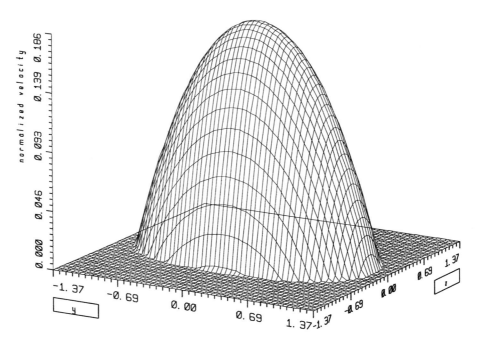

FIGURE 10.16   Axial velocity distribution in the cross-section of a uniformly collapsed tube ($k_0 = 1.005$) conveying a steady fully-developed laminar flow for the normalized transmural pressure $\tilde{p}_p$. The flow velocity is normalized with the scale $a_0^2 f_v / \mu$.

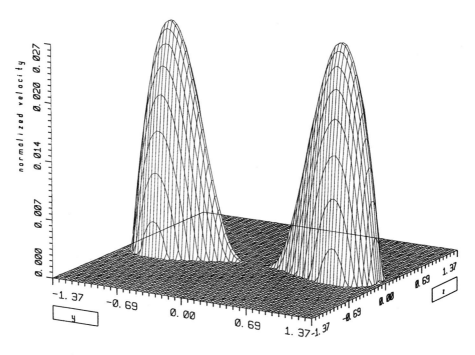

FIGURE 10.17   Axial velocity distribution in the cross-section of a uniformly collapsed tube ($k_0 = 1.005$) conveying a steady fully-developed laminar flow for $\tilde{p}_c$.

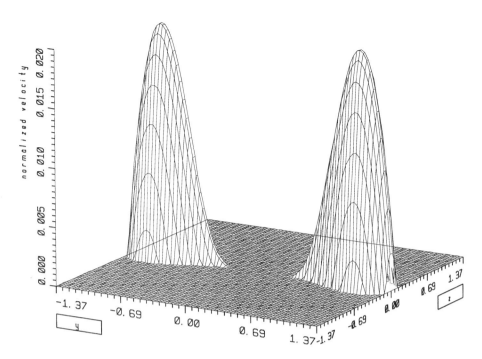

FIGURE 10.18 Axial velocity distribution in the cross-section of a uniformly collapsed tube ($k_0 = 1.005$) conveying a steady fully developed laminar flow for $\tilde{p}_l$.

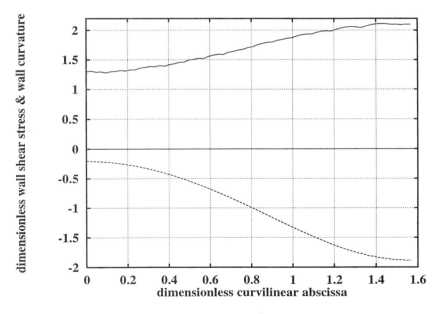

FIGURE 10.19 Variation of the dimensionless wall shear stress $\bar{\tau}_w$ (top-continuous line) along the quarter of the wetted perimeter of the tube cross-section and of the wall dimensionless curvature (bottom-dotted line) at $\tilde{p}_p$ in laminar fully developed flow ($k_0 = 1.005$). The normalization is based on constant flow rate (stress scale: $\mu q / a_0^3$).

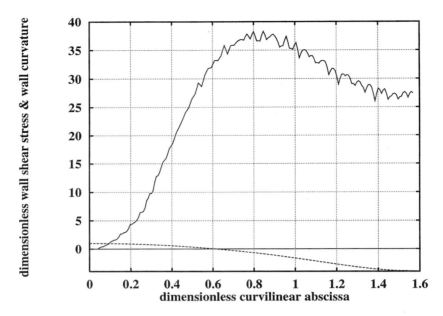

FIGURE 10.20 $\tilde{\tau}_w$ (top-continuous line) and wall dimensionless curvature changes (bottom-dotted line) at $\tilde{p}_c$ in laminar fully developed flow ($k_0 = 1.005$).

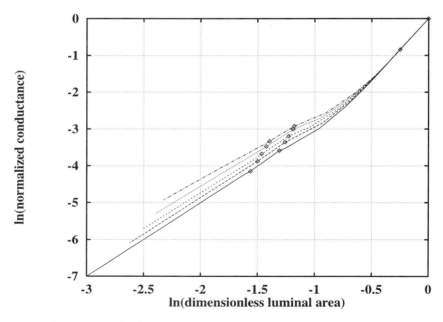

FIGURE 10.21 ln $(\bar{\sigma}/\bar{\sigma}_0)$ vs. ln($\bar{A}_i$) relationships for different ellipticities ($k_0 = 1.005, 2, 3, 5,$ and 10 from bottom to top). The points on the curves define the characteristic values ($\bar{A}_{i0}, \bar{A}_{ip}, \bar{A}_{ic}, \bar{A}_{il}$), The conductivity σ is normalized with the scale a_0^4/μ; $\tilde{\sigma}_0 = \pi/[4k_0(1 + k_0^2)]$.

Kinetic Energy and Momentum Coefficients

In order to take into account the transverse distribution of the flow velocity, kinetic energy α_k[†] and momentum α_m coefficients should be introduced in the one-dimensional model:

$$\alpha_m = \int_\Omega \left(u/U\right)^2 \left(dA_i / A_i\right), \quad \alpha_k = \int_\Omega \left(u/U\right)^3 \left(dA_i / A_i\right)$$

The velocity distribution must then be known in order to compute these coefficients. Analytical expressions are obtained for both coefficients in the range $p \leq p_l$ and interpolated when $p_l \leq p \leq 0$.[63]

10.5 The One-Dimensional Model

One-dimensionality is based upon several hypotheses: (1) the longitudinal area gradient is sufficiently small,[††] (2) the tube axis remains straight during the test; and (3) the wall inertia and the fluid motion due to the collapsing process are negligible. The tube wall material is usually assumed to be homogeneous and purely elastic.

The set of equations is composed from the mass and momentum conservation equations associated with the tube law. The straight tube is inclined with an angle β with respect to the horizontal (sin β = $-dZ/dx$; x is the streamwise coordinate and Z is the upward vertical direction). For the sake of simplicity, the tube is supposed to be subjected to a uniform and constant external pressure p_e. Moreover, the flowing fluid is assumed to be incompressible (Mach number $Ma < 0.2$). The wave speed depends mainly on the tube compliance.

In steady tests, mass conservation gives the relationship between the steady cross-sectional average of the velocity U and the luminal area A_i:

$$\frac{dU}{U} + \frac{dA_i}{A_i} = 0 \tag{10.17}$$

The differential expression of the speed index is given by

$$\frac{dS^2}{S^2} = -M \frac{dA_i}{A_i} \tag{10.18}$$

The space gradient of the cross-sectional area may be written in a differential form typical of the one-dimensional steady model.[1][†††]

$$\left(1 - S^2\right) \frac{dA_i}{dx} = \frac{A_i}{\rho c^2} \left(\rho g \sin \beta - f_\upsilon\left(x, q\right)\right) \tag{10.19}$$

[†]In collapsible tube flows, because of the axial cross-sectional area changes, the following coefficient η may be used:[64]

$$\eta = \alpha_k - \frac{A_i}{2} \frac{d\alpha_k}{dA_i}$$

[††]The no-cross flow hypothesis is open to criticism, especially in the downstream collapsible tube segment. It is in conflict with the critical flow concept associated with sudden variation in cross-sectional area ($dA_i/dx \to \pm \infty$). The axial space derivative of A_i is generally studied in the collapsible tube literature and may be found to be large, at least locally.

[†††]Two terms always appear in the differential equation of the one-dimensional model of flow in inclined pipe: a propagation-dependent term $1 - S^2$ and a gravity friction term $\rho g \sin \beta - f_\upsilon$. Because the luminal area depends on the axial coordinate $A_i[p(x)]$, contrary to the uniformly-collapsed tube situation (see section 10.4), the head loss is denoted here as $f_\upsilon(x, q)$.

The longitudinal configuration of the flexible pipe depends then on the respective magnitude of the viscous and hydrostatic terms (Table 10.3). Elastic jump may occur in a cross-sectional expansion ($dA_i/dx \rightarrow +\infty$), downstream from a super-critical ($S > 1$) flow segment, or choking in a throat ($dA_i/dx \rightarrow +-\infty$).[†] A peculiar configuration corresponds to the uniform longitudinal shape $dA_i/dx = 0$; the viscous forces balance the hydrostatic pressure forces whatever the cross-section. Examples of A_i longitudinal variations for given flow rates are illustrated in the case of subcritical flows for both horizontal (Fig. 10.22) and downward inclined tubes (Figs. 10.23 and 10.24).

TABLE 10.3   Sign of dA_i/dx from Eq. 10.19

	$S < 1$	$S = 1$	$S > 1$
$(\rho g \sin \beta - f_v) < 0$	< 0	∞	> 0
$(\rho g \sin \beta - f_v) = 0$	0	$0/0$	0
$(\rho g \sin \beta - f_v) > 0$	> 0	∞	< 0

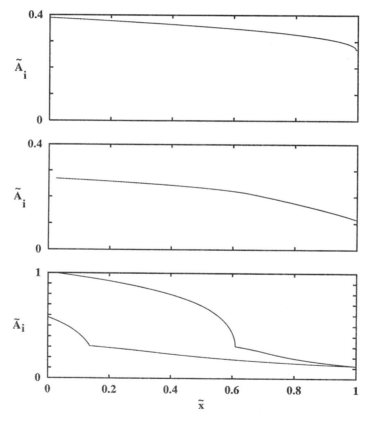

FIGURE 10.22 Dimensionless cross-sectional area $\tilde{A}_i$ vs. dimensionless axial length $\tilde{x}$. Horizontal flow ($\lambda/\gamma = 0.0073$) Top: $k_0 = 1.005$, $\tilde{p}_2 = \tilde{p}_c$, $\tilde{q} = \tilde{q}^*_{c+}$; mid: $k_0 = 1.005$, $\tilde{p}_2 = 2.5\ \tilde{p}_l$, $\tilde{q}^*_{c+} < \tilde{q} < \tilde{q}^*_{c-}$; bottom panel with constant downstream transmural pressure $\tilde{p}_2 = 5\ \tilde{p}_l$, $k_0 = 10$, $\tilde{q} = 0.105$ and $\tilde{q}^*_{c+}$. The parameters λ, γ, and δ are defined at the end of section 5 (paragraph similarity conditions).

[†]The spatial derivative of the speed index dS/dx can be obtained using Eq. 10.18. The sign of dS/dx depends on the sign of $-MdA_i/dx$.

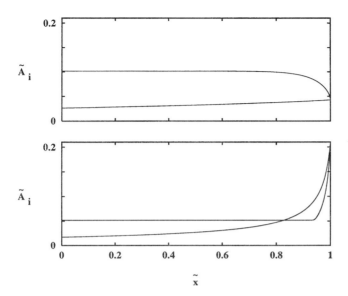

FIGURE 10.23 Dimensionless cross-sectional area $\tilde{A}_i$ vs. dimensionless axial length $\tilde{x}$. Downward flow with constant downstream transmural pressure $\tilde{p}_2$ in a compliant tube ($k_0 = 1.005$, $n = 3/2$, $\mathcal{M} > 0$). Top: $\lambda/\gamma = 0.035$, $\delta \sin \beta = 50$, $\tilde{p}_2 = 10\ \tilde{p}_l$, $\tilde{q} = 0$ and $\tilde{q}_2^\star$; Bottom: $\lambda/\gamma = 0.014$, $\delta \sin \beta = 216$, $\tilde{p}_2 = \tilde{p}_l$, $\tilde{q} = 0$ and $\tilde{q}_1^\star$.

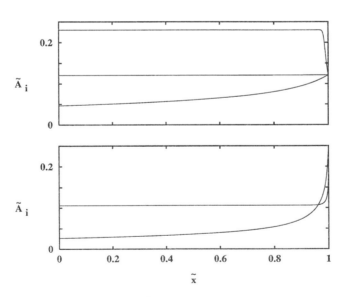

FIGURE 10.24 Dimensionless cross-sectional area $\tilde{A}_i$ vs. dimensionless axial length $\tilde{x}$. Downward flow with constant downstream transmural pressure $\tilde{p}_2$ in a compliant tube ($k_0 = 10$, $n = 9/4$, $\mathcal{M} < 0$, $\lambda/\gamma = 0.014$). Top: $\delta \sin \beta = 112$, $\tilde{p}_2 = 5\ \tilde{p}_l$, $\tilde{q} = 0$, $\tilde{q}^+$, and $\tilde{q}_1^\star$; Bottom: $\delta \sin \beta = 431$, $\tilde{p}_2 = \tilde{p}_l$, $\tilde{q} = 0$ and $\tilde{q}_2^\star$. In some traces, the wholly (superscript +) or partial uniform configuration is observed.

In horizontal tubes, Eq. 10.19 becomes: $dA_i/dx = -(A_i/\rho c^2)(f_v/(1 - S^2))$. The sign of dA_i/dx depends only on the sign of $1 - S^2$, because f_v is positive. If $S < 1$, the tube is convergent streamwise and conversely. Due to the discontinuity, $dA_i/dx \to -\infty$ in the mid-tube segment, while, in the absence of any discontinuity in the tube law, $dA_i/dx \to -\infty$ only at the tube exit, where c reaches its lowest values.

In inclined tubes, the hydrostatic pressure distribution induces a divergent or convergent tube shape at rest, whether the flow is downward ($\beta > 0$) or upward ($\beta < 0$). The speed-index dependence of the tube axial configuration convecting an ascending flow is similar to the dependence in horizontal tubes. In a descending flow, dA_i/dx depends on the respective magnitude of hydrostatic and viscous effects. Stationary flow simulations are performed with constant p_2 and increasing flow rates up to critical conditions $S = 1$ (Figs. 10.23 and 10.24). The head loss is maximum in the most collapsed tube segment. Hence, the hydrostatic effect balances the viscous effects first on the tube entry. With further increases in flow rate, an upstream uniform segment occurs and lengthens. The axial shape of the collapsed tube can become uniform over its entire length, as long as the flow remains subcritical. Then, the uniform tube disappears, but a uniform upstream segment still exists. Whereas the flow rate continues to rise, the uniform segment shortens and expands transversely, p_2 remaining constant. When the numerical experiments are carried out with a constant inlet cross-section, an entry uniform segment cannot appear. However, the tube can become uniform along its entire length for a given flow rate as soon as the unconstrained pressure p_2 reaches the imposed entry pressure p_1.

The one-dimensional model predicts not only the occurrence of a critical section but also the existence of critical uniform pipe segments ($dA_i/dx = 0$ and $S = 1$). The critical conditions are reached at once either in the upstream uniform segment or at the downstream end, whether the inlet cross-sectional area A_{i1} is greater than A_{i2} or not, and whether $\mathcal{M}$ is positive or not.

Mass and momentum conservations in the case of the one-dimensional unsteady flow give the following system of equations:

$$\frac{\partial A_i}{\partial t} + \frac{\partial A_i U}{\partial x} = 0$$
$$\frac{\partial A_i U}{\partial t} + \frac{\partial A_i U^2}{\partial x} = -\frac{A_i}{\rho}\left(\frac{\partial p}{\partial x} + f_v(x,q)\right) \tag{10.20}$$

Similarity Conditions

The similarity conditions between two experiments are given by dimensionless governing parameters. The tube law must be of the same type, i.e., same k_0; the flow regime is either laminar or turbulent. A set of dimensionless parameters[†] can be actually defined: (1) an aspect ratio $\lambda = L/a_0$, which can also be considered as an axial viscous effect parameter; (2) $\gamma = a_0(\rho K)^{1/2}/\mu$, which is a combination of elastic, inertial, and viscous forces; and (3) for gravity-friction flows, a hydrostatic effect parameter $\delta = \rho g L/K$. A flow-pattern dependent group of the two first quantities λ/γ^m ($m = 1$ or $m = 1/4$, whether the regime is laminar or turbulent) has been found to give a suitable similarity parameter in the description of the $\Delta p - q$ curves. High values of λ/γ indicate that, for given tube properties, the viscous forces affect the flow significantly. When gravity and friction are combined, the hydrostatic similarity parameter $\delta \sin \beta$ must be added. This parameter is needed to adjust the tube inclination angle to the space available for the set-up.

[†]Using the unstressed perimeter $\chi_0/4$ as the transverse characteristic length, the tube's aspect ratio is expressed by $\lambda = 4 L/\chi_0$ and $\lambda/\gamma = 16 \mu L/\chi_0^2(\rho K)^{1/2}$.

10.6 Collapsed Tube Flow Experiments

Steady Flow Experiments

Horizontal Tube

Experiments have been performed on a Starling resistor in order to investigate the behavior of the tube–flow couple at constant outlet pressure p_2.[65] The flow is induced by a constant head reservoir. The rigid chamber is filled with air so that the pressure is adjustable. The passage through the chamber is made with elastic membranes (Fig. 10.25) in order to reduce end effects as much as possible. The inlet internal pressure was measured in the collapsible pipe via a small tube bonded on the flexible tube wall, and the outlet was in a small box located at the exit of the test section. Note that the water-filled tube was maintained horizontally in the air-filled chamber by means of thread as in a hammock.

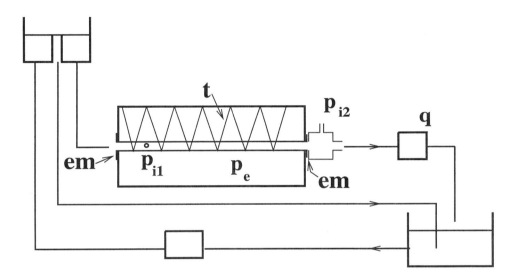

FIGURE 10.25 Bonis experimental set-up. The compliant tube is attached at both ends to elastic membranes (em). The water-filled tube is maintained horizontally in the air-filled chamber with a set of suspension threads (t), responsible for changes in contact conditions (see text).

The similarity criterion λ/γ was validated by subcritical laminar flow experiments performed under different conditions.[66] In a first set of experiments, the physical quantities were measured in two tubes, A and B, of different properties (Fig. 10.26). The parameter γ is imposed by both fluid and tube properties; λ is adjusted in order to satisfy the similarity condition λ/γ = constant in laminar flow. Experimental iso-p_2 $\Delta p - q$ curves (Fig. 10.26) are similar when the similarity criterion is satisfied. A second series of experiments was carried out with a rubber tube B, through which a water-glycerol mixture was flowing (Fig. 10.27). the experimental results show that when the similarity condition λ/γ = constant is respected, the data obtained for given dimensionless p_2 are similar, except for the critical conditions. This criterion is therefore able to take into account variations in both tube and fluid properties, whatever the collapsed tube configuration, with or without contact.

Numerical tests were carried out with the assumption of a subcritical laminar flow.[56†] A reasonable agreement was found for $p_2 > p_l$ between dimensionless numerical and experimental iso-p_2 $\Delta p - q$ curves. When $p_2 < p_l$, the numerical predictions do not fit the experimental results, in particular the slopes of the curves at very low flow are very different. The discrepancy can be imputed only to the wetted tube perimeter χ_i (Eq. 10.22), i.e., to experimental contact artifacts (Fig. 10.28). Indeed, during the collapse, in the absence of a foam bed, a single contact point exists in the folded cross sections and χ_i remains

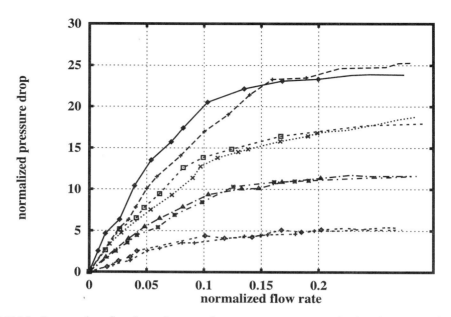

FIGURE 10.26   Pressure drop $\tilde{p}_1 - \tilde{p}_2$ vs. flow rate $\tilde{q}$ curves at constant normalized outlet pressure $\tilde{p}_2$ for eight series of experimental values performed from top to bottom at $\tilde{p}_2 = -26.4$ ($\Diamond$ and +), -19.8 ($\square$ and $\times$), -13.2 ($\triangle$ and $\star$), and -6.6 ($\Diamond$ and +). Two tubes A ($\Diamond$, $\times$, $\triangle$) and B (+, $\square$, $\star$) with the following properties and corresponding values of the dimensionless parameters:

Tube	a_0 (mm)	h_0 (mm)	k_0	K (Pa)	λ	γ	λ/γ
A	7.2	0.27	1.29	13.9	36	944	0.038
B	5.1	0.32	1.31	72.5	58	1521	0.038

constant until the transmural pressure becomes sufficiently small to set upright the lobes on the threads. The occurrence of the line of contact is thus delayed. Therefore, when computed from the tube law, the absolute value of the line-contact pressure p_l and the wetted perimeter χ_i are both underestimated. The suspension threads induce a stiffening effect; the tube law no longer applies. Another set of experiments was performed with a horizontal tube lying both on a foam bed and on a set of V-shaped plates to minimize any artifact bearing on the contact phenomenon; the lobes of the collapsed tube may lie without folding. Good agreement is then found between numerical and experimental data, in particular between the slopes at the origin.

[†] The axial gradient in p is derived from Eq. 10.19, noting that $A_i/\rho c^2 = dA_i/dp$:

$$\left(1 - S^2\right) dp/dx = \rho\left[g \sin\beta - \left(\Lambda/8\right)\left(\chi_i/A_i^3\right)q^2 \right] \tag{10.21}$$

The normalized governing equation for horizontal flow (b = 0), using the dimensionless variables defined in section 10.3 and after some algebra, becomes:

$$\frac{\tilde{A}_i^3 \left(\tilde{q}^{*2} - \tilde{q}^2\right)}{4\tilde{q}^{*2} \left(f_L/f_{L0}\right)\left(1 + k_0^2\right)\left(\lambda/\gamma\right)\left(\tilde{\chi}_i/\tilde{\chi}_{i0}\right)^2 \tilde{q}} \, d\tilde{p}/d\tilde{x} = -1 \tag{10.22}$$

Note that for a given flow rate $\tilde{q}$, for a given $\tilde{q}_2$, and for a given λ/γ ratio, the pressure gradient depends on $\tilde{\chi}_i/\tilde{\chi}_{i0}$.

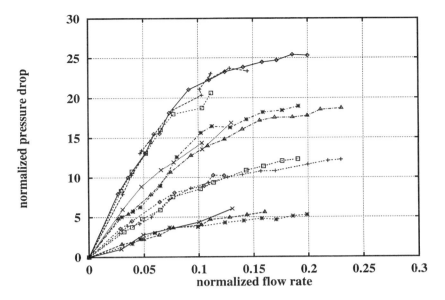

FIGURE 10.27 Pressure drop $\tilde{p}_1 - \tilde{p}_2$ vs. flow rate $\tilde{q}$ curves in tube B (see Fig. 10.26) at constant outlet pressure $\tilde{p}_2$ for 12 series of experimental values from top to bottom at $\tilde{p}_2 = -26.4$ ($\Diamond$, $+$, $\square$), -19.8 ($\triangle$, $*$, $\times$), -13.2 ($+$, $\square$, $\Diamond$), and -6.6 ($*$, $\triangle$, $\times$) for 20, 40, and 50% glycerol-water mixtures respectively, with the following properties and values of the dimensionless parameters:

$\rho\,(kg/m^{-3})$	$\mu\,(mPa/s)$	$L\,(mm)$	λ	γ	λ/γ
1048	1.45	187	37	966	0.038
1096	2.91	95	19	489	0.038
1120	4.30	65	13	334	0.038

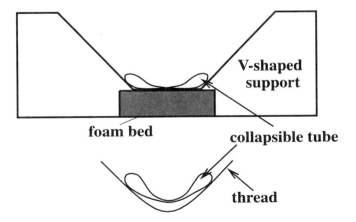

FIGURE 10.28 Top: foam beds are inserted between perspex regularly spaced V-shaped supports in order to prevent the tube folding (i.e., a curvature of the opposite walls in the same direction). Bottom: thread-induced collapse with folding of a water-filled flexible tube in an air-filled chamber. The occurrence of the line of contact is delayed. In such tests, the previously described model of the tube deformation is not suitable.

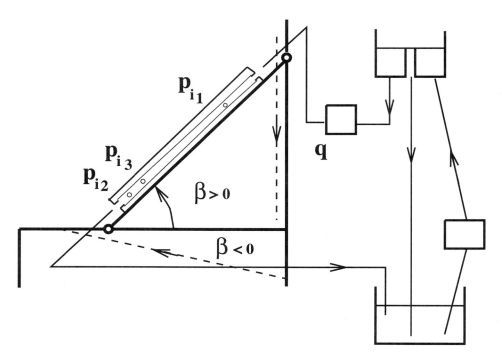

FIGURE 10.29   Experimental set-up. The flow may be either descending ($\beta > 0$) from the vertical position or ascending ($\beta < 0$) from −20°. The three pressure measurement stations are labeled p_{i1} near the tube entry and p_{i2} and p_{i3} in the downstream segment. Pressure taps are located at the tube lateral edges.

Inclined Tube

Gravity-friction flows have been explored in thin-walled long collapsible tubes put in an air-filled rigid box.[6] The silicon-rubber tube, attached to rigid ducts at both ends, lays on a foam bed and on a set of V-shaped plates (Fig. 10.29) to ensure the alignment of rigid and water-filled floppy tube axes and to prevent the folding of the cross-section (see comments in the previous section). Both entry (p_1) and exit (p_d) pressures are measured. The water flow is induced by a constant head tank (head of 2.5 m with respect to the tube axis in a horizontal position). The flow rate is adjusted by a downstream valve.[†] Downward (0 to 90°) and upward (−20 to 0°) flows were investigated.

In upward pipes, a flow-limiting throat appears quickly and the iso-p_2 $q − \Delta p$ curves do not exhibit any plateau. The downward inclined tube is divergent at rest, because of the hydrostatic effects. The downstream transmural pressure p_d is set to negative values. The highly collapsed tube opens gradually while the flow rate is increased. When the pressure drop Δp is equal to zero (the measurement stations are sufficiently remote from the tube attachments), the tube configuration is uniform between the pressure taps p_1 and p_d. Flow data in uniformly collapsed inclined tubes[††] are given in Fig. 10.30.

Circular and elliptical attachments were used. The downstream pressure (p_d) was measured at 150 mm (p_3) or 15 mm (p_2, i.e., possibly in the experimentally imposed divergent) using circular attachments and 60 mm upstream from the collapsible tube outlet at a single station using elliptical attachments. Three modes of operation for each kind of flow (horizontal, ascending, and descending) were defined: (1) circular attachment and p_2 constant; (2) circular attachment and p_3 constant; and (3) elliptical attachment and p_2 constant. The attachment condition plays an important role, because it affects the outlet divergent shape and the longitudinal deformation.[†††] Results in a downward inclined tube (inclination angle of 45° with the three modes of operation are illustrated by iso-p_d $\Delta p − q$ curves (Figs. 10.31

[†]The downstream valve entry pressure is given by 190.5 $q^{-0.12}$, where the pressure and flow rate units are *Pa* and *cm³/s⁻¹*, respectively.

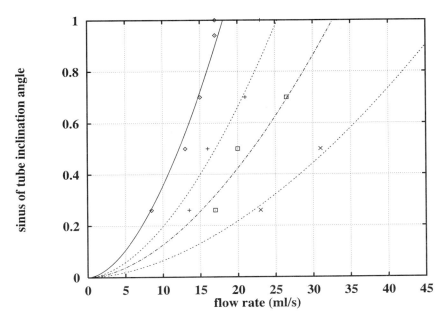

FIGURE 10.30 Experimental data in uniformly collapsed tube flow conditions (superscript +) for an inclination angle ranging from 0° to 90° at $p^+ = -1$ (×), -1.5 (□), -2 (+), and -3 kPa (◊), and computational results from the force balance equation in a turbulent pattern (q in cm^3)/s^{-1}).

to 10.33). When mode 1, which is the usual mode, is operated, the experimental curves are stopped when the flow rate reaches the threshold set by the experimental bench. Fluctuations and oscillations are observed in the plateau region for a flow rate more or less close to the curve knee. With mode 2, a disruption is observed after the uniform state conditions; a downstream throat appears. Stability is thus tightly linked with location of the p_d measurement station; the greater the local wall axial tension the steadier the flow for modes 1 and 2. The mode 3 iso-$p_2 \, \Delta p - q$ curves resemble the mode 1 plots. However, strong fluctuations appear at high flow rates when the downstream divergent angle is smaller.

The bulge displayed in Fig. 10.34 is located near the downstream end of an inclined collapsed tube. The tube axial configuration may be decomposed into three segments. Segment I is the highly collapsed part ($p \leq p_c$) where the line of contact shortens in the streamwise direction to reach a distal point of

---

[†] In uniform tubes (superscript +), computations from the force balance $\rho g \sin \beta^+ = f_v^+$ were performed both for laminar and turbulent flows. Computational and experimental results are in agreement. The equilibrium equation, for example, in turbulent flow, is given by:

$$\sin \beta^+ = \kappa f_T^+ \left(v^{1/4} / g \right) p^{+(11-n)/4n} q^{+7/4}$$

where $\kappa = \chi_{i0}^{5/4} / \left(8\sqrt{2} A_{il}^3 p_l^{(11-n)/4n} \right)$ is a tube-dependent coefficient. In this set of experiments, the above equation becomes, with SI units:

$$\sin \beta^+ = 0.2606 \, 10^5 \, f_T^{+} \, p^{+1.24} \, q^{+1.75}$$

The experimental pressure-dependent shape factors are given below for the investigated pressures:

$p + (kPa)$	-1	-1.5	-2	-3
f_T^+	0.27	0.31	0.34	0.37

[†††] The collapsible tube flow behavior depends strongly on the location of the measurement site of the downstream internal pressure. In other experiments in horizontal tubes, p_{i2} is measured in the rigid hydraulic circuit downstream from the collapsible tube.

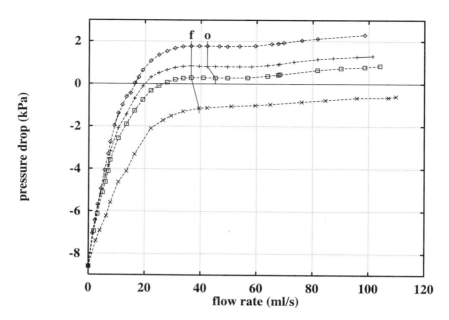

FIGURE 10.31   Pressure drop $p_1 - p_2$ *(kPa)* vs. descending flow rate q (cm³/s⁻¹) curves in iso-p_2 conditions in a 45°
inclined tube. Mode of operation 1: circular attachment, $L = 1.2$ m, $p_3 = -3$ (◊), -2 (+), -1.5 (□), and -0.2 *kPa* (×).
Peculiar points are connected by labeled lines which define the beginning of fluctuations (**f**) and the beginning of
low-frequency small-amplitude oscillations (**o**), which grow with further increase in flow rate. The governing parameters are $\gamma = 3584$, $\lambda = 50$, and $\delta = 369$. The tube properties are the following:

a_0 (mm)	k_0	h_0 (mm)	n	$K\,(N/m^{-2})$
21.3	2.98	0.76	1.845	28.31

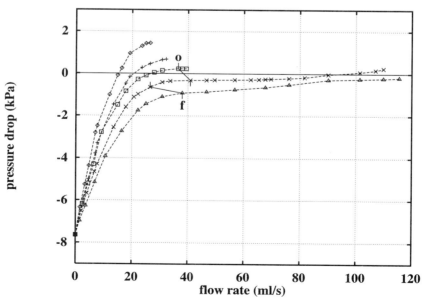

FIGURE 10.32   Pressure drop $p_1 - p_3$ *(kPa)* vs. descending flow rate q (cm³/s⁻¹) curves in iso-p_3 conditions in a 45ᵒ
inclined tube. Mode of operation 2: circular attachment, $L = 1065$ mm, $p_2 = -3$ (◊), -2 (+), -1.5 (□), -1 (×), and
-0.4 *kPa* (△). For definitions of labeled lines, see Fig. 10.31. The governing parameters are $\gamma = 3584$, $\lambda = 56$, and $\delta = 416$.

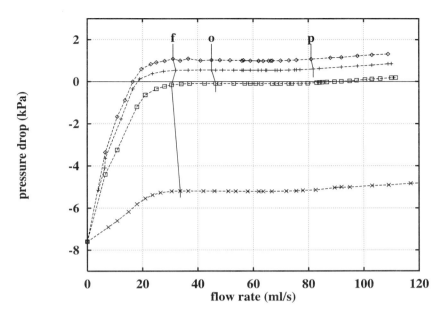

FIGURE 10.33 Pressure drop $p_1 - p_2$ (kPa) vs. descending flow rate q (cm³/s⁻¹) curves in iso-p_2 conditions in a 45° inclined tube. Mode of operation 3: elliptical attachment, $L = 1065$ mm, $p_2 = -2$ (◊), -1.5 (+), -1 (□), and $+4.2$ kPa (×). For definitions of labeled lines, see Fig. 10.31; label **p** means a tube pinch without significant change in p_2. The governing parameters are $\gamma = 3584$, $\lambda = 50$, and $\delta = 369$.

contact at its exit (dA_i/dx increases progressively). Segment II is shaped like a bulge: dA_i/dx rises, becomes nul, and decreases rapidly. Segment III is the downstream throat (locally, $dA_i/dx \rightarrow -\infty$) at the entrance of the exit divergent. This peculiar configuration[67] may be the consequence of a non-stationary process.[18]

Flow Limiter

A specific device, the flow limiter, has been used to investigate the flow limitation in horizontal flows with constant inlet transmural pressure ($p_1 = 0$). The test section is composed of a compliant pipe enclosed in two coaxial transparent rigid cylinders (Fig. 10.35). The fluid simultaneously enters the collapsible tube and fills the space between the two chambers, so that p_1 is held constant at zero ($\Delta p = -p_2$). Water glycerol mixtures were used. The liquid leaves the flexible pipe through a short rigid bend[†] to avoid air trapping in the flexible pipe. Both p_e and q are measured.

Experimental flow rate $-p_2$ vs. q curves can be subdivided into six successive phases (Fig. 10.35). During phase 1, the tube collapses progressively while q is increasing. Phase 2 is characterized by an unsteady collapse; both p_2 and q decrease without oscillations. When the collapsing wave reaches the downstream end, a sudden additional narrowing is observed, which induces a backward-running pressure wave. Phase 3 corresponds to a new steady state, defined by a very small decrease in flow rate. Phases 4 through 6 belong to the inverse path of the test, displaying a hysteresis. An unsteady re-opening of the conduit is observed during phase 5. In phase 6, the flow is again steady.

In Fig. 10.36, the greater μ, the higher the slope of the curves and the lower the flow rate at which phase 2 is initiated. The steady flow is bounded by the theoretical curve based on the critical flow rate associated with p_2. Flow limitation (beginning of phase 2) seems to appear with the onset of tube instability: the flow-rate is the maximum obtainable flow rate ($q = q_2^\star$). In phases 3 and 4, the flow rate

---

[†]The head loss (in *Pa*) of the rigid segment downstream from the compliant test section depends on the flow rate for a given flowing fluid, according to the following formulae: $1.11\,q + 2.96$, with q in *cm³/s⁻¹*, for a fluid viscosity μ of 1.3×10^{-3} *Pa/s* and of $3.41\,q + 4.80$ for μ of 15.9×10^{-3} *Pa/s*. The internal pressure at the flexible pipe exit can thus be estimated.

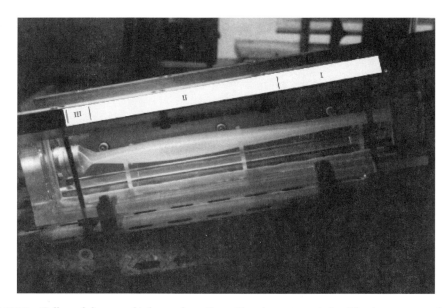

FIGURE 10.34 Collapsed downward tube configuration with a downstream bulge. The downstream tube segment exhibits three segments: segment I, the highly collapsed part, segment II, the bulge, and segment III, the distal constriction. The tube properties are given in Fig. 10.31.

variations follow a path parallel to the critical outlet flow rate. For the highest viscosity (plot d of Fig. 10.36), the flow rate q remains lower than q_2^* during phases 1 and 2, does not follow the q_2^* traces during phase 3, and does not exhibit hysteresis.

Unsteady Flow Experiments

Unsteady flow experiments have been performed in collapsible tubes subjected to pressure ramps in either a Starling resistance type model[51,52] or in a two-serial element model, composed of a balloon and a flexible straight tube of elliptical unstressed cross-section.[68] The latter model is briefly presented (Fig. 10.37). the model was placed in a rigid box. A negative pressure was initially set in the box to determine the initial volume of the compliant structure. The pressure in the box was then raised to a maximum in a given time span.

The flow rate time evolution can be subdivided into three successive stages (Fig. 10.38): (1) a phase of flow acceleration characterized by a rapid increase in flow rate up to a maximum reached for $p_c < p_2 < 0$; (2) a brief phase of flow limitation, during which q decreases quickly whereas $\Delta p = p_1 - p_2$ and p_e continue to rise; and (3) a drainage phase. The speed indices (S_1 and S_2) can be estimated from pressure and exit flow rate measurements at the pressure-measurement sections when the tube law is known, neglecting the collapse flow due to the wall motion.[†] The flow becomes critical ($S_2 = 1$) simultaneously with the peak flow, before the opposite walls come into contact at the tube exit.[††] When contact between opposite walls occurs at the tube outlet, S_2 reaches its maximum. Due to the contact discontinuity, S_2 then drops abruptly. Soon afterwards, p_1 becomes negative. The critical section travels upstream, followed by the point-contact section, to reach the pipe inlet in a very short time. The critical section migrates with a decaying critical transmural pressure $p^*(x)$ while q is decreasing.[70] The final period of phase 2 is

---

[†]This hypothesis is questionable in the range $p_c < p < 0$, where the tube compliance is important. However, $\partial A_i / \partial p$ becomes negligible when $p < p_l$.

[††]The numerical results, in good agreement with experimental data, show that the peak flow is critical, with $p_c < p_2 < 0$.[69]

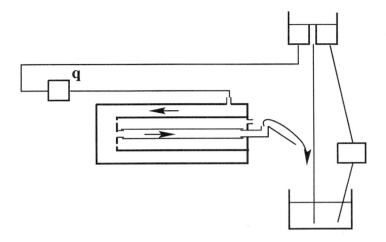

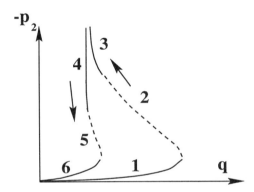

FIGURE 10.35 Top: the flow limiter device. The inlet transmural pressure p_1 is nul because of negligible entry effects (the collapsible tube chamber senses p_1; the fluid inside is at rest during the stationary state). Bottom: flow rate vs. pressure drop relationship of the flow limiter device. Six phases can be defined (see text). Hysteresis is observed between the increasing flow rate (from zero, "forth test") and decreasing flow rate ("back test") experimental curves. In forth tests, instabilities arise from critical conditions.

characterized by a slightly supercritical flow in the entire tube.[†] The flow then returns to a subcritical state throughout the entire conduit, the flow velocity being too low; the viscous forces are then dominant in the fully collapsed tube.[††]

Variations in the rate of pressurization were investigated, too. The stronger the initial rate of pressurization, the shorter the time required for the critical section to appear at a given station of the pipe, the greater the peak flow, and the higher the value of the critical transmural pressure p^*. The flow rate depends, therefore, on the pressure history. However, the timespan of critical section motion along the pipe is almost constant. Similar findings were observed experimentally[51] and in human healthy subjects.

[†]Flow-rate changes during this stage may be understood with the help of the following equation, derived from the mass and momentum conservation principles:

$$\rho / A \left(\partial q / \partial t + 2U \partial q / \partial x \right) + \left(1 - S^2 \right) \partial p / \partial x + f_\upsilon = 0 \qquad (10.23)$$

During the entire test, the axial pressure gradient $\partial p / \partial x$ is negative, whereas the terms $\partial q / \partial x$ and f_υ are positive. The third term of Eq. 10.23 is positive when $S > 1$; $\partial q / \partial t$ must be negative.

[††]When $S_1 < 1$, q decreases with time when the viscous effects are predominant (see Eq. 10.23).

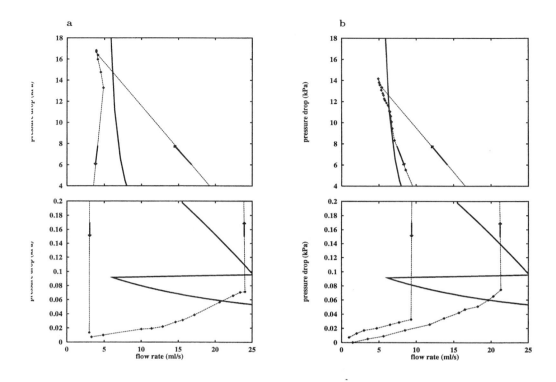

FIGURE 10.36 Relationships between the pressure drop $-p_2$ (kPa) and the volume flow rate q (ml/s) curves in the flow limiter device for four water–glycerol mixtures of increasing viscosity from a to d. Direction of flow rate changes is indicated by arrows. The critical flow rate q^* (A_{i2}) is also plotted (thick line). The four plots, a to d, are decomposed in order to display the high pressure range [4, 18] (kPa) including the contact condition (top), and the low pressure range [0, 0.2] (kPa) including the flow-limitation region (bottom). The tube and flow properties are given below, with the similarity-condition paramater λ/γ:

a_0 (mm)	k_0	χ_0 (mm)	h_0 (mm)	$K\,(N/m^{-2})$		test	$\mu\,(mPa/s)$	λ/γ
7.70	1.30	43.0	0.415	31.35		1	1	0.033
						2	3	0.073
						3	17	0.406
						4	34	0.813

The external pressure maximum was also modified. The flow-limiting site does not reach the pipe inlet when the external pressure maximum is below a given threshold. A decrease in initial negative external pressure p_e (0) induces an increase in the total exhaled volume and a higher peak flow, which appears later, p_2 remaining nearly constant. In experiments with smaller pressurization rates, the flow was found to remain subcritical during the entire operation.[†]

10.7 Physiological Applications

Collapsible tube flow investigations can either be focused on the flow behavior with applications to physiological flows, directed toward the fluid–solid coupled stability with applications to vascular and

---

[†]Eq. 10.23 shows that a peak flow may occur while the flow remains subcritical; indeed, $\partial q/\partial t = 0$ when $S < 1$ because $\partial q/\partial x > 0$ and $\partial p/\partial x < 0$.

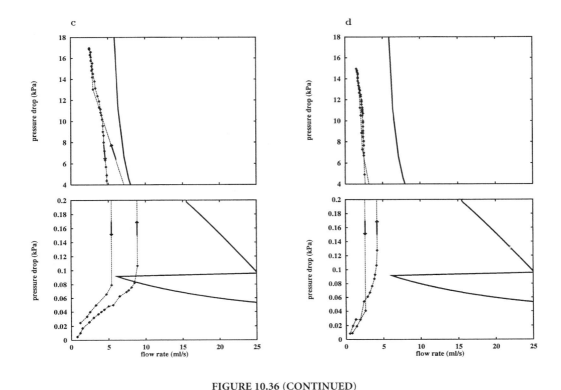

FIGURE 10.36 (CONTINUED)

pulmonary auscultation sounds,[71,72] or aimed at describing the fluid-dependent displacement of the solid boundary with applications to cardiac valve motions.[73]

Physiological conduits vary both in size and structure along their length. Axial changes in bending stiffness have been taken into account in some works.[74] Yet, the sigmoidal shape of the $A_i(p)$ relationship is observed in every compliant duct, whether the wall is relatively homogeneous in size and in mechanical properties or is made of a composite material of non-uniform geometry.[75] As a consequence, the cross-sectional area decreases very quickly for slightly negative transmural pressures down to its opposite wall contact value in test sections as well as *in vivo*.[76]

The venous circulation is characterized by the vein compliance, which allows the storage of blood volume (up to 70%) in the venous network; veins are thus modeled by thin-walled floppy tubes. Flows with combined friction and gravity have been investigated in order to model the downward blood flow in superficial veins supposed to be subjected, at least partially, to the atmospheric pressure. The external jugular veins of the giraffe provides a useful physiological model to explore the gravity effects because of the large range in the pressure gradient. Additionally, the veins of test pilots experience an added acceleration term.[77]

The time-mean exit pressure in proximal veins is controlled by the heat pump to fit the physiological requirements. However, at the exit of any capillary network remote from permanent sources of unsteadiness, the pressure can be supposed to be constant, especially in the rigid skull. Whether the inlet of an isolated vein, like the external jugular vein, is located downstream from a capillary network or the outlet of a distal vein is the junction with an in-parallel capillary network, the venous blood can be assumed to flow either under inlet or outlet constant conditions, respectively. Blood in superficial veins, in their natural environment or slightly pressurized by support stockings, is conveyed under constant external pressure.

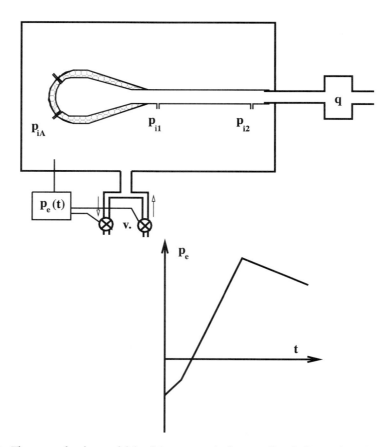

FIGURE 10.37 The monoalveolar model (top) is composed of a compliant balloon whose inner wall is covered with soft foam and a flexible pipe in series. Internal pressures are measured in the balloon p_{iA}, near the tube inlet p_{i1} and near the outlet p_{i2}. The deformable model, placed in a transparent rigid box on a foam bed, is first inflated and then submitted to a pressure ramp (bottom). The flow rate is measured at the tube exit.

Many scenarios in terms of tube axial profiles and critical conditions can occur (see section 10.4). Critical conditions may be attained either at the tube entrance or exit or at an intermediate site because of the tube law discontinuity. Such phenomena provide a mechanical process liable to produce a quick transient regulation of blood flow. This physical mechanism can be understood as the first step of a local regulation, which can be taken over later by a biochemical regulation of the vasomotor tone set by flow-induced stresses on the endothelial cells. The biochemical regulation is associated with the local release of substances like endothelin, prostacyclins, nitric oxide, etc. The nitric oxide, which induces a relaxation of the vascular tone, might require the shortest possible response time, the magnitude being of order of 100 ms. The low time-constant mechanical flow regulation can be set for a faster response.

The shear stress distribution is affected by the cross-sectional shape and the wall curvature changes. A set of shear stress magnitudes has been computed for different veins of known diameters and given flow rates, assuming a steady laminar developed flow (Table 10.4). For a given flow rate and azimuth, the wall shear stress can increase up to 30 times from the open section ($p = 0$) to the point-contact section in steady flow (see Figs. 10.19 and 10.20). Furthermore, at a given station, the more collapsed the cross-section, the greater the circumferential variation in wall shear stress. Consequently, the endothelium, subjected to huge changes in tangential stress, might limit the stress range by releasing substances which, acting on the adjoining media cells, might locally modify the lumen size.

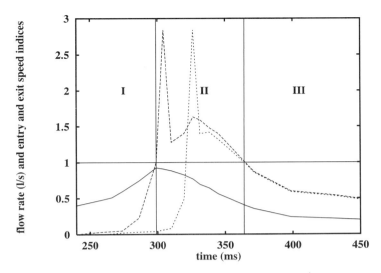

FIGURE 10.38   Time variations of the volume flow rate q and of the speed indices computed at the measuring stations of the downstream (S_2, dashed line) and upstream (S_1, dotted line) transmural pressure p in a monoalveolar compliant model. Three phases are defined: phase I of flow acceleration, phase II of wave-speed flow limitations ($S \geq 1$), and phase III of viscous drainage.

TABLE 10.4   Estimated Maximal Shear Stresses at the Venous Wall from the Normalized Value $\tau_w(\tilde{p}_c)$ (see Fig. 10.20), using the Scale $\mu q/l_0^3$ and Assuming a Blood Dynamic Viscosity of 4mPa/s

a_0 (mm)	q (cm³.s⁻¹)	τ_w (Pa)
12.5	75.00	5.68
5	1.9	2.25
1.5	0.13	5.70

The confluence of vein tributaries is located at the lateral edges of the veins, while the venous valvules, aimed at avoiding backflow during transient external compression, are inserted on the flat sides. These sides might come into contact temporarily but repeatedly during the venous lifespan, and, undergoing strong reaction forces, experience tissue fatigue and possible damage. Such contact reactions might, in the long run, induce a functional deficiency of valvules, without locking the venous return.

Unsteady flow experiments have been performed in order to model venous flow under muscle contraction, external cuff compression,[51] or forced expiration maneuvers.[68] The latter test is commonly performed in lung function testing to quantify the pulmonary function loss especially in patients with chronic obstructive lung disease. The forced expiration test corresponds to the quickest exhalation of the vital capacity, which is the maximal volume of air that can be expelled from the lungs after a complete inspiration. Flow limitation has been demonstrated from iso-p_1 $\Delta p - q$ curves (i.e., in iso-volume conditions),[78] when the lung volume is lower than 75% of the vital capacity.

Flow limitation may be linked to a critical condition (see section 10.6). The results obtained from simple models of the respiratory network have been validated by *in vivo* experiments, during which both the flow rate and the tracheal cross-section areas were measured.[76] The trachea collapses very early and quickly, the peak flow occurring during the first 300 ms of the maneuver. During this period, the tracheal area is reduced to 40% of its initial size. All conditions required for critical conditions are set up with such catastrophic changes associated with the wave of collapse. Estimates of the local velocity and wave speed were of the same order of magnitude. A critical flow is highly probable, which could explain the occurrence of the peak flow and the flow pattern, at least for a short period after the peak flow.

10.8   Conclusion

The one-dimensional model has been very useful in the investigation of flow in collapsible tubes. However, the one-dimensionality of fluid flow is based on a set of hypotheses which are incompatible with the complexity of flow behavior. Criticisms may be directed toward the fluid mechanics side as well as to the solid mechanics aspects.

In a one-dimensional model, momentum and kinetic energy coefficients should be introduced in order to take into account the non-uniform velocity distribution in any collapsed pipe cross-section. However, these coefficients are usually supposed to be equal to unity. Furthermore, as soon as the opposite walls come into contact, the flow splits into two separate streams. The upstream and downstream effects of the stream division on the flow field are ignored. Transverse velocities involved during the wall motion are also neglected. Above all, the axial change in cross section area may be found to be experimentally important (Fig. 10.34). Downstream from the constriction designed by the collapsed tube, an induced divergent segment can be associated with jet and flow separation. Flow separation strongly affects the pressure distribution and the tube mechanics. Additionally, the three-dimensional flow through the elastic jump has been experimentally demonstrated.[7]

In the standard one-dimensional model, the physical agents of the longitudinal flexural deformation of the tube are not taken into account. They can be incorporated in a more complex way than in the standard model via the local axial tension.[11,12,16,17] These factors are important because they affect the stability of the fluid–tube couple. However, a detailed analysis of the actual mechanics of the fluid–tube couple will need future three-dimensional studies both in fluid mechanics and in solid mechanics. At the present time, three-dimensional models, oversimple in fluid mechanics but more complex in solid mechanics and incorporating contact, are in progress.[19] Time-dependent velocity profiles measured by laser Doppler velocimetry in an oscillating collapsible tube[79] confirm the need to venture deeper. However, advanced models do not necessarily enhance the knowledge of the physiological system of interest.

Nomenclature

Tube-Dependent Variables

A	:	area	a_0	:	major semi-axis
c	:	wave speed	D	:	flexural rigidity
E	:	Young modulus	$\mathcal{F}$	:	polynomial function
f	:	friction shape factor	f	:	external forces
$\mathcal{G}$	:	mapping coefficient	h	:	wall thickness
K	:	pressure scale	k_0	:	tube ellipticity
L	:	tube length	$\mathbf{M}$	:	bending moment
$\mathcal{M}$	:	tube law parameter	N	:	lobe number
$\mathbf{n}$	:	normal	n	:	similarity tube law exponent
p	:	transmural pressure	$\mathbf{R}$	:	contact reaction
r	:	reaction distribution	s	:	curvilinear abscissa
$\mathbf{T}$	:	stress resultant	$\mathbf{t}$	:	tangent
β	:	tube inclination angle	θ	:	(j,t) angle
ν	:	Poisson coefficient	ξ	:	correction factor
χ	:	tube perimeter			

$\alpha_{po}, \beta_{po}, \gamma_{po}$: tube law coefficients for $p_p < p \leq 0$

$\alpha_{cp}, \beta_{cp}, \gamma_{cp}$: tube law coefficients for $p_c < p \leq p_p$

$\alpha_{lc}, \beta_{lc}, \gamma_{lc}$: tube law coefficients for $p_l < p \leq p_c$

Fluid-Dependent Variables

f_υ	:	head loss per unit length	M_a	:	Mach number
m	:	flow pattern exponent	q	:	volume flow rate
R	:	flow resistance	S	:	speed index
Re	:	Reynolds number	u	:	fluid velocity
U	:	cross-sectional average velocity	α_k	:	kinetic energy coefficient
υ	:	dimensionless velocity	γ	=	$a_0(\rho K)^{1/2}/\mu$
α_m	:	momentum coefficient	Δp	:	axial pressure drop
δ	=	$\rho g L/K$	λ	=	L/a_0
Λ	:	head loss coefficient	ρ	:	fluid density
μ	:	fluid dynamic viscosity	τ	:	shear stress
σ	:	conductivity			

Other Variables

g	:	gravity	t	:	time
Z	:	ascending vertical axis			
$\{x, y, z\}$	:	cartesian coordinates with unit vector set $\{i, j, k\}$			

Sub- and Superscripts

$\tilde{\cdot}$	:	normalized	$\bar{\cdot}$	:	time mean
$\cdot^\star$	:	critical	$\cdot_0$	:	unstressed
$\cdot_b$	:	buckling	$\cdot_p$	:	for first parallel (wall edges)
$\cdot_c$	:	point-contact at $p = p_c$	$\cdot_l$	:	line-contact at $p = p_l$
$\cdot_L$	:	laminar	$\cdot_T$	:	turbulent
$\cdot_1$	:	tube entry	$\cdot_2$	:	tube exit
$\cdot_d$	:	downstream tube segment	$\cdot_u$	:	upstream (head tank)
$\cdot_e$	:	external	$\cdot_i$	:	internal, luminal
$\cdot_+$	:	right of discontinuity	$\cdot_-$	:	left of the discontinuity
$\cdot_g$	:	driving (pressure)	$\cdot_w$	:	wall
$\cdot_A$	:	alveolar (balloon)	$\cdot^+$	:	uniform axial shape

References

1. Shapiro, A.H., *ASME J. Biomechanical Eng.*, 99, 126, 1977.
2. Pedley, T.J., *Fluid Mechanics in Large Blood Vessels*, Cambridge University Press, Cambridge, 1980.
3. Comolet, R., *Biomecanique Circulatoire*, Masson, Paris, 1984.
4. Comolet, R., *Cardiovascular and Pulmonary Dynamics*, INSERM, Paris, 1977, 451.
5. Nahmias, J., *Thèse de Troisieme Cycle VI*, University of Paris, 1980.
6. Mazghi, M., *Thèse de Troisieme Cycle XII*, University of Paris, 1986.
7. Kicecioglu, I. et al., *J. Fluid Mech.*, 109, 367, 1981.
8. Griffiths, D.J., *Med. Biol. Eng.*, 9, 581, 1971.
9. Cowley, S.J., *J. Fluid Mech.*, 116, 459, 1982.
10. Bertram, C.D. and Pedley, T.J., *J. Biomechanics*, 15, 39, 1982.
11. Cancelli, C. and Pedley, T.J., *J. Fluid Mech.*, 157, 375, 1985.
12. Jensen, O.E. and Pedley, T.J., *J. Fluid Mech.*, 206, 339, 1989.
13. Rubinow, S.I. et al., *J. Theor. Biol.*, 35, 299, 1972.
14. Wild, R. et al., *J. Fluid Mech.*, 81, 273, 1977.
15. Hayashi, S. et al., *JSME Int. J.*, Ser. B, 37, 349, 1994.
16. Matsuzaki, Y. and Matsumoto, T., *ASME J. Biomechanical Eng.*, 111, 180, 1989.

17. Matsuzaki, Y., *ASME J. Biomechanical Eng.*, 116, 469, 1994.
18. Jensen, O.E., *J. Fluid Mech.*, 220, 623, 1990.
19. Heil, M. and Pedley, T.J., *J. Fluids Structures*, 10, 565, 1006.
20. Knowlton, F.P. and Starling, E.H., *J. Physiol.*, 44, 206, 1912.
21. Holt, J.P., *Am. J. Physiol.*, 134, 292, 1941.
22. Rodbard, S. and Saiki, H., *Am. Heart J.*, 46, 715, 1953.
23. Conrad, W.A., *IEEE Trans. Biomed. Eng.*, 16, 284, 1969.
24. Katz, A.I. et al., *Biophys. J.*, 9, 1261, 1969.
25. Bonis, M. et al., *J. de Méca. Appl.*, 5-2, 123, 1981.
26. Flaherty, J.E. et al., *SIAM J. Appl. Math.*, 23, 446, 1972.
27. Kresch, E., *J. Biomechanics*, 12, 825, 1979.
28. Ribreau, C. et al., *ASME J. Biomechanical Eng.*, 115, 432, 1993.
29. Ribreau, C. and Bonis, M., *J. Fr. Biophysique Médecine Nucléaire*, 3, 153, 1978.
30. Palermo, T. and Flaud, P., *J. Biophys. Bioméca.*, 11, 105, 1987.
31. Dion, B. et al., *Med. Biol. Eng. Comput.*, 33, 196, 1995.
32. Thiriet, M. et al., *Innov. Technol. Biol. Med.*, 8, 99, 1987.
33. Rosenberg, N. et al., *Photogrammetric Eng. Remote Sensing*, 56, 1273, 1990.
34. Bertram, C.D. and Ribreau, C., *Med. Biol. Eng. Comput.*, 27, 357, 1989.
35. Ribreau, C. et al., *J. Biophys. Biomecanique*, 10, 57, 1986.
36. Brower, R.W. and Scholten, C., *Med. Biol. Eng.*, 13, 839, 1975.
37. Brower, R.W. and Noordergraaf, A., *Ann. Biomed. Eng.*, 1, 333, 1973.
38. Bonis, M. and Ribreau, C., *La Houille Blanche* 34, 165, 1978.
39. Lyon, C.K. et al., *Circ. Res.*, 47, 68, 1980.
40. Lyon, C.K. et al., *Circ. Res.*, 49, 988, 1981.
41. Bertram, C.D., *J. Biomechanics*, 15, 201, 1982.
42. Bertram, C.D., *J. Biomechanics*, 19, 61, 1986.
43. Bertram, C.D. et al., *J. Fluids Structures*, 4, 125, 1990.
44. Fry, D.L., *Med. Phys. Biol.*, 3, 174, 1958.
45. Ohba, K. et al., *Technol. Rep. Kansai Univ.*, 25, 1984.
46. Ohba, K. et al., Progress and New Directions of Biomechanics, Fung, Y.C. et al., Eds., Mita Press, 1989, 213.
47. Bertram, C.D. et al., *J. Fluids Structures*, 5, 391, 1991.
48. Low, H.T. and Chew, Y.T., *Med. Biol. Eng. Comput.*, 28, 217, 1991.
49. Beyar, R. et al., *Med. Biol. Eng. Comput.*, 22, 499, 1984.
50. Bendaoud, M., Doctoral thesis, University of Compiegne, 1983.
51. Kamm, R.D. and Shapiro, A.S., *J. Fluid Mech.*, 95, 1, 1979.
52. Jan, D.L. et al., *ASME J. Biomechanical Eng.*, 105, 12, 1983.
53. Pedley, T.J. and Stephanoff, K.D., *J. Fluid Mech.*, 160, 337, 1985.
54. Tadjbakhsh, I. and Odeh, F., *J. Math. Anal. Appl.*, 18, 59, 1967.
55. Bonis, M. and Ribreau, C., ASME *J. Biomechanical Eng.*, 103, 27, 1981.
56. Ribreau, C., Doctoral thesis, University of Paris, 1991.
57. Lambert, R.K. and Wilson, A.T., *J. Appl. Physiol.*, 33, 150, 1972.
58. Elad, D. et al., ASME *J. Biomechanical Eng.*, 109, 1, 1987.
59. Kimmel, E. et al., ASME *J. Biomechanical Eng.*, 110, 292, 1988.
60. Taylor, L.A. and Gerrard, J.H., *Med. Biol. Eng. Comput.*, 15, 11, 1977.
61. Oates, G.C., *Med. Biol. Eng.*, 13, 780, 1975.
62. Olsen, J.H. and Shapiro, A.H., *J. Fluid Mech.*, 6, 513, 1967.
63. Ribreau, C. et al., *J. Fluids Structures*, 8, 183, 1994.
64. Langlet, A., Doctoral thesis, University of Paris, 1992.
65. Bonis, M., Doctoral thesis, University of Technology of Compiegne, 1979.
66. Bonis, M. and Adedjouma, S.D., personal communication.

67. Holt, J.P., *Circ. Res.*, 7, 342, 1959.

68. Thiriet, M. and Bonis, M., *Med. Biol. Eng. Comput.*, 21, 681, 1987.

69. Thiriet, M. et al., *Med. Biol. Eng. Comput.*, 25, 551, 1987.

70. Thiriet, M. et al., *ASME J. Biomechanical Eng.*, 111, 9, 1989.

71. Shimizu, *J. Fluid Mech.*, 158, 113, 1985.

72. Grotberg, J.B. and Gavriely, N., *J. Appl. Physiol.*, 66, 2262, 1989.

73. Bitbol, M. et al., *J. Fluid Mech.*, 114, 187, 1982.

74. Elad, D. et al., *J. Appl. Physiol.*, 65, 14, 1988.

75. Begis, D. et al., *J. Appl. Physiol.*, 64, 1359, 1988.

76. Thiriet, M. et al., *J. Appl. Physiol.*, 67, 1032, 1989.

77. Gaffie, X. and Guillaume, O., *Archiv. Physiol. Biochem.*, 104, 693, 1996.

78. Fry, D.L. and Hyatt, R.E., *Am. J. Med.*, 29, 672, 1960.

79. Ohba, K. et al., *Technol. Rep. Kansai Univ.*, 31, 1, 1989.

11

Techniques in the Modeling and Simulation of Blood Flows at the Aortic Bifurcation with Flexible Walls

Wen-Jei Yang
University of Michigan

Paul P. T. Yang
Southeast Permanente Medical Group

11.1   Introduction..11-1
11.2   Physical Models for Arterial Bifurcations................ 11-2
        Geometry • Blood and Its Rheology • Arterial Wall Models •
        Fluid Dynamic Parameters • Experimental Methods
11.3   Numerical Models 11-6
11.4   Aortic Bifurcation with Flexible Walls..........................11-7

This article presents a computer simulation of blood flow through the aortic bifurcation with flexible walls. A two-dimensional numerical model is employed to describe the blood flow and the wall behavior is delineated by a linear viscoelastic constitutive equation. The arbitrary Lagrangian-Eulerian method is adopted to deal with the moving boundaries. The successive-over-relaxation method is employed to solve both the vorticity and Poisson equations. It is disclosed that flexible walls do not significantly affect the overall flow patterns but tend to facilitate flow reversals or eddies during expansion but restrict them during contraction. A flexible bifurcation experiences shear stresses approximately 10% lower than those of a rigid one.

11.1   Introduction

An artery consists of a tube lined by endothelium. The tube wall consists of three layers: tunica intima, tunica media, and tunica adventitia. The flattened surface cells of the tunica intima are longitudinally oriented. The tunica intima includes subendothelial connective tissue and an elastic tissue layer designated the lamina elastica interna. Surrounding these inner layers is a relatively thick tunica media composed of smooth muscle and elastic tissue in varying proportions. The outer covering of an artery is the tunica adventitia, composed mainly of collagenous fibers.

The representative disease of arteries is atherosclerosis. Curvatures, junctions, and bifurcations of the large and medium arteries are severely affected by atherosclerosis. In the aorta, the abdominal segment is more severely affected than the arch or the thoracic segment. The blood flows in these regions are disturbed and the arterial walls are exposed to either high or low shear stresses. Many experimental and theoretical studies have been conducted to investigate the role of fluid dynamics in atherogenesis. Comprehensive reviews pertinent to the subject are available.[1-3] Various studies[2] favor the association of a low shear region with atherosclerosis.

Lou and Yang[3] reviewed and critiqued biofluid studies at various arterial bifurcations. They found that the coronary arteries are highly vulnerable to atherosclerosis and that a strong association exists between the sites of atherosclerosis and the inner walls of the curved coronary arteries. In order to understand the mechanism of localization of atherosclerosis, Tokuda et al.[4] and Kajiya et al.[5] conducted studies comparing the relationship between the flow structure and the morphology function of endothelium at a location of favorable artherogenesis with the relationship at an unfavorable location. They found that in the region of unfavorable atherogenesis:

1. Stress fibers are distributed mainly near the base of endothelium but also extend into the tunica media. Like cell nuclei, they are aligned perfectly parallel with the flow direction.
2. Acetyl LDL (lipoprotein) is distributed mainly in the endothelial layer and practically none is distributed in the tunica media.

In contrast, in the region of favorable atherogenesis, shear flow velocity is low, flow separation exists, and flow is more oscillatory. Most importantly, the alignment of stress fibers is nonhomogeneous with lower density and alignment of nuclei is also random.

In order to determine the fluid dynamic aspects of the mechanism of atherosclerosis, modeling and simulation techniques are employed to analyze flow structure at arterial bifurcations.

11.2 Physical Models for Arterial Bifurcations

The modeling and simulation of blood flows are the most popular and well-documented subjects in the field of biofluid mechanics. Fig. 11.1[6] illustrates how they can be modeled, from the simplest combination of a steady, one-dimensional, laminar viscous flow of a Newtonian fluid in a long straight tube of constant cross-section to the most complicated model of an unsteady, three-dimensional, pulsatile flow of a non-Newtonian fluid in a short, tapered tube with a flexible wall, imbedded in a flexible bed. Fig. 11.1 can be used as a reference in the study of flows at an arterial bifurcation.

Geometry

Many bifurcations exist in the arterial system and they can be classified into two general shapes. One type is Y-shaped, and has two branches of comparable sizes originating from the end of a trunk. The aortic, carotid, iliac, neural, and coronary bifurcations belong to this category. The other group is T-shaped, with a single side branch growing from a relatively continuous trunk. The renal, femoral, celiac, and mesenteric bifurcations belong to this class. However, when two bifurcations are close to each other, hydrodynamic interference exists between them, and it is necessary to treat them as a single unit. For example, the left and right renal bifurcations are almost perpendicular to the abdominal aorta and on opposite sides, thus forming a cross shape. The superior mesenteric and celiac bifurcations are situated next to each other on the same side of the abdominal aorta, forming a π shape.

Fig. 11.2 depicts schematics of idealized Y- and T-bifurcations. The two most employed geometric parameters in the former case are the bifurcation angle, β, defined as the angle between two branch axes, and the area ratio *(AR)* which is the ratio of the sum of the two branch areas to the trunk area. In general, β is equal to the sum of individual bifurcation angles β_1 and β_2, and *AR* is equal to the sum of individual

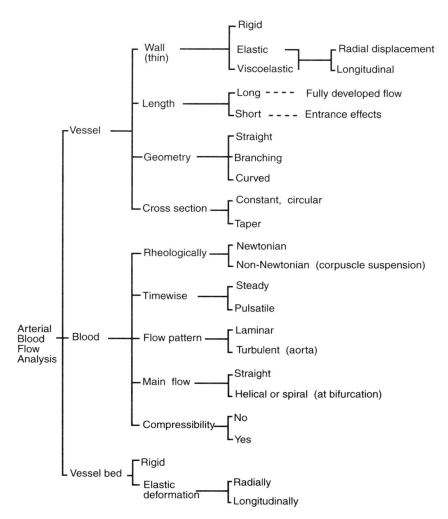

FIGURE 11.1 Selection of models for arterial flow analyses. (*Source:* Stettler, J.C., Niederer, P., Anliker, M., and Casty, M., *Ann. Biomed. Eng.*, 9, 1665, 1981. With permission from AIP.)

area ratios AR_1 and AR_2. In contrast, an idealized T-shaped bifurcation has only a (total) bifurcation angle, β, and three area ratios, AR, AR_1, and AR_2, as defined for an asymmetric Y-shaped bifurcation. Branch 1 is the side branch and branch 2 is a straight continuation of the trunk. Physiologically, AR_2 is close to unity while AR_1 is much less than unity. In general, AR ranges from 0.52 to 1.39 in human arteries and changes with age. The average AR for the aortic bifurcation varies from 1.1 in early infancy to 0.75 in the fifth decade. The AR is related to the shear stress level, flow field, and pressure wave reflection at bifurcation. In the aortic bifurcation, an increase in AR leads to a reduction in shear level in the branches and a noticeable increase in the duration of reverse flow distal to the lateral junctions. In the abdominal bifurcation, however, a reduction in AR values from 1.15 leads to an increase in the pulse wave reflection.

Blood and Its Rheology

It is commonly believed that the non-Newtonian effect is small in large vessels where the shear rate is high. However, low shear spots are also found near bends and bifurcations. Lou and Yang[3] used the

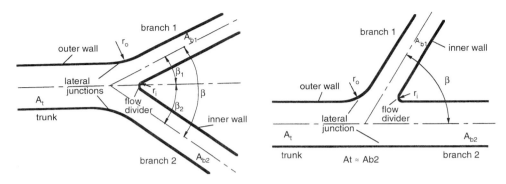

FIGURE 11.2   Schematics of idealized bifurcation shapes, (a) Y-shaped and (b) T-shaped.

Casson model, which is known to adequately represent non-Newtonian blood rheology. The Casson model expresses the blood stress (τ) strain rate ($\dot{\gamma}$) relation as

$$\tau^{1/2} = \eta^{1/2} \dot{\gamma}^{1/2} + \tau_o^{1/2} \tag{11.1}$$

where τ_o denotes the yield stress and η is the Casson viscosity or the limiting apparent viscosity. A power law non-Newtonian fluid model was employed to study flow in renal and general T-shaped bifurcations. It has the form of

$$\tau = b \dot{\gamma}^n \tag{11.2}$$

where b and n are constants. The model lacks the yield stress to address the low shear rate region. Another power law expression was derived by Walburn and Schneck which includes the hematocrit and two adjustable parameters. The Herschel-Bulkley equation is a power law model with an additional yield stress term, as

$$\tau = b \dot{\gamma}^n + \tau_o \tag{11.3}$$

It fails to fit the flow rate-pressure relationship for blood over a wide range of shear rates. All the models described above are for steady flow and do not address the time-dependent behaviors that arterial blood shows.

Numerical Bifurcation Studies

Numerous investigators have employed the Casson model in numerical studies of arterial bifurcation. For example, Lou and Yang[1,10] used a range of the yield stress, 0.0477, 0.1, and 0.143 dyn/cm^2, with a dynamic viscosity of 0.048 poise. The non-Newtonian effect on the fluid dynamics in arterial bifurcations was found to be insignificant.

Experimental Bifurcation Studies

The Separan solutions of various compositions with the power law type rheological behavior were used in experimental studies of arterial bifurcations. One example is

$$\tau = 0.031 \dot{\gamma}^{0.68} \tag{11.4}$$

Experimental results showed dramatic non-Newtonian influence globally compared with numerical simulations with the Casson model.

Arterial Wall Models

As noted in the introduction section, the arterial walls are made of composite materials consisting of three major layers: the intima, media, and adventitia. From the viewpoint of material strength, the intima contributes little to the mechanical strength of the walls and has a layer of endothelial cells in direct contact with blood. The media, consisting of smooth muscle and elastin embedded in collagen and an amporphous gel, provides the main strength of the wall. The adventitia consists of smooth muscles and connective tissues, such as collagen, elastic, and amorphous material. Its structure is loose and irregular.

The complicated structure causes arterial walls to exhibit nonlinear and viscoelastic rheological behavior. Different layers are so structured that the wall becomes stiffer as it is stretched. In addition, the arteries are subject to large initial deformation. Thus, the formulation of a complete artery model becomes a difficult task.

Three basic artery models are classified based on their geometric arrangements. The types are independent rings, thin membranes, and thick membranes. The thick membrane model is the best, while the thin membrane model is more realistic than the independent rings model. Each of these three basic models can be further classified elastic and viscoelastic models, based on constitutive behavior. Elastic and viscoelastic models can be further divided into linear, incrementally linear, and nonlinear models.

For wall modeling, potential simplifications can be made on longitudinal movement, external restraint on radial expansion, inertial effect from wall mass, linearity, independent rings, and viscoelasticity. The viscoelastic wall model yields results closer to normal physiological conditions than the elastic wall model, while a dissipative wall is more effective than a dissipative fluid in eliminating the high frequency oscillations.

As the material for arterial walls, silicon rubber has been used in all *in vitro* experimental simulations for distensible fiburcations, with a glass (rigid) model to determine the effect of the wall distensibility. In general, the major shortcomings of the experimental studies include lack of data or characterization of the viscoelasticity; lack of identical rigid models, especially those with the same curvatures at the flow divider and laterial junctions, for comparison; and lack of a precise means of directly measuring the shear rate at walls.

Fluid Dynamic Parameters

In dealing with a vessel bifurcation into two branches, two flow ratios are needed to determine the distribution of the flow in the branches:

QR_1 = flow in branch 1/total flow
QR_2 = flow in branch 2/total flow

The flow is said to be balanced if the ratio of two flow ratios is equal to that of two area ratios, i.e., $QR_1/QR_2 = AR_1/AR_2$.

For a physiologically pulsatile flow in the cardiovascular system, a complete description demands an entire time history of the flow or one representative Reynolds number *(Re)* plus a time history of a normalized flow. This represents the best approach to obtain more accurate results than the use of the Reynolds number at the peak, minimum, and average flow rates in a cycle, the Womersley number *(α)* of the base harmonic, or the Strouhal number *(St)*. Here,

$$\mathrm{Re} = \frac{DV}{\nu}, \ \alpha = \left(\frac{D}{2}\right)^2 \left(\frac{\omega}{\nu}\right)^{1/2}, \ St = \frac{\omega D}{V}$$

D denotes the hydraulic diameter; V, characteristic velocity; ν, kinematic viscosity; and ω, angular frequency. Hence the three dimensionless numbers are interrelated as

$$St = 4\alpha^2 / Re$$

In an arterial bifurcation, secondary flow is a spiral flow pattern from the outer wall to the inner wall in the bifurcation plane, and in the opposite direction along the upper and bottom walls. Flow depends on the interplay among the viscous forces, Coriolis force owing to the change of flow direction, other inertial forces, and the axial pressure gradient. It is harder to detect secondary flow in a pulsatile flow than to detect it in a steady flow. However, it induces a higher shear rate along the inner wall and either enhances or induces flow separation along the outer wall in a bifurcation.

Experimental Methods

Experimental approaches in biofluid dynamics at arterial bifurcations include flow visualization, velocity measurement, and shear rate or stress measurement. Flow visualization derives qualitative information from the overall flow field that can then be converted into quantitative data through the application of an image processing technique. In the study of arterial bifurcations, streaming birefringence, dye injection, particle tracers, hydrogen bubbles, and MRI (magnetic resonance image) may be used. Invasive hot-wire probes and noninvasive LDV (laser Doppler velocimetry) are available for *in vitro* bifurcation studies. Noninvasive UDV (ultrasonic Doppler velocimetry) and MRI can be adopted for *in vivo* studies. LDV with an optical fiber is feasible for an *in vivo* study, but the technique is invasive. In addition, digitized cineangiography and a combination of particle tracers and cinephotography have also been used.

One of the main pursuits in arterial bifurcation studies is the measurement of wall shear stresses. Three methods can be used for indirect wall shear rate evaluations:

(1) Using an opaque coating layer on the surface, an electrochemical probe, or a hot-film probe.
(2) Using the velocity profile derived from a nonlinear theory or wave equation together with measured pressures at two neighboring stations, or a flow rate.
(3) Extrapolating the wall shear rate from the measured point velocity at desired sites.

The difficulty of precisely locating an arterial wall, especially a distensible wall, causes errors in velocity and position values, resulting in errors in the shear rate values. Another important source of errors is the curve-fitting procedure because the spatial resolutions for most commercial LDVs and the distance limit required to avoid light scattering at the liquid-wall interface are of the same order of magnitude as the boundary layer thickness of a physiologically pulsatile flow at a bifurcation.

11.3 Numerical Models

Numerical models have evolved from two-dimensional (2D) to three-dimensional (3D) geometries, from Newtonian to non-Newtonian flows, and from rigid to distensible walls. In the case of arterial bifurcations, various simplifications have been imposed on complex wall geometry by using straight lines, round-up area ratios, round-up bifurcation angles, and the adoption of coordinate transformations and finite element methods. It has been reported that the use of round corners can avoid unrealistic wall shear stress peaks and exaggerated flow separations that are obtained from sharp corners at the flow divider and lateral junctions.

Two sets of variables have been in use in computational fluid dynamics: the velocity and pressure set and the stream function and vorticity set. The first set evaluates the wall shear rate indirectly by taking the derivative of the resulting velocity. As a consequence, errors in the shear rate and the velocity introduced by a mesh of low resolution can be significant in dealing with arterial bifurcations, that have thin boundary layers. In contrast, with the use of the stream function and vorticity set, the error arises directly from the process of solving a system of equations, because the wall shear rate is equal to the wall vorticity.

Due to the geometric complexity of a bifurcation, the 3D computational model has begun to surface only in recent years. The majority of the numerical models for arterial bifurcations are two-dimensional. Hence, the validity of 2D models needs to be examined.

The 2D model has three disadvantages or problems. First is its inability to provide 3D information. It is fortunate that major atherogenic sites around arterial bifurcations are more concentrated in bifurcation planes, which coincide with 2D planes, especially in early stages of atherosclerosis. The second problem is the difficulty of keeping the branch-to-trunk Womersley number ratio in a 2D model identical to that of the physiological bifurcation. The third problem with a 2D model is its lack of mechanism to accommodate secondary flow. Nevertheless, for the carotid bifurcation, a 2D model can provide basic information on the complexity of the flow field similar to that provided by a 3D model. For the aortic bifurcation, the results for a 2D model show a good agreement with those from the MRI for pulsatile flows.

In general, a 3D model is better than a 2D model in exhibiting overall flow patterns. However, a 3D model without proper flow pulsation, *Re*, *AR*, corner curvature, and spatial resolution may not give good results. The complexity of 3D modeling makes it difficult to produce a relatively fine mesh. Hence, a careful 2D model can provide a basic flow pattern without substantial errors, although it fails to accommodate secondary flow.

11.4 Aortic Bifurcation with Flexible Walls

In order to demonstrate the technique of modeling, this section presents a 2D numerical model for blood flow at the aortic bifurcation with flexible walls.[8,9] The model is a Y-shaped bifurcation, as shown on the left side of Fig. 11.2. It is symmetric with respect to the trunk center line. Flow rates into the two branches are equal, i.e., $QR_1 = QR_1 = 0.5$. Because of symmetry, the problem can be solved in one half of the flow domain.

Fig. 11.3 depicts a physiologically pulsatile flow rate over one cycle, adopted from Stettler et al.[7] A Reynolds number is defined in the trunk, with the flow rate at peak 1 in a phasic flow rate profile calculated as

$$Re_{pt} = LV / \nu \tag{11.5}$$

where L is the characteristic length, equal to the trunk hydrodynamic diameter, chosen to be 15 mm, and V is the characteristic velocity, equal to the mean velocity across a hydrodynamically equivalent circular pipe. At the kinematic viscosity ν of 4.5×10^{-6} m²/s, the Re_{pt} is equal to 1300. The time-average Reynolds number in the trunk over a cardiac cycle is 181, a laminar flow. Blood is treated as a Newtonian fluid in this section; the non-Newtonian effect has been presented elsewhere1.[10,11] The area ratio *AR* varies from 0.75 to 1.0 and 1.1, while the bifurcation angle β is 75 degrees, which is within the physiological range. In pulsatile flow, Lou[1] derived the Womersley number for a channel as

$$\alpha = 1.531 H_l \left(2\pi f / \nu \right)^{1/2} \tag{11.6}$$

where H_l is the half width of a channel. At H_l of 3.75 mm (15 mm ÷ 4), α is 8.8623 for a heart beat of 60 beats/min.

The vorticity transport and Poisson equations read

$$\frac{\partial \omega}{\partial t} \left(\psi_\eta' \right) \frac{\partial \omega}{\partial \xi} - \psi_\xi' \right) \frac{\partial \omega}{\partial \eta} / C_0 \left(C_1 \frac{\partial^2 \omega}{\partial \xi^2} - 2C_2 \frac{\partial^2 \omega}{\partial \xi \partial \eta} \right.$$
$$\left. + C_3 \frac{\partial^2 \omega}{\partial \eta^2} + C_4 \frac{\partial \omega}{\partial \eta} + C_5 \frac{\partial \omega}{\partial \xi} \right) / \left(C_0^2 Re_{pt} \right) \tag{11.7}$$

$$C_1 \frac{\partial^2 \psi}{\partial \xi^2} - 2C_2 \frac{\partial^2 \psi}{\partial \xi \partial \eta} + C_3 \frac{\partial^2 \psi}{\partial \psi^2} + C_4 \frac{\partial \psi}{\partial \eta} + C_5 \frac{\partial \psi}{\partial \xi} = -C_0^2 \omega \tag{11.8}$$

Here,

$$\psi'_{\eta} = \psi_{\eta} + \psi_{\eta m} \tag{11.9a}$$

$$\psi'_{\xi} = \psi_{\xi} + \psi_{\xi m} \tag{11.9b}$$

and

$$\psi_{\eta m} = y_t x_{\eta} - x_t y_{\eta} \tag{11.10a}$$

$$\psi_{\xi m} = -\left(x_t y_{\xi} - y_t x_{\xi}\right) \tag{11.10b}$$

where (x, y) denotes the spatial domain; (ξ, η), referential domain; t, time; ω, vorticity, ψ, stream function; and x_t and y_t, mesh velocities in the x and y directions, respectively. C_i's for $i = 0, 1, 2, 3, 4,$ and 5 are some coordinate transformation coefficients.

The arbitrary Lagrangian Eulerian (ALE) method is employed for the Lagrangian viewpoint to take care of moving boundaries and for the Eulerian viewpoint to avoid large grid distortions.

No-slip conditions apply along the solid walls, which are constant ξ lines, with fluid particles moving with the boundary. The walls are allowed to move only in the normal direction but not in the tangential. A potential flow region exists along the center line of the trunk, which is a constant η line, where ω is equal to zero. No flow is allowed across the line of symmetry, and ψ is constant along the boundary. The inlet is at some constant ξ line, with a fully developed flow. The pulsatile blood flow profile (Fig. 11.3) proximal to the aortic bifurcation is imposed at the inlet. Much of the flow activity is limited to the systole with a period of 1 sec in a normal case, typical in the descending aorta, iliac arteries, and femoral arteries. The exit of the branch corresponds to another constant ξ line, at a nearly fully developed flow. The second derivative of stream function and the first derivative of vorticity are set equal to zero in the axial direction.

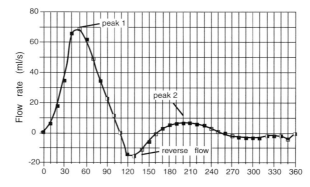

FIGURE 11.3 Physiologically pulsatile flow rate over one cycle. (*Source:* Stettler, J.C., Niederer, P., Anliker, M., and Casty, M., *Annuals Biomed. Eng.*, 9, 1665, 1981. With permission.)

A linear incremental model is used to describe the wall flexibility. A wall stiffness k is defined as the ratio of pressure change Δp to area change ΔA, all in the dimensionless versions. The Voight model, one of the linear viscoelastic models, is employed to depict wall viscoelasticity. In the computer program, the wall displacement, instead of the area change, is directly calculated using

$$p = \cong 2kD_n + 2C\frac{dD_n}{dt} \tag{11.11}$$

Here, p denotes the dynamic pressure; D_n, normal wall displacement (positive for an outward moving wall); and C, wall damping constant.

The pressure is calculated only along its solid walls for the determination of wall movements. Along a solid wall with no movement in the axial or tangential direction, the Navier-Stokes equations can be simplified to

$$\frac{\partial p}{\partial s} = -\frac{1}{Re_{pt}}\frac{\partial \omega}{\partial n} \tag{11.12}$$

Here, n is a coordinate normal to the wall and s denotes a coordinate along the wall, positive downstream and negative upstream; s is zero at the vertex and hip for the inner and outer walls, respectively. The pressure along the walls can be obtained by integrating the above equation. In order to start computing p from the inlet, Lou[1] developed a simple model that delivers a physiologically pulsatile pressure profile at the inlet.

The finite difference method is used in numerical analysis. The first derivatives of vorticity in the convective terms in the vorticity transport equation are treated by a hybrid upwind scheme. All other spatial derivatives in the governing equations are discretized by a central difference scheme for interior nodes. The vorticity at a solid wall is calculated using Wood's approximation. The stream function at a solid wall is derived from the inlet flow rate using continuity. Both the vorticity transport and Poisson equations are solved using the successive over-relaxation (SOR) technique. The enforced inlet flow rate data are smoothed with the cubic spine, which ensures continuous pressure and rate of change, thus yielding stable wall motion and numerical results.

Fig. 11.4 shows a computational mesh with 41 nodes across a channel and 87 nodes along the axial direction. It is nonuniform to give the best resolution around the hip and the vertex and near the walls. The convergence in vorticity with respect to the grid size is within 1% in both the axial and width directions. The time integration of the vorticity is performed implicitly to ensure stability, while the pressure is calculated explicitly in order to reduce the computing time. The time step size is 0.5 degrees in a cardiac cycle of 360 degrees, small enough to induce practically no error in the time integration.

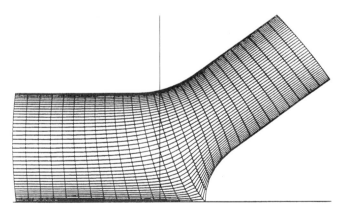

FIGURE 11.4 Mesh network for numerical computations, 41 nodes across the branch and 87 nodes along the axis (showing only a part around the bifurcation).

Typical results are presented in Figs. 11.5 through 11.9 for $RE_{pt} = 1300$, $\alpha = 8.8623$, $\beta = 75$ degrees, $r_i/L = 0.15$, and $r_i/L = 0.2$. The velocity profiles and moving wall positions are presented at six phase angles in Fig. 11.5. The dotted lines are zero-p (or reference) wall positions. Overall flow patterns for flexible wall cases are similar to those for rigid wall cases. In Figs. 11.6 and 11.8, the sign of the vorticity along the inner wall is reversed. The vorticity is positive when the flow is forward, and negative when the stream is reversed. The absolute vorticity value is used to evaluate the average absolute vorticity. Fig. 11.7 illustrates reverse flow patterns along the walls. The dotted areas show reverse flows. A point is dotted if the flow is reversed at that particular position *(s)* and time (phase angle). It is observed that dots are clustered around the hip and vertex, because of a high concentration of computational nodes, and between 40 and 140 degrees, because the results are screened more frequently (once every 1 degree). No permanent eddy appears due to the pulsatile nature of a physiological flow and the vertex wedge effect. A temporary eddy exists downstream of the hip under certain conditions, such as a large area ratio. The tiny vertex point is characterized by a low shear stress. Two high shear stress regions exist, one on the outer wall peaking at a point immediately distal to the hip and the other on the inner wall peaking at a point distal to the vertex. The peak value on the inner wall is higher than that on the outer wall. They are induced by the wedge effect of the vertex.

Fig. 11.9 shows time histories of the flow rate, Q, outer wall pressure, p, normal wall displacement, D_n, and vorticity, ω at $s = -0.9$, 0, and 0.9. The maximum values and corresponding phase angles for Q are listed in the figure. It is revealed that[8]

1. The pressure drop through a bifurcation is small compared with the pressure itself.
2. The flow resistance and consequent pressure drop increase with the wall stiffness and the damping ratio. The effect of the damping ratio is relatively small in the physiological range.

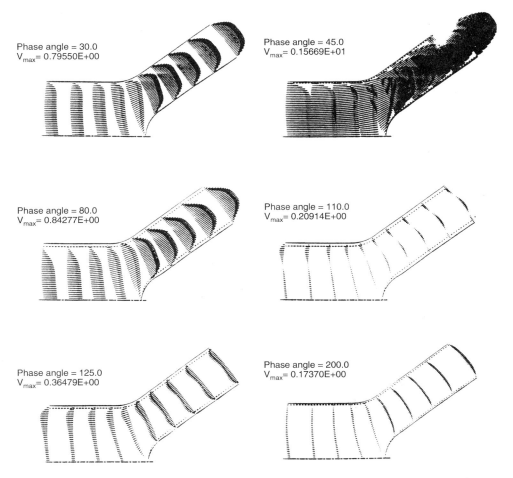

FIGURE 11.5   Velocity profiles and moving wall positions, with the dotted lines showing zero-p (or reference) wall positions.

3. Flexible walls smooth out the peaks of a pulsatile flow rate as the flow travels downstream.
4. The time history of wall movement resembles that of pressure. The wall movement lags behind the pressure in the artery and the magnitude of the phase increases with the damping ratio.
5. The wall expansion tends to induce flow reversals or eddies during the decelerating systole, while contraction tends to restrict them during the diastole.
6. The wall shear stress leads the flow rate in time. The wall flexibility enhances the phase advance.
7. A flexible bifurcation experiences lower shear stresses and has lower shear stress concentration factors than its rigid counterpart. This effect is amplified for a bifurcation with a smaller area ratio.

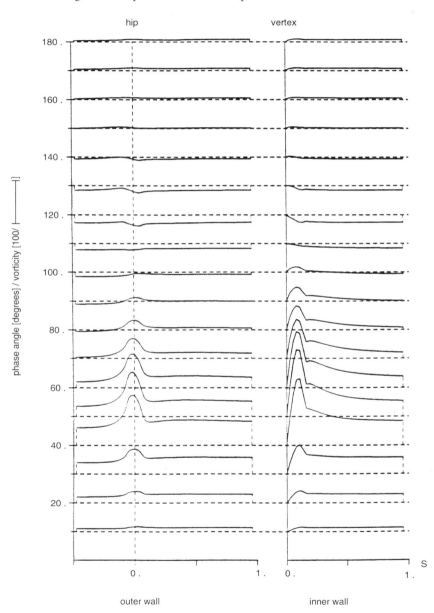

FIGURE 11.6 Velocities along the walls at various phase angles, with the shear stress in dyn/cm² which is equal to 1.24 times the dimensionless vorticity.

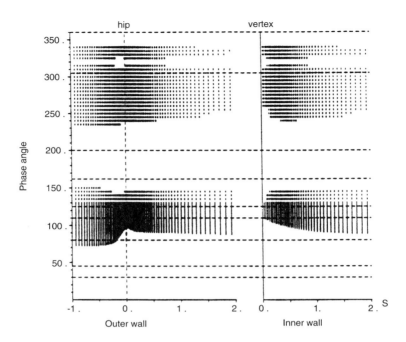

FIGURE 11.7 Reversed flow patterns along the walls, with the dotted areas showing reverse flows. A point is dotted if the flow is reversed at that particular position (*s*) and time (phase angle).

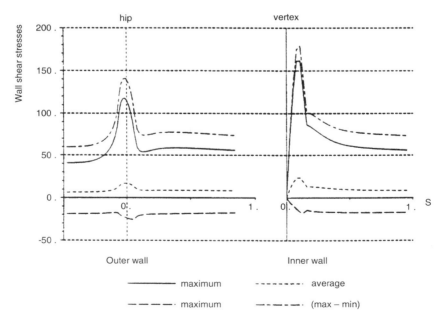

FIGURE 11.8 The maximum, minimum, average absolute, and maximum-minimum vorticities along the walls, with the shear stress in dyn/cm^2 which is equal to 1.24 times the dimensionless vorticity.

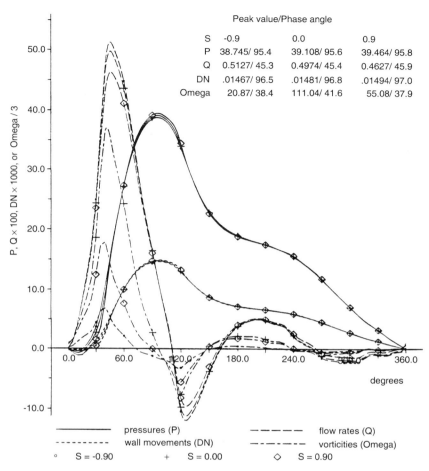

Peak value/Phase angle			
S	-0.9	0.0	0.9
P	38.745/ 95.4	39.108/ 95.6	39.464/ 95.8
Q	0.5127/ 45.3	0.4974/ 45.4	0.4627/ 45.9
DN	.01467/ 96.5	.01481/ 96.8	.01494/ 97.0
Omega	20.87/ 38.4	111.04/ 41.6	55.08/ 37.9

FIGURE 11.9 Time history of flow rate Q, outer wall pressure p, normal wall displacement D_n, and vorticity ω at position $s = -0.9$, 0, and 0.9.

References

1. Lou, Z., A computer simulation of the flow field at the aortic bifurcation, Ph.D. thesis, University of Michigan, Ann Arbor, 1990.
2. Nerum, R.M. and Levesque, in *Handbook of Bioengineering*, Skalab, R. and Chien, S., Eds., McGraw-Hill, New York, 1987, 21.1.
3. Lou, Z. and Yang, W.J., *Crit. Rev. Biomed. Eng.*, 19, 455, 1992.
4. Tokuda, C., Tsujioka, K., Ogasawara, Y., and Kajiya, F., in *Proceedings of 15th Meeting of the Laser Microscope Research Society*, 1995, 39.
5. Kajiya, F., Tokuda, S., and Yamamoto, T., *Opt. Tech. Contact*, 24, 286, 1996.
6. Yang, W.J., *Biothermal Fluid Sciences*, Hemisphere, Washington, 1986.
7. Stettler, J.C., Niederer, P., Anliker, M., and Casty, M., *Ann. Biomed. Eng.*, 9, 1665, 1981.
8. Lou, Z. and Yang, W.J., *J. Biomechanical Eng.*, 115, 306, 1993.
9. Lou, Z. and Yang, W.J., in *Fluid Mechanics and Biomechanics*, Vol. 1, Adeli H. and Sierakowski, R.L., Eds., The Society, New York, 1991, 544.
10. Lou, Z. and Yang, W.J., *J. Biomechanics*, 26, 37, 1993.
11. Lou, Z. and Yang, W.J., *Biomed. Matl. Eng.*, 1, 173, 1991.

12

Airway Dimensions in the Human Determined by Non-Invasive Acoustic Imaging

12.1   Introduction...**12**-1
12.2   Principle of the Acoustic Reflection Method **12**-2
        Acoustic Propagation Wave in a Duct • Reflection upon a
        Singular Site of Reflection • Reflection upon Multiple Sites of
        Reflection • Reflection upon a Tube with a Continuous
        Longitudinal Area Profile Variation
12.3   Physical Idealizations Required by the Acoustic
        Reflection Method **12**-7
        One-Dimensional Wave Propagation • Linear Behavior
        (Acoustic Linear Theory) • Lossless Wave Propagation • Rigid
        Wall • Homogeneous Gas • Single Duct
12.4   Impulse Response Measurement and Digital Signal
        Processing..**12**-12
        The One-Microphone Strategy • The Two-Microphone
        Strategy • Signal Processing
12.5   Nasal Applications **12**-22
12.6   Pulmonary Applications..**12**-24
        Validation of the Method • Applications
12.7   Conclusion **12**-26

B. Louis
INSERM

P. Drinker
Hood Laboratories, Inc.

G.M. Glass
Hood Laboratories, Inc.

D. Isabey
INSERM

J.J. Fredberg
Harvard School of Public Health

12.1   Introduction

Whenever possible, tools used to evaluate the structures and mechanical properties of the respiratory system should be noninvasive and require little cooperation from the subject. In 1956, Dubois et al.[1] were among the first to successfully meet this challenge. They determined mechanical properties of the lung from its response to external pressure oscillations imposed at the airway opening. Their work was the first of many studies of frequency response of the normal and abnormal lung.[2] Early studies focused on low frequencies (< 100 Hz), where the wavelength sound waves are much longer than the dimensions of typical respiratory structures. In that case, lumped parameter models are sufficient to describe the frequency-dependent behavior of the respiratory system.[3] Lumped parameter models are characterized by the fact that they have no spatial extent. At higher frequencies (250 Hz through 10 to 12 kHz), however, wavelengths become as small as airway dimensions or smaller, and it therefore becomes necessary to take into account the spatial properties of the airway tree, including local airway geometry, local wall properties, and branching. In these circumstances spatially distributed models along the system have to be used.[3] This also suggests that the high frequency response of the respiratory system could be used to infer geometrical information about the airway tree.

The acoustic reflection method gives the longitudinal cross-sectional area profile along the airway, and is a good example of spatial consideration in oscillation mechanics of the lungs. Knowledge of the area profile is potentially useful in understanding the structure and the function associated with oral airway, larynx, glottis, trachea, pulmonary airways, and nasal airway in health and in disease. The pathologies that can be studied with this method, and for which the method could be a diagnostic tool, include obstructive sleep apnea, obstructive pulmonary disease, asthma, tracheal stenosis, and nasal obstruction. The acoustic reflection method has also been used to study musical instruments in the wind group, such as brasses and alphorns.[4]

In the next section of this chapter, we examine the basic physical principles of the acoustic reflection method and its associated computational algorithms. We then describe the underlying assumptions and the various limits imposed by those assumptions. We also point out the different methods that have been proposed to overcome those limits. In the following section, we review the different strategies proposed in the literature for fabrication of a working apparatus. In the last two sections, we present some of the principal applications of the acoustic reflection method that have been established in the literature.

12.2 Principle of the Acoustic Reflection Method

The acoustic reflection method is based upon the analysis of planar acoustic wave propagating in rigid ducts.

Acoustic Propagation Wave in a Duct

For small amplitude, one-dimensional wave propagation of a nonviscous gas in a rigid duct under adiabatic conditions, the equations for conservation of momentum and mass[5,6] take the forms:

$$A \cdot \frac{\partial p}{\partial x} = -\rho_0 \cdot \frac{\partial q_v}{\partial t} \tag{12.1}$$

$$\gamma \cdot p_0 \cdot \frac{\partial q_v}{\partial x} = -A \cdot \frac{\partial p}{\partial t} \tag{12.2}$$

where A is the cross-sectional area of the duct, q_v is the flow rate, x is the spatial coordinate, t is the time, ρ_0 is the density of the fluid at equilibrium, p_0 is the equilibrium pressure, p is the pressure variation (the acoustic pressure), γ is the ratio of the specific heat at constant pressure of the gas to specific heat at constant volume, and $\partial/\partial t$ is the Eulerian time derivative.

Differentiating Eq. 12.1 by x and Eq. 12.2 by t allows the elimination of qv, and thus gives the classical wave equation:

$$\frac{\partial^2 p}{\partial x^2} = \frac{1}{c^2} \cdot \frac{\partial^2 p}{\partial t^2} \quad \text{with} \quad c^2 = \frac{\gamma \cdot p_0}{\rho_0} \tag{12.3}$$

By defining $\psi = t - x/c$ and $\varphi = t + x/c$, the general solution to Eq. 12.3 can be given as:

$$p = f\{\psi\} + g\{\varphi\} = f\{t - x/c\} + g\{t + x/c\} \tag{12.4}$$

Eqs. 12.4 and 12.1 give the general solution for the flow rate:

$$q_v = \frac{A}{\rho_0 \cdot c} \cdot \left(f\{t - x/c\} - g\{t + x/c\} \right) \tag{12.5}$$

Both p and q_v are the sum of two waves that propagate with the same wave speed, c, but in opposite directions; f propagates in the positive x direction while g propagates in the negative x direction. Moreover, for waves propagating in the same direction, the ratio of the pressure to the flow is a constant, $\rho_0 \cdot c/A$. This ratio, called characteristic (or specific)[7] acoustic impedance,[5] is a pure real number, which means that any pressure wave is in phase with its associated flow rate wave. The values of f and g are determined from the boundary and initial conditions of the problem.

Such a description of wave propagation can be interpreted as well in the frequency domain. Consider the harmonic time dependence of all time varying quantities. Pressure and flow rate can then be written under the form: $W\{\omega,x\} \cdot e^{j\omega t}$ with ω being the radian frequency and $j^2 = -1$. $W\{\omega,x\}$ is a function of both x and ω. Eqs. 12.1 and 12.2 may be recast in form of the classic transmission line equations:

$$\frac{\partial p}{\partial x} = -j \cdot \omega \cdot \frac{\rho_0}{A} \cdot q_v \tag{12.6}$$

$$\frac{\partial q_v}{\partial x} = -j \cdot \omega \cdot \frac{A}{\rho_0 \cdot c^2} \cdot p \tag{12.7}$$

Any elemental segment of the duct can be represented by an electrical analog model (see Fig. 12.1). This model is composed of series impedance, Zs, and shunt admittance, Ys, per unit duct length. Zs is a pure self-inductance, whose value is given by Eq. 12.6 ($I_g = \rho_0/A$), taking into account the gas inertia, for a flat velocity profile; 1/Ys is the impedance due to a pure compliance inductance, whose value is given by Eq. 12.7 ($C_g = A/[\rho_0 \cdot c^2]$), taking into account the gas compressibility, for a pure adiabatic transformation.

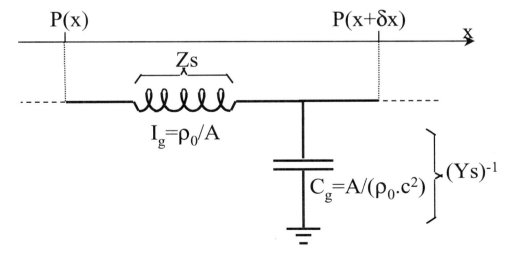

FIGURE 12.1   Equivalent transmission line element for an infinitesimal length of tube, δx. The associated model assumes a one-dimensional lossless wave propagation and rigid wall duct. The series impedance per unit duct length, Zs, takes into account gas inertia while the shunt admittance per unit duct length, Ys, takes into account the compressibility of the gas. Zs and Ys are not frequency dependent.

Reflection upon a Singular Site of Reflection

Consider two semi-infinite, straight, gas-filled, rigid-walled uniform tubes with two distinct cross-sectional areas connected to their ends (see Fig.12.2). Within the first tube (on the left side), a planar pressure wave, $p_i(t,x)$, propagates to the right and becomes incident upon the second tube (on the right side). As this incident wave impinges upon the interface between the two tubes, a reflected wave propagating in the first tube arises while a transmitted wave propagates in the second tube. The reflected wave, $p_r(t,x)$, propagates to the left while the transmitted wave, $p_t(t,x)$, propagates to the right. Each pressure wave is associated with a flow rate wave propagating in the same direction and with the same velocity. Writing the continuity of both pressure and flow rate at the connection of the two tubes, $x = x_0$, gives two relations:

$$p_i(t,x_0)+p_r(t,x_0)=p_t(t,x_0) \quad \text{and} \quad q_{vi}(t,x_0)-q_{vr}(t,x_0)=q_{vt}(t,x_0) \qquad (12.8)$$

Using Eqs. 12.4 and 12.5 permits elimination of the flow rate:

$$p_i(t,x_0)+p_r(t,x_0)=p_t(t,x_0) \quad \text{and} \quad \frac{A_0}{\rho_0 \cdot c}\Big[p_i(t,x_0)\big(-p_r(t,x_0)\big)\Big]=\frac{A_1}{\rho_0 \cdot c}\Big[p_t(t,x_0)\Big] \qquad (12.9)$$

Using the first relation to eliminate p_t in the second relation, it follows that:

$$\frac{A_1}{A_0}=\frac{p_i(t,x_0)-p_r(t,x_0)}{p_i(t,x_0)+p_r(t,x_0)} \quad \text{and} \quad \frac{p_r(t,x_0)}{p_i(t,x_0)}=\frac{A_0-A_1}{A_0+A_1} \qquad (12.10)$$

Eq. 12.10 shows that by measuring the amplitude of the incident and reflected pressure wave, the value of A_1 can be determined if A_0 is known. Since the wave propagation is assumed to be nonviscous, incident and reflected pressure waves can be measured at any point of the first tube and corrected for the propagation delay without loss of information.

Therefore,

$$p_i(t,x_0)= p_i(t-\tau,0) \text{ and } p_r(t,x_0)=p_r(t+\tau,0) \text{ with } \tau \text{ propagation delay } =\frac{x_0-0}{c}=\frac{x_0}{c}.$$

Measuring the round trip propagation delay, $2.\tau$, also permits determination of the position of the first tube end, i.e.,

$$x_0 =\frac{1}{2}\cdot\frac{2\cdot\tau}{c}.$$

Reflection upon Multiple Sites of Reflection

Consider, now, n + 1 straight, gas-filled, rigid-walled uniform tubes with distinct cross-sectional areas, $A_0, A_1, \ldots, A_{n+1}$ connected each one to the next in series at $x = x_0, x_1, \ldots, x_{n+1}$. At each site of reflection, the same analysis as that made for a single site of reflection can be applied, except for two aspects. First, the incident wave for the kth site is the transmitted wave through the $k-1$ sites ordered before this k^{th} site. Secondly the reflected wave for the kth site includes also the reflected waves due to the sites ordered beyond. The reflected wave at $x = x_0$ can be written under the form:

$$p_r(t, x_0) = r_0 \cdot p_i(t, x_0) + r_1(1 - r_0^2) \cdot p_i\left(t - 2 \cdot \frac{x_1 - x_0}{c}, x_0\right)$$

$$+ \left[r_2 \cdot (1 - r_1^2) \cdot (1 - r_0^2) - r_1^2 \cdot r_0 \cdot (1 - r_0^2)\right] \cdot p_i\left(t - 2 \cdot \frac{x_2 - x_0}{c}, x_0\right)$$

$$+ \ldots\ldots\ldots\ldots\ldots\ldots\ldots\ldots\ldots\ldots\ldots\ldots\ldots\ldots\ldots$$

$$+ \left[r_i \cdot (1 - r_{i-1}^2)\ldots\ldots\ldots(1 - r_0^2) + k_i\left\{r_0, r_1, \ldots\ldots\ldots, r_{i-2}, r_{i-1}\right\}\right] \cdot p_i\left(t - 2 \cdot \frac{x_i - x_0}{c}, x_0\right)$$

$$+ \ldots\ldots\ldots\ldots\ldots\ldots\ldots\ldots\ldots\ldots\ldots\ldots\ldots\ldots\ldots$$

$$+ \left[r_n \cdot (1 - r_{n-1}^2)\ldots\ldots\ldots(1 - r_0^2) + k_n\left\{r_0, r_1, \ldots\ldots\ldots, r_{n-2}, r_{n-1}\right\}\right] \cdot p_i\left(t - 2 \cdot \frac{x_n - x_0}{c}, x_0\right) \quad (12.11)$$

where r_i designates the reflection coefficient at the distance $x = x_i$. This coefficient would give the ratio between incident and reflected wave at this distance only if the reflection site at $x = x_i$ was a single site of reflection:

$$r_i = (A_i - A_{i+1})/(A_i + A_{i+1}) \quad (12.12)$$

Note that k_i is a function of both $r_0, r_1, \ldots, r_{i-2}$ and r_{i-1}. This problem was studied by Ware and Aki,[8] who showed that $k_i\{\}$ may be computed using a recursive procedure. This property allowed them to develop an efficient algorithm to determine the different reflection coefficients when the incident wave, $p_i(t, x_0)$, is of the form $\delta(t)$ and $\delta(t)$ is the well-known Dirac delta function ($\delta(t) = 1$ if $t = 0$; $\delta(t) = 0$ if t # 0). In this case, the reflected wave is called the impulse response. This algorithm is based upon the property that k_i is a function of both $r_0, r_1, \ldots, r_{i-2}$ and r_{i-1} but not of r_i. The reflection coefficient may then be computed incrementally. First r_0 is determined with the reflected wave measured at the initial time, i.e., the time ($t = 0$) where the incident wave is in $x = x_0$. Indeed, due to the properties of Dirac delta function, Eq. 12.11 gives $p_r(t = 0; x_0) = r_0$. Then, using the predetermined value of r_0, r_1 is calculated with the reflected wave measured at $t = 2(x_1 - x_0)/c$. Using the values of r_1 and r_0, r_2 is determined; the procedure continues until r_n is calculated.

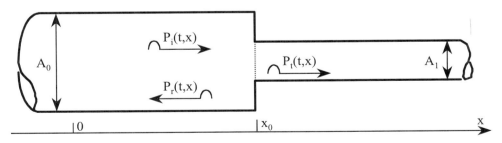

FIGURE 12.2 Schematic of two semi-infinite, straight, gas-filled, rigid-walled uniform tubes with two distinct cross-sectional areas connected to their ends. An incident pressure wave, $p_i(t, x)$, propagating in the first tube impinges upon the tube connection, giving rise to a transmitted wave in the second tube, $p_t(t, x)$, and to a reflected wave in the first one, $p_r(t, x)$. Axial position = x. Distance of the connection = x_0.

Thus, the measurement of the impulse response permits determination of the sequence of reflection coefficients. Introducing those coefficients in Eq. 12.12 allows estimation of the different unknown cross-sectional areas. Moreover, the round-trip propagation delay, Ti, permits a straightforward estimation of the locations of all reflection sites:

$$x_i - x_0 = \frac{c \cdot Ti}{2} \tag{12.13}$$

Reflection upon a Tube with a Continuous Longitudinal Area Profile Variation

This problem is a generalization of the discrete problem addressed in the previous section. In this case, the cross-sectional area, A, in Eqs. 12.1 and 12.2 is a continuous function of x, and Eq. 12.3 should be replaced by:

$$\frac{1}{c^2} \cdot \frac{\partial^2 \phi}{\partial t^2} - \frac{\partial^2 \phi}{\partial x^2} = A^{-1/2} \cdot \frac{\partial}{\partial x}\left[-\frac{1}{2} \cdot A^{-1/2} \cdot \frac{\partial A}{\partial x}\right] \cdot \phi \tag{12.14}$$

where

$$\phi = A^{1/2} \cdot p(t, x)$$

that can be recast as:

$$\frac{\partial^2 \phi}{\partial \varsigma^2} - \frac{\partial^2 \phi}{\partial t^2} = q\{\varsigma\} \cdot \phi \text{ with } \varsigma = x/c \text{ and } q\{\varsigma\} = A^{-1/2} \cdot \frac{\partial}{\partial \varsigma}\left[-\frac{1}{2} A^{-1/2} \cdot \frac{\partial A}{\partial \varsigma}\right] \tag{12.15}$$

Eq. 12.15 is recognized as a one-dimensional Schroedinger equation with a potential, $q(\zeta)$. Ware and Aki[8] showed that this equation can be recast in a form of a Gel'fand-Levitan integral equation, and that $q(\zeta)$ can be estimated from the knowledge of the impulse response, as defined previously. This potential is a real number which is a function of the cross-sectional area of the duct, $A(\zeta)$. If this potential is known, it can be integrated in order to infer A as a function of ζ, which is simply x/c, so the cross-sectional area of the duct can be inferred as a function of the spatial coordinate.

In summary, the longitudinal gradient of internal cross-sectional area, $A\{x\}$, along a singular duct can be inferred from the impulse response if both the wave speed and the initial section are known.

Independently, more or less similar theories were developed by Sondhi and Gopinath[9–11] and by Schroeder.[12] Nevertheless, beyond validation of these theories in models, these methods were never used to successfully study airway dimensions in animals or humans. For example, Schroeder's method,[12] based upon the measurement of singularities (poles and zeroes) of the lip impedance function, required some extremely constraining boundary conditions not really compatible with physiological and clinical applications. The vocal tract had to be assumed to be closed at the glottis and the length of the vocal tract had to be assumed to be constant at some estimated value. The methods developed subsequently avoided these shortcomings. Based upon the analysis of Ware and Aki,[8] Jackson et al.[13] studied excised canine lungs, and Fredberg and coworkers obtained the first successful measurements in human subjects, and described the roles of the carrier gas, the physiologic sources of error, the wall elasticity, loss, and concomitant flow.[7,14–17]

12.3   Physical Idealizations Required by the Acoustic Reflection Method

The main assumptions of the previous formulation are one-dimensional wave propagation, linear behavior of the system, lossless wave propagation, rigid wall, and homogeneous gas in a single duct.

One-Dimensional Wave Propagation

The one-dimensional propagation of pressure wave along the longitudinal direction of a duct implies that the variations of acoustic pressure along the radial direction of the duct are negligible. This fails to be so[5,6,15] when acoustic cross modes, i.e., the nonplanar modes, begin to propagate. This occurs when the wave-length, $2\pi/\omega$, is less than twice the value of the radial dimension of the duct. Thus, the higher bound of the frequency to ensure one-dimensional propagation assumption is roughly: $f_{max} \approx c/2d$, where d is the local duct diameter. Considering a maximal radial dimension of about 3 cm, the frequency of the acoustic pressure should not be higher than 6 kHz in air.

The first consequence of this limit was in the initial embodiment of the working apparatus. The authors used a casted, custom-made mouthpiece to fill the oral cavity between the lips and the hard palate.[14,17–22] Indeed, the transverse dimension of the mouth cheek-to-cheek (4 to 8 cm) was the most susceptible location for the generation of acoustic cross-modes. Subsequently, Rubinstein et al.[23] showed that this precaution was not required; replacement of the casted mouthpiece by a scuba-diving mouthpiece did not introduce significant artifacts in estimates of pharyngeal, glottic, or tracheal areas. It seems that when the lips are pushed forward and tightly sealed around the scuba-diving piece, the distances between the cheeks are naturally reduced so that the one-dimensional assumption is not violated.

The second consequence is that the higher bound of the frequency determines the spatial resolution of the method. The spatial resolution is the distance over which the inferred area accommodates to step changes of area. In the case of lossless wave propagation in rigid ducts, the resolution can be crudely approximated as one-sixth of the shortest wavelength of the pressure wave.[7,14,15] Of course, this resolution is altered both by viscosity and wall nonrigidity. Practically, the resolution obtained is in the range of 1 cm.[7]

Linear Behavior (Acoustic Linear Theory)

In the above wave propagation description we have used an equation of motion in which the pressure variations are assumed to be small enough that the acoustic approximation holds and variations in pressure to the first order are retained. The amplitude of the acoustic pressure used with the acoustic reflection technique is very small (about 1 cm $H_2O \approx 1$ hPa) compared with atmospheric pressure. Thus, the assumption of acoustic linear theory seems to be reasonable to describe the wave propagation used with the acoustic reflection method.

Lossless Wave Propagation

In Eqs. 12.1 and 12.2, we have assumed nonviscous gas and adiabatic conditions. In fact gases like air are viscous, and isothermal wall conditions seem to be more realistic for the boundary condition. When the duct is circular, several authors[24–29] described an analytical solution, taking into account viscosity and realistic thermodynamic boundary conditions. In this case, the schematic of Fig. 12.1 can still be used to represent the oscillatory flow but the series impedance and the shunt admittance have to be modified. Zs is replaced by Zp which is composed by a resistance, Rv, in series with a self inductance, Iv. Ys is replaced by Yp that is composed by a resistance, Rt, in parallel with a compliance, Ct. Rv takes into account viscous losses, whereas Iv takes into account the gas inertia. Rt represents the thermal losses, while Ct represents the gas compression. The parameters Rv and Iv depend on the radial profile of flow velocity and appear as functions of the Womersley number,

$$\alpha \cdot \alpha = \frac{d}{2}\sqrt{\frac{\omega}{\nu}}$$

where d is the diameter and ν is the kinematic viscosity of the gas. The parameters Rt and Ct depend on the radial profile of the temperature and appear to be functions of both α and of the Prandtl Number *Pr*.

In Fig. 12.3 we plotted Zp and Yp as a function of α when the gas is air (*Pr* = 0.707). At the higher α values, Iv tends toward Ig and Ct tends toward Cg as defined in Eqs. 12.6 and 12.7. Moreover, for these higher α values, the series impedance Zp tends toward Zs; Rv can be neglected in front of the impedance associated to Iv (j.ω.Iv). In the same way, for the higher α values, (Yp)$^{-1}$ tends to (Ys)$^{-1}$; Rt can be neglected for the computation of (Yp)$^{-1}$. For the lower α values, Rt and Rv cannot be neglected and the model associated with the acoustic reflection method is no longer valid. It means that to use the acoustic reflection method there is a low limit on α. It introduces a low limit on the frequency that depends on the diameter, the viscosity, and the Prandtl number of the gas. For example, considering a 1.2 cm diameter duct filled with air, using the lossless model (Eqs. 12.6 and 12.7), introduces errors in Zs and in Ys that are less than 5% if the frequency is above 60 Hz. The lower bound of the frequency to ensure the lossless propagation assumption is roughly:

$$f_{\min \nu} \approx \frac{15.\nu}{\pi^2 \cdot d^2}$$

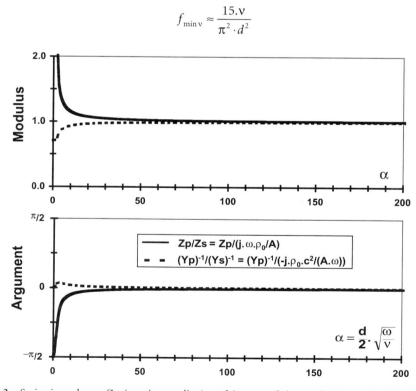

FIGURE 12.3 Series impedance, Zp (continuous line), and inverse of shunt admittance, Yp^{-1} (dashed line), for viscous gas (Prandtl number for air = 0.707) and isothermal thermodynamic wall condition plotted as function of the Womersley number, α. The series impedance and the shunt admittance were normalized by their values obtained in nonviscous and adiabatic transformation model, i.e., with the model used for the acoustic reflection method. Top: modulus; bottom: argument.

Rigid Wall

Propagation in a Nonrigid Walled Tube

When the walls are not rigid, the term $\partial A/\partial t$ is no longer equal to zero. The temporal variation of area should be introduced in the equations for conservation of mass (Eq. 12.2). This equation can be recast as:

$$\frac{\partial q_v}{\partial x} = -\left[\frac{A}{\rho_0 \cdot c^2} + \frac{\partial A}{\partial p}\right]\cdot\frac{\partial p}{\partial t} \text{ or in the frequency domain } \frac{\partial q_v}{\partial x} = -j\cdot\omega\cdot p\cdot\left[\frac{A}{\rho_0 \cdot c^2} + \frac{\partial A}{\partial p}\right] \quad (12.16)$$

while Eqs. 12.1 and 12.6 (for conservation of momentum) are not modified by the temporal variation of the area. The solutions of Eqs. 12.6 and 12.16 are:

$$p = \left(a\cdot e^{-j\frac{\omega}{c}\left(1+\frac{\rho_0 \cdot c^2}{A}\cdot\frac{\partial A}{\partial p}\right)^{1/2}\cdot x} + b e^{+j\frac{\omega}{c}\left(1+\frac{\rho_0 \cdot c^2}{A}\cdot\frac{\partial A}{\partial p}\right)^{1/2}\cdot x}\right)\cdot e^{j\cdot\omega t}$$

$$q_v = \frac{A}{\rho_0 \cdot c}\cdot\left(1+\frac{\rho_0 \cdot c^2}{A}\cdot\frac{\partial A}{\partial p}\right)^{1/2}\cdot\left(a\cdot e^{-j\frac{\omega}{c}\left(1+\frac{\rho_0 \cdot c^2}{A}\cdot\frac{\partial A}{\partial p}\right)^{1/2}\cdot x} - b e^{+j\frac{\omega}{c}\left(1+\frac{\rho_0 \cdot c^2}{A}\cdot\frac{\partial A}{\partial p}\right)^{1/2}\cdot x}\right)\cdot e^{j\cdot\omega t}$$

Both p and q_v represent the sum of two waves which propagate with the same wave speed, c_w, but whose amplitude is attenuated by a factor $(1/e)$ within a distance da;

$$e = 2.718, c_w = c/\text{real part}\left\{\left(1+\frac{\rho_0 \cdot c^2}{A}\cdot\frac{\partial A}{\partial p}\right)^{1/2}\right\}, \text{ and da } = 1/\text{imaginary part}\left\{\frac{\omega}{c}\left(1+\frac{\rho_0 \cdot c^2}{A}\cdot\frac{\partial A}{\partial p}\right)^{1/2}\right\}.$$

A segment of the conduct is represented by introducing in the electrical schematic a shunt admittance Y_d in parallel to Ys (see Fig. 12.4). This pathway is represented by a compliance, $C_d = \partial A/\partial p$, called local dynamic compliance of the duct. C_d is not necessarily a real number because it depends on the elastic, resistive, and inertial properties of the airway wall.[6,14,30–35]

The consequences of assuming nonrigid walls upon the area inferred with the acoustic method are complicated. For example, consider two semi-infinite, straight, gas-filled, uniform tubes with two distinct cross-sectional areas connected to their ends (see Fig. 12.2). The first tube is rigid-walled. The wall of the second tube is purely elastic ($C_{d1} = \partial A_1/\partial P$ is a real number). Using the same procedure described earlier, it is easy to show:

$$\frac{P_i(t,x_0)-P_r(t,x_0)}{P_i(t,x_0)+P_r(t,x_0)} = r_0 = \frac{A_1}{A_0}\cdot\left(1+\frac{\rho_0 \cdot c^2}{A_1}\cdot C_{d1}\right)^{1/2} > \frac{A_1}{A_0}$$

Thus, the area inferred from Eq. 12.10 ($A_1 = r_0 \cdot A_0$), when the wall rigidity is assumed, would overestimate the area of the second tube. On the contrary, if the second tube was rigid-walled and the wall of the first tube was elastic, the area inferred from Eq. 12.10 would underestimate the area of the second tube. Finally, the wall nonrigidity modifies the wave speed which introduces an error on the determination of distance of the reflection site.

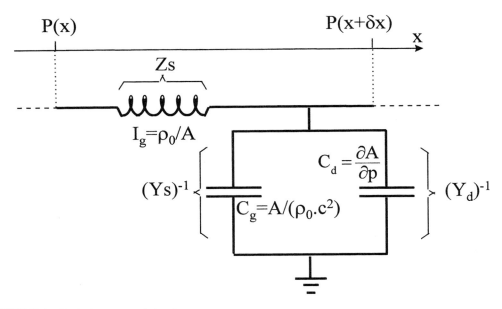

FIGURE 12.4 Equivalent transmission line element for an infinitesimal length of tube with nonrigid wall. Zs and Ys are defined by the rigid wall model. The local compliance of the duct, C_d, is not necessarily a real number and it could be frequency dependent in the frequency range used with the acoustic reflection method.

Limits of the Wall Rigidity Assumption in Airway

For simplicity, let us consider a model of locally reacting airway wall responses; the area of a section depends only on the local parameter occurring in this section. Theoretically, it is clear that this model may represent an oversimplification of the dynamic airway behavior.[14] Nevertheless, measurements of wave propagation characteristics in excised trachea showed that such a model represents the dynamic behavior of canine trachea[33] and calf trachea[36] fairly well at the high frequencies used in the acoustic reflection method.

The wall of the duct has a non-negligible mass. This mass confers inertia per unit length, I_d, to the duct wall.[33] When the duct is circular, the volumetric inertia per unit of tube length is given by

$$I_d = \frac{t_h}{\pi \cdot d} \cdot \rho_d,$$

where d is the internal diameter, t_h is the thickness of the wall, and ρ_d is the density of the wall tissue. This inertia should be taken into account in the determination of the local dynamic compliance of the duct. The impedance due to inertia ($I_d \cdot \omega$) increases with ω, while the impedance due to the gas compliance decreases as $1/\omega$. The wall rigidity assumption is justified if the former is much larger than the latter. The lowerbound of the frequency to ensure wall rigidity assumption is roughly

$$f_{\min w} = \frac{1}{\pi} \cdot \sqrt{\frac{\gamma \cdot P_0}{d \cdot t_h \cdot \rho_d}}.$$

This lower bound of frequency decreases as the mass wall per unit length ($d \cdot t_h \cdot \rho_d$) increases. This fact was experimentally corroborated by Brooks et al.[17] in excised canine trachea. They found that adding wall mass with petroleum jelly significantly enhanced the accuracy of the trachea area inferred by acoustic reflection.

To overcome this limit, the initial idea was to increase the working frequency range. Unfortunately, this procedure tended to be in conflict with the one-dimensional propagation assumption. To overcome this dilemma, Fredberg et al.[14] proposed use of a 78% He–22% O_2 mixture rather than air. Indeed, the wave speed of this mixture is 1.8 times higher than that of air. For this mixture, the higher bound of the frequency to ensure one dimensional propagation assumption ($f_{max} \approx c/2d$) is almost twice as high as the bound of air, which allows the bandwidth to be extended up to 12 kHz. These authors showed that such a mixture permits accurate inferences of the airway area in a healthy adult from the mouth to the carina. By comparison, using air under the same conditions permits accurate inferences only up to the glottis, but strongly overestimates the area of the trachea by a factor as large as 250%. The airway wall rigidity assumption for such a bandwidth is corroborated by Rice's measurement of sound speed in upper airways of anesthetized dogs.[37] Indeed, this author finds that the wave speed at frequencies above 10 kHz was the wave speed in free field, i.e., the wave speed, c, defined in Eq. 12.3.

Recently, another procedure was proposed by Louis et al.[38] to de-emphasize the effects of the wall nonrigidity. Rather than extending the bandwidth up to 12 kHz, those authors proposed to introduce a numerical high pass filter of the impulse response. This is intended to reduce that part of the response spectrum (low frequencies) where the influence of local wall dynamics is the greatest. A 250 Hz cutoff frequency associated with a linear phase finite impulse response (FIR) filter with a 125-point hamming window, was found sufficient to minimize the effects of wall nonrigidity in estimating pulmonary airway response. Fig. 12.5 illustrates this procedure. The area-distance functions of a human healthy subject obtained (1) with the He and O_2 mixture, (2) with air without a high-pass filter, and (3) with the high-pass filter are plotted graphically. When the high-pass filter is used, the results obtained in air are in close agreement with the results obtained with He and O_2 mixture. When the high-pass filter was not used with air, the area of the subglottic region was clearly overestimated. This procedure permits accurate inferences of the airway area from the mouth up to the carina using air as breathing gas. This is important because it permits considerable simplification of the apparatus and the measurement procedures.

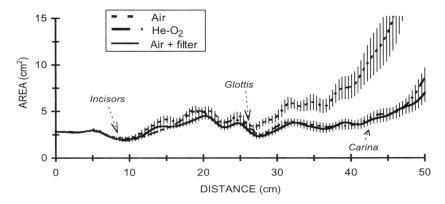

FIGURE 12.5   Pulmonary airway area-distance function from the mouth to the carina of a healthy subject. Effect of the high-pass filter procedure used to de-emphasize the wall nonrigidity artifact. The area vs. distance curve was inferred using the two-microphone strategy with (1) He and O_2 mixture, (2) with air and without impulse-response high-pass filter procedure, and (3) with air and a high pass filter procedure of the impulse response. Results obtained with He and O_2 mixture are considered "gold standard" results. The error bars are the standard deviations associated with the areas inferred by 10 repeated pressure acquisitions. The position of different landmarks are indicated by the arrows

Homogeneous Gas

The acoustic reflection method requires a homogeneous gas in the duct. Indeed, the reflection sites are determined by the longitudinal gradient of the characteristic impedance. Consider the two semi-infinite uniform tubes discussed in Section 12.2 (see Fig. 12.2) that were filled with two distinct gases. The gases had distinct densities (ρ_0, ρ_1) and distinct wave speeds (c_0 and c_1). In this case, Eq. 12.8 is still valid, but Eq. 12.10 should be replaced by:

$$\frac{p_i(t,x_0) - p_r(t,x_0)}{p_i(t,x_0) + p_r(t,x_0)} = \frac{A_1}{\rho_1 \cdot c_1} \cdot \frac{\rho_0 \cdot c_0}{A_0} \quad \text{and} \quad \frac{p_r(t,x_0)}{p_i(t,x_0)} = \frac{\dfrac{\rho_0 \cdot c_0}{A_0} - \dfrac{A_1}{\rho_1 \cdot c_1}}{\dfrac{\rho_0 \cdot c_0}{A_0} + \dfrac{A_1}{\rho_1 \cdot c_1}} \tag{12.17}$$

The analysis of the reflection gives the variation of the characteristic impedance ($\rho \cdot c/A$) which can give the area variation only if both density and wave speed are constant. Moreover, the variation of wave speed introduces an error on distance of the reflection sites.

Gas is not exactly homogeneous in the pulmonary airway. Under usual conditions, the inspiratory gas is generally dry air at 20°C while the expiratory gas is humid air with CO_2 at 37°C. Nevertheless, those differences are slight and in first approximation they will not modify the density and the wave speed. The effect of accumulation of carbon dioxide on the method was investigated by D'Urzo et al.[18] who measured the area of a glass model with different concentrations of CO_2 up to 10%. Those authors found that the variations introduced by the different concentrations of CO_2 are well within the inherent run-to-run variation of this technique.

Single Duct

The algorithm is written for a duct without bifurcations. Sidell and Fredberg,[7] who simulated different branching networks, showed that the Ware and Aki algorithm[8] cannot be expected to infer the summed cross-sectional area of asymmetric branching networks of ducts. From the clinical point of view, this means that the acoustic reflection method does not rest on firm theoretical footings when applied to distal regions in the lung. Nevertheless, if all parallel pathways had identical properties, then the reflective waves in all parallel pathways would be identical and would simultaneously emerge from the duct network. In this case, the algorithm permits accurate inferences for the summed cross-sectional area.

This probably explains why Jackson et al.[13] found in airway casts that the method was able to correctly infer the summed cross-sectional area of the main stem bronchi up to 6 cm past the carina. The single-duct assumption also requires that the velum be closed in order to infer the geometries of more distal structures. Generally the velum closes if a subject breathes through the mouth without wearing nose-clips.[14,15,17] When a subject wears noseclips, the velum is not so systematically closed, and this invalidates the measurement. Fortunately, the open velum is easily detected on the area-distance plots.[15,23] Indeed, the open velum introduces a clear overestimation of the sub-velum airway area and a dramatic increase in the run-to-run variation (see Fig. 12.6).

12.4 Impulse Response Measurement and Digital Signal Processing

The cross-sectional area can be inferred from the impulse response, $h(t, x_0)$, via the Ware and Aki algorithm.[8] To measure the impulse response, different strategies were proposed in the literature.

The One-Microphone Strategy

This strategy attempts to obtain the impulse response from the separate values of both incident and reflected pressure waves. By definition, the impulse response is the reflected wave when the incident wave is of the form of the Dirac delta function, $\delta(t)$. In a real system a pressure wave of the form $\delta(t)$ cannot be achieved . Nevertheless, there is a relation between $h(t, x_0)$, $p_i(t, x_0)$, and $p_r(t, x_0)$ whatever the form of $p_i(t, x_0)$ is.

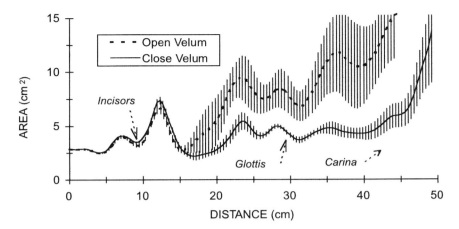

FIGURE 12.6   Representative example of unacceptable pulmonary airway area-distance plots arising when the velum is open (dashed line). The results are compared to those obtained with a closed velum (continuous line). The results were obtained using the two-microphone strategy with air and using the impulse-response high-pass filter procedure.

$$p_r\!\left(t,x_0\right)=p_i\!\left(t,x_0\right)\otimes h\!\left(t,x_0\right)=\int\limits_{s=0}^{s=t} h\!\left(t-s,x_0\right)\cdot p_i\!\left(s,x_0\right)\cdot \partial s \qquad (12.18)$$

The $\otimes$ denotes the convolution integral. If both $p_r(t,x_0)$ and $p_i(t,x_0)$ are known, $h(t,x_0)$ can be determined using a variety of deconvolution algorithms.

The Time Window Approach

In 1971 Sondhi and Gopinath[9,10] suggested a technique to separate $p_r(t,x_0)$ and $p_i(t,x_0)$ using only a single pressure transducer. Consider a wave tube with one of its ends connected to a sound source and the other end connected to the airway opening (see Fig. 12.7, top). An incident pressure wave of short but finite duration is generated by the sound source. This incident wave is recorded by the pressure transducer at a fixed distance, L_1, from the airway opening. The incident wave impinges upon the airway opening and creates a reflected wave that is also recorded by the transducer. If the duration of the incident wave, T_i, is less than the round-trip propagation delay between the pressure transducer and the airway opening, i.e., if $T_i < 2\,L_1/c$, the incident wave and the reflected wave are non-overlapping in time. These two waves are easy to separate and identify in the recording of the transducer (see Fig. 12.7, bottom). The main constraint of this technique is to obtain a sufficiently short incident wave to avoid a too-long wave tube. This strategy was successfully developed by Fredberg et al.[39–41] It was applied to infer areas of excised dog lungs and pulmonary cast,[13,42] and later used to infer nasal airway areas[43–47] and upper airway area up to the glottis.[48]

The Single Calibration Approach

A variant of this first approach was proposed by Fredberg et al. in 1980.[14,16] Rather than using time window technique, this approach relied upon a highly reproducible sound source with a calibration procedure to effect the separation of overlapping incident and reflected waves. Consider a horndriver connected to a wave tube with a pressure transducer. The calibration procedure consists of connecting the wave tube to a tube of the same diameter also called the calibration tube (see Fig. 12.8, top). Since there is no reflection site on the wave tube or the calibration tube, the recorded pressure, $p_{cal}(t,0)$, is the incident wave, $p_i(t,0)$. With the calibration tube removed and the airway connected (see Fig. 12.8, middle)

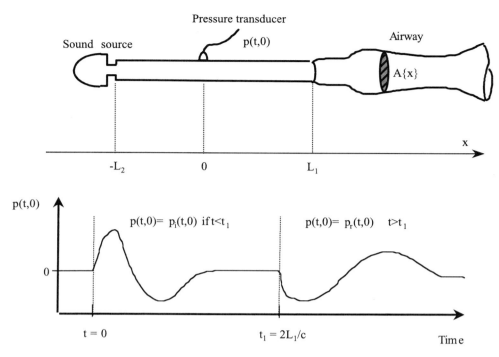

FIGURE 12.7 Top: Schematic of the wave tube for measuring the area by the one-microphone strategy using the time window approach. This schematic corresponds also to the set-up used by the one-microphone strategy with the dual calibration procedure. Bottom: Schematic of the pressure vs. time recorded by the transducer if the incident wave is short enough to avoid overlapping in time between the incident wave and the reflected wave. In such a case, the pressure recorded before t_1 is the incident wave while the pressure recorded after t_1 is the reflected wave.

the recorded pressure, $p(t,0)$ is the superposition of the incident wave, $p_i(t,0)$, and of the reflected wave, $p_r(t,0)$. The reflected wave is then extracted by simple subtraction:

$$p_r(t,0) = p(t,0) - p_{cal}(t) \quad \text{with} \quad \mathrm{p_i}(t,0) = p_{cal}(t,0). \tag{12.19}$$

This approach permitted accurate and reproducible *in vivo* airway area reconstruction in humans.[14,17–23,49–52] In the first attempts using this approach, the calibration tube was replaced by a slide valve considered an ideal reflector. The pressure recorded during the calibration was: $p_{cal}(t,0) = 2p_i(t,0)$.

A hidden constraint pertains to both approaches. The associated theory assumes implicitly that the wave tube has no reflection site. The horndriver is a major and unavoidable reflection site. The reflected wave that comes from the airway opening impinges upon the horndriver and creates a secondary reflected wave. Let us take the time origin at the moment when the incident wave is created by the sound source. The secondary reflected wave reaches the pressure transducer after $t_{max} = L_2/c + 2(L_1 + L_2)/c$ (see Fig. 12.8, bottom). After that, the secondary reflected wave included in the recorded pressure acts as a new incident wave that is not taken into account by the time window technique or by the calibration procedure. Incident and reflected waves cannot be separated after t_{max}, so the area cannot be inferred for distance beyond $x_{max} = (c \cdot t_{max} - L_2/c)/2 = L_1 + L_2$. As a result of this constraint, the distance separating the loudspeaker from the pressure transducer, L_2, must be larger than the maximum airway penetration depth of interest, $x_{max} - L_1$.

Opening the second extremity of the wave tube to permit spontaneous breathing during measurement (as in Fig. 12.8) introduces another length constraint. The sound source creates a wave that propagates toward this tube extremity. This wave impinges upon this end and creates a secondary reflected wave

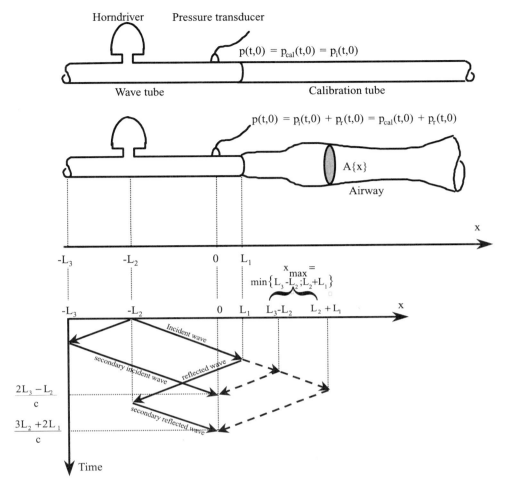

FIGURE 12.8   Schematic of the wave tube for measuring the area by the one-microphone strategy using the singular calibration approach. Top: The calibration tube is connected; there is no reflection site; the recorded pressure, $p_{cal}(t,0)$ is the incident pressure. Middle: The airway is connected to the wave tube; the recorded pressure is the sum of the incident and reflected waves; the reflected wave is obtained by subtracting $p_{cal}(t,0)$ from the recorded pressure. Bottom: Schematic diagram (time vs. axial position) of reflections resulting from the impulse generated by the horndriver. The maximum distance which can be investigated, x_{max}, depends (1) on the distance between the horndriver and the airway opening, and (2) on the distance between the horndriver and the wave tube extremity opened to the atmosphere; x_{max} is the smallest value of these two distances.

propagating back to the sound source. This secondary reflection acts as a secondary incident wave (Fig. 12.8, bottom). This secondary incident wave arrives at the pressure transducer after a propagation delay of $t = (2(L_3 - L_2) + L_2)/c$. The secondary incident wave is then superposed to the first incident and reflected wave, which prohibits the use of the time window technique (Sondhi's approach). In this case, the distance separating the end of the wave tube and the loudspeaker, $L_3 - L_2$, should be larger than the distance of maximum penetration in the airway, x_{max}, to ensure that the new incident wave arrives only after that the data acquisition has been completed. The single calibration procedure (Fredberg's approach) does not have this constraint. As a result of these length constraints, instruments for pulmonary application (from the mouth to the carina) described in the literature were 2 m or more in length which is hardly consistent with standard clinical applications.

The Dual Calibration Approach

To de-emphasize this length constraint, Marshall[53] proposed another variation of this one-microphone strategy. This variation did not explicitly separate incident and reflected waves, but used the Fourier transform and its inverse function. This variation is based upon a double calibration procedure associated with a measurement. The first calibration gives the incident wave. The second one gives the impulse response of the horndriver. Then, the measurement gives the impulse response of the airway. Consider the geometry in Fig. 12.7. The recorded pressure is given by:

$$p(t,0) = \overbrace{p_i(t,0)}^{\substack{\text{incident} \\ \text{wave}}} + \overbrace{p_i(t,0) \otimes h(t,0)}^{\substack{\text{reflected} \\ \text{wave}}} + \overbrace{p_i(t,0) \otimes h(t,0) \otimes h_h(t,0)}^{\substack{\text{secondary reflected} \\ \text{wave}}}. +$$

$$+ \overbrace{p_i(t,0) \otimes h(t,0) \otimes h_h(t,0) \otimes h(t,0)}^{\substack{\text{third reflected} \\ \text{wave}}} + \ldots\ldots$$

Then,

$$p(t,0) = p_i(t,0) \otimes \left[\delta(t) + h(t,0)\right] \otimes \left[\sum_{k=0}^{k=\infty} \delta(t) \otimes \left[h_h(t,0) \otimes h_h(t,0)\right]^k\right] \qquad (12.20)$$

$h(t,0)$ designates the impulse response of the airway including the part of wave tube located between the pressure transducer and the airway opening; $h_h(t,0)$ designates the impulse response of the source sound including the part of wave tube located between the pressure transducer and the source sound. Using the Fourier transform, Eq. 12.20 becomes:

$$P(\omega,0) = P_i(\omega,0) \cdot \left[1 + H(\omega,0)\right] \cdot \left[\sum_{k=0}^{k=n} \left[H(\omega,0) \cdot H_h(\omega,0)\right]^k\right] \qquad (12.21)$$

where F (capital letter), function of (ω, x), designates the Fourier transform of f (lowercase letter) function of (t, x). Using the properties of the binomial series Eq. 12.21 becomes

$$P(\omega,0) = \frac{P_i(\omega,0) \cdot \left[1 + H(\omega,0)\right]}{1 - H(\omega,0) \cdot H_h(\omega,0)} \qquad (12.22a)$$

$$H_h(\omega,0) = \frac{P(\omega,0) - P_i(\omega,0) \cdot \left[1 + H(\omega,0)\right]}{P(\omega,0) \cdot H(\omega,0)} \qquad (12.22b)$$

$$H(\omega,0) = \frac{P(\omega,0) - P_i(\omega,0)}{P_i(\omega,0) + P(\omega,0) \cdot H_h(\omega,0)} \qquad (12.22c)$$

The first step of the calibration consists of connecting to the wave tube a calibration tube with the same dimension as in the single calibration approach. There is no site of reflection; $h(t,0) = 0$; $H(\omega,0) = 0$.

The recorded pressure is the incident pressure (Eq. 12.22a). The second step of the calibration is closing the wave tube by a slide valve. This slide valve is considered as an ideal reflector,

$$\left(h(t,0) = \delta(t - 2L_1/c); \; H(\omega,0) = e^{j\omega \frac{2L_1}{c}} \right).$$

Associated with the knowledge of $P_i(\omega,0)$ determined at the first calibration step, this permits us to obtain $H_h(\omega,0)$ via Eq. 12.22b. The measurement then involves connecting the airway to the wave tube and measuring the pressure. The recorded pressure immediately shows $H(\omega,0)$, via Eq. 12.22c where both $P_i(\omega,0)$ and $H_h(\omega,0)$ were determined during the calibration steps. Performing the inverse Fourier transform gives the impulse response. This strategy permits an arbitrarily short wave tube. It was developed and used by Marshall[53] to infer the area from glass tube models up to three times the wave tube length but it was not validated in living subjects. Indeed, to our knowledge, this approach will not allow the subject to breathe freely during the measurement. A convective flow engendered by breathing flow modifies the wave speed and thus modifies $H_h(\omega,0)$ which would thereby not be the same between the calibration time and the measurement time.

The Two-Microphone Strategy

This strategy attempts to obtain the impulse response without explicitly separating incident and reflected pressure waves. In principle, simultaneous pressure recording at two distinct sites of a wave tube is sufficient to determine the impulse response of the airway. Such a strategy, developed by Louis et al.[54] to infer airway area, was inspired by Schroeder's work and followed a similar two-transducer approach used by Seybert et al., Chung et al., and Krishnappa and Bodén et al.[55-61] to measure duct acoustic input impedance (but not duct dimensions).

Principle of the Two-Microphone Approach

Consider an horndriver connected to a wave tube with two distinct pressure transducers (Fig.12.9). The pressure recorded in the two distinct sites are given by:

$$p(t,-L) = p_i(t,-L) + p_r(t,-L), \text{ i.e., in the frequency domain } P(\omega,-L) = P_i(\omega,-L) + P_r(\omega,-L) \quad (12.23a)$$

$$p(t,0) = p_i(t,0) + p_r(t,0), \text{ i.e., in the frequency domain } P(\omega,0) = P_i(\omega,0) + P_r(\omega,0) \quad (12.23b)$$

The propagation delays, τ, of the two waves between the transducer sites are given by:

$$p_i(t,0) = p_i(t - \tau, -L), \text{ i.e., in the frequency domain } P_i(\omega,0) = P_i(\omega,-L) \cdot e^{j\omega\tau} \quad (12.23c)$$

$$p_r(t,-L) = p_r(t - \tau, 0), \text{ i.e., in the frequency domain } P_r(\omega,-L) = P_r(\omega,0) \cdot e^{j\omega\tau} \quad (12.23d)$$

$$\text{with } L = c \cdot \tau \quad (12.23e)$$

The impulse response of the airway including the part of the wave tube located between the second transducer ($x = 0$) and the attached airway is given by:

$$p_r(t,0) = h(t,0) \otimes p_i(t,0), \text{ i.e., in the frequency domain } P_r(\omega,0) = H(\omega,0) \cdot P_i(\omega,0) \quad (12.23f)$$

The impulse response of the second end of the wave tube including horndriver and part of the wave tube up till the first transducer is given by:

$$p_i(t,-L) = h_h(t,-L) \otimes p_r(t,-L) \quad \text{or in the frequency domain} \quad P_i(\omega,-L) = H_h(\omega,-L) \cdot P_r(\omega,-L) \quad (12.23g)$$

Using 12.23c and 12.23d, Eqs. 12.23f, 12.23a, and 12.23b can be recast in the frequency domain as:

$$P_r(\omega,-L) = P_i(\omega,-L) \cdot e^{j\omega\tau} \cdot H(\omega,0) \qquad (12.24a)$$

$$P(\omega,-L) = P_i(\omega,-L) + P_r(\omega,-L) \cdot e^{j\omega\tau} \qquad (12.24b)$$

$$P(\omega,0) = P_i(\omega,0) + P_r(\omega,0) \cdot e^{j\omega\tau} \qquad (12.24c)$$

Adding Eqs. 12.24a through 12.24c, we get:

$$\begin{cases} P(\omega,-L) = P_i(\omega,-L) \cdot \left(1 + H(\omega,0) \cdot e^{j\omega 2\tau}\right) \\ P(\omega,0) = P_i(\omega,-L) \cdot \left(e^{j\omega\tau} + \cdot e^{j\omega\tau} \cdot H(\omega,0)\right) \end{cases} \qquad (12.25b)$$

From Eq. 12.26a and 12.26b, it is easy to show

$$\frac{P(\omega,-L)}{P(\omega,0)} = \frac{e^{-j\omega\tau} + H(\omega,0) \cdot e^{j\omega\tau}}{1 + H(\omega,0)}$$

which can be recast as

$$H(\omega,0) \cdot \left[P(\omega,-L) - P(\omega,0) \cdot e^{-j\omega\tau}\right] = P(\omega,0) \cdot e^{j\omega\tau} - P(\omega,-L)$$

or in the time domain:

$$h(t,0) \otimes \left[p(t,-L) - p(t-\tau,0)\right] = p(t+\tau,0) - p(t,-L) \qquad (12.26)$$

If τ and the pressures in $x = 0$ and $x = -L$ are known, $h(t,x_0)$ can be determined from Eq. 12.26 using a variety of deconvolution algorithm. Because of the additive length of wave tube between the second microphone ($x = 0$) and the airway opening without any reflection site, the early portions of the recorded pressure at the two measurement sites are identical except for the propagation delay. The comparison of the two-microphone outputs in the early parts of their respective transients permits us to determine the propagation delay, τ. The wave speed is obtained from τ: $c = L/\tau$ which permits us to determine, via Eq. 12.13, the distance at which the reflection sites are located.

    No assumptions concerning non-overlapping time windows or secondary reverberation from the sound source, were made. Acoustic reverberation within the wave tube is implicitly permitted by the two-microphone method since Eq. 12.26 does not contain the quantity $h_h(t,-L)$. Moreover, there is no length constraint upon the wave tube with this approach, so that the wave tube length can be considerably reduced by comparison with the one-microphone strategy. A 10 cm long wave tube and a 20 cm long wave tube were tested for nasal applications[54] and for pulmonary applications.[37] The results obtained were substantially equivalent to the results obtained by the classical one-microphone method. Remarkably, although the two-transducer system was potentially noisier, the coefficient of variation associated with the dual-transducer system was barely increased in comparison with the single transducer system.

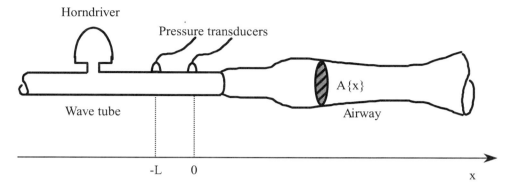

FIGURE 12.9   Schematic of the wave tube used for measuring area by the two-microphone strategy. The pressure, sum of the incident, and the reflected waves, are recorded in two loci of the wave tube to infer the area, A, vs. the axial position, x. The 0 and -L represent the distances of the two pressure transducers. This strategy does not require preliminary calibration procedure.

Convective Flow

A convective flow, such as the one engendered by breathing, slightly modifies the associated theory of the two-microphone strategy. Propagation depends on the direction of the propagation. Eqs. 12.23c, 12.23d, and 12.23e are replaced by

$$p_i(t,0) = p_i(t-\tau_i, -L), \text{ i.e., in the frequency domain } P_i(\omega,0) = P_i(\omega,-L)\cdot e^{j\omega\tau_i} \qquad (12.27a)$$

$$p_r(t,-L) = p_r(t-\tau_r, 0), \text{ i.e., in the frequency domain } P_r(\omega,-L) = P_r(\omega,0)\cdot e^{j\omega\tau_i} \qquad (12.27b)$$

with

$$L = \tau_i \cdot (c+u) = \tau_r \cdot (c-u) \qquad (12.27c)$$

The τ_i is the propagation delay of the incident wave and τ_r is that of the reflected wave; u is the mean velocity of the steady flow in the duct, which is taken to be positive in the x direction ($\rightarrow +x$). In this case Eq. 12.26 should be replaced by:

$$H(\omega,0)\cdot\left[P(\omega,-L) - P(\omega,0)\cdot e^{j\omega\tau_r}\right] = P(\omega,0)\cdot e^{j\omega\tau_i} - P(\omega,-L)$$

or in the time domain:

$$h(t,0)\otimes\left[p(t,-L) - p(t-\tau_r,0)\right] = p(t+\tau_i,0) - p(t,-L) \qquad (12.28)$$

The τ_i can be directly measured by comparing the two-microphone outputs in the early part of their respective transients, while The τ_r can be inferred from Eq. 12.27c if u is known, and $h(t,0)$ can be inferred from Eq. 12.28 using a variety of deconvolution algorithms. Using simulation technique,

Louis et al.[37,62] showed at least to a first approximation that the effects of the convective flow due to quiet breathing may be neglected. When the convective flow is neglected, i.e., when we assume $\tau_r = \tau_i$, τ_r is by a factor,

$$\frac{2L}{c} \cdot \left(\frac{\mathcal{M}}{1 - \mathcal{M}^2} \right),$$

where $\mathcal{M}$ is the Mach number, $\mathcal{M} = u/c$, the $\tau_r = \tau_i$ assumption is valid if $\mathcal{M} \ll 1$. In the present embodiment of the two-transducer system (bandwidth, wave tube dimension, gas, investigation distance < 60 cm), the authors found that if $\mathcal{M}$ was less than 0.045, the relative error on the inferred area was less than 10%. Moreover for the higher Mach number values a correction is always possible if we know the value of the convective flow.

The two-microphone strategy has advantages in that it permits (1) miniaturization of the wave tube (small dead space), (2) absence of a calibration procedure, and (3) the ability for the subject to breathe air during measurement.[37,54,63] This allows *in vivo* measurement under varied and difficult conditions as encountered in intensive care units,[64,65] with a sleeping patient during obstructive sleep apnea, or during parabolic flight.

Signal Processing

Deconvolution Algorithm

With the time window approach, the single calibration approach, and the two-microphone method, the impulse response can be obtained after the deconvolution of an equation of the form:

$$y(t) = h(t) \otimes x(t) \tag{12.29}$$

$h(t)$ is the unknown impulse response, $x(t)$ is the input function, and $y(t)$ is the output function. For the one-microphone strategy, $x(t) = p_i(t,0)$ and $y(t) = p_r(t,0)$. For the two-microphone strategy, $x(t) = p(t,-L) - p(t-\tau,0)$ and $y(t) = p(t+\tau,0) - p(t,-L)$. Once Eq. 12.29 is discretized by the Riemann sum approximation, $h(t)$ can be computed in a manner analogous to the standard time domain deconvolution algorithm:[66]

$$h(n\Delta t) = \frac{y(\cdot \Delta t)}{x(0 \cdot \Delta t)} - \sum_{k=1}^{k=n} \frac{h([n-k] \cdot \Delta t) \cdot x(k \cdot \Delta t)}{x(0 \cdot \Delta t)} \tag{12.30}$$

Δt is the sampling period of the discretization, while $x(0 \cdot \Delta t)$ denotes the first non-zero discrete value of $x(t)$, and n is an integer. The deconvolution problem, known in the acoustic field as an inverse problem, is numerically ill-posed. This means that small perturbations in the input and output functions can lead to wild oscillations of the numerical solutions of the impulse responses that do not make sense. To de-emphasize this instability, a regularization procedure was used. The computational algorithm implied by Eq. 12.30 remains stable only if $x(0 \cdot \Delta t)$ was larger than some minimal value.[54] The regularization procedure introduces a threshold to define the first non-zero value of $x(t)$ (see Fig. 12.10). The data corresponding to earlier times are neglected and that gives a first approximation of the impulse response.

$$h_1(n\Delta t) = \frac{y((n+g) \cdot \Delta t)}{x(g \cdot \Delta t)} - \sum_{k=1}^{k=n} \frac{h((n+g-k) \cdot \Delta t) \cdot x((k+g) \cdot \Delta t)}{x(g \cdot \Delta t)} \tag{12.31}$$

Because this regularization procedure introduces an error upon the determination of h, a correction time sequence was added. This correction was generated by convoluting the early part of the neglected time sequence with the first approximation of the impulse response, $h_1(t)$. Indeed, Eq. 12.29 can be recast as:

$$y(t) = h(t) \otimes \left[x_n(t) + x_t(t) \right]$$
(12.32)

where $x_n(t)$ is the early part of $x(t)$ neglected by the threshold procedure and $x_t(t)$ is the truncated part of $x(t)$ defined by this threshold procedure (see Fig. 12.10):

$$
\begin{aligned}
x_n(t) &= x(t) \quad \text{if} \quad t < g \cdot \Delta t \quad \text{and} \quad x_n(t) = 0 \quad \text{if} \quad t \geq g \cdot \Delta t \\
x_t(t) &= 0 \qquad \text{if} \quad t < g \cdot \Delta t \quad \text{and} \quad x_t(t) = x(t) \quad \text{if} \quad t \geq g \cdot \Delta t
\end{aligned}
$$

Eq. 12.32 can be recast as:

$$h(t) = \left[y(t) - \left(x_n(t) \otimes h(t) \right) \right] \varnothing x_t(t)$$
(12.33)

where $\varnothing$ is the deconvolution operator. If we consider that the impulse response is approximated by h_1 well enough to compute the effect of the early part of the time sequence previously neglected, the corrected second approximation is given by:

$$h_2(t) = \left[y(t) - \left(x_n(t) \otimes h_1(t) \right) \right] \varnothing x_t(t)$$
(12.34)

This procedure of correction can be theoretically repeated as long as we wish:

$$h_m(t) = \left[y(t) - \left(x_n(t) \otimes h_{m-1}(t) \right) \right] \varnothing x_t(t)$$

Nevertheless, such correction re-introduces some instability in the solution, so the procedure is generally repeated once or twice. Regularization and correction secure both adequate stability and accuracy of the method.[37,54,64,65] Eq. 12.29 can also be deconvoluted in the frequency domain.[13,14,16,42] In this case $h(t)$ is the inverse Fourier transform function of the ratio $Y(\omega)/X(\omega)$. The time domain algorithm was chosen because it offers substantial saving in computation time compared with the Fourier transform technique.[17] Time domain computation requires only the signal up to the time corresponding to the deepest penetration of interest. Use of the deconvolution algorithm with the Fourier transform method requires the entire pressure signal (time truncated waveforms yield incorrect results) and generates the entire impulse response signal of which only the earliest portion is used to infer area. This time saving was especially important in the 1980s[17] because it permitted area reconstruction in quasi-real time (one area reconstruction per second). Twelve years later, considering the great increases in microcomputer power, the interest in time computation savings offered by the time domain deconvolution is perhaps more questionable.

Data Filtering

Pressure data and impulse response are also moderately band pass filtered. The low pass tends to de-emphasize the instability of the impulse response and tends to eliminate artifacts associated with acoustic cross-mode. The high pass de-emphasizes artifacts associated with airway wall nonrigidity. It also tends to eliminate physiologic noise associated with breathing and tends to ensure the lossless propagation assumption.

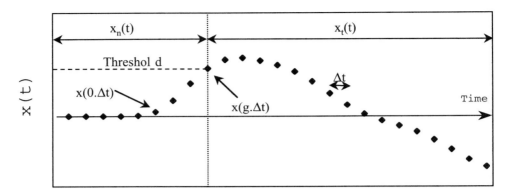

FIGURE 12.10 Representative example of the discrete record of input function used in time domain deconvolution. Δt is the sampling frequency; $x(0 \cdot \Delta t)$ is the true first non-zero value of the input function; $x(g \cdot \Delta t)$ is the first non-zero value of the input function determined by the threshold procedure; $x_n(t)$ and $x_t(t)$ are successively the parts of the input function neglected and truncated by the threshold procedure.

12.5 Nasal Applications

Measurements of the geometry of the nasal cavity via the acoustic reflection method were recently introduced by Hilberg et al.[43] Those authors validated the method by comparing acoustic measurement results with the results obtained by CT scan in cadavers and by water displacement in healthy humans. Since their study was done, acoustic reflection technology has proven to be a useful tool to detect changes in the nasal cavity produced by decongestants, posture, nasal functioning, intramuscular steroid injections of nasal polyps, or nasal histamine challenges.[67–70] The method was also used to evaluate nasal obstruction due to chronic rhonchopathies[46] and obstruction caused by septal deviation before and after septoplasty.[44,71]

The authors found satisfactory correlation between subjective assessments of nasal patency and acoustic rhinometry results. The acoustic reflection method, called acoustic rhinometry for nasal applications, presents the advantage of being noninvasive and producing objective documentation of the nasal cavity with a quantitative evaluation and topical data on mucosal and skeletal changes in response to physiological states or artificial manipulation (decongestion, surgery, etc.). By comparison, if CT scanning is suitable for a three-dimensional reconstruction of the nasal tract, it only shows a static picture. Classical rhinomanometry shows the resistance of the nasal tract, but it cannot furnish details about the location of an obstruction.

Generally, acoustic rhinometry results are clinically analyzed by focusing upon the cross-sectional areas at specific locations of the area distance curve. Indeed, this curve is usually described on the basis of two or three relative minima, called notches (see example in Fig. 12.11, top). Considering that the first notch corresponded to the minimal cross-sectional area (MCA) in 134 normal subjects, Lenders and Pirsig[45] concluded that the first notch corresponded to the functional isthmus nasi. They also concluded that the second notch corresponded to the head of the inferior turbinate. Similar results were reported by Grymer et al.[44] and by Kunkel et al.[72] Nevertheless, other studies without changing the interpretations of the different notches, showed that MCA was not systematically in the first notch. The bottom section of Fig. 12.11 presents an example where the MCA is in the second notch due to hypertrophy of the inferior turbinate in the left nostril.

The different studies comparing results obtained with acoustic rhinometry and those obtained with CT scan or magnetic resonance imaging revealed satisfactory correlations among the methods applied to the anterior part of the nose cavity, i.e., from the nostril to the second notch. Acoustic rhinometry results seemed to overestimate the area of the posterior part of the nasal cavity, i.e., the area of the region lying after the second notch.[43,70,73] Using a nasal cavity cast made by a stereolithographic method, Hilberg et al.[47] showed that this overestimation was introduced by the presence of the maxillary sinuses. The

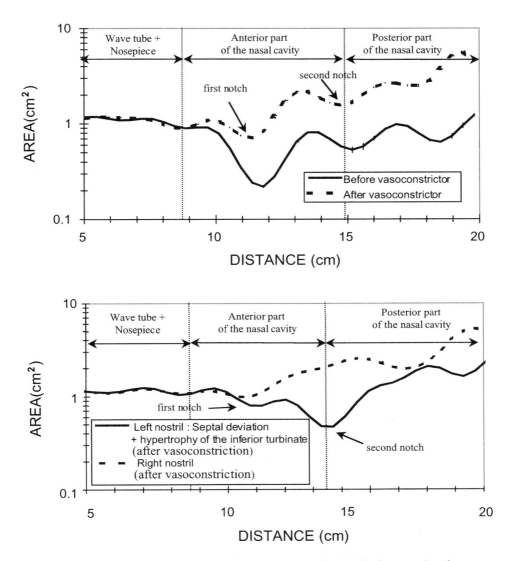

FIGURE 12.11 Typical acoustic rhinometry area-distance curves inferred with the two-microphone strategy in patients of the Hospital Henri-Mondor, Créteil, France. The interpretation of the different characteristic points of the curves (notches) is indicated in the text. Top: Example of the effect of a vasoconstrictor in a patient with rhinitis. Bottom: Example of a patient with an obstruction in his left nostril due to turbinate hypertrophy. The curve of the left nostril is compared to the normal curve obtained in the right nostril. Using a vasoconstrictor prior to measurement permits de-emphasis of the artifacts linked with nasal cycle.

presence of sinuses when the ostia is open tends to violate the single duct assumption. Part of the acoustic wave would be transmitted in the sinuses via the ostia, which would introduce an overestimation of the inferred area as in the case of the open velum (see "Single Duct" in Section 12.3). Therefore, the degree to which acoustic rhinometry can be used to evaluate airway area of the posterior part of the nasal cavity is questionable. On the other hand, such a property may present some positive aspects. First of all, it may suggest that the acoustic technique may be potentially for obtaining information about sinus and ostium pathology. Further studies have to be done to reject or confirm such speculation. Second, even if sinuses engender an overestimation of the area of the posterior part of the nasal cavity, the method still permits comparisons in this region. Indeed, in a group of 20 children, the change in the volume of the nasopharynx after adenoidectomy, i.e., in a nasal part far after the second notch as inferred by acoustic rhinometry, was found highly proportional to the volume of the adenoid removed.[74]

For those applications, the investigation distances of interest are short (< 15 cm). That almost erases the disadvantage of the wave tube length constraints encountered with the one-microphone strategy. Nasal breathing during measurement generates positive or negative intranasal pressure and noise that can alter the shape of the nasal cavity. The recommendation is to perform measurement during a brief breathing pause.[75] Other types of error are acoustic leaks[76] and the distortion of the vestibulum nasi by a nose-piece.[77] Special care in the choice of nose-piece size will minimize the distortion of the vestibulum nasi. The use of gel between nose-piece and nostril can ensure a good seal.[77] Moreover, using head support and wave tube support[77] permitted enhancement of the reproducibility of the measurement, a reduction of interoperator variability, and standardization of the angle between wave tube and nostril that can be another source of error.[78]

12.6 Pulmonary Applications

Pulmonary applications follow three distinct stages: (1) a validation stage in which the results of the acoustic reflection method were compared *in vitro* and *in vivo* to the results obtained with different techniques considered as reference methods, (2) a second stage at which the method was used to better understand the physiopathology of the upper airway; (3) a third stage in which the method was used for direct clinical application.

Validation of the Method

The first validation of the method was realized *in vitro* through use of a 16 cm depth test cavity by Sondhy and Gopinath in 1971.[9] Later, Jackson et al.[42] compared the area of dried excised dog lung inferred with acoustic method to the area inferred by radiographic bronchograms. These early *in vitro* studies were followed by the measurement of a cast of a human airway,[13] a rigid glass model simulating pharyngeal-laryngeal-tracheal regions,[17,48,53,79] and a combination of compliant tube and rigid glass model simulating the upper airway from the mouth to the carina.[37] In all of these studies, acoustic measurements were in good agreement compared with measurements obtained by radiographic or direct methods.

The first *in vivo* validation of the method was made by Fredberg et al.[14] who compared the tracheal areas of healthy subjects inferred with acoustic method to areas measured by bilateral radiography. The authors found a satisfactory correlation between the two methods. Their findings were confirmed by Brooks et al.[17] and D'Urzo et al.[19] who found differences of less than 10% between the results measured by acoustic reflection and results from CT scan or lateral radiography. The measurement of glottis area by acoustic reflection was validated by d'Urzo et al.[80] with similar degree of correlation and much later by Zhou et al.[81] who followed vocal cord movement. Marshall et al.[48] compared the acoustic and the magnetic resonance imaging (MRI) methods of assessing pharyngeal areas in ten normal subjects. They observed that only the inference of the maximal hypopharyngeal area was statistically different in comparing the two methods. To explain this difference, the authors pinpointed three facts: (1) the uncertainty of the MRI associated to the pixel size was $\pm 19\%$ for an area of 1 cm² and it was $\pm 14\%$ for an area of 2 cm²; (2) the acoustic method shows an instantaneous image of the upper airway while MRI shows an image corresponding to an average over many breaths including effects of possible swallowing; and (3) the MRI image can be affected by a non-ideal tissue/air contrast.

Applications

The first field of application of the acoustic reflection method concerned the structure and the function of the upper airway and especially of the pharyngeal regions in patients with obstructive sleep apnea

(OSA). In 1984 Rivlain et al.[82] showed that the area of the pharyngeal region, defined as the region from the hard palate up to the glottis, was significantly smaller in nine awake patients with OSA than in normal controls. Later, Brown et al.[83] found that by measuring upper airway areas at different pharyngeal pressures, patients with OSA showed larger pharyngeal compliance than those in the normal control group. These results were extended to the snorer population. Snoring is considered a precursor of OSA.[84] The authors found that snorers without OSA and patients with OSA had similar pharyngeal areas at functional residual capacity; at residual volume, the pharyngeal areas of snorers without OSA were larger than those of patients with OSA. This finding was confirmed for obese women by Rubinstein et al.[85] Using acoustic reflection to evaluate the changes in pharyngeal area with lung volume in awake patients, Hoffstein et al.[86] showed that the lung volume associated with changes in pharyngeal area is related to the severity of sleep apnea.

The acoustic reflection method also showed that the pharyngeal area changed from the supine to the upright position and that those changes are quantitatively different in patients with OSA and normal subjects.[51,87,88]

Another application of the acoustic reflection method concerns the structure and the dynamic of the trachea. Hoffstein and Zamel[20] showed that acoustic technique was a rapid and noninvasive method capable of identifying and quantifying tracheal stenosis prior the development of characteristic abnormalities in the flow volume curve. Many investigations involved healthy subjects. Hoffstein[89] systematically analyzed the relationship between the tracheal area and maximal expiratory flow. They found that tracheal area varies with lung volume.

Hoffstein and co-authors[21,90] also showed that hysteresis of airways and tracheal distensibility can be noninvasively assessed via the acoustic reflection method. This last application is a clear example of the possibility of providing noninvasive dynamic images of the airways. In Hoffstein's study, measurements made during inspiration and expiration were separated and airway areas were reconstructed at a rate slightly above 3 per second. Brooks et al. showed that men have larger trachea size areas than women and that these differences are not functions of body size.[91] The method was also used to evaluate the tracheal response during isocapneic hypoxia.[92] The acoustic reflection technique was used to assess airway tone in patients with asthma.[93,94] The effects of different phamacologic agents such as methacholine, histamine, LTC_4,[95] procaterol hydrochloride, and ipratropium bromide[96] upon the upper airway were also tested with the help of the acoustic reflection method.

Due to the improvement of the method associated with the two-microphone strategy, it was possible to apply the method at bedside in an intensive care unit. Van Surrel et al.[64] have shown that the method permits assessment of the patency of the adult endotracheal tube (ETT) in mechanically ventilated patients by assessing the extent and locus of the accumulation of mucus deposited at the wall of the ETT. It was later shown that this method did not even require interruption of the artificial ventilation during the measurement.[65] Routine assessment of the ETT may help reduce the consequences of ETT obstruction with regard to the risk of emergency extubation. The method also permits a noninvasive estimate of the work of breathing due to the ETT,[65] assuming that the pressure drop through an adult mucus-lined ETT is determined by the Blasius resistance formula.[97] This formula relates pressure drop and inner geometry of the tube in fully developed and hydraulically smooth turbulent flow. Postulating that this formula would remain locally applicable in any segment of the ETT, Heyer et al.[65] showed that the determination of the inner ETT geometry by acoustic reflection allowed accurate estimation of the pressure drop as well as the work of breathing due specifically to the ETT.

The acoustic reflection method was also adapted to the assessment of the neonatal pediatric ETT.[98] During this study, the authors observed another exciting possibility. The method allowed us to characterize and detect early the well-known problem of ETT obstruction due to head position[99] when babies are ventilated in womb posture. To avoid some muscular atrophy and bone development problems, head position is regularly modified (from left to right and inversely). Occasionally the distal extremity of the ETT abuts the tracheal wall, which produces a dramatic obstruction that can compromise the benefits of mechanical ventilation in newborns. Generally, this problem is suspected after an episode of hypercapnia. In their study,

Jarreau et al.[98] observed that such a situation is clearly visible and thus detectable on the acoustic profiles which exhibit a major decrease in area of the region immediately following the distal extremity of the ETT (see Fig. 12.12). These area profiles are easy to distinguish from normal situations.

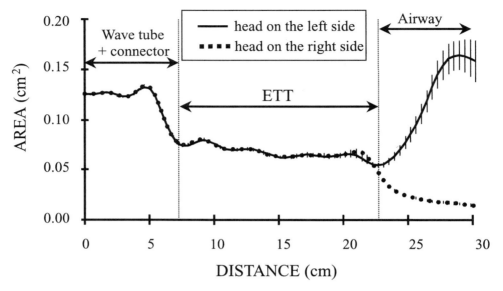

FIGURE 12.12 Longitudinal profile area measured with acoustic reflection in an intubated baby. The ETT was 3 mm in diameter. The curves were obtained in air with the two-microphone method in the neonatal intensive care unit of the Hospital Port-Royal in Paris. Left and right positions of the baby head were tested. On the right side the ventilation was dramatically compromised, probably because the extremity of the ETT was abutting the tracheal wall. It produced, by comparison with the left side, a major decrease on the portion of the area-distance curve corresponding to the region immediately following the distal extremity of the ETT (airway). By contrast, there was no obstruction inside the ETT.

12.7   Conclusion

The acoustic reflection method is a powerful tool for studying the physiopathological properties of the upper airways. It now permits an accurate noninvasive and reproducible measurement of the upper airway area at bedside and in clinical situations. The short duration of the measurement ($\cong 5$ ms) in comparison with the breathing period also allows clinicians to estimate the upper airway geometry under both static and dynamic conditions. Such a tool should lead to a better understanding of the structure and function of the upper airway in health as well as in disease.

The method is robust and easy to use. The recent improvements of the method (miniaturization and use of air as a breathing gas) should lead to new applications of the method, especially in the severe clinical conditions of an intensive care unit.

References

1. Dubois, A.B., Brody, A.W., Lewis, D.M., and Burgers, B.F., *J. Appl. Physiol.*, 8, 587, 1956.
2. Peslin, R. and Fredberg, J.J., in *Handbook of Physiology*, Geiger, S.R., Ed., American Physiological Society, Bethesda, MD, 1986, vol. 3, 145.
3. Fredberg, J.J., *Fed. Proc.* 39, 2747, 1980.
4. Drinker, P.A. and Bowsher, J.M., *Historic Brass Soc. J.*, 5, 107, 1993.
5. Morse, P.M. and Ingard, K.U., in *Theoretical Acoustics*, Princeton University Press, Princeton, NJ, 1968, 227, 467.
6. Comolet, R., in *Biomecanique Circulatoire*, Masson, Paris, 1984, 155.

7. Sidell, R.S. and Fredberg, J.J., *J. Biomechanical Eng.*, 100, 131, 1978.
8. Ware, J.A. and Aki, K., *J. Acoust. Soc. Am.*, 45, 911, 1969.
9. Sondhi, M.M. and Gopinath, B., *J. Acoust. Soc. Am.*, 49, 1867, 1971.
10. Gopinath, B. and Sondhi, M.M., *Bell System Tech. J.*, 49, 1195, 1972.
11. Sondhi, M.M., *J. Acoust. Soc. Am.*, 55, 1070, 1974.
12. Schroeder, J., *J. Acoust. Soc. Am.*, 41, 1002, 1967.
13. Jackson, A.C. and Olson, D.E., *J. Appl. Physiol.*, 48, 896, 1980.
14. Fredberg, J.J., Wohl, M.E.B., Glass, G.M., and Dorkin, H.L., *J. Appl. Physiol.*, 48, 749, 1980.
15. Fredberg, J.J., *Ann. Biomed. Eng.*, 9, 463, 1981.
16. Fredberg, J.J., Acoustic pulse response measuring, U.S. Patent 4,326,416, 1982.
17. Brooks, L.J., Castille, R.G., Glass, G.M., Griscombe, N.T., Wohl, M.E.B., and Fredberg, J.J., *J. Appl. Physiol.*, 57, 777, 1984.
18. D'Urzo, A.D., Rebuck, A.S., Lawson, V.G., and Hoffstein, V., *J. Appl. Physiol.*, 60, 398, 1986.
19. D'Urzo, A.D., Lawson, V.G., Vassal, K.P., Rebuck, A.S., Slutsky, A.S., and Hoffstein, V., *Am. Rev. Respir. Dis.*, 135, 392, 1987.
20. Hoffstein, V. and Zamel, N., *Am. Rev. Respir. Dis.*, 130, 472, 1984.
21. Hoffstein, V., Castille, R.G., O'Donnell, C.R., Glass, G.M., Strieder, D.J., Wohl, M.E.B., and Fredberg, J.J., *J. Appl. Physiol.*, 63, 2482, 1987.
22. Brown, I.G., Zamel, N., and Hoffstein, V., *J. Appl. Physiol.*, 61, 890, 1986.
23. Rubinstein, I., MacLean, P.A., Boucher, R., Zamel, N., and Fredberg, J.J., *J. Appl. Physiol.*, 63, 1469, 1987.
24. Iberall, A.S., *J. Res. Natl. Bur. Stan.*, 45, 85, 1950.
25. Brown, F.T., *J. Basic Eng.*, 84, 547, 1962.
26. Benade, A.J., *J. Acoust. Soc. Am.*, 44, 616, 1968.
27. Franken, H., Clement, J., Cauberghs, M., and Woestijne, K.P.V.d., *IEEE Trans. Biomed. Eng.*, 28, 416, 1981.
28. Keefe, D.H., *J. Acoust. Soc. Am.*, 75, 58, 1984.
29. Louis, B., Dynamique des ecoulements gazeux oscillants de faible amplitude: application á la mesure de l'impedance du système respiratoire, Thesis, University of Paris, 1990.
30. Begis, D., Delpuech, C., Le Tallec, P., Loth, L., Thiriet, M., and Vidrascu, M., *J. Appl. Physiol.*, 64, 1359, 1988.
31. Bertram, C.D. and Raymond, C.J., *Med. Biol. Comput.*, 29, 493, 1991.
32. Cremer, L. and Heckl, M., in *Structure-Borne Sound*, Springer-Verlag, New York, 1973, 466.
33. Guelke, R.W. and Bunn, A.E., *Acustica*, 48, 101, 1981.
34. Ribreau, *J. Mal. Vasc.*, 14, 287, 1989.
35. Thiriet, M. and Bonis, M., *Med. Biol. Comput.*, 21, 681, 1983.
36. Bela, S., Habib, R.H., and Jackson, C., *J. Appl. Physiol.*, 75, 2755, 1993.
37. Rice, A.D., *J. Appl. Physiol.*, 49, 326, 1980.
38. Louis, B., Glass, G.M., and Fredberg, J.J., *J. Appl. Physiol.*, 76, 2234, 1994.
39. Fredberg, J.J., Sidell, R.S., Wohl, M.E., and DeJong, G., *J. Biomechanical Eng.*, 100, 67, 1978.
40. Fredberg, J.J., Jackson, C., and Dawson, S.V., *J. Acoust. Soc. Am.*, 58, S128, 1975.
41. Fredberg, J.J., Jackson, C., and Dawson, S.V., *Med. Instrum.*, 9, 64, 1974.
42. Jackson, A.C., Butler, J.P., Millet, E.J., Hoppin, F.G., and Dawson, S.V., *J. Appl. Physiol.*, 43, 523, 1977.
43. Hilberg, O., Jackson, A.C., Swift, D.L., and Pedersen, O.F., *J. Appl. Physiol.*, 66, 295, 1989.
44. Grymer, L.F., Hilberg, O., Elbrond, O., and Pedersen, L.F., *Laryngoscope*, 99, 1180, 1989.
45. Lenders, H. and Pirsig, W., *Rhinology*, 28, 5, 1990.
46. Lenders, H. and Pirsig, W., *Rhinology*, 31, 101, 1993.
47. Hilberg, O. and Pedersen, O.F., *J. Appl. Physiol.*, 80, 1589, 1996.
48. Marshall, I., Maran, N.J., Martin, S., Jan, M.A., Rimmington, J.E., Best, J.K.K., Drummond, G.B., and Douglas, N.J., *Clin. Phys. Physiol. Meas.*, 14, 157, 1993.

49. Fouke, J.M. and Strohl, K.P., *J. Appl. Physiol.*, 63, 375, 1987.

50. Jan, M.A., Marshall, I., and Douglas, N.J., *Am. J. Respir. Crit. Care Med.*, 149, 145, 1994.

51. Martin, S.E., Marshall, I., and Douglas, N.J., *Am. J. Respir. Crit. Care Med.*, 152, 721, 1995.

52. Martin, T.R., Castille, R.G., Fredberg, J.J., Wohl, M.E.B., and Mead, J., *J. Appl. Physiol.*, 63, 2042, 1987.

53. Marshall, I., *J. Acoust. Soc. Am.*, 91, 3558, 1992.

54. Louis, B., Glass, G.M., Kresen, B., and Fredberg, J.J., *J. Biomechanical Eng.*, 115, 278, 1993.

55. Seybert, A.F. and Ross, D.F., *J. Acoust. Soc. Am.*, 61, 1362, 1977.

56. Chung, J.H. and Blaser, D.A., *J. Acoust. Soc. Am.*, 68, 907, 1980.

57. Chung, J.H. and Blaser, D.A., *J. Acoust. Soc. Am.*, 68, 914, 1980.

58. Chung, J.H. and Blaser, D.A., *J. Acoust. Soc. Am.*, 68, 1570, 1980.

59. Krishnappa, G., *J. Acoust. Soc. Am.*, 69, 307, 1981.

60. Boden, H. and Abom, M., *J. Acoust. Soc. Am.*, 79, 541, 1986.

61. Abom, M. and Boden, H., *J. Acoust. Soc. Am.*, 83, 2429, 1988.

62. Louis, B., Isabey, D., and Fredberg, J.J., Influence of a constant flow on the measurement of airway area by the two-microphone acoustic reflection method, in Jan, L.B. and Kooloos, G.M., Eds., Proceedings, Second World Congress of Biomechanics, Amsterdam, 1994.

63. Fredberg, J.J., Gloass, G.M., Lehr, J., and Louis, B., Airway geometry imaging, U.S. Patent 02/808,907, 1993.

64. Van Surrel, C., Louis, B., Lofaso, F., Beydon, L., Brochard, L., Fredberg, J.J., Harf, A., and Isabey, D., *Am. J. Respir. Crit. Care Med.*, 149, 28, 1994.

65. Heyer, L., Louis, B., Isabey, D., Lofaso, F., Brochard, L., Fredberg, J.J., and Harf, A., *Anesthesiology*, 85, 1324, 1996.

66. Claerbout, J.F., *Fundamentals of Geographical Data Processing with Application to Petroleum Prospecting*, McGraw-Hill, New York, 1976.

67. Elbrond, O., Felding, J.U., and Gustavsen, K.M., *J. Laryngol. Otolaryngol.*, 105, 178, 1991.

68. Fouke, J.M. and Jackson, A.C., *J. Lab. Clin. Med.*, 19, 371, 1992.

69. Fisher, E.W., Scadding, G.K., and Lund, V.J., *Rhinology*, 31, 57, 1993.

70. Hilberg, O., Grimer, L.F., and Pedersen, O.F., *Allergy*, 50, 166, 1995.

71. Marais, J., Murray, A.M., Marshall, I., Douglas, N., and Martin, S., *Rhinology*, 32, 1994.

72. Kunkel, M. and Hochban, W., *J. Cranio-Maxillo-Facial Surg.*, 22, 244, 1994.

73. Min, Y.G. and Jang, Y.J., *Laryngoscope*, 105, 757, 1995.

74. Elbrond, O., Hilberg, O., Felding, J.U., and Andersen, O.B., *Clin. Otolaryngol.*, 16, 84, 1991.

75. Tomkinson, A. and Eccles, R., *Rhinology*, 33, 138, 1995.

76. Tomkinson, A., Phil, M., and Eccles, R., *Am. J. Rhinol.*, 10, 77, 1996.

77. Roth, N. et al., *Am. J. Rhinol.*, 10, 83, 1996.

78. Fischer, E.W., Morris, D.P., Biemans, J.M.A., Palmer, C.R., and Lund, V.J., *Rhinology*, 33, 219, 1995.

79. Sondhi, M.M. and Resnick, J.R., *J. Acoust. Soc. Am.*, 73, 985, 1983.

80. D'Urzo, A.D. et al., *J. Appl. Physiol.*, 64, 367, 1988.

81. Zhou, Y. and Daubenspeck, J.A., *Ann. Biomed. Eng.*, 23, 85, 1995.

82. Rivlain, J. et al., *Am. Rev. Respir. Dis.*, 129, 355, 1984.

83. Brown, I.G., Bradley, T.D., Phillipson, E.A., Zamel, N., and Hoffstein, V., *Am. Rev. Respir. Dis.*, 132, 211, 1985.

84. Bradley, T.D. et al., *New Engl. J. Med.*, 315, 1327, 1986.

85. Rubinstein, I., Hoffstein, V., and Bradley, T.D., *Eur. Respir. J.*, 4, 344, 1989.

86. Hoffstein, V., Wright, S., Zamel, N., and Bradley, T.D., *Am. Rev. Respir. Dis.*, 143, 1294, 1991.

87. Brown, I., McLean, P., Boucher, R., Zamel, N., and Hoffstein, V., *Am. Rev. Respir. Dis.*, 136, 628, 1987.

88. Yildirim, N. et al., *Am. Rev. Respir. Dis.*, 144, 845, 1991.

89. Hoffstein, V., *Am. Rev. Respir. Dis.*, 134, 956, 1986.

90. Katz, I., Zamel, N., Slutsky, A.S., Rebuck, A.S., and Hoffstein, V., *J. Appl. Physiol.*, 65, 2390, 1988.

91. Brooks, L.J., Byard, P.J., Helms, R.C., Fouke, J.M., and Strohl, K.P., *J. Appl. Physiol.*, 64, 1050, 1988.

92. Julià-Serdà, G. et al., *J. Appl. Physiol.*, 75, 1728, 1993.

93. Julià-Serdà, G. et al., *J. Appl. Physiol.*, 73, 2328, 1992.

94. Molfino, N.A. et al., *Am. Rev. Respir. Dis.*, 148, 1238, 1993.

95. Molfino, N.A. et al., *Am. Rev. Respir. Dis.*, 146, 577, 1992.

96. Hoffstein, V., Zamel, N., McLean, P., and Chapman, K.R., *Am. J. Respir. Crit. Care Med.*, 149, 81, 1994.

97. Lofaso, F., Louis, B., Brochard, L., Harf, A., and Isabey, D., *Am. Rev. Respir. Dis.*, 146, 974, 1992.

98. Jarreau, P. et al., Surveillance des sondes d'intubation endotracheales (sit) par reflexion acoustique en reanimation neonatale, *Arch. Pediatrie*, 1997, in press.

99. Brasch, R.P., Heldt, G.P., and Hecht, S.T., *Radiology*, 141, 387, 1981.

Index

A

ABAQUS, **5**-2, **5**-10, **5**-12, **5**-14, **5**-15–**5**-20, **5**-23, **5**-25, **5**-26
Abdominal aorta, flow and stenosis relationships, **7**-7–**7**-8
Acoustic linear theory, **12**-7–**12**-12
Acoustic reflection method, **12**-1–**12**-26
   assumptions, **12**-7
     acoustic linear theory, **12**-7–**12**-12
     homogeneous gas, **12**-11–**12**-12
     one-dimensional wave propagation, **12**-7
     single duct, **12**-13
     wall rigidity, **12**-9–**12**-11
   helium-oxygen mixture, **12**-11
   impulse response measurement, **12**-12–**12**-20
     convective flow, **12**-17, **12**-19–**12**-20
     dual calibration approach, **12**-16–**12**-17
     one-microphone strategy, **12**-12–**12**-17
     single calibration approach, **12**-13–**12**-15
     time window approach, **12**-13
     two-microphone strategy, **12**-17–**12**-20
   in vivo validation, **12**-24
   multiple reflection sites, **12**-4–**12**-6
   nasal applications, **12**-22–**12**-24
   signal processing, **12**-20–**12**-22
   singular reflection site, **12**-4
   tube with continuous longitudinal area profile variation, **12**-6
   wave propagation in duct, **12**-2–**12**-3
Acoustic rhinometry, **12**-22–**12**-24
Adhesive agents, **3**-9
Airway structure, **12**-1–**12**-2
   acoustic rhinometry, **12**-22–**12**-24
   helium-oxygen mixture, **12**-11
   noninvasive acoustic imaging, **12**-2–**12**-26, *See* Acoustic reflection method
Angiography technologies, **7**-11–**7**-12
   pressure measurement method, **8**-5, **8**-23, **8**-24
Angioplasty, **1**-2, **5**-23, **7**-12, **7**-14
ANSYS, **5**-12, **5**-22
Aorta, wall shear stress and pulsatile pipe flow model, **2**-2–**2**-9
Aortic bifurcation modeling, **11**-1–**11**-13, *See also* Arterial bifurcations
Apexcardiogram, **7**-11
Apnea, **12**-2, **12**-24–**12**-25
Arbitrary Lagrangian Eulerian method (ALE), **11**-8
Arterial bifurcations, *See also* Y-shaped bifurcation, numerical model
   atherosclerosis and, **3**-17–**3**-18, **11**-2
   computer-aided design, *See* Branching blood vessels, optimal design considerations

electrochemical shear stress measurement method
   application, **3**-21–**3**-30
   flow visualization, **3**-21, **3**-22–**3**-23
   generating system of pulsating flow, **3**-22
   physiological implications, **3**-29
   pulsating flow, **3**-29
   steady flow, **3**-24–**3**-28
experimental models, **11**-4, **11**-6
flexible-walled aortic bifurcation model, **11**-1–**11**-13
flow measurement technology, **1**-11
fluid dynamic patterns, **11**-5–**11**-6
   results, **1**-24–**1**-27
   validation run, **1**-19–**1**-21
hemodynamics models for optimal design, **1**-1, *See* Branching blood vessels, optimal design considerations
non-Newtonian flow effects, **9**-31
numerical models, **9**-11, **9**-18–**9**-24, **11**-4, **11**-6–**11**-7, *See also* Y-shaped bifurcation, numerical model
physical models, **11**-2
   experimental methods, **11**-6
   fluid dynamic parameters, **11**-5–**11**-6
   geometry, **11**-2–**11**-3
   non-Newtonian fluid, **11**-3–**11**-4
three-dimensional pulsatile flow simulations, **1**-11
two-dimensional model for Y-shaped bifurcation with flexible walls, **11**-7–**11**-13
   arbitrary Lagrangian Eulerian method, **11**-8
   3D vs. 2D models, **11**-7
   vorticity transport and Poisson equations, **11**-7, **11**-9
vortex formation, **7**-3, **7**-4
wall models, **11**-5
wall shear stress and, **1**-26–**1**-27, **3**-17–**3**-18, **7**-6, **7**-7, **7**-8, **11**-6
wall stiffness relationship, **11**-8, **11**-10–**11**-11
Arterial collapse, **7**-10, *See also* Collapsible tube flow
Arterial compliance, *See* Wall compliance
Arterial disease, *See* Atherosclerosis
   hemodynamic severity parameters, **1**-7–**1**-8
   hemodynamics simulation and optimal blood vessel design, **1**-5–**1**-11, *See also* Branching blood vessels, optimal design considerations
   impacts on branching blood vessels, **1**-2–**1**-5
Arterial fluid dynamics, **7**-1–**7**-15, *See also* Computational fluid dynamics; Wall shear stress; specific parameters
   atherosclerosis relationships, *See also* Atherosclerosis
   diagnostic applications, **7**-11–**7**-15
   endothelial cell behavior relationship, **7**-8–**7**-9
   in vitro studies, **7**-8–**7**-9
   in vivo studies, **7**-9–**7**-10

investigations, 7-5–7-10
basics of, 7-2–7-5
bifurcating flows with flexible walls, 11-5–11-6, *See also*
     Arterial bifurcations
finite element models, *See* Arterial structural mechanics,
     finite element models
in vitro studies of artery wall effects, 7-8–7-9
in vivo measurement considerations, 7-9–7-10, 8-1,
     8-4–8-6, 8-10, 8-23, 8-24
investigations, 7-5–7-10
numerical simulations, *See* Numerical models
pressure drop-velocity change relationships, 8-2–8-25,
     See Arterial pressure drop,
     measurement/calculation study for velocity
     relationship
pressure wave propagation, 7-4–7-5
Reynolds number and, 7-2–7-3
Arterial pressure drop, measurement/calculation study for
     velocity relationship, 8-2–8-25
animal hemodynamic data, 8-13–8-14
assumptions, 8-2
blood rheology, 8-2–8-3, 8-12–8-13, 8-16
calculation for consecutive pulse cycles, 8-22–8-23
diagnostic application, 8-23
experimental method, 8-4–8-13
     arterial geometry, 8-7–8-8
     boundary conditions, 8-10–8-11
     formulation, 8-8–8-10
     in vivo measurement considerations, 8-4–8-6, 8-10,
          8-23, 8-24
hysteresis effects, 8-13
parabolic inlet flow calculation, 8-14–8-20
     non-Newtonian blood viscosity, 8-16
     pressure drop, 8-19–8-20
     shear rate, 8-15
     shear stress, 8-14
     velocity distribution, 8-15
summary and conclusion, 8-23–8-24
uncertainties, 8-4
Arterial structural mechanics, finite element models,
     5-1–5-28
applications, 5-12
CFD models and, 5-2, 5-22–5-23
future research directions, 5-2, 5-24
hyperelastic (HE) models, 5-1, 5-12–5-13, 5-24
     diseased arteries and, 5-21–5-22
normal artery applications, 5-2, 5-12–5-21
     cardiac cycle simulation, 5-19
     hyperelastic FEMs, 5-12–5-13
     PHETS FEMs, 5-20–5-21
     porohyperelastic FEMs, 5-13–5-19
other arterial structures, 5-23
PHETS model, 5-1–5-10, *See*
     Porohyperelastic-transport-swelling (PHETS)
     model
porohyperelastic (PHE) model, 5-1, 5-3, 5-10, 5-13–5-19
     biphasic mixture model equivalence, 5-3, 5-10
Arterial structure, 5-2–5-3, 11-1, 11-4, 11-5
Arterial temperature distribution, 6-3–6-14, *See also*
     Temperature distribution, three-dimensional
     vascular model

Arterial tissue remodeling, FEM applications, 5-13, 5-25,
     See also Branching blood vessels, optimal design
     considerations
Arteriole blood flow, 4-1, *See* Perfusion measurement
Arterio-venous fistula, 7-10
Asthma, 12-2
Atherosclerosis, 1-2
     arterial bifurcations and, 1-2, 3-17–3-18, 11-2
     arterial wall FEMs, 5-2
     blood flow patterns and, 7-2
     combined CFD and FEM models, 5-25
     endothelial cell behavior and, 7-9
     fluid mechanics and
          diagnostic applications, 7-11–7-15
          endothelial cell behavior relationship, 7-8–7-9
          in vitro studies, 7-8–7-9
          in vivo studies, 7-9–7-10
          investigations, 7-5–7-10
     hemodynamics and optimal blood vessel design,
          1-5–1-11, *See also* Branching blood vessels, optimal
          design considerations
     historical studies, 7-5
     hyperelastic FEM model, 5-21–5-22
     indicator equations for localization, 1-5–1-7
     methods to determine flow limitations, 7-14–7-15
     post-operative effects, 1-2–1-4
     vulnerable arteries, 7-2, 11-2
     wall shear stress and, 1-2, 2-1, 3-17–3-18, 3-29, 7-6–7-7
     wall shear stress gradient and, 1-6–1-8
Auscultation, 7-11, 10-37

B

Ballon angioplasty, 1-2, 5-23, 7-12
Bifurcating flows, *See* Arterial bifurcations
Bilateral radiography, 12-24
Bioheat transfer models, 6-1–6-2
     bioheat equation, 4-2–4-3, 6-1, 6-14
          old versus new forms, 6-1–6-2
     blood-tissue temperature equilibrium, 6-2
     efficiency function approach, 6-14–6-18, 6-23
     Pennes approach, 4-2–4-3, 4-12, 6-1, 6-14–6-15
          efficiency function approach and, 6-14–6-18, 6-23
          reliability under different conditions, 6-17–6-18
     three-dimensional model, 6-3–6-14, *See* Temperature
          distribution, three-dimensional vascular model
Biphasic mixture models, poroelastic model equivalence,
     5-3, 5-10
Blood flow measurement, 1-11, *See also* Flow visualization;
     specific methods
     branching blood vessel studies, 1-1
     diagnostic applications, 7-11–7-15
     Doppler methods, *See* specific methods
     electrochemical method for wall shear stress, *See*
          Electrochemical method
     instantaneous *in vivo* measurement for pressure drop
          calculation, 8-4–8-6, 8-10, 8-24, *See also* Arterial
          pressure drop, measurement/calculation study for
          velocity relationship

LDA, *See* Laser Doppler anemometry
microvascular vessels, *See* Perfusion measurement
MRI velocimetry, 7-8, *See also* Magnetic resonance
    imaging
Ultrasonic Doppler, 1-11, **4**-24–**4**-25, **7**-11, 7-14–**7**-15,
    9-16, 9-30
Blood flow regulation, **6**-23
Blood particle trajectories, 1-2
Blood perfusion, *See* Perfusion measurement
Blood pressure measurement, 7-11
Blood-tissue temperature equilibrium, **6**-2
Blood viscosity, 7-2, **8**-2–**8**-3
    Carreau model, **8**-12
    Casson model, 1-9, 1-10, 2-24–**2**-29, 9-30–**9**-31, 11-4
    non-Newtonian rheology, 2-1, 7-4, **8**-3
        arterial taper and, **8**-23
        bifurcation effects, 9-31, **11**-3–**11**-4
        experimental-calculation approach for pressure
            drop-velocity change relationship, **8**-12–**8**-13
        numerical simulations and, 9-30–**9**-31
        parabolic inlet velocity profile and, **8**-16
        self-excited oscillation, 2-24
    wall shear rates and, 7-4
Brain damage, 4-1
Branching blood vessels, optimal design considerations,
    1-1–**1**-2, *See also* Arterial bifurcations
    arterial disease impacts, **1**-2–**1**-5
    clinical implications and recommendations, **1**-51–**1**-53
    discretized equations, **1**-13–**1**-15
    flow measurement technology, 1-11
    model validation, **1**-15–**1**-23
        carotid artery bifurcation validation run, **1**-19–**1**-21
        error estimation, **1**-15–**1**-17
        inlet length, 1-15
        LDA measurements, 1-18
    numerical method, **1**-11–**1**-12
        discretized equations, **1**-13–**1**-15
        mesh generation, 1-12
    optimal computer-aided design, 1-1
    post-operative problems, **1**-2–**1**-4
    results, carotid artery bifurcation, **1**-24–**1**-27
        mid-plane velocity vector profiles, 1-26
        wall shear stress distributions, 1-26
        wall shear stress gradients, 1-26
    results, graft-to-artery anastomosis, **1**-27–**1**-56
        distal end anastomotic connector design, **1**-40–**1**-46
        flow fields, **1**-32–**1**-35
        flow input waveform effects, **1**-46–**1**-49
        input flow waveforms, 1-28
        suture line position effects, 1-37
        Taylor patch connector, **1**-38–**1**-40
        wall shear stress calculation, 1-30
        wall shear stress gradient and oscillatory shear index
            comparison, **1**-36–**1**-37
        wall shear stress vectors, **1**-35–**1**-37
    theory, **1**-5–**1**-11
        flow waveform parameters, 1-8
        indicator equations, **1**-5–**1**-7
        inlet and outlet conditions, **1**-10–**1**-11
        severity parameters, **1**-7–**1**-8

transport equations auxiliary condition, **1**-8–**1**-10
wall boundary conditions, 1-11
Bursting frequency, 2-7
Bypass grafts, *See* Graft-to-artery anastomoses

C

Capillary blood flow, 4-1, *See* Perfusion measurement
Captopril, 5-14
Cardiac cycle simulation, finite element model, 5-19, 5-25
Cardiac flow pulse, 7-14
Cardiac output, 7-14
Carotid artery, atherosclerosis susceptibility,
    7-2
Carotid artery bifurcation, model for optimal design,
    1-1–**1**-24, *See* Branching blood vessels, optimal
    design considerations
    results, **1**-24–**1**-27
    validation run, **1**-19–**1**-21
Carotid endarterectomy, **1**-1–**1**-5, 1-24, 1-27
Carotidogram, 7-11
Carreau model, **8**-12
Casson model, 1-9, 1-10, 11-4
Catheter atherectomy, 1-2
Cell culture, 1-11, 7-8
Clotting responses, 9-25
Cold stress, classical bioheat approach and, 6-17
Collapsible tube flow, **10**-1–**10**-41, *See also* Wall compliance
    bidimensional Cartesian system, 10-7
    critical conditions, **10**-2–**10**-3
    early theoretical models, **10**-3
    ellipticity and, **10**-4
    flow experiments, **10**-5–**10**-6
    lumped parameter model, 10-3, 10-6
    one-dimensional model, **10**-23–**10**-26, 10-40
    physiological applications, **10**-36–**10**-39
    pressure-area relationship, **10**-4–**10**-5
    Starling resistor, **10**-3–**10**-4, 10-27, 10-34
    steady flow experiments
        flow limiter, **10**-33–**10**-34
        horizontal flow, **10**-27–**10**-29
        inclined flow, **10**-30–**10**-33
    tube law, **10**-6–**10**-14
        analytical expressions, **10**-11–**10**-12
        modes of collapse, **10**-7–**10**-9
        normalization, 10-10
        normalized similarity law, **10**-10–**10**-11
        opposite wall-edge contact, 10-10
        positive transmural pressure, **10**-13–**10**-14
    tube law dependent data, **10**-14–**10**-23
        critical flow rate, **10**-15–**10**-16
        head losses, 10-17
        kinetic energy and momentum coefficients, **10**-23
        Reynolds number, **10**-16–**10**-17
        shape factors, **10**-17–**10**-18
        wall shear stress, **10**-18–**10**-22
        wave speed, 10-14
    unsteady flow experiments, **10**-34–**10**-36, 10-39
    unstressed configuration dependent collapse, 10-6
Composite prosthetic vein grafts, 1-4

Computational fluid dynamics (CFD) models, *See also*
    Numerical models
    combined structural FEM models, 5-2, 5-12, 5-22–5-23,
        5-25
    grid generation, 1-12
    optimal branching blood vessel design, *See* Branching
        blood vessels, optimal design considerations
    pressure drop-velocity change relationships, 8-2–8-25,
        See Arterial pressure drop,
        measurement/calculation study for velocity
        relationship
Computer-aided design, branching blood vessels, *See*
    Branching blood vessels, optimal design
    considerations
Conjugate gradient squared (CGS) method, 2-12
Continuous wave Doppler ultrasound methods, 4-24
Contrast agents, 4-23–4-24, 7-12
Control volume method (CVM), 1-11, 1-13
Coronary artery
    atherosclerosis susceptibility, 7-2, 11-2
    flow and stenosis relationships, 7-7, 9-11, 9-25–9-31
        assessing hemodynamic-severity associations,
            9-29–9-30
        particle tracking, 9-28–9-29
        shear data, 9-26–9-28
        velocity data, 9-26
Coronary artery bypass graft (CABG), 7-14, *See also*
    Graft-to-artery anastomoses
Coronary artery disease, arterial fluid mechanics diagnostic
    applications, 7-11–715
Countercurrent arterio-venous network, 6-3–6-5

D

Dacron grafts, 5-23
Diffuser, energy conversion rate of, 2-27–2-29
Distensibility, *See* Wall compliance
Doppler flow meter, 8-4, 8-5–8-6, 8-23, 8-24
Doppler ultrasound velocimetry, 1-11, 4-24–4-25, 7-11,
    7-14–7-15
Drug delivery systems, 5-26

E

Efficiency function approach, 6-14–6-18
    hyperthermia treatment application, 6-18–6-23
Electrochemical method, 2-2, 3-2–3-8
    effective electrode surface area, 3-11–3-14
    45° bifurcation model, 3-21–3-30
        flow visualization, 3-21, 3-22–3-23
        generating system of pulsating flow, 3-22
        physiological implications, 3-29
        pulsating flow, 3-29
        steady flow, 3-24–3-28
    frequency characteristics and limitations, 3-14–3-17
    Laser Doppler velocimeter measurement comparison,
        3-8
    physiological perspective, 3-17–3-18
    stenosis model, 3-18–3-20

oscillation shear index, 3-20
pulsating flow, 3-20
steady flow, 3-19–3-20
test electrode, 3-9–3-10
Electrode, 3-9–3-10
    effective surface area, 3-11–3-14
End-to-end anastomoses, 1-27
    FEM model, 5-23
End-to-side anastomoses, 1-27, *See also* Graft-to-artery
    anastomoses
    hemodynamics model for optimal design, 1-27–1-56
    model for different angles and flow rate, 9-11, 9-12–9-18
    three-dimensional pulsatile flow simulations, 1-11
Endothelial cells
    in vitro flow effects, 7-8–7-9
    tissue culture, 1-11, 7-8
Endothelial permeability, 1-2, 1-8, 7-10
Endothelin, 7-9
Endotracheal tube (ETT), 12-25–12-26
Epoxy, 3-9
E-selectin, 7-9
Exercise, blood temperature distribution model, 6-9

F

Femoral artery
    atherosclerosis susceptibility, 7-2
    experimental-calculation approach for pressure
        drop-velocity change relationship, 8-2–8-25
Finite difference methods
    aortic bifurcation model, 11-9
    bioheat transfer and vascular temperature distribution,
        6-2, 6-5, 6-9
    shear stress model for oscillating wall, 2-12
Finite element models (FEMs), 5-1–5-3, 5-10–5-12, 8-9,
    9-35–9-39, 9-5–9-7
    ABAQUS program, 5-2, 5-10, 5-12, 5-14, 5-15–5-20,
        5-23, 5-25, 5-26
    arterial grafts and stents, 5-23
    arterial tissue remodeling, 5-13, 5-25
    arterial wall structure, 5-1–5-3
    balloon angioplasty, 5-23
    cardiac cycle simulation, 5-19
    experimental-calculation approach for pressure
        drop-velocity change relationship, 8-8–8-10
    hyperelastic (HE) models, 5-1, 5-12–5-13, 5-21–5-22,
        5-24
    local drug delivery systems, 5-26
    normal artery applications, 5-12–5-21
    porohyperelastic (PHE) model, 5-1, 5-3, 5-10, 5-13–5-19
    porohyperelastic-transport-swelling (PHETS) model,
        5-20–5-21
        combined CFD simulations, 5-2, 5-12, 5-22–5-23,
            5-25
        diseased arteries and, 5-2, 5-21–5-22
        future applications, 5-2, 5-24–5-26
        normal artery applications, 5-2, 5-12–5-21
        tissue remodeling, 5-25
    viscous flow calculation, 2-26
    whole-limb heating model, 6-23

Finite volume method (FVM), 1-11, 1-13, 1-15
Flow choking, 7-10
Flow visualization, 3-21, 3-22–3-23, *See also* specific
    methods
    arterial bifurcations, 11-6
    model for evaluating MRI system, 9-22–9-24
FLOW3D, 1-11
Fluorescent microspheres, 4-22
Flush-mounting probe method for shear stress
    measurement, 2-2
Forced expiration test, 10-39

G

Gadolinium diethylenetriamine, 4-24
Galerkin method, 1-12, 8-9
Graft-to-artery anastomoses,
    FEM models, 5-23
    flow and angle considerations for surgeons, 9-17–9-18
    hemodynamics model for optimal design results,
        1-27–1-56
        design applications and implications, 1-51–153,
            9-14–9-18
        distal end connector design, 1-40–1-46
        flow field, 1-32–1-35
        input flow waveforms, 1-28, 1-46–1-49
        secondary flow, 1-33
        suture line position and, 1-37–1-38
        Taylor patch connector, 1-38–1-40
        wall shear stress calculation, 1-30
    numerical model for different angles and flow rates,
        9-11, 9-12–9-28
        in vitro experiment correlations, 9-15–9-16
        Reynolds number, 9-12
        surgical decision-making application, 9-14–9-18
        velocity vector plots, 9-12–9-13
        wall shear rates on artery, 9-13–9-14
        wall shear rates on graft, 9-13
        wall shear stress mapping, 9-16–9-17
    restenosis, 1-2–1-4, 1-50, 9-17
    S-connector designs, 1-41, 1-43–1-46, 1-48
    vortex formation, 9-12
    wall shear stress effects, 1-30, 1-35–1-37, 9-13–9-14
Gravity-friction flows, in collapsible tubes,
    10-30

H

Head losses, collapsed tubes, 10-2, 10-17
Heart rate, canine, 8-4
Heat transfer models, *See also* Bioheat transfer models
    intrinsic tissue conductivity, 4-9, 4-11
    Pennes equation, 4-2–4-3, 4-12
    thermal clearance techniques, 4-2–4-21
Helium-oxygen mixture, 12-11
Hematocrit, 1-9, 1-10
Hemodynamics simulations, *See* Computational fluid
    dynamics; Numerical models; specific applications,
    methods, models

optimal branching blood vessel design, 1-1–1-56, *See*
    Branching blood vessels, optimal design
    considerations
Herschel-Bulkley equation, 11-4
Hoods, 1-4
Hot film probe, 3-1
Hot wire probe, 3-1
Hydralazine, 5-14
Hyperelastic (HE) models, 5-1, 5-12–5-13, 5-24
    diseased arteries and, 5-21–5-22
Hyperplasia, 1-2
Hyperthermia therapy, 6-3, 6-17, 6-18–6-23
Hysteresis, pressure drop-velocity change relationship
    study, 8-13

I

ICAM-1, 7-9
Inclined collapsed tube flows, 10-26, 10-30–10-33
Indicator dilution techniques, 4-21–4-22
Indicator imaging, 4-1
Infrared photoplethysmography, 4-26
Instantaneous left ventricular pressure, 7-12, 7-14
Intimal hyperplasia, 1-1, 1-2, 1-4, 1-27, *See also*
    Atherosclerosis; Stenosis
    bypass graft distal anastomosis and, 9-17
    indicator equations for localization, 1-5–1-7
    wall shear stress gradient indicator, 1-50
Intrinsic tissue conductivity, 4-9, 4-11, 4-19

L

L and M Doppler flow meter, 8-5–8-6
Laser ablation, 1-2
Laser Doppler anemometry (LDA), 1-11, 3-2, 9-2
    arterial bifurcation visualization, 11-6
    arterial flow and damage studies, 7-6–7-7
    branching blood vessel model validation, 1-8
    electrochemical wall shear stress measurement
        comparison, 3-8
    perfusion measurements, 4-25
Laser photocoagulation, 4-1
Limb blood flow measurement, 4-25
Liquid metal strain gauge, 4-25
Local drug delivery systems, 5-26
Low density lipoprotein (LDL), 1-2, 1-8, 5-24
Lower extremity bypass, 1-1
Lumped parameter model, 10-3, 10-6, 12-1

M

Magnetic resonance imaging (MRI), 4-24
    acoustic reflection method validation, 12-24
    angiography applications, 7-11
    arterial bifurcation visualization, 11-6
    computational simulation application for evaluating,
        9-22–9-24
    velocimetry, 7-8
MARC, 5-12, 5-23

Mass transfer
    combined CFD and heat transfer applications, **7**-8
    electrochemical method for wall shear stress
        measurement, **3**-2–**3**-6
    indicator techniques, **4**-21–**4**-22
Mechanography, **7**-11
MENTAT, **5**-23
Mesh generation, **1**-12
Microvascular blood flow, **4**-1, *See* Perfusion measurement
Moens-Korteweg equation, **7**-5
Monkey aortic coarctation model, **7**-10
Monocyte recruitment, **7**-9
Muscle vasculature, **6**-2, **6**-3–**6**-4

N

Nasal cavity, **12**-2, **12**-22–**12**-24
Nitric oxide, **10**-38
Non-Newtonian fluids, **7**-4, *See also* Blood
        viscosity
    assumptions for experimental-calculation model for
        velocity change-pressure drop relationships, **8**-3
    bifurcation effects, **9**-31, **11**-3–**11**-4
    Casson model, **1**-9, **1**-10, **2**-24–**2**-29, **9**-30–**9**-31, **11**-4
    diffuser energy conversion rate and, **2**-27–**2**-29
    experimental-calculation approach for pressure
        drop-velocity change relationship, **8**-12–**8**-13
    numerical simulations and, **9**-30–**9**-31
    self-excited oscillation, **2**-24, **2**-27
    wall shear stress evaluation and application, **2**-24–**2**-29
    wall shear stress for flow through axisymmetric diffuser,
        2-24-**2**-29
Numerical models, **9**-1–**9**-3, *See also* Computational fluid
        dynamics; specific approaches, methods, models,
        parameters
    accuracy limitations, **9**-2
    applications, **9**-11
    bifurcations, **9**-11, **9**-18–**9**-24, **11**-6–**11**-13, *See also*
        Arterial bifurcations; Branching blood vessels,
        optimal design considerations; Y-shaped
        bifurcation, numerical model
    distal graft anastomoses, **9**-11, **9**-12–**9**-18, *See also* under
        Graft-to-artery anastomoses
    finite element formulation, **9**-5–**9**-7, **9**-35–**9**-39, *See also*
        Finite element models
    general applications, **9**-2
    global coordinates, **9**-8, **9**-42–**9**-43
    governing equations for blood flow, **9**-3–**9**-5
    local coordinates, **9**-8, **9**-41–**9**-43
    measurement/calculation study for pressure drop
        relationship, **8**-2–**8**-25, *See* Arterial pressure drop,
        measurement/calculation study for velocity
        relationship
    non-Newtonian flow effects, **9**-30–**9**-31
    optimal branching blood vessel design, **1**-11–**1**-12, *See
        also* Branching blood vessels, optimal design
        considerations
    pulsatile flow considerations, **9**-9–**9**-10
    shape function, **9**-7–**9**-8, **9**-39–**9**-41

stenosed artery, **9**-11, **9**-25–**9**-38, *See also* Stenosed
        coronary artery model
stream function, **9**-7
vessel distensibility effects, **9**-31
vorticity, **9**-7
Nusselt number, **6**-2, **6**-5, **6**-11, **6**-13

O

Obstructive pulmonary disease, **12**-2
Obstructive sleep apnea, **12**-24–**12**-25
One-dimensional law, collapsible tube flow, **10**-23–**10**-26,
        10-40
Opening angles, **5**-2, **5**-13, **5**-25
Optical Doppler technique, **4**-25
Optical stereoscope, **10**-4–**10**-5
Oscillating wall
    parabolic inlet velocity profile shear rate, **8**-15
    pressure drop and, **2**-16–**2**-20
    separation bubble structure, **2**-20–**2**-23
    shear stress evaluation, **2**-9–**2**-23
Oscillation, non-Newtonian fluids, **2**-24, **2**-27
Oscillatory shear index (OSI), **1**-5, **7**-7
    bypass graft model, **1**-32, **1**-36–**1**-37, **1**-40
    electrochemical wall shear stress measurement and, **3**-20
    intimal thickness relationship, **7**-7, **7**-8
Osmotic effects, FEM model, **5**-25

P

Parabolic inlet velocity profiles, **8**-14–**8**-20,
        8-23
    non-Newtonian blood viscosity, **8**-16
    pressure drop, **8**-19–**8**-20
    shear rate, **8**-15
    shear stress, **8**-14
    velocity distribution, **8**-15
Partial differential equations, **9**-1, *See also* Numerical
        models
Particle tracking, **11**-6
    branching blood vessel model validation, **1**-18
    stenosed artery model and, **9**-28–**9**-29
Pennes bioheat transfer approach, **4**-2–**4**-3, **4**-12, **6**-1,
        6-14–**6**-15
    efficiency function and, **6**-14–**6**-18, **6**-23
Percutaneous transluminal coronary angioplasty (PTCA),
        7-14
Perfusion heat transfer, **6**-1–**6**-2, **6**-14, *See also* Perfusion
        measurement, thermal clearance
Perfusion measurement, **4**-1–**4**-27
    Doppler techniques, **4**-24–**4**-25
    imaging techniques, **4**-22–**4**-24
    indicator techniques, **4**-21–**4**-22
    plethysmography, **4**-25–**4**-26
    thermal clearance, **4**-2–**4**-21
        bioheat transfer equation, **4**-2–**4**-3
        self-heated thermistor, **4**-1, **4**-9–**4**-21, *See also*
            Self-heated thermistor method
        thermal pulse decay, **4**-1, **4**-3–**4**-8

Peripheral vascular imaging, **4**-24–**4**-25
Pharyngeal structure, **12**-24–**12**-25
Photocardiography, **7**-11
Photochromic tracer, **1**-11, **1**-18
Pitot tube, **3**-1
Plethysmography, **4**-1, **4**-25–**4**-26
Poiseuille, Jean, **7**-1
Poisson equations, aortic bifurcation model, **11**-7, **11**-9
Polytetrafluoroethylene (PTFE) grafts, **5**-23
Porohyperelastic (PHE) model, **5**-1, **5**-13–**5**-19
   biphasic mixture model equivalence, **5**-3, **5**-10
Porohyperelastic-transport-swelling (PHETS) model,
   5-1–**5**-10
   biphasic mixture models and, **5**-3, **5**-10
   charged mobile species, **5**-24
   combined CFD simulations, **5**-2, **5**-12, **5**-22–**5**-23, **5**-25
   diseased arteries and, **5**-2, **5**-21–**5**-22
   drug delivery systems, **5**-26
   equivalence of poroelastic and mixture-based models,
      5-3, **5**-10
   Eulerian initial boundary problem, **5**-5–**5**-6
   fundamental fields, **5**-3
   future research and applications, **5**-2, **5**-24–**5**-26
   intimal layer scale extension, **5**-25
   kinematics, **5**-3–**5**-4
   Lagrangian initial boundary problem, **5**-7–**5**-9
   material properties, **5**-9–**5**-10
   osmotic effects, **5**-25
   phenomenological equations, **5**-5
   quasi-static problem solution method, **5**-12
   virtual velocities, **5**-7
Positron emission tomography (PET), **4**-23
Post-operative thrombosis, **1**-1, **1**-2, **1**-3
Pressure drop
   collapsible tube flow experiments, **10**-5–**10**-6,
      10-27–**10**-39, *See also* Collapsible tube flow
   in vivo measurement considerations, **8**-4–**8**-5, **8**-23, **8**-24
   measurement/calculation study for velocity relationship,
      See Arterial pressure drop,
      measurement/calculation study for velocity
      relationship
   oscillating wall and, **2**-16–**2**-20
   parabolic inlet condition, **8**-19–**8**-20
   porohyperelastic FEM model, **5**-18
   velocity profile prediction, **9**-9–**9**-10
   wall shear stress evaluation for pulsatile pipe flow,
      2-2–**2**-9
Pressure pulses, **7**-11
Preston tube, **3**-1
Principles of virtual velocities, **5**-7
Prostacyclin, **7**-9
Pulmonary mechanical properties, **12**-1–**12**-2
   acoustic applications, **12**-1–**12**-2, **12**-24–**12**-26, *See also*
      Acoustic reflection method
Pulsatile flows
   general discussion, **7**-4
   numerical simulation considerations, **9**-9–**9**-10
   simulation for pressure drop-velocity change
      relationship study, **8**-10
   three-dimensional bifurcation flow simulation, **1**-11

wall shear stress model for transitional Reynolds
   number, **2**-2–**2**-9
Pulse evaluation technologies, **7**-11
Pulsed wave Doppler ultrasound methods, **4**-24

R

Radio-labeled indicators, **4**-22
Renal arteries, **7**-8
Respiratory system structure, **12**-1–**12**-2
   acoustic rhinometry, **12**-22–**12**-24
   noninvasive acoustic imaging, **12**-2–**12**-26, *See* Acoustic
      reflection method
Reverse flow
   bypass graft model, **1**-33
   pressure drop-velocity change relationship study, **8**-13,
      8-21
   velocity and pressure gradient studies, **8**-3
Reynolds number, **2**-12, **7**-2–**7**-3, **8**-3, **9**-10
   bifurcation and, **9**-19, **11**-5
   collapsible tube flow, **10**-16–**10**-17
   distal graft model, **9**-12
   oscillating wall and, **2**-21
Rhie-Chow algorithm, **1**-14
Rhinomanometry, **12**-22–**12**-24
Round-off error, **1**-15–**1**-16

S

Schroedinger equation, **12**-6
Self-heated thermistor method, **4**-1, **4**-9–**4**-21
   approach, **4**-9–**4**-12
   calibration, **4**-12, **4**-14, **4**-15
   measurement volume, **4**-15–**4**-18
   sensitivity analysis, **4**-12–**4**-15
   simultaneous tissue conductivity-perfusion
      measurement, **4**-19–**4**-20
   thermistor properties, **4**-15
   tissue intrinsic conductivity, **4**-9, **4**-11, **4**-19
   transient measurements, **4**-18–**4**-19
Shear stress, *See* Wall shear stress
SIMPLE, **1**-14
SIMPLEC, **1**-14
Simple washout model, **4**-20–**4**-21
Single photon emission computed tomography (SPECT),
   4-23
Sleep apnea, **12**-2, **12**-24–**12**-25
Smooth muscle, FEM model applications, **5**-25
Starling resistor, **10**-3–**10**-4, **10**-27, **10**-34
Stenosed coronary artery model, **9**-11, **9**-25–**9**-31
   assessing hemodynamic-severity associations, **9**-29–**9**-30
   particle tracking, **9**-28–**9**-29
   shear data, **9**-26–**9**-28
   velocity data, **9**-26
Stenosis, *See also* Atherosclerosis
   arterial flow and intimal thickening relationships,
      7-6–**7**-7
   arterial fluid mechanics diagnostic applications,
      7-11–715

clinically significant effects, **7**-10
collapsed walls, **7**-10, *See* Collapsible tube flow
combined CFD and FEM models, **5**-23
coronary artery model, **9**-11, **9**-25–**9**-31
electrochemical wall shear stress measurement method
    application, **3**-18–**3**-20
hemodynamic severity parameters, **1**-7–**1**-8
hyperelastic FEM model, **5**-22
methods to determine flow limitations, **7**-14–**7**-15
optimal branching blood vessel design and, *See*
    Branching blood vessels, optimal design

Tissue temperature distribution, *See* Temperature
    distribution, three-dimensional vascular model
Tracheal stenosis, **12**-2
Tracheal structure, **12**-25
Truncation error, **1**-15–**1**-16
T-shaped bifurcation, **11**-2
Turbulence effects, **7**-3, **7**-10
Turbulent shape factor, for collapsed tube flow, **10**-18

U